D1561224

ELECTRONIC CIRCUIT BEHAVIOR

Other books by the same author:

ELECTRONIC COMPONENTS, INSTRUMENTS, AND TROUBLESHOOTING

Bridges the gap between electronics as it is taught in the classroom and electronics as it is practiced in the industry.

ELECTRONICS POCKET HANDBOOK

Most-needed formulas, charts, and component data; unit conversions, codes, test procedures, and glossary of terms—all in a convenient pocket size.

Second Edition

ELECTRONIC CIRCUIT BEHAVIOR

DANIEL L. METZGER

Monroe County Community College
Monroe, Michigan

PRENTICE-HALL, INC., *Englewood Cliffs, New Jersey 07632*

Library of Congress Cataloging in Publication Data

METZGER, DANIEL L., date
Electronic circuit behavior.

Includes index.
1. Electronic circuits. 2. Electronic apparatus and appliances. I. Title.
TK7867.M43 1983 621.3815'3 82-16174
ISBN 0-13-250241-0

Editorial/production supervision by *Virginia Huebner*
Manufacturing buyer: *Gordon Osbourne*
Cover design by Photo Plus Art, *Celine Brandes*

Printed in the United States of America

10 9 8 7 6 5 4

ISBN 0-13-250241-0

PRENTICE-HALL INTERNATIONAL, INC., *London*
PRENTICE-HALL OF AUSTRALIA PTY. LIMITED, *Sydney*
EDITORA PRENTICE-HALL DO BRASIL, LTDA., *Rio de Janeiro*
PRENTICE-HALL CANADA INC., *Toronto*
PRENTICE-HALL OF INDIA PRIVATE LIMITED, *New Delhi*
PRENTICE-HALL OF JAPAN, INC., *Tokyo*
PRENTICE-HALL OF SOUTHEAST ASIA PTE. LTD., *Singapore*
WHITEHALL BOOKS LIMITED, *Wellington, New Zealand*

For Susie

CONTENTS

Asterisks (*) indicate sections which are especially important or are prerequisite for later sections.

3 BASIC POWER SUPPLIES 24

4 TRANSISTOR CHARACTERISTICS 56

5 TRANSISTOR SWITCHING CIRCUITS 76

6 THE COMMON-EMITTER AMPLIFIER 94

7 SPECIAL AMPLIFIER TYPES AND CIRCUITS 122

8 FIELD-EFFECT TRANSISTORS 146

17 CIRCUIT EXAMPLES 336

APPENDICES 366

INDEX 397

PREFACE

This text has been prepared to help you understand electronic devices and circuits in an intuitive yet quantitative way. The emphasis is placed squarely on the behavior of actual circuits commonly encountered in electronic systems, rather than on transistor theory and generalized mathematical techniques. To this end the text contains over 150 schematic diagrams of functional circuits *with all component values specified.* These circuits have been built and tested by the author to ensure that they work as predicted. By building them yourself you can confirm your understanding of circuit-behavior concepts and gain valuable experience with testing and troubleshooting procedures.

Approximations have been employed freely to obtain formulas for circuit analysis that are simple enough to be worked out right at the bench, but which nevertheless produce answers that are accurate enough for troubleshooting and preliminary-design applications. Mathematics has therefore been used only as a tool for obtaining numerical answers. Elementary algebra is an adequate math prerequisite for all but a few optional sections. In the electrical area, a prior course in basic dc and ac electricity has been assumed.

The text has been specifically designed for basic-electronics courses as commonly offered in the second semester (or second and third quarters) of an electronics-technology program. It is also appropriate for individual study and as a reference for practicing technicians and designers.

Topical coverage includes the devices and circuits most likely to be encountered in a broad range of electronic applications. Sections which contain essential topics, or material that is prerequisite for later sections, are marked with an asterisk (*) in the Table of Contents to permit an abbreviated coverage if time is limited.

Chapter 5 on Transistor Switching Circuits is logically placed after transistor characteristics and before transistor amplifiers, because switching circuits are the simplest transistor applications. However, common practice places this material in a separate course on digital circuits; therefore the text has been written so that Chapter 5 can be skipped without loss of continuity. Many users of the First Edition indicated that they preferred to cover zener regulators with basic power supplies, so Sections 15.1 and 15.2 have been written to allow them to be transplanted to Chapter 3 without prerequisite problems. The various sections of Chapter 16 (Special Diodes and Triggered Devices) may also be covered earlier if desired. Chapter 17 (Circuit Examples) is a fun chapter—it would be a shame to go through all the work and then miss the fun.

Many persons have contributed to making this Second Edition thoroughly complete and up-to-date, but particular recognition must be given to Professor William H. Mowbray of Community College of Rhode Island. It is rare good fortune when such a competent individual is able to devote an adequate portion of his time to improving a manuscript.

Daniel L. Metzger

Temperance, Michigan

1

AN OVERVIEW OF ELECTRONICS

1.1 A CAPSULE HISTORY

Like many of our present technologies, electronics has evolved from a rather crude experimental art at the turn of the century to a highly organized scientific pursuit requiring a division of specialties within its own ranks and often intertwined with outside specialties such as optics, chemistry, metallurgy, and mathematics. Thomas Edison was undoubtedly the greatest of the nineteenth-century experimentalists, and his work on electric lighting and power distribution laid much of the foundation for the later development of electronics. Of the same breed were men like Morse (telegraph, 1844), Bell (telephone, 1876), and Marconi (radiotelegraph, 1895). In each case, the basic principles for these devices had been discovered earlier by men of a theoretical or mathematical turn of mind. It was left to the experimentalists to try the original crude laboratory device, modify it, and try it again, sometimes for years, until a practical commercial product resulted.

The device which started the age of electronics was developed in precisely this way. Lee De Forest's notebooks show a long series of "try–try again" experiments, leading to the development of the triode tube, or "audion" as he called it, in 1907.

The importance of the triode to the development of electronics cannot be overstated. Prior to the triode, radio signals were developed by dumping large bursts of current into a tuned circuit amid great crashing sparks. Minute fractions of this energy were then intercepted by huge receiving antennas and used to produce feeble clicks in a pair of headphones. With vacuum tubes it became possible to generate continuous radio waves and to superimpose voice or music on them. The feeble signal at the receiver could also be amplified to fill a room or a concert hall if desired.

It is an interesting illustration of the character of experimental development that these potentials of the vacuum tube lay unused for 10, or in some cases, 15 years. Conflicting and entirely incorrect theories of how the triode operated abounded in this period, and this lack of understanding allowed the deficiencies of the original tubes to obscure the possibilities that could so easily have been realized.

The decades of the 1920s and 1930s saw the rise of a multimillion-dollar radio industry. Competition was fierce, and countless new circuits and improvements to the basic triode were developed. By 1929 the long struggle for a workable television system was well under way, and by 1939 the British had in operation a primitive radar to warn of Nazi air raids. These devices were of a different order of complexity from the first radio sets, however; and their development cannot be attributed to any one man, or even to any one company. In spite of many developments, by 1945, electronics was essentially limited to the field of communications (i.e., radio, telephone, phonograph, and radar). Although radar had produced some rather basic changes in its form, all thought still centered around the vacuum tube. In fact it was considered a good joke when someone said that there was "nothing wrong with electronics that getting rid of the vacuum tube couldn't cure," for at that time getting rid of the vacuum tube would have gotten rid of electronics as well.

The device which was in fact destined to replace the vacuum tube was invented in 1947 by a team of Bell Laboratories scientists. Conceived in a somewhat different form by William Shockley as early as 1940, the transistor did not make any appreciable impact on the electronics industry until several years after a 1952 symposium at which Bell Labs released its transistor secrets to a number of large firms and development work began on a large scale. It was not until 1963 that the transistor actually overtook the vacuum tube in units sold.

Meanwhile a number of other developments were taking place which were changing the character of electronics. In 1946 the first electronic digital computer containing some 19 000 vacuum tubes was developed for the U. S. Army at the University of Pennsylvania. By the mid-1950s computers were in common use, and a few years later transistorized computers had reached such a level of efficiency that few large companies could afford to operate without one. In other fields jet aircraft, nuclear reactors, guided missiles, and automated industrial plants were calling upon electronic systems not only for *communications* between controlling and activating portions of their structure but also for automatically *monitoring* and *controlling* some of the processes.

The use of electronics in complex computational and control functions has been made possible by new techniques in microminiaturization, of which the integrated circuit is the prime example. Using photoreduction techniques the function of 10 000 vacuum tubes can be duplicated in a single package smaller than a postage stamp. Because transistors do not "burn out," large computers containing 1 000 000 transistors in integrated circuit form can be built and operated reliably, whereas the early vacuum-tube computers required a special maintenance staff just for replacement of burned-out tubes. In addition, vacuum-tube costs cannot be brought much lower than $1 per tube, while integrated circuits have brought costs down to much less than 1 cent per transistor. These improvements in size, reliability, and cost have made possible the application of electronics to countless areas beyond the original radio-communications field.

1.2 "WHAT'S ELECTRONICS LIKE?"

Student technicians are usually curious about what it is like to work in electronics. The questions is a fair one, but an answer is almost impossible to give, because electronics technicians do such a wide variety of things. Nevertheless, the following grab bag of observations is offered, perhaps more to raise questions than to answer them.

1. What the general public sees is far less than the tip of the iceberg of electronic devices. Enter an elevator and you see only a few lighted buttons. Upstairs may be an entire rack of electronic devices controlling floor stops, motor runs, and door closures. Back at the elevator plant there will be switches and motors undergoing life tests, assembly lines, research labs, test towers, and service-training centers—all necessary to the existence of that elevator, but all unseen by the casual rider in it.
2. The production process usually involves electronics, even if the product does not. To produce shampoo in plastic bottles we would probably call on electronic instruments to control the pressure and temperature in the plastic extrusion machine, scan for defective bottles, meter the filling process, and package the final product.
3. Probably the majority of research and development projects never result in a successful product. Estimates are made, prototypes constructed, vendors contacted, components stocked and tested, and production lines set up with their specially designed test equipment, flow-solder machines, automatic wire-stripping machines, inspection and calibration devices—hundreds of man-years are invested—and in the end the project is canceled or achieves only marginal success before being superseded by something new.
4. Working in industry is very much like working in the school or home lab. The projects are more sophisticated, but you stay on them longer; the pay is better but the hours are longer; concern for grades is replaced by concern for prices—but the "feel" is most often the same. If you enjoy one, you should enjoy the other.
5. Electronics in the United States is highly regionalized. California—from San Diego to San Francisco—has the lion's share. The East Coast, Chicago, Texas, and parts of the Southwest and Florida have strong concentrations. Many cities—even large cities—have almost no job opportunities in electronics.
6. Most employers encourage and support continuing education for their technical employees. Successful employees with associate (two-year) degrees are usually

encouraged to go another two years for a bachelor's degree and engineering status within the company.

7. You will have to get used to that "snowed" feeling. If you ever learn enough to feel "on top of things" in electronics it will not be for many, many years. The thing to remember is that you are valuable for the skills that you do have. Other skills will be required that you do not have, but it is up to the project director to hire individuals for these tasks. When the team is assembled it will be able to accomplish things that no one person could accomplish alone.

1.3 HOW TO LEARN ELECTRONICS

Students often complain about the amount of mathematics in electronics: "I don't want to be an engineer, I want to be a service technician. Can't you skip the math and just teach me how to troubleshoot?" Unfortunately, the answer to that is a simple "no."

The idea that such an approach is possible probably got started with TV servicing as it was in the 1950s and 1960s; "If the picture blooms, change the 1B3; if the horizontal-output plates glow red, change the horizontal-oscillator tube; shrinking picture—change the 5U4." Television receivers were such standard items that a collection of such maxims was enough to make a person into a passable field-service technician.

Industrial electronics was never this routine. Consumer electronics has become very much less so, and the total victory of solid-state has made tube swapping a lost art. Today, the game plan for troubleshooting (except for some rather obvious problems) goes like this:

1. Examine the circuit diagram to determine what voltage waveform *should* appear at a selected point.
2. Observe this voltage waveform on the oscilloscope and compare what you see with what you expected to see.
3. If expectation and observation match, move on to another selected point in the circuit. If they do not match, investigate that area of the circuit more closely until the defective component is located.

This procedure requires that you be able to predict, *in numerical terms*, what a given circuit will do if it is working properly. Power supplies are supposed to put out a big dc voltage with a little ac ripple on it. Maybe the dc is suspiciously low, and the ripple seems unusually high. How low is too low, and how high is too high? There is no maxim to handle that question. Eight volts dc with 4 V ripple is fine in some circuits, but 80 V dc with 0.4 V ripple is not good enough in some others. You must be able to analyze the circuit mathematically to determine what is fair and what is foul.

You don't often see service technicians filling up reams of paper solving math problems, because they have learned—after doing it on paper several times—to do it in their heads. It might take you 20 minutes the first time you try to calculate the voltage gain of a loaded common-emitter amplifier, but if you have to do it every day on the job for a year you'll have the answer (to within $\pm 20\%$) in 15 seconds without touching a pencil or a calculator.

Theories, equations, and math problems are important only because they lead to an ability to deal with electronic hardware. For this reason most of the circuits in this book include component values. It is recommended that these circuits actually be constructed and that the propositions presented in the text be verified in the lab.

Many people tend to confuse reading with studying. This is a mistake, because most of us do not learn technical material by reading; we learn it by doing. Of course you must read the text, but this is not yet studying; its primary purpose is to acquanit you with the general ideas of the topic at hand—to get the lay of the land, so to speak. The next step is to attempt some of the problems at the end of the chapter. Undoubtedly, you will encounter some difficulty, but you will remember having read something about it in the text, and you will go back and give it a closer look. *Now* you are studying. Next you will take the circuit presented in the problem and attempt to verify in the laboratory that it really behaves as the text says it should. At this stage you are learning things of practical value. And you are not limited by the text or the problems at hand; you are free to investigate as far as your curiosity can take you.

1.4 THE INTERNATIONAL SYSTEM OF UNITS (SI)

Quantities are physical characteristics capable of being expressed numerically. Examples are length, time, and electric current. **Units** are arbitrary amounts which form the basis for measurement of quantities. Examples are meters, seconds, and amperes. The *Systeme International d'Unites* (SI) has been developed to provide a single, well-defined, and universally accepted unit for each quantity.

1. Quantity symbols consist of a single letter of the English or Greek alphabet, modified by subscripts and/or superscripts where appropriate.
2. When printed, quantity symbols (and mathematical variables) appear in italic (slanted) type. Subscripts which are quantity symbols or variables (in their own right) are also italic. Other subscripts are roman (upright). Examples: V_R, V_s, I_x, I_CEO.
3. Because of the limited number of characters available, two quantities may be assigned the same letter symbol. To avoid confusion the quantities may be differentiated by subscripts, or upper- and lowercase letters may be defined differently by the writer. Several subscripts may be attached to a single quantity symbol, separated by a comma, hyphen, or parentheses if necessary for clarity.
4. Uppercase (capital) letters are used to designate dc, rms, average (av), maximum (m), or minimum (min or n) values of such quantities as voltage, current, and power. (The appropriate subscript may be added where clarification is needed.) In this case uppercase subscripts indicate dc values, or, with the

additional subscripts M, N, or AV, maximum, minimum, or average value of the total waveform. Lowercase subscripts indicate the rms value of the ac component or, with the additional subscript m, maximum value of the ac component. See Fig. 1-1(a).

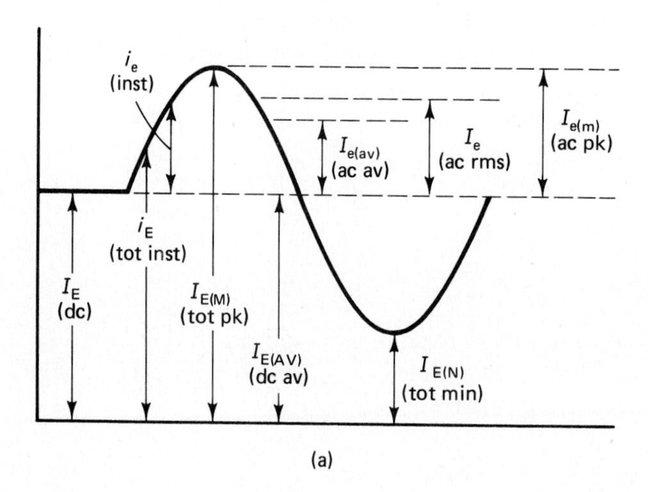

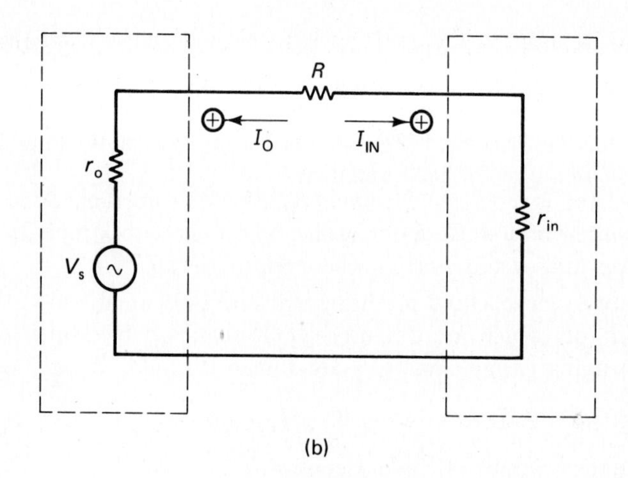

Figure 1-1 (a) Correct use of upper- and lower-case in quantity symbols and subscripts. (b) Conventional current flows into a terminal if positive, out of the terminal if negative.

5. Lowercase (small) letters are used to designate the *instantaneous* value of a time-varying quantity. Here, uppercase subscripts indicate instantaneous value of the total (dc + ac) and lowercase subscripts indicate the instantaneous value of the ac component only.
6. Double uppercase subscripts designate the dc supply for the element indicated. Example: V_{CC} (dc voltage supply for collector).

7. Two-letter subscripts to the voltage symbol designate voltage from the first point to the second point as a reference. Single-letter subscripts to the voltage symbol may indicate voltage across the device, or from the point designated to circuit ground as a reference. Examples: V_{CE} (collector to emitter), V_Z (across the zener), V_C (collector to ground), V_o (output to ground, ac)
8. Hyphenated subscripts may be used where two elements have the same name: Examples: $V_{B1\text{-}B2}$ (base one to base two of a UJT), $V_{1B\text{-}2B}$ (base of the first transistor to base of the second).
9. Conventional current (positive to negative) is regarded as flowing into the terminal indicated by a subscript to the current symbol. Conventional current out of the terminal gives the quantity a negative value. Of course, this *convention* is opposite to the actual direction of *electron flow* in a wire, but in practice each system works equally well, and the *conventional* direction was established before electrons were discovered. In troubleshooting we usually trace conventional current in negative-ground circuits and electron flow in positive-ground circuits, so we always follow a path from supply to ground. Therefore, it is a good idea to learn to handle either system. Where the system in use is not specifically stated, the conventional direction is assumed. See Fig. 1-1(b).
10. Resistance as a circuit element is designated by uppercase R. Dynamic resistance of devices such as transistors and diodes, and internal source resistance, are designated by lowercase r.

Unit symbols consist of a letter or group of letters from the English and Greek alphabets, plus a few special symbols.

1. Unit symbols are printed in roman (upright) type. They are never given subscripts or superscripts.
2. Lowercase letters are used except where the symbol was derived from a proper name, in which case the first letter is capitalized.

QUESTIONS AND PROBLEMS

1-1. How many years elapsed from the invention of the telegraph to the invention of the triode vacuum tube? How many from the invention of the triode to the invention of the transistor?

1-2. When was the first electronic digital computer developed? Did it use tubes or transistors?

1-3. Name three areas of the United States that have heavy concentrations of electronics industries.

1-4. What is the recommended procedure for troubleshooting electronic circuits?

1-5. Why is mathematical analysis necessary in electronic servicing?

1-6. The quantity symbol for the instantaneous total voltage across a resistor is v_R. Distinguish clearly between capital and lowercase letters and write the symbols for:
(a) dc voltage across the resistor
(b) rms ac voltage across the resistor
(c) peak ac voltage across the resistor
(d) instantaneous ac voltage across the resistor, exclusive of dc

1-7. Figure 1-2(a) shows the flow of electrons in a circuit and the polarity of the voltage across each resistor as determined by the fact that electrons flow from negative to positive through a resistance. Copy Fig. 1-2(b) and mark the direction of electron flow in each resistor. Mark the polarity of the voltage across each resistor.

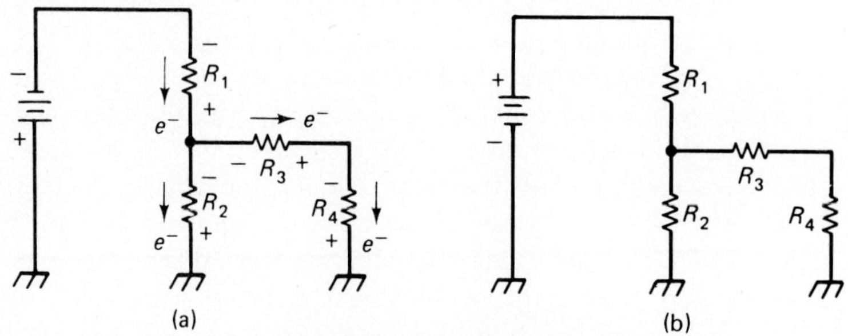

Figure 1-2 Problems 1-7 and 1-8.

1-8. Copy Fig. 1-2(b) again and mark the direction of conventional current in each resistor. Use the symbol →⊕. Mark the polarity of the voltage across each resistor using the fact that the direction of conventional current is from positive to negative through a resistor. Is the polarity the same or different from what you determined in Problem 1-7?

1-9. Draw a resistor in a box. Bring out the two leads and mark them A and B where they leave the box. If $I_A = -120$ mA, which end (A or B) of the resistor is positive? What is I_B?

1-10. Use Appendix H to identify the quantity and unit symbols below. Examples: I: quantity symbol for electric current; F: farad; unit of capacitance.

(a) R	**(b)** V	**(c)** P	**(d)** V	**(e)** L
(f) Hz	**(g)** H	**(h)** C	**(i)** Ω	**(j)** W
(k) A	**(l)** s	**(m)** S	**(n)** Q	**(o)** C

2

DIODE CHARACTERISTICS

2.1 DIODE PROPERTIES AND APPLICATIONS

The diode provides an excellent starting point for the study of electronic circuits because it leads immediately to practical applications in power supplies (Chapter 3) and also because it serves to introduce concepts and terminology which will be carried on to later chapters.

Diodes are *two-terminal* devices like resistors, capacitors, and inductors, and they are likewise used to control the current from one terminal to the other. Unlike resistors, capacitors, and inductors, however, diodes are *unilateral* rather than bilateral devices. This means that they do not exhibit the same response to currents in either direction. While there are special-purpose diodes which conduct in both directions, the basic diode, constituting perhaps 95% of all diodes used, is a one-way street for electric current. Thus we have resistors to limit the passage of current, capacitors to pass alternating current only, inductors to pass direct current only, and diodes to pass current in one direction only. The particular direction of allowed current can be changed by simply interchanging the two leads of the diode in the circuit.

Several applications for a device which conducts in only one direction suggest themselves immediately. One is a safety device to prevent the mistaken application of the wrong polarity voltage to a circuit, as shown in Fig. 2-1(a). Note that the direction of allowed *electron* flow is *against* the arrow of the diode symbol; the diode arrow points in the direction of the allowed *hole* current. Such circuits are sometimes used in portable radios to prevent damage to the transistors from improperly inserted batteries.

The "band" side of the diode symbol is called the *cathode*. It is the side to which the negative voltage is applied to make the diode conduct. The "arrowhead" side of the diode is called the *anode*. In general, a positive electrode is an *anode* and a negative electrode is a *cathode*. Diodes usually have a band painted on their cathode end, or sometimes they are bullet-shaped, with the bullet pointing in the direction of the arrowhead.

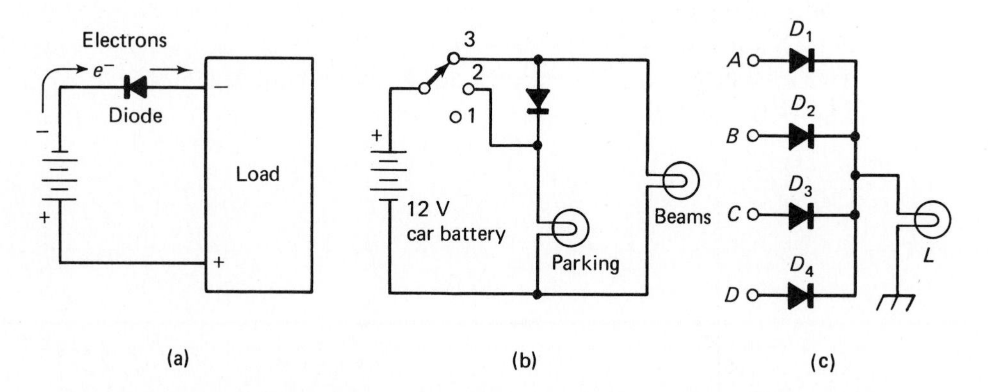

Figure 2-1 Basic diode applications: (a) reverse-voltage protection; (b) current steering; (c) diode decoupling.

A second example, called current steering, is shown in Fig. 2-1(b). The diode has been connected to allow current through the parking lights whenever the beams are switched on (position 3) but current cannot go back through the diode to turn on the beams when the parking lights only are switched on (position 2).

A final example, called diode decoupling, is shown in Fig. 2-1(c). A positive signal at any of the four inputs will cause lamp L to light, but if A only is positive, for example, diodes D_2, D_3, and D_4 will prevent A from "backing up" current into B, C, and D. This trick is often used in computing circuits.

The most common use of the diode is in converting alternating current (from the 60-Hz ac line, for example) into the direct current which is needed for powering most electronic systems. This is the subject of Chapter 3, and further discussion is deferred until then.

2.2 LINEAR VS. NONLINEAR ELEMENTS

Although we often think of the diode as being simply a short circuit in one direction and an open circuit in the other, it is good to keep in mind that this is not strictly true. Solid-state diodes do have some finite resistance in the reverse direction (generally a high number of megohms), but, more important, they do exhibit some resistance in the forward or "turned-on" direction. The character of this forward resistance is not at all the same as what we have been used to in resistors, as Fig. 2-2 illustrates.

Figure 2-2 shows the voltage vs. current curves for (a) a 1-kΩ resistor, (b) a nonexistent "perfect" diode, and (c) a real diode. Notice that the "curve" for the resistor is actually a straight line; we say that the resistor is a *linear* device. For this type of device, if 0.1 V will cause 0.1 mA, then twice that much voltage will cause twice that much current, 0.3 V will cause 0.3 mA, and so on. In the reverse direction, −0.1 V will cause −0.1 mA of current (the minus signs simply indicate that the polarity of the applied voltage is reversed and that the direction of current is reversed). For a linear resistive device, the ratio of voltage to current is always the same. This ratio, of course, is called resistance, and in the example of the resistor of Fig. 2-2(a), it is always 1000 Ω.

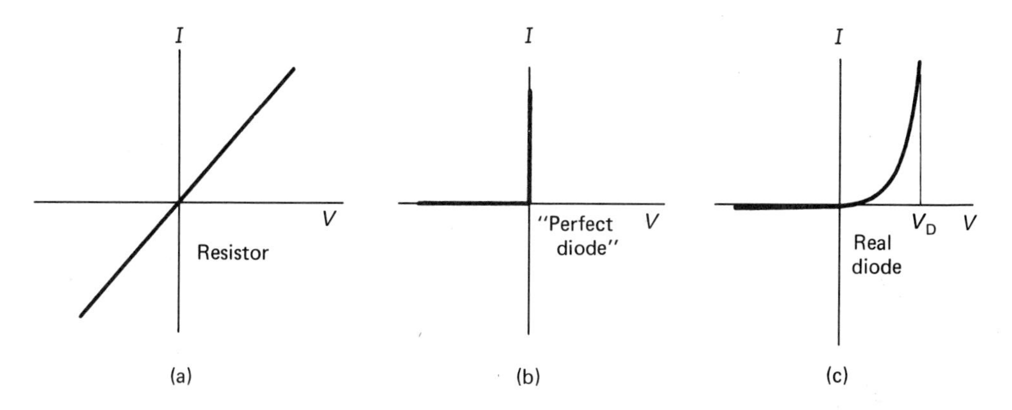

Figure 2-2 Linear and nonlinear elements.

The curve of the idealized diode in Fig. 2-2(b) consists of two linear segments. In the forward direction (vertical line segment) a very large current, say 1 A, could be caused by nearly zero voltage, so the resistance is

$$R = \frac{V}{I} = \frac{0}{1} = 0\,\Omega$$

In the reverse direction (horizontal line segment) a fairly large voltage, say 10 V, will produce no current at all, and the resistance is

$$R = \frac{V}{I} = \frac{10}{0} \longrightarrow \infty$$

in other words, an open circuit.

This ideal diode does not exist, of course, but the foregoing does make the point that the resistance of a device is indicated by the *slope* of its *V* vs. *I* curve. Very low resistances are indicated by a nearly vertical line, and very high resistances by a nearly horizontal line.

For the real diode whose characteristics are given in Fig. 2-2(c), the reverse resistance is very nearly infinite, but the forward part of the curve is decidedly *nonlinear*. Its slope changes from nearly horizontal at low forward voltages to nearly vertical at some forward voltage V_D. This means that the forward resistance of the diode remains quite high until the forward voltage reaches a *turn-on* region where current begins to increase rapidly and the resistance drops to a low value.

2.3 DIODE VOLTAGE LIMITS

Peak inverse voltage (PIV), also called peak reverse voltage (PRV), is the maximum voltage that can be applied across the diode in the back direction with assurance that it will not break down and begin to conduct heavily. Breakdown in a circuit usually means the destruction of the diode (and possibly a great amount of other hardware) since the power dissipation increases tremendously for currents in the reverse direction. For example, compare 100 mA at 1 V in the forward direction (0.1 W) with 100 mA at 200 V in the reverse direction under breakdown (20 W).

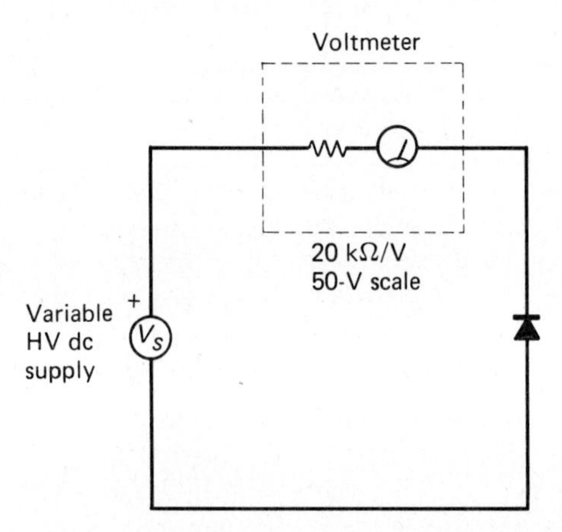

Figure 2-3 Experimental setup for nondestructive determination of the reverse breakdown voltage of a diode.

The circuit of Fig. 2-3 can be used to test diode PIV nondestructively, since the reverse current is limited by the high resistance of the voltmeter. The supply voltage is advanced until the voltmeter just begins to show a reading above zero. The diode is then at the breakdown point, and the supply voltage equals the diode PIV.

Series stringing: When an application requires a peak inverse voltage greater than that of any available diode, a number of diodes of the same type may be connected in series as shown in Fig. 2-4(a). The PIV of the string is then equal to the *sum* of the individual diode PIV ratings. For low-current applications the diodes alone can usually withstand the back voltage, but for higher-power systems it is advisable to shunt each diode with an identical capacitor as in Fig. 2-4(b).

Turn-off time: The reason for the capacitors is illustrated in Fig. 2-4(c). When the diode is turned on by a forward voltage, there is a forward current which is determined by the value of the voltage and the circuit resistance in series with the voltage source and diode. If the polarity of the voltage source is suddenly reversed, however, the diode will remain turned on for a very short time and will pass current

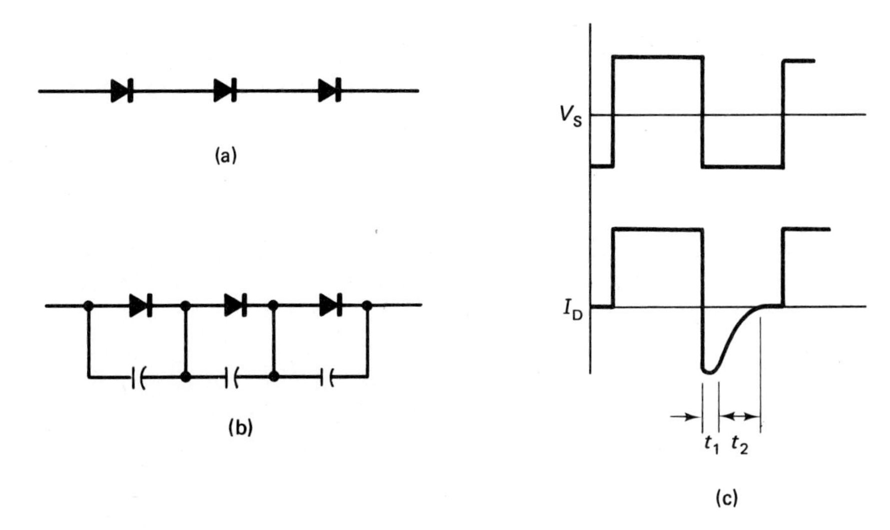

Figure 2-4 (a) Series stringing to increase diode PIV; (b) equalizing capacitors to protect "fast" diode; (c) diode turn-off time.

in the reverse direction. This recovery time is longer for higher forward currents and may be several microseconds for power diodes. The problem arises because some of the diodes in the string will turn off faster than others. The first diode to turn off will then be required to handle the *entire* reverse voltage since the other diodes will still be conducting. The capacitors ensure a path around the fast diode, in effect making all the diodes appear slow. The value of these equalizing capacitors should be in the range 0.005 to 0.02 μF. If the applied voltage is a 60-Hz sine wave which has been filtered against transient spikes, the diodes will have ample time to recover before the reverse voltage rises to a dangerous level, and the capacitors should not be necessary. The availability of silicon diodes with PIVs of 1 kV and more makes series stringing of diodes unnecessary in most cases.

2.4 SOLID-STATE-DIODE TYPES

Germanium point contact diodes are the direct descendents of the "cat's whisker" radio receivers of the early part of this century and have been available since the early 1950s. Although they are still popular for detecting small signals, they are not suited for high-current applications because of the delicate nature of the fine wire contact. Notice from the characteristic curve of Fig. 2-5 that the forward voltage drop becomes quite large at high currents.

Germanium junction diodes have a low forward voltage drop (around 0.2 V for low currents), but the material germanium is, unfortunately, rather temperature sensitive. Most germanium devices are designed to operate at case temperatures no higher than 80°C. This limits the current-carrying capability of germanium diodes.

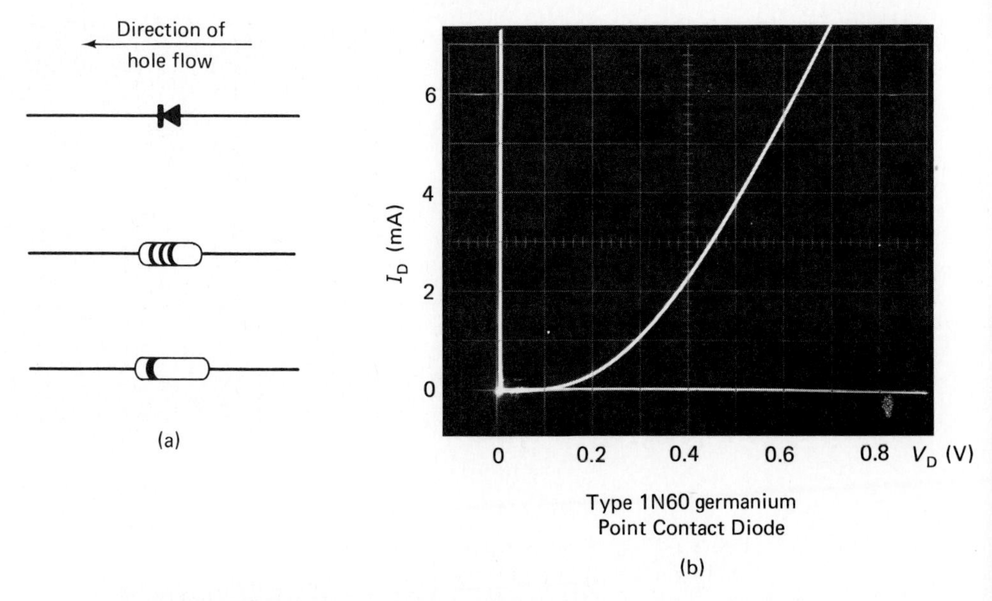

Figure 2-5 Germanium point contact diodes: (a) physical appearance; (b) characteristic curves.

Furthermore, at high temperatures, germanium diodes tend to leak current excessively in the reverse direction, a problem which is much less severe in silicon diodes.

Silicon junction diodes are extremely popular both for small-signal and power applications, with voltages up to the kilovolt range and currents up to the 100-A range. They can operate at temperatures up to 200°C, making them generally more suited to high-current applications than germanium diodes, even though their forward voltage drop is greater (around 0.6 V for low currents). Even with this relatively low voltage drop, a diode carrying 50 A of current can generate a considerable amount of heat, so the higher current units are often constructed with a threaded stud which is used to attach the diode to a finned heat sink.

Figure 2-6 shows the relative characterstics of silicon and germanium junction diodes.

Schottky or hot-carrier diodes have a forward drop of about 0.3 V at low currents. They can switch from the forward *on* state to the reverse *off* state much more quickly than can silicon or germanium junction diodes. Low-power Schottky diodes can switch at frequencies in the range of 1 GHz (10^9 Hz), while high-power Schottkys can switch several amperes at 100 kHz. Their disadvantages include lower reverse breakdown voltage, higher reverse leakage current, and higher cost.

Light-emitting diodes (LEDs) are used as indicator lamps, although in principle they are much the same as other junction diodes. The most popular LEDs employ gallium arsenide or gallium phosphide semiconductor and emit a red glow when forward biased at about 3 to 15 mA. Their turn-on voltage is typically about 1.2 to 1.5 V. Other types of LED emit yellow, green, or invisible infrared radiation. LEDs

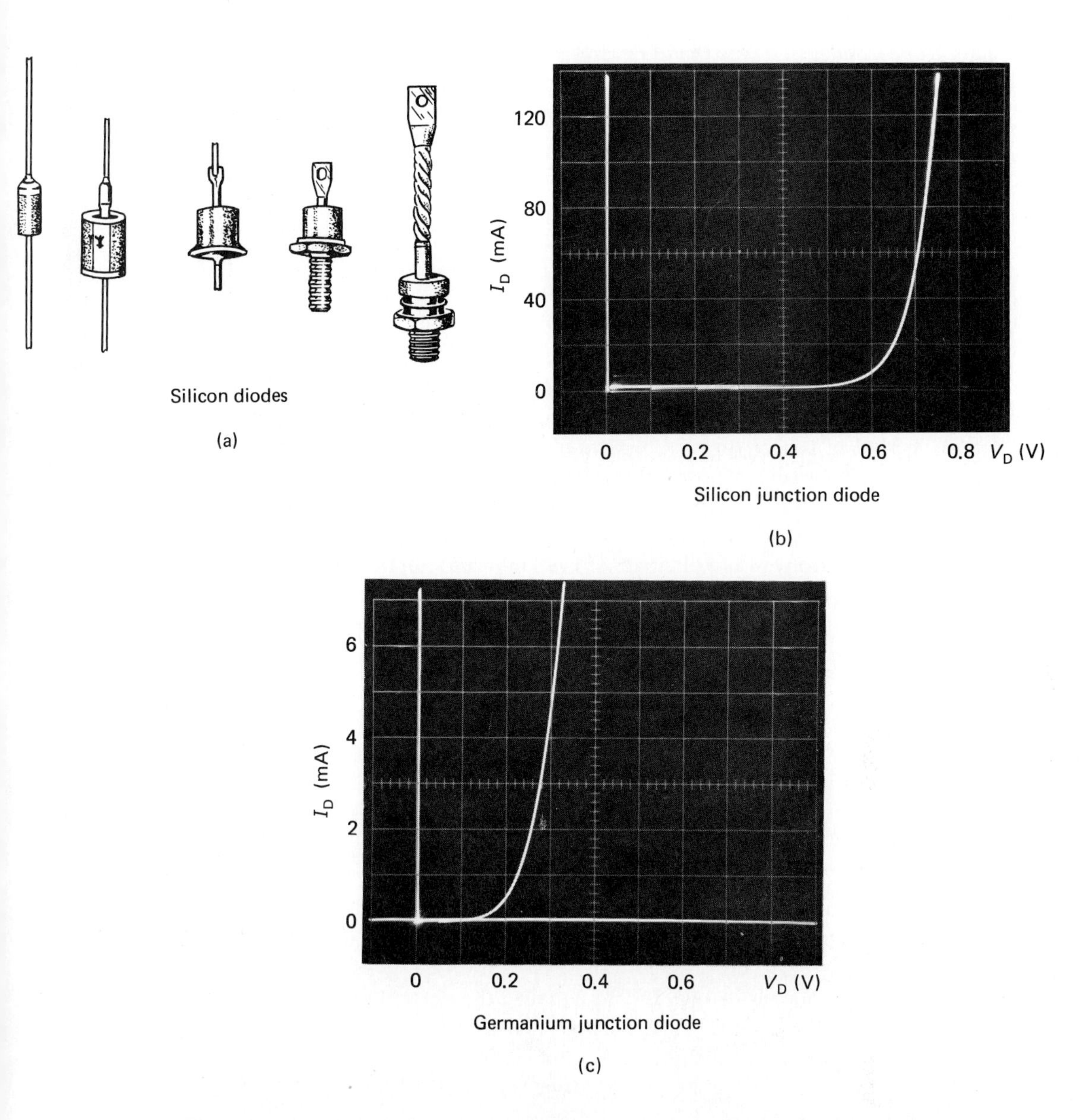

Figure 2-6 Junction diodes: (a) low- and high-power silicon diodes; (b) silicon junction diode curves; (c) germanium junction diode curves.

have all but replaced incandescent lamps as indicators because they have a longer life, are more rugged, less expensive, and require less voltage and current. They can be switched on and off in a fraction of a microsecond.

Diode-drop approximations: For the remainder of this text we will use the following approximations for diode forward-voltage drop, realizing that the actual drop varies continuously with current, with temperature, and from unit to unit.

Diode	10 mA or less	Above 10 mA
Silicon junction	0.6 V	0.9 V
Schottky hot-carrier	0.3 V	0.6 V
Germanium junction	0.2 V	0.5 V
Light-emitting	1.4 V	1.4 V

2.5 DIODE CURRENT LIMITS

Two diode current limits are generally specified: the *average* and the *peak* (or surge) current limits. The *average current* can be maintained indefinitely without damage to the diode, whereas the peak current can be maintained only for some specified short period of time (typically 20 ms). The peak current is typically 10 times the average value for a silicon power diode. Although there is no simple nondestructive test to check the current capability of a diode, some indication can be obtained from the physical size of the diode. For example, a silicon diode carrying 100 mA (at 0.9 V forward drop) would dissipate only 90 mW and might be as small as a grain of wheat, but a 10-A diode would have to dissipate more than 9 W and would be quite a bit larger, probably operating with a heat sink.

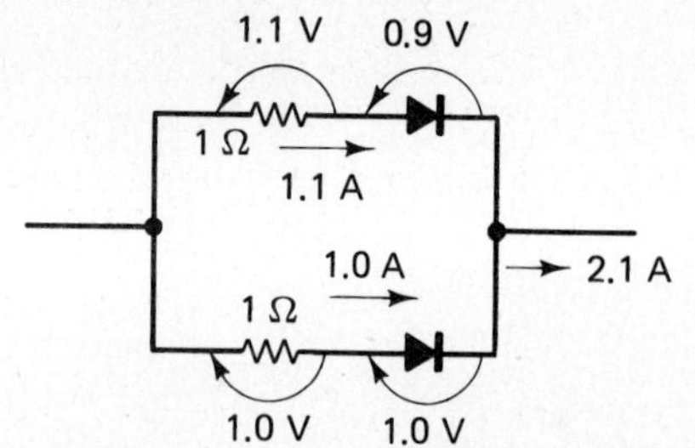

Figure 2-7 Series resistors equalize currents when paralleling diodes.

Diodes cannot be simply connected in parallel to provide higher current ratings, since one diode will usually turn on at a lower voltage than the others and hog most of the current. This problem can be circumvented by using a series resistor with each diode, as shown in Fig. 2-7, to equalize the current in each branch. However, this increases the forward voltage drop, since the resistor must be calculated to drop about 1 V to swamp out the effects of diode differences. Since diodes are readily available with average forward-current ratings of several hundred amperes, it is not often desirable to parallel diodes.

2.6 TEMPERATURE EFFECTS ON DIODES

Silicon diodes are typically designed to operate with maximum case temperatures from 100 to 200°C (212 to 392°F). To keep their temperature down, the larger diodes are generally bolted to a finned heat sink or to the metal chassis with a thin mica wafer providing electrical isolation. Small diodes are sometimes cemented to the chassis with a special thermally conductive adhesive to provide heat sinking. Germanium diodes are especially temperature sensitive and are consequently used mostly in low-power applications.

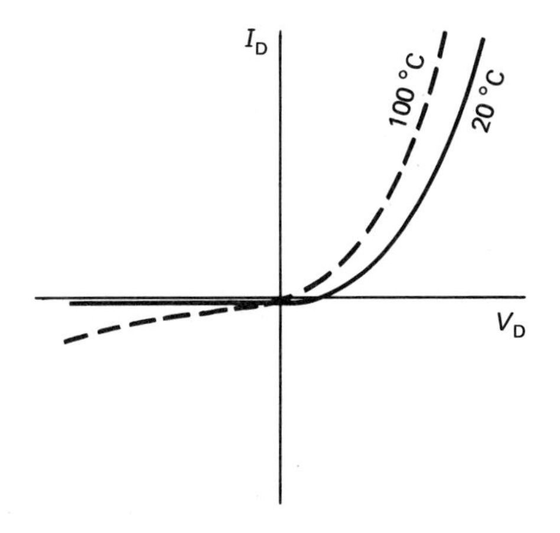

Figure 2-8 Solid-state diode curves at room temperature (solid line) and elevated temperature (dashed line).

The two major effects of an increase in temperature on diode characterstics are shown in Fig. 2-8. First, higher temperatures make the diode conduct more readily, so the turn-on voltage decreases by as much as a tenth of a volt. Second, the resistance of the diode in the reverse direction decreases. This reverse leakage becomes serious at relatively low temperatures for germanium diodes but is not generally as noticeable in silicon types.

2.7 SEMICONDUCTOR CRYSTAL STRUCTURE

Up to this point we have been discussing the external behavior of solid-state diodes without much regard for the internal structure which gives these devices their properties. This topic, usually called semiconductor physics, can become quite involved, but a simplified and abbreviated treatment is given in Sections 2.7 and 2.8. Numerous texts are available for those who wish to study semiconductor physics in greater detail.

Silicon and germanium atoms have a structure which places them in the category of materials called semiconductors. Silicon atoms have 14 electrons and germanium atoms have 32 electrons per atom, but both types of atoms have 4 electrons in their *valance shell*, the outside layer which gives the atom its basic electronic properties. A complete shell would have 8 electrons in it, and there is a tendency for atoms with incomplete shells to cling to other atoms with similar shortages in order to produce a filled shell by sharing electrons.

In a piece of pure silicon, each silicon atom is surrounded by four neighbors, as represented in Fig. 2-9.

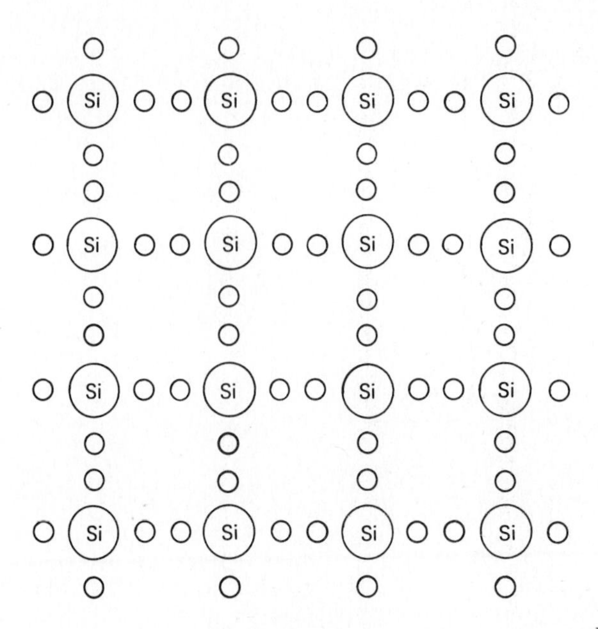

Figure 2-9 A perfect crystal lattice has no free electrons to carry current.

It can be seen that each inner atom is surrounded by 8 electrons in this structure; 4 of its own and 1 from each of its four neighbors. This structure is called a crystal lattice and tends to be quite stable, allowing little freedom of movement for the electrons. This is reflected by the fact that pure silicon is almost insulative at low temperatures.

Impurity doping: It is only when impurities are added to the lattice structure that the material becomes a semiconductor. For example, if one atom in every million is not silicon but arsenic, there will be a certain number of extra free electrons in the lattice, since arsenic has five electrons in its outer shell (see Fig. 2-10). These extra electrons have no place to fit into the lattice structure and are not tightly bound to any particular atom. We say that the semiconductor has been *doped* with an *N*-type impurity since free *negative* electrons have been added to the lattice. These free, mobile electrons are termed *charge carriers* since they enable the material to conduct electric current.

It is important to note that the material has not actually been given a net electric charge. The arsenic impurity atoms bring an extra proton for every extra electron in the lattice structure. The point is that the electrons are mobile while the protons are fixed in the lattice. Therefore, the *N*-type silicon does not have a negative charge; it has negative charge carriers. It also bears repeating that impurities on the order of 1 part per million are actually capable of transforming the material from a near insulator to a semiconductor.

P-type doping: If a pure crystal lattice is doped with an impurity element having three electrons in the outer shell (such as boron) a similar but distinctly different phenomenon occurs (see Fig. 2-11). Now instead of a free electron we have a hole or

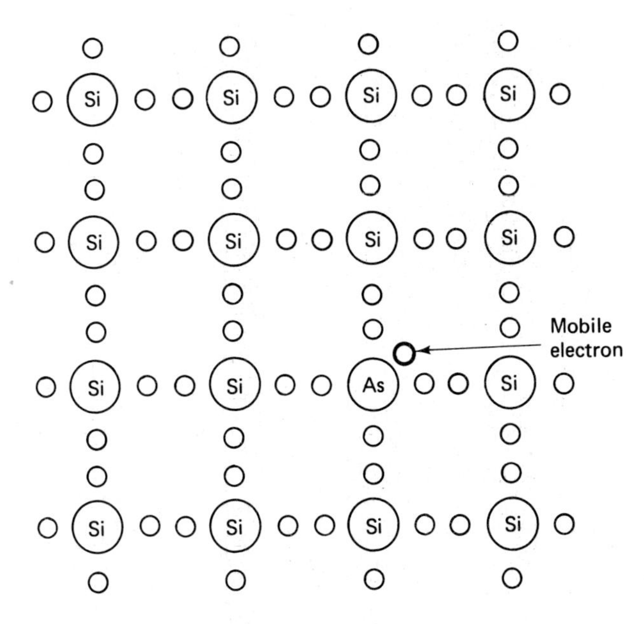

Figure 2-10 An impurity element in the lattice can provide an extra (free) electron, providing a means of carrying current (*N*-type semiconductor).

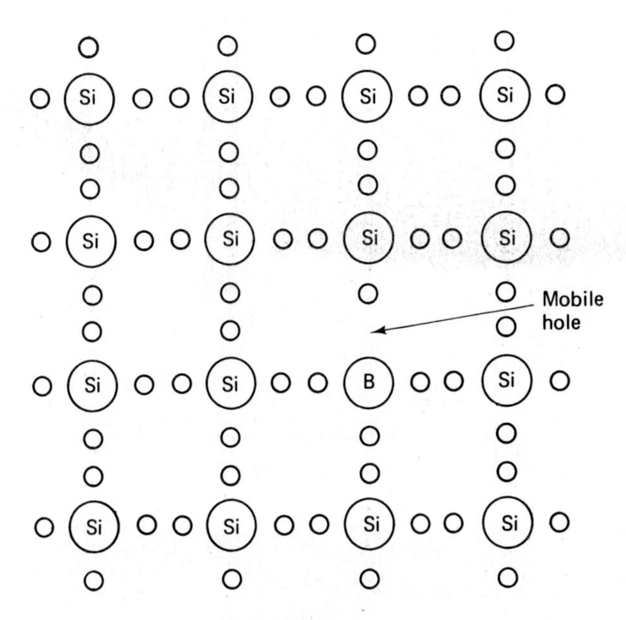

Figure 2-11 An impurity which introduces holes (electron missing) also provides a means of carrying current (*P*-type semiconductor).

vacant spot where there should be an electron, but no electron is present. This hole is an invitation for an electron from a neighboring atom to move in, leaving a new hole behind. If a voltage is applied across the semiconductor material, there will be a migration of electrons toward the positive side made possible by using the holes as stepping stones. A little reflection will show that as the electrons move toward the positive side, the holes will move toward the negative side of the material.

The conduction in this case is made possible by the holes. Since the holes move from positive to negative, they are termed positive charge carriers, and the material is called a *P-type semiconductor*.

2.8 P-N JUNCTION BEHAVIOR

When *P*-type and *N*-type semiconductor materials are joined, a diode is formed. The basic action of this diode can be explained with reference to Fig. 2-12.

With no external voltage applied, the *P-N* device has hole charge carriers in the *P* side and free-electron charge carriers in the *N* side. Notice that only the mobile charge carriers are depicted in the figure; the atoms and outer-shell electrons are not active and therefore not shown.

Reverse bias: When an external voltage is applied, positive to the *N* side and negative to the *P* side of the diode, the free electrons are pulled by the positive charge

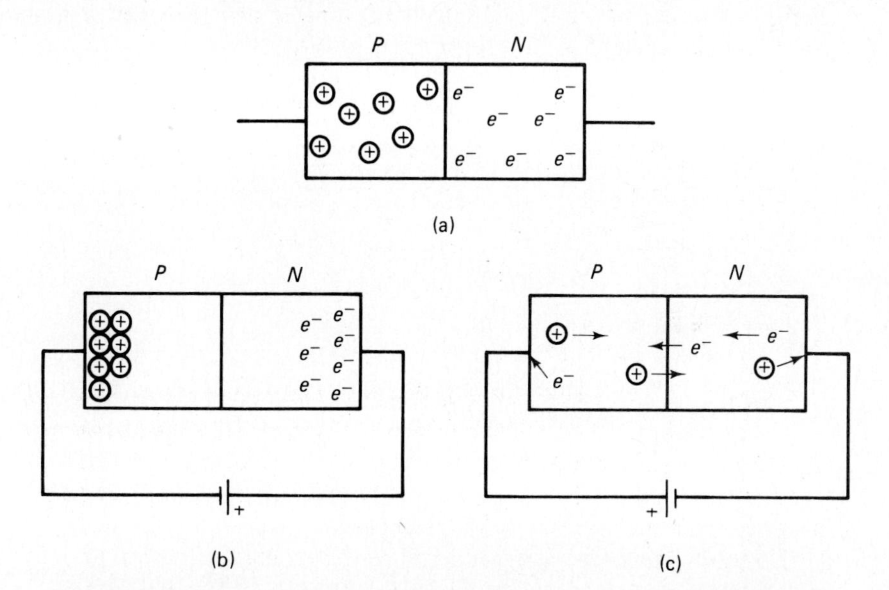

Figure 2-12 (a) A diode is formed by the junction of *P*-type (mobile holes) and *N*-type (mobile electrons) semiconductor material. (b) Reverse bias: charge carriers are attracted away from the junction, leaving an insulating depletion zone. (c) Forward bias: charge carriers are attracted across the junction, resulting in current through the diode.

of the battery to the extreme right-hand end of the diode, and the holes are pulled by the negative charge of the battery to the opposite end [see Fig. 2-12(b)]. This leaves a wide region around the junction of the *P* and *N* materials which is depleted of charge carriers. With the voltage applied in this way, no current can pass across the *P-N* junction, and the diode is said to be *reverse biased.*

Forward bias: If the external battery is connected positive to the *P* side and negative to the *N* side of the diode, as shown in Fig. 2-12(c), holes from the *P* side are attracted across the junction to the negative terminal of the battery. Similarly, electrons from the *N* side of the diode are attracted across the junction to the *P* side, where they make their way to the positive terminal of the battery. The diode is now *forward biased,* and current passes freely.

Forward voltage drop: This leaves unanswered the question of why it takes a certain voltage (0.2 V for germanium or 0.6 V for silicon) before the diode begins to conduct at all. This phenomenon can be explained by taking a closer look at what happens in the vicinity of the *P-N* junction before any voltage is applied.

Figure 2-13 shows that some of the free electrons in the *N*-doped material diffuse across the junction to fill in some of the holes in the *P* material near the other side of the junction. In a similar manner, some of the holes in the *P* material near the junction diffuse across to combine with the electrons on the other side. This represents a migration of charges (not just charge carriers), positive to the right and negative to the left in Fig. 2-13. The result is that a potential difference or voltage barrier is formed at the *P-N* junction by diffusion of charge carriers. This voltage must be overcome by the externally applied voltage before current can begin.

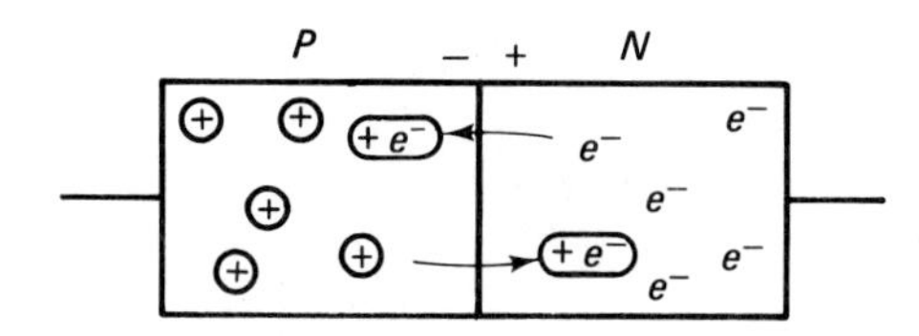

Figure 2-13 Migration of electrons to *P* side and holes to *N* side creates charge imbalance (− on *P* side, + on *N* side), resulting in potential barrier, V_D.

CHAPTER SUMMARY

1. Diodes pass current in one direction but block it in the other direction.
2. The arrow of the diode symbol points the direction of allowed *conventional current* (positive to negative). Electron flow is against the direction of the arrow.
3. Most diodes have a band or group of bands at one end. This represents the bar or band to which the arrow points in the symbol.
4. Silicon junction diodes, which are the most common, drop about 0.6 V when passing currents less than 10 mA and about 0.9 V for forward currents greater than 10 mA. These values are approximate and vary from unit to unit and with temperature.
5. Diodes may be called upon to pass high forward currents (50 A), withstand high reverse voltages (1 kV), or switch from conduction to nonconduction very quickly

(0.01 μs). No one diode can do all of these things, so diodes must be selected for each application.

6. The basic diode is formed as a junction of *P*- and *N*-type silicon. *N*-type silicon has an impurity element which leaves unbound (mobile) electrons in the crystal structure. *P*-type silicon has an impurity element which leaves mobile holes in the crystal structure.

QUESTIONS AND PROBLEMS

2-1. In Fig. 2-1(b), is the diode forward biased (ON) or reverse biased (OFF) with the switch in position 2? With the switch in position 3?

2-2. In Fig. 2-1(c), assume that four voltage sources are connected to the inputs: $V_A = +2$ V, $V_B = -3$ V, $V_C = +5$ V, and $V_D = -7$ V. Which diodes are ON and which are OFF?

2-3. On a graph of I (vertical) vs. V (horizontal), what resistance is indicated by (a) slight upward slope; (b) steep upward slope; (c) horizontal line; (d) vertical line?

2-4. How can a 2500-V reverse voltage be withstood if the best diodes available have a PRV rating of 1000 V?

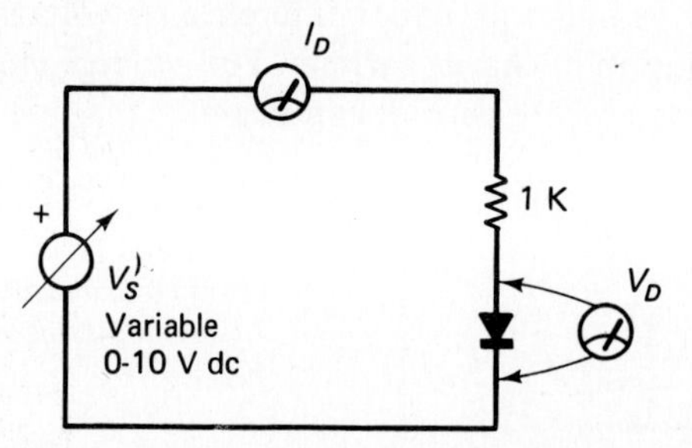

Figure 2-14 Problem 2-5.

2-5. In Fig. 2-14, V_S is set to 4.0 V and the diode is a silicon junction type. Determine V_D and I_D. Use the approximation given in Section 2.4.

2-6. What is the power dissipated in a silicon junction diode carrying a forward current of 1 A?

2-7. Two silicon junction diodes are connected in parallel. The combination carries 1 A. Does each one carry $\frac{1}{2}$ A? Why or why not?

2-8. Increased temperature causes silicon-diode forward voltage drop to _________ and reverse leakage current to _________.

2-9. What is the purpose of shunting each power diode in a series string with a capacitor?

2-10. How many electrons are there in the outer shell of each silicon atom? How many in the outer shell of each germanium atom?

2-11. How many electrons are there in the outer shell of a *P*-type impurity element? How many in the outer shell of an *N*-type impurity element?

2-12. What is a "hole" in semiconductor terminology?

2-13. If the negative side of a source is connected to the *N*-doped side of a diode, the diode is __________ (forward or reverse) biased.

2-14. Migration of charge carriers across a *P-N* junction leaves a __________ charge on the *P* side.

3

BASIC POWER SUPPLIES

3.1 FUNCTION OF A POWER SUPPLY

Electronic systems such as amplifiers, radio receivers, temperature controllers, and oscilloscopes perform their functions by using small voltages and currents to control much larger sources of current. The success of this technique depends on the availability of sources of pure (unvarying) direct current at the desired voltage and current. For solid-state systems, the voltage required is commonly in the range 5 to 50 V. The voltage available at the electric utility outlet is generally ac at 110 or 220 V. Thus we are faced with the problem of taking the readily available ac line voltage and converting it to direct current at the desired voltage level. This is the function of a power supply.

Almost every electronic device utilizes at least one power supply, and because they are relatively uncomplicated in their basic form, they often consititute part of a new technician's initial assignment on a project. Another reason they are frequently encountered by technicians is that they are, in a sense, the "workhorse" part of the system, since all the power used by the system must be fed through the power supply. Hence a high percentage of system failures tends to be concentrated in the power supply.

3.2 THE DIODE RECTIFIER

A diode used to change ac to dc is called a rectifier. An elementary rectifier circuit is shown in Fig. 3-1.

R_L is the load resistance and is not generally a resistor as such, but simply represents the system which the power supply is feeding. Since most electronic systems pass current and dissipate power like resistors, it is helpful to represent them as a resistor, with a value equal to V/I, that is, the voltage they require divided by the current they pass.

D_1 is, of course, the diode rectifier, and it may be a silicon, germanium, or hot-carrier type. In most new designs, D_1 will be silicon, and we must remember that it will, in that case, take the first 0.9 V from the transformer just to turn it on; any voltage beyond that will appear across the resistor. The transformer in Fig. 3-1 delivers a voltage of 10 V, but it must be remembered that this is a sine wave with a 10-V rms value. Its *peak* value will be 1.41×10 V or 14.1 V. The silicon diode drops 0.9 V, so the peak voltage at the load is $14.1 - 0.9$, or 13.2 V. The diode allows no current through the resistor on the negative half-cycle, so the total voltage across R_L is a pulsating dc, as shown in Fig. 3-1.

Know I Flow
Know Diode Direction

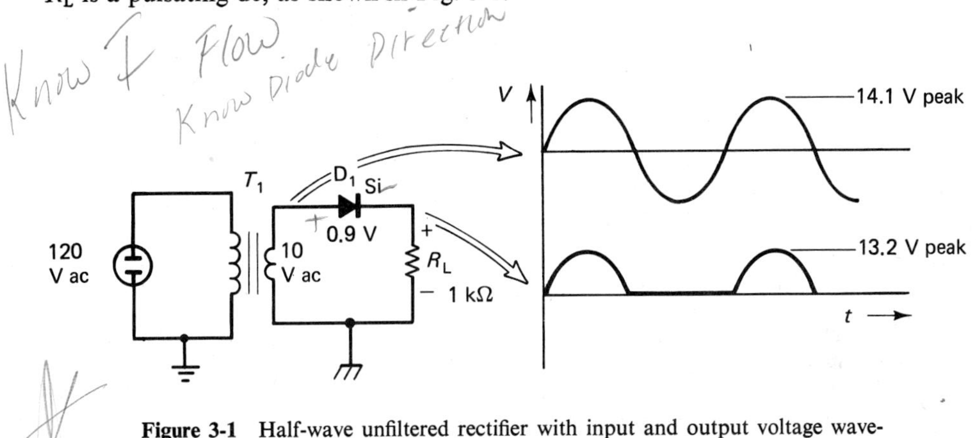

Figure 3-1 Half-wave unfiltered rectifier with input and output voltage waveforms.

This circuit is called a *half-wave* rectifier, since current is allowed to pass for only one-half of the incoming waveform.

Since hole current passes in the direction of the diode arrow, and holes flow from positive to negative, it can be seen that the polarity of the voltage output is plus at the diode's cathode end as marked. If a negative output voltage is required, it is only necessary to reverse the diode in the circuit (point it to the left).

3.3 ISOLATION TRANSFORMERS

The transformer in Fig. 3-1 is used to step the voltage up or down to the level required by the load, but it serves the additional purpose of isolating the circuit chassis from earth ground. This is an important safety feature and is illustrated in Fig. 3-2.

One side of the ac line (the so-called neutral) is generally connected to earth ground. Thus it is possible to be connected to one side of the ac line by simply standing on a damp floor or touching a water pipe, a faucet, an electrical panel, or heating duct. If the ac plug is inserted in such a direction that the chassis of a piece of transformerless gear is "hot" with respect to ground, a severe shock can be obtained by simply touching the chassis, if a path to earth ground is present.

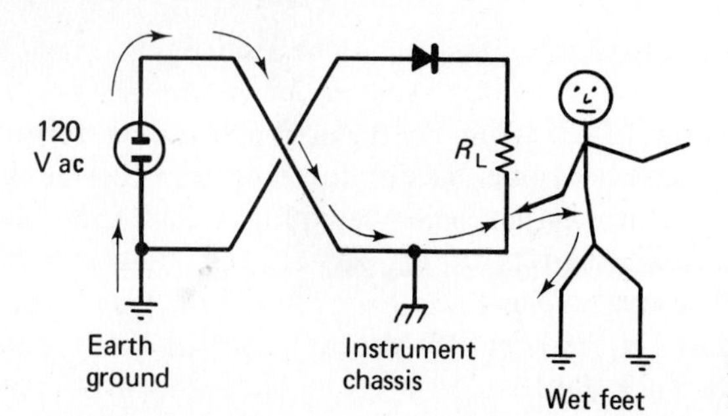

Figure 3-2 Shock hazard presented by transformerless instruments.

The 120-V line will give a peak voltage of 170 V when rectified, and it is very tempting to omit the transformer when voltages in that range are required, especially in view of the weight and cost added to a piece of gear by the transformer. However, this is *not* good practice and is to be strongly discouraged. Transformerless, or so-called ac-dc, power supplies have been quite common in low-cost entertainment equipment because of the fiercely competitive nature of this high-volume market, but quality equipment for commercial or industrial use invariably uses isolating transformers.

3.4 THE SIMPLE CAPACITOR FILTER

The rectifier of Fig. 3-1 is suitable for a few applications which do not require pure dc (electroplating is one such application), but power supplies are more often required to deliver, not pulses of dc, but pure unfluctuating dc. This requires that the rectifier circuit be followed by a filter circuit.

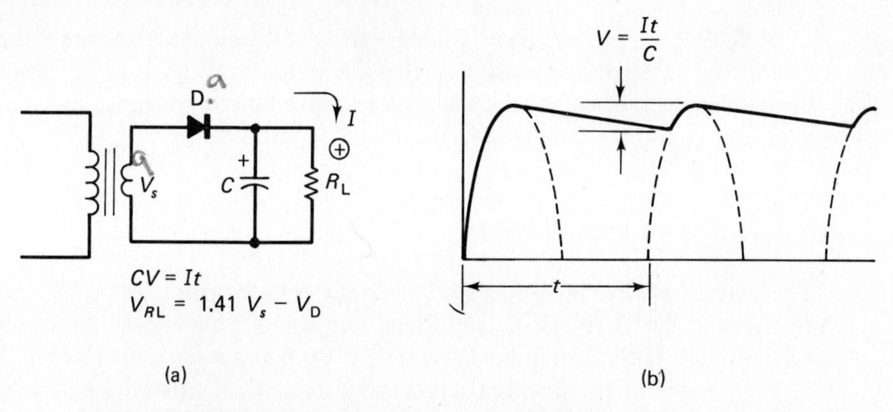

Figure 3-3 Half-wave rectifier with single capacitor filter: (a) circuit with design formulas; (b) output waveform showing ripple amplitude.

The simplest filter is shown in Fig. 3-3. It consists of a single large capacitor connected across the load. The capacitor will store a charge,

$$Q = CV \tag{3-1}$$

where C is the capacitance in farads and V is the peak transformer secondary voltage minus the diode's voltage drop. As soon as the transformer voltage begins to drop below its peak value (at the 90° point of the cycle) the diode will see a lower positive voltage on its left side than on its right side, since the capacitor tends to remain charged at its peak voltage. The diode then turns off, leaving a circuit consisting of the capacitor and the load resistance. The current drawn by the load resistance will start to discharge the capacitor, resulting in a lowering of the output voltage. If the capacitor were allowed to discharge to any great extent, we might attempt to find the value of filtering capacitance required by the familiar time-constant discharge formula, $\tau = RC$. However, in practice the capacitor is seldom allowed to discharge to less than 80% of its peak voltage, and a more convenient formula, which assumes a constant rate of discharge, can be employed: The definition of charge is $Q = It$. But the capacitor charge is $Q = CV$. Combining the two expressions for Q, we obtain

$$CV = It \tag{3-2}$$

where C is the value of the filter capacitor in farads, V is the voltage drop during discharge, I is the dc current drawn by the load, and t is the time between successive charging pulses. The pulses occur at a rate of 60 per second for a half-wave rectifier on a 60-Hz ac line, so the time, t, is

$$t = \frac{1}{f} = \frac{1}{60} = 16.6 \text{ ms}$$

Example 3-1

Design a power supply which will deliver −9 V dc to a load at 30 mA. The output voltage must drop no more than 0.5 V between charging pulses. Use the circuit of Fig. 3-4.

Solution If the output is 9 V and the diode drops 0.9 V, the peak transformer secondary voltage must be

$$V_{s(\text{pk})} = 9 + 0.9 = 9.9 \text{ V}$$

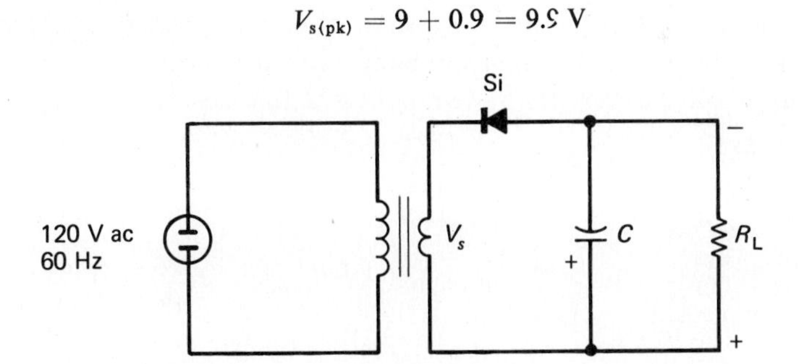

Figure 3-4 Negative-voltage power supply design example.

The rms transformer voltage is then

$$V_s = \frac{V_{s(pk)}}{1.41} = \frac{9.9}{1.41} = 7.0 \text{ V}$$

The capacitor value is determined next:

$$CV = It \tag{3-2}$$

$$C = \frac{It}{V}$$

$$= \frac{(30 \times 10^{-3})(16.6 \times 10^{-3})}{0.5}$$

$$= 1000 \times 10^{-6}$$

$$= \mathbf{1000\ \mu F}$$

Notice that the polarity of the capacitor has been correctly indicated. This is important, since the only capacitors available with such high values are electrolytic types and these will short circuit if reverse voltage is applied.

3.5 RIPPLE-PERCENTAGE SPECIFICATIONS

The sawtooth-shaped voltage component which appears in the output of the power supply is given the graphic term *ripple*. This ripple is usually the cause of the low-pitched buzz or hum that is sometimes heard in the speaker of a low-cost piece of audio equipment. To compare power supplies in their ability to suppress ripple, a percentage-ripple specification is used:

$$\text{ripple factor} = \frac{\text{rms ripple voltage}}{\text{dc output voltage}} \tag{3-3}$$

Multiplying the ripple factor by 100 will, of course, give the *percentage ripple*.

Because the power content of the ripple voltage is generally what is of concern, the peak-to-peak value of the ripple voltage must be divided by 2.82 before it is used in the above formula. There is some error in using the factor 2.82 to convert from peak-to-peak to rms values for a sawtooth waveform, since that factor applies only to sine waves, but the error is not serious and we are interested in approximate results in most cases. In fact, the electrolytic filter capacitors typically used in the kinds of circuits under discussion have tolerances in the range of −20 and +80%, so highly accurate calculations are quite pointless in this case.

Example 3-2

Find the percent ripple for the circuit of Fig. 3-4:

Solution

$$\text{rms ripple} = \frac{\text{p-p ripple}}{2.82} = \frac{0.5}{2.82} = 0.177 \text{ V}$$

$$\text{ripple factor} = \frac{\text{rms ripple}}{\text{dc output}} = \frac{0.177}{9} = \mathbf{0.0197}$$

The ripple specification is approximately **2%**.

3.6 LOAD REGULATION

The output of an elementary power supply drops somewhat from no-load to full-load currents. This is generally undesirable and a load-regulation specification is used to compare supplies in their ability to deliver a constant output voltage regardless of load current.

$$\text{load regulation} = \frac{V_{NL} - V_{FL}}{V_{FL}} \tag{3-4}$$

where V_{NL} is the no-load voltage and V_{FL} is the voltage at full-rated load current. Ideal load regulation is zero percent. Typical values for elementary power supplies range from 2 to 50%. A fair comparison of load regulation requires that all supplies being compared be rated for the same full-load current. A supply with 20% regulation when rated at 200 mA would have a 10% regulation if rerated for 100 mA full load.

There are three causes for the drop in output voltage at full load. We will discuss each one in turn.

Diode drop: First, the drop across the rectifier diode is greater at full-load than at no load. We will use the approximations $V_D = 0.9$ V full load and $V_D = 0.6$ V no load for silicon junction rectifiers. Thus we will estimate $V_{O(FL)}$ to be 0.3 V less than $V_{O(NL)}$ due to the diode. In practice we may find this difference to be closer to 0.5 V in the case of a 1-A supply, and more like 0.2 V for a 10-mA supply, but we will stick with the 0.3-V approximation for the sake of convenience.

Capacitor drop: The second reason for the voltage drop at full load is the discharging of the filter capacitor between charging pulses. At no load, the filter capacitor simply charges up to $V_{s(pk)} - V_D$ and stays there. Under load the capacitor discharges (by a voltage $V_{rip(p\text{-}p)} = I_L t/C$) to some valley voltage ($V_{pk} - V_{rip(p\text{-}p)}$). The dc voltage is the *average* between the peaks and valleys of the rippling output voltage, as shown in Fig. 3-5.

In this Figure there is a 20-V p-p ripple across the first filter capacitor under full load. That is, the capacitor discharges from its peak of 100 V to a minimum of 80 V because of the heavy load current drawn from it between charging pulses. The

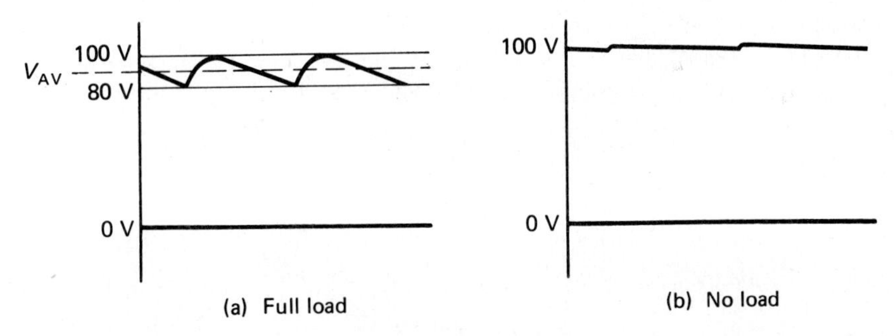

Figure 3-5 The average output voltage of a supply may drop to 90 V under load (a) and rise to 100 V at no load (b) because of ripple across the first filter capacitor.

average dc voltage of 90 V appears at the output under full load. Under no-load conditions the filter capacitor does not discharge appreciably and the output voltage rises to essentially 100 V dc. Thus the full-load output voltage is less than the no-load output by $\frac{1}{2}V_{\text{rip(p-p)}}$ because of the discharging of the filter capacitor. It should be apparent that larger filter capacitors will produce not only less ripple, but also lower (better) load regulation.

Transformer drop: The third component of load regulation is the drop in voltage from $V_{s(\text{pk})}$ because of the capacitor's charging current in the winding resistance of the transformer. We can calculate this drop $V_{r(\text{w})}$ from Ohm's law (IR), but neither the I nor the R are as straightfowared as might have been expected. I in this case is the capacitor *charging* current, which flows only during the rising part of the rippling capacitor voltage. The *average* charging current must equal the discharging current I_L, but charging time may be anywhere from $\frac{1}{3}$ to $\frac{1}{20}$ of the total time between charging pulses (see Fig. 3-6). We will settle on the approximation that $t_{\text{chg}} \approx \frac{1}{7}t_{\text{total}}$. Thus $I_{\text{chg(pk)}} \approx 7I_L$. This approximation is *for half-wave rectifiers only* and is liable to errors on the order of $\pm 50\%$ for normally encountered circuit values.

The winding resistance through which these pulses flow is the secondary *plus reflected primary* resistance of the transformer.

$$r_w = r_S + n^2 r_P = r_S + \left(\frac{N_S}{N_P}\right)^2 r_p \tag{3-5}$$

For 60-Hz transformers r_S and r_P can be measured with an ohmmeter, but for high-frequency transformers, the skin effect will cause the actual resistance to be many times the dc value indicated on the ohmmeter. Most commercial 60-Hz transformers having only a single secondary are wound so as to balance the primary and secondary losses, so it is a fairly safe assumption that $n^2 r_P$ will approximately equal r_S, and $r_w \approx 2r_S$. Transformer-winding resistance will thus cause $V_{O(FL)}$ to drop by approximately $7I_L r_w$, or $14I_L r_S$.

The complete formula for predicting V_O is then

$$\begin{aligned} V_{O(FL)} &= V_{s(\text{pk})} - V_{r(\text{w})} - V_D - \tfrac{1}{2}V_{\text{rip(p-p)}} \\ &= 1.41V_{s(\text{rms})} - 7I_L r_w - 0.9 - \frac{I_L t}{2C} \end{aligned} \tag{3-6}$$

With no load, $I_L = 0$, $V_D = 0.6$ V, $V_{r(\text{w})}$ and $\frac{1}{2}V_{\text{rip(p-p)}}$ are zero, and the expression for no-load output becomes

$$V_{O(NL)} = 1.41V_{s(\text{rms})} - 0.6 \tag{3-7}$$

Example 3-3

What is the load regulation for a supply which delivers 24 V at the full-rated current of 1.8 A and 24.1 V at zero load current?

Solution

$$\begin{aligned} \text{load regulation} &= \frac{V_{NL} - V_{FL}}{V_{FL}} = \frac{24.1 - 24.0}{24.0} \\ &= \frac{0.1}{24} = \mathbf{0.4\%} \end{aligned}$$

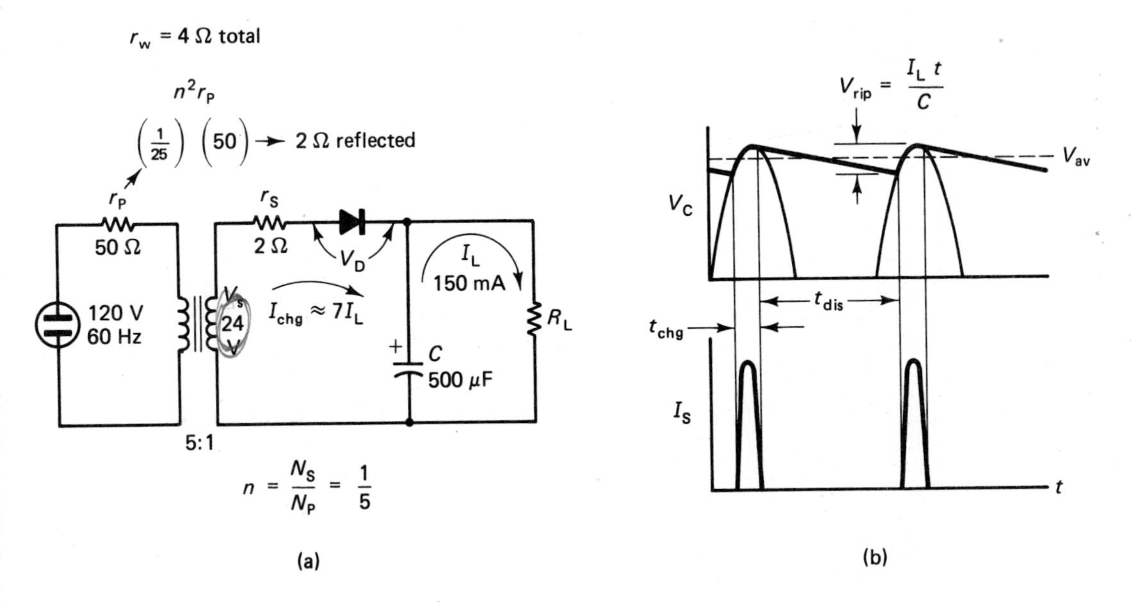

Figure 3-6 (a) Winding resistance r_w is secondary plus reflected-primary resistance. (b) Capacitor-charging current comes in pulses which are about $7I_L$ for a half-wave rectifier.

Example 3-4

Predict the load regulation for the supply of Fig. 3-6.

Solution

$$V_{O(NL)} = 1.41\,V_{s(rms)} - 0.6 = (1.41)(24) - 0.6 = 33.2 \text{ V} \tag{3-7}$$

$$V_{O(FL)} = 1.41\,V_{s(rms)} - 7I_L r_w - 0.9 - \frac{I_L t}{2C} \tag{3-6}$$

$$= (1.41)(24) - (7)(0.15)(4) - 0.9 - \frac{(0.15)(0.0167)}{(2)(500 \times 10^{-6})}$$

$$= 33.8 - 4.2 - 0.9 - 2.5 = 26.2 \text{ V}$$

$$\text{load regulation} = \frac{V_{O(NL)} - V_{O(FL)}}{V_{O(FL)}} = \frac{33.2 - 26.2}{26.2} = \mathbf{26.7\%} \tag{3-4}$$

3.7 SURGE-CURRENT LIMITING

When a power supply with a capacitor filter is first turned on, the capacitor voltage is at zero and the entire transformer voltage is applied to the diode. (Recall that the voltage across a capacitor cannot change instantly, so the capacitor behaves like a short circuit for a few milliseconds until it charges.) The only thing limiting this surge current is the resistance of the transformer winding. In many cases this small resistance is enough to protect the diode, since surge current ratings are generally quite high, but an additional surge-current-limiting resistor must sometimes be placed in series with the diode.

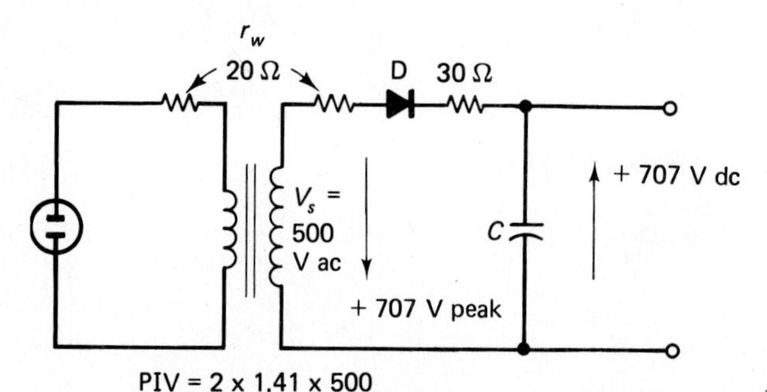

Figure 3-7 Illustration of diode surge-current limiting (Section 3.7) and peak-inverse-voltage calculation (Section 3.8).

Example 3-5

The diode in the circuit of Fig. 3-7 has a surge-current rating of 10 A. The transformer winding resistance r_w is 40 Ω. Find the value of surge-limiting resistor, if one is required.

Solution The peak secondary voltage is $1.41 \times 500 = 705$ V. To limit the surge current to 10 A, the resistance in series with the diode must be

$$R = \frac{V_s}{I_{\text{surge}}} = \frac{705}{10} = 70.5\ \Omega$$

The transformer winding provides only 40 Ω, so an addition **30 Ω** is required to limit the current to a safe value when the supply is first turned on.

It should be emphasized that the calculations above give only a general idea of what additional surge-limiting resistance is needed and are not highly accurate. The size of the filter capacitor has a great deal to do with the question also since it determines how long the diode will be asked to endure the surge current. For small capacitors the surge resistor will not be needed at all, since the high current will exist for a considerably shorter time than the approximately 20 ms normally specified as the surge-current time.

There is an upper safe limit on capacitor size, above which the addition of a surge-limiting resistor will do little good because increasing the series resistance will also lengthen the charging time of the capacitor beyond the time limit of approximately 20 ms generally specified in surge-current ratings. An approximation of the largest safe value of filter capacitor is given by

$$C_{\max} = \frac{0.02 I_{\text{surge}}}{V_{s(\text{pk})}} \tag{3-8}$$

Example 3-6

Using the data of Example 3-5, calculate the largest capacitor which may safely be used with the rectifier system shown.

Solution

$$C_m = \frac{0.02(10)}{705} \tag{3-8}$$
$$= \mathbf{284\ \mu F}$$

3.8 PEAK INVERSE-VOLTAGE REOUIREMENTS

Few diodes manufactured for power-supply applications have PIV ratings below 50 V, so diode back voltage need not be a concern in designing supplies in the 10-V range. For higher-voltage supplies, such as the one in Fig. 3-7, it is a vital concern. The diode PIV required will be higher than the output voltage of the supply because the capacitor holds a charge equal to the peak value of the transformer secondary voltage (the 0.9-V diode drop is negligible in the face of such large voltages). Notice that on the negative half-cycle the transformer peak secondary voltage appears in series with the voltage across the capacitor. This total voltage, equal to twice the transformer peak secondary voltage, appears as a reverse voltage across the diode. In Fig. 3-7, for example, the diode must be able to withstand a reverse voltage of 1414 V. The ac line is notorious for having transient spikes of voltage on it, and the consequences if a diode should break down in an unfused supply may be quite severe: First the filter capacitor shorts, and then the power transformer burns up. It is therefore good practice to use diodes with a PIV *twice* the reverse voltage actually expected. This safety factor of 2 will be used throughout the text.

For a half-wave rectifier, diode PIV required is then given by

$$\text{PIV} = 4V_{s(pk)}$$
$$= 4(500 \times 1.41) \tag{3-9}$$
$$= 2820\ \text{V}$$

Three 1000-PIV diodes in series would be suitable for Fig. 3-7.

3.9 THE FULL-WAVE RECTIFIER

The amount of ripple voltage present in a power-supply output is directly dependent on the time between successive charging pulses from the rectifier. If the pulses can be made to occur twice as fast (doubling the ripple frequency), the ripple voltage can be cut in half, or looking at it from an economy point of view, a filter capacitor one-half the present size can be used to produce the same ripple reduction. A scheme for obtaining a pulse every *half*-cycle, and thus two pulses per cycle, is shown in Fig. 3-8. This is termed a full-wave rectifier, since current is supplied to the filter on both halves of the input waveform. This is a negative supply, but of course it can be made positive by simply reversing the diodes and capacitor.

The full-wave circuit is simply a combination of two half-wave circuits. Only one diode and one-half of the transformer secondary winding are in use at any one

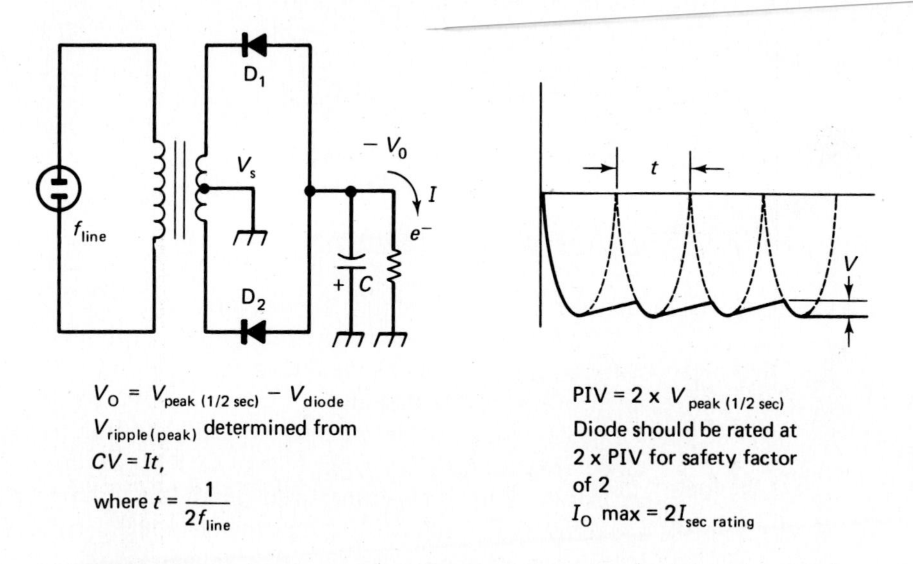

Figure 3-8 Full-wave center-tapped rectifier circuit with single capacitor filter.

time. When the top of the transformer secondary is negative, turning D_1 on, the bottom is positive, turning D_2 off, and vice versa. The output voltage is determined by subtracting the diode drop from the peak value of *one-half* of the secondary voltage. Thus a transformer with an output of 20 VCT (volts center-tapped) would produce an output of $(10 \times 1.41) - 0.9 = 14.1 - 0.9 = 13.2$ V dc. The current output can be twice the average rated current of the transformer secondary, since each half of the secondary supplies one-half of the output current. The diode PIV required is found as in Section 3.8, using the voltage of one-half the secondary.

In calculating ripple voltage, it must be remembered that the ripple frequency is twice the ac line frequency. For a 60-Hz line, $f_{rip} = 120$ Hz and

$$t = \frac{1}{f} = \frac{1}{120} = 8.3 \text{ ms}$$

Winding-resistance drop: In calculating the effective winding resistance r_w, remember to measure only *one-half* of the secondary for r_S, and that n is the ratio of *one-half* of the secondary voltage to the primary voltage.

With full-wave rectifiers the capacitor-charging pulses come more frequently, so the ratio of I_{chg}/I_L is generally lower than for half-wave rectifiers. We will use the approximation $I_{chg} \approx 4\, I_L$ for full-wave rectifiers, again realizing that this is subject to error of $\pm 50\%$.

Filtering limitations: Figure 3-9 illustrates that it is quite useless to follow a full-wave rectifier with a large filter capacitor in an attempt to bring ripple to a low level immediately because the two diodes used may have forward voltage drops which differ by 0.1 V or so. One side of the transformer will thus charge the capacitor to a slightly higher voltage than the other side. If the capacitor is not permitted to discharge to the peak level of the "weak" side, the diode on this side will never conduct. If the

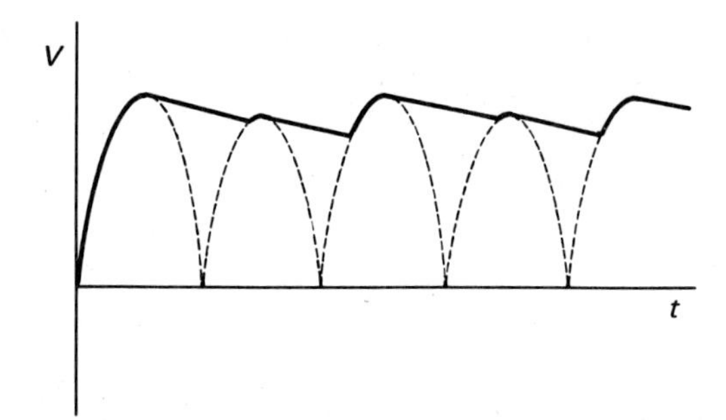

Figure 3-9 Slightly unequal pulses on alternate half-cycles can defeat the advantages of the full-wave rectifier if too much filtering is attempted.

imbalance is less extreme, one diode will conduct more heavily than the other, resulting in unequal ripple peaks.

As a rule of thumb for full-wave rectifiers, it is therefore best to allow the ripple across the first capacitor to be at least 0.5 V p-p. This ripple can be further reduced by additional filter stages, as described in Section 3.10.

Example 3-7

Analyze the circuit of Fig. 3-10 for V_O, diode PIV and surge current, and ripple voltage. Make note of any design errors. Winding-resistance drop $V_{r(w)}$ is negligible.

Solution

1. $V_O = (\frac{1}{2} \times 100 \times 1.41) - 0.9 = \mathbf{70\ V}$.
2. **PIV** $= \frac{1}{2} \times 100 \times 1.41 \times 4 = \mathbf{280\ V}$. The 200-V diodes are in danger of being destroyed by line transients.
3. $$r_w = \left(\frac{V_S}{V_P}\right)^2 r_P + r_S = \left(\frac{50}{120}\right)^2 (5) + 2 = 2.9\ \Omega$$

$$I_{surge} = \frac{70\ \text{V}}{2.9\ \Omega} = \mathbf{24\ A}$$

The surge-current rating of the diodes is not given, but if it is not 24 A or greater, the diodes may burn out when turning on the supply.

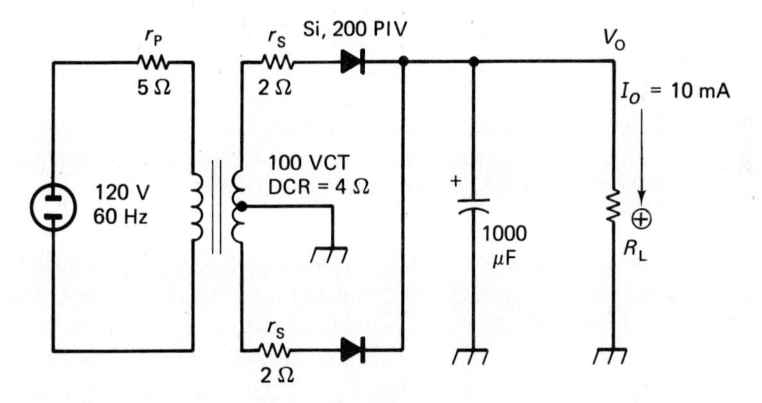

Figure 3-10 Circuit-analysis Example 3-7, containing several design errors.

4. Ripple voltage is found next:

$$V = \frac{It}{C} = \frac{(10 \times 10^{-3})(8.3 \times 10^{-3})}{1000 \times 10^{-6}} \tag{3-2}$$

$$= \mathbf{83\ mV\ p\text{-}p}$$

This circuit attempts to do too much filtering in one stage. The result is almost certain to be unequal division of current between the two diodes, as pictured in Fig. 3-9.

3.10 THE PI-SECTION FILTER

The output of the power supplies discussed thus far consists of an ac ripple voltage riding on a dc voltage level. We have seen that it is desirable to reduce the ripple content to the lowest possible level in order to have a pure source of dc to power the rest of the system. We know that a capacitor presents a low impedance to ac but will not pass dc and that an inductor passes dc quite readily but presents a high impedance to ac. Using this knowledge, we can block the ac portion of the power-supply output with a series inductor and shunt any ac that does get by to ground through a capacitor. This scheme, along with the filter capacitor already discussed, is shown in Fig. 3-11(a). The three components are arranged in the shape of the Greek letter Π(pi)--hence the name of the circuit.

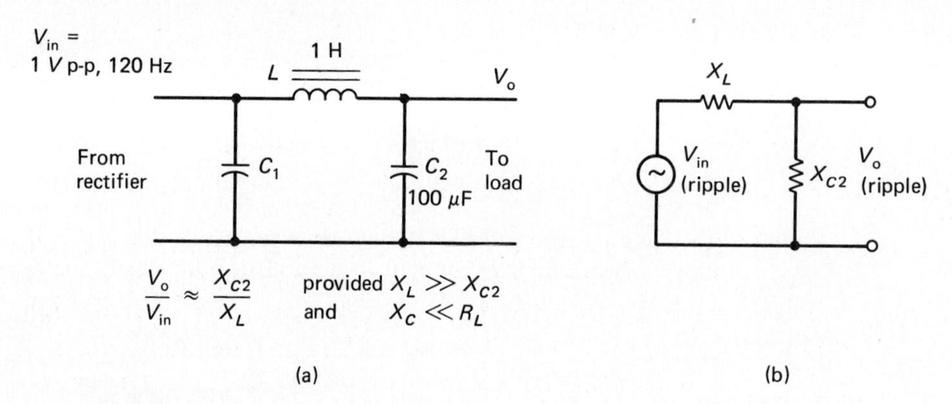

Figure 3-11 Pi-section filter for use after half- or full-wave rectifiers. (a) Circuit diagram and design equation [the first capacitor (C_1) is analyzed as in the previous sections]. (b) L and C_2 may be analyzed as a voltage divider to determine ripple reduction.

To be successful, this filter must use large capacitors and inductors, so that X_L is many times larger than X_{C2}. In that case, L and C_2 form a voltage divider, as shown in Fig. 3-11(b). If X_{C2} is negligibly small compared to X_L, the total impedance seen by the ripple input will be very nearly equal to X_L. If X_{C2} is also much smaller than the load resistance, the output ripple will appear across an impedance equal to X_{C2}. The output ripple voltage can then be determined according to the formula

$$\frac{V_o}{V_{in}} = \frac{X_{C2}}{X_L} \tag{3-10}$$

Remember that the ripple across the *first capacitor*, C_1, is determined as $V_{C1(p\text{-}p)} = I_L t/C_1$. The ripple across the *second capacitor*, C_2, is then found as $V_{C2} = V_{C1}(X_{C2}/X_L)$.

It must be cautioned that inductors and capacitors cannot be treated like resistors in this way when X_L and X_{C2} are of the same order of magnitude or when there is appreciable resistance present. There is also some error in applying the concept of reactance to the sawtooth waveforms which the input ripple voltage takes, since this concept is valid only for sine waves. However, in normal power-supply practice, the formulas just presented are entirely adequate, and they have the decided advantage of being about 10 times easier than a strict analysis. Actual ripple will be about 30% less than calculated by this technique, and will tend to be sinusoidal rather than sawtooth shaped. Again, the tolerances of the capacitors and inductors generally employed in power supplies will mask this error completely.

Example 3-8

Determine the output ripple for the filter of Fig. 3-11(a).

Solution

$$X_L = 2\pi fL = (6.28)(120)(1) = 750\ \Omega$$

$$X_{C2} = \frac{1}{2\pi fC_2} = \frac{0.159}{120 \times 10^{-4}} = 13.2\ \Omega$$

$$V_o = V_{in}\frac{X_{C2}}{X_L}$$

$$= (1)\frac{13.2}{750}$$

$$= \mathbf{0.018\ V\ p\text{-}p}$$

DC choke drop: Power-supply inductors (called filter chokes) necessarily have a certain dc resistance because of their wire windings, and this resistance will cause a voltage drop because of the direct current through it. The voltage drop across this resistance is $I_L r_{ch}$, and it must be considered when calculating $V_{O(FL)}$.

$$V_{O(FL)} = V_{s(pk)} - V_{r(w)} - V_D - \tfrac{1}{2} V_{C1(p\text{-}p)} - I_L r_{ch} \tag{3-11}$$

3.11 PI FILTER USING A RESISTOR

Filter chokes cost from 10 to 100 times as much as resistors and generally weigh more than 100 times as much as resistors. The practice of replacing chokes with resistors wherever possible is therefore widely followed. Performance must suffer, of course, and if a filter choke is replaced with a resistor, we have the option of

1. Using a high-value resistor and letting load regulation suffer,
2. Using a low-value resistor and letting ripple voltage increase,

3. Using a high-value resistor and attempting to restore load regulation by raising input capacitor C_1, or
4. Using a low value of resistor and raising the value of the output capacitor to restore ripple reduction.

In low-current supplies (below 50 mA) it is quite common to use a filter resistor in place of a filter choke, since good load regulation is easy to obtain at low currents.

Analysis of a resistor pi filter is exactly the same as for an inductor pi filter, except that X_L is replaced by R, the value of the filter resistor. The power dissipated in the filter resistor should always be calculated. Since it is frequently greater than $\frac{1}{2}$ W, a power resistor may be required.

Decoupling networks: Pi-section filters using resistors and capacitors are also very commonly used as *decoupling networks* to keep voltage from one amplifier stage from feeding through the power-supply line to another amplifier stage powered by the same supply. This practice is illustrated in Fig. 3-12.

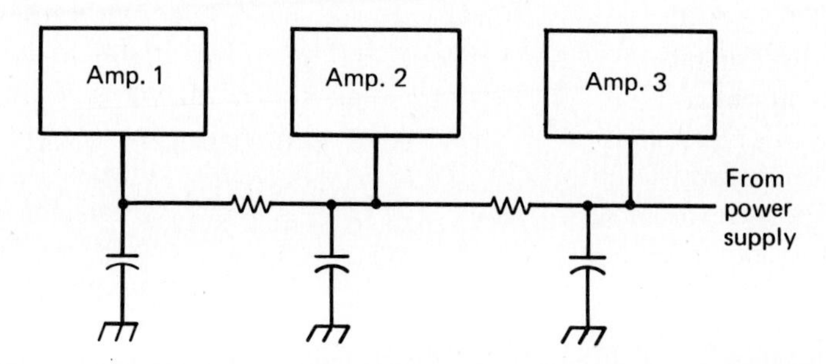

Figure 3-12 Pi-section filters to reduce interaction between parts of a system.

Example 3-9

Analyze the circuit of Fig. 3-13 for $V_{o(\text{rip})}$, $V_{O(\text{NL})}$, and $V_{O(\text{FL})}$.

Solution

1. $V_{C1(\text{rip})} = \dfrac{I_L t}{C_1} = \dfrac{(25 \text{ mA})(16.7 \text{ ms})}{500\ \mu\text{F}} = 0.84 \text{ V p-p}$

$$X_{C2} = \frac{1}{2\pi f C_2} = \frac{1}{2\pi(60)(1 \times 10^{-3})} = 2.65\ \Omega$$

$$V_{B(\text{rip})} = V_{A(\text{rip})}\frac{X_{C2}}{R_f} = (0.84)\frac{2.65}{100} = \mathbf{0.022\ V\ p\text{-}p}$$

2. $V_{O(\text{NL})} = 1.41V_s - V_D = (1.41)(12) - 0.6 = \mathbf{16.3\ V}$
3. $V_{O(\text{FL})} = 1.41V_s - V_{r(w)} - V_D - \frac{1}{2}V_{C1(\text{p-p})} - V_{R(f)}$

$$r_w = \left(\frac{V_S}{V_P}\right)^2 r_P + r_S = \left(\frac{12}{120}\right)^2(200) + 2 = 4\ \Omega$$

$$V_{r(w)} = 7I_L r_w = 7(0.025)(4) = 0.7 \text{ V}$$

$$V_{R(f)} = I_L R_f = 0.025 \times 100 = 2.5 \text{ V}$$

$$V_{O(\text{FL})} = (1.41)(12) - 0.7 - 0.9 - 0.4 - 2.5 = \mathbf{12.4\ V}$$

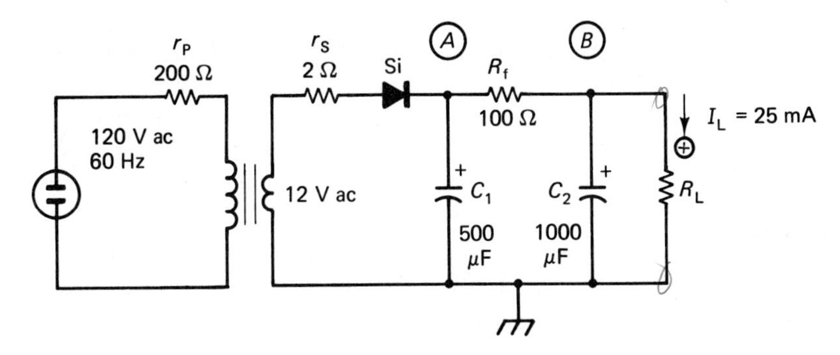

Figure 3-13 Circuit-analysis Example 3-9.

3.12 BRIDGE RECTIFIER CIRCUIT

The bridge rectifier provides a full-wave output without the use of a center-tapped transformer. The bridge circuit is most frequently drawn in the standard diamond-bridge configuration of Fig. 3-14(a), but it is more readily understood by redrawing it as in Fig. 3-14(b). In this form it can be visualized as a scheme for alternately reversing the polarity of the transformer secondary connections on each half-cycle of ac, thus always providing the same polarity output. Diodes D_1 and D_2 conduct on the positive half-cycle, while D_3 and D_4 conduct on the negative half-cycle. The voltage pulses on alternate half-cycles will be of the same amplitude, except for the small difference in diode voltage drops. Two diodes operate on each half-cycle,

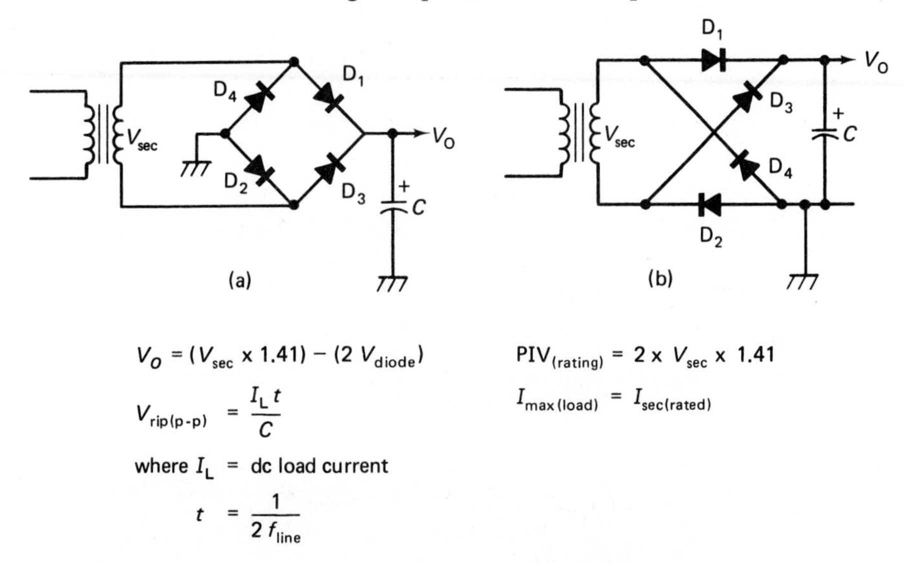

Figure 3-14 Full-wave bridge rectifier circuit with design equations: (a) as commonly drawn; (b) alternate drawing clarifying the circuit function.

so this inequality may be on the order of 0.2 V. The first stage of filtering can efficiently bring the ripple voltage down to about 1 V without unequal ripple pulses.

The polarity of the output voltage can, of course, be reversed by simply reversing all the diodes and the filter capacitor. This circuit has become so popular that encapsulated bridge rectifier units are available in a wide variety of current and voltage ratings. These packages have four terminals, two for ac input and a plus and minus for dc output. They have the additional advantage that the diode forward voltages are matched to typically 10 mV, permitting ripple filtering to less than 0.1 V p-p on the first filter capacitor.

A disadvantage of the bridge circuit is that neither end of the transformer secondary can be grounded. Often we wish to operate several rectifiers (say, a positive and a negative) from a single secondary winding. This is possible with the half-wave and center-tapped designs, but not with the bridge circuit.

Example 3-10

Design a bridge rectifier with a single capacitor filter for +40 V at 100 mA with 2% ripple and 10% load regulation.

Solution

1. If $V_O = 40$ V, then $V_{s(pk)}$ must be 40 plus two diode drops, or 41.8 V. This is an rms voltage of $41.8/1.41 = 29.6$ V. A 30-V transformer would be ordered.
2. 2% of 40 V is 0.8 V rms ripple. Expressed in peak-to-peak values, the ripple is

$$V_{rip} = (0.8)(2.82) = 2.26 \text{ V p-p}$$

3. For a full-wave rectifier on a 60-Hz line, t between pulses is 8.3 ms, so C is calculated as

$$C = \frac{It}{V} = \frac{(100 \times 10^{-3})(8.3 \times 10^{-3})}{2.26} \tag{3-2}$$

$$= 367 \times 10^{-6} = 367\ \mu\text{F}$$

A 500-μF capacitor would be used to ensure adequate filtering, since capacitor tolerances are often quite broad.

4. PIV $= 2V_{s(pk)} = (2)(41.2) = 82.4$ V. Diodes with at least 100-V PIV would be used.
5. $V_{O(NL)} = V_{s(pk)} - 0.6$ (3-7)

$V_{O(FL)} = V_{s(pk)} - V_{r(w)} - 0.9 - \frac{1}{2}V_{rip(p\text{-}p)}$ (3-6)

$V_{O(NL)} - V_{O(FL)} = V_{r(w)} + 0.3 + \frac{1}{2}V_{rip(p\text{-}p)}$

$$\frac{V_{O(NL)} - V_{O(FL)}}{V_{O(FL)}} = 0.10 = 10\% \tag{3-4}$$

$$V_{O(NL)} - V_{O(FL)} = 0.10V_{O(FL)}$$

$$= 0.10(40) = 4 \text{ V}$$

$$4 = V_{r(w)} + 0.3 + \tfrac{1}{2}(2.26)$$

$$V_{r(w)} = 4 - 0.3 - 1.13 = 2.57 \text{ V}$$

$$r_w = \frac{V_{r(w)}}{4I_L} = \frac{2.57}{4(0.1)} = 6.4\ \Omega$$

The transformer secondary resistance should be 3.2 Ω. The primary resistance would then be $3.2(120\text{ V}/30\text{V})^2$ or 51 Ω.

3.13 VOLTAGE-DOUBLING RECTIFIERS

Two popular circuits exist for obtaining a dc output voltage of approximately twice the peak value of ac input voltage. The first [Fig. 3-15(a)] is a full-wave circuit, since the output contains ripple pulses at a frequency of twice the line frequency. The second [Fig. 3-15(b)] is a half-wave circuit and can be used without a transformer.

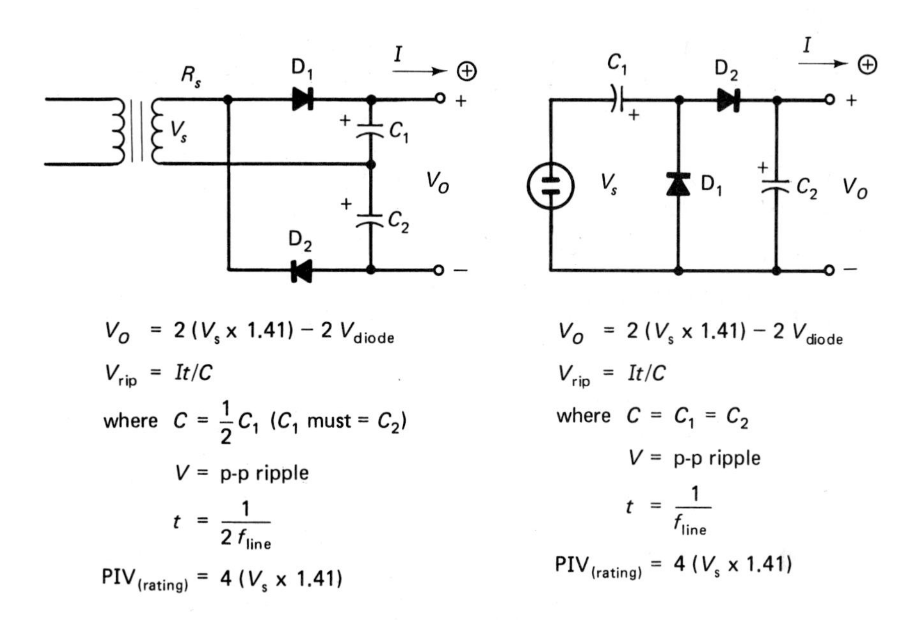

Figure 3-15 Voltage-doubling power-supply circuits with design formulas: (a) full-wave; (b) half-wave.

In both circuits, C_1 is charged through D_1 on the positive half-cycle. In the full-wave circuit the voltages on C_1 and C_2 simply appear in series. In the half-wave circuit the voltage stored on C_1 during the negative half-cycle appears in series with the line voltage, V_s, on the positive half-cycle. In the half-wave circuit, C_1 does not necessarily have to be equal to C_2, but a low value of C_1 will cause poor load regulation. In both circuits, the capacitors must be considered an integral part of the rectifier circuit. These circuits will not function unfiltered, as will all the others discussed so far.

Example 3-11

Compare the output voltage and ripple for the two circuits of Fig. 3-15. $C_1 = C_2 = 100\ \mu F$. $V_s = 120$ V ac and $I_L = 40$ mA.

Solution

1. The peak output voltages are the same for each circuit:

$$V_O = 2(120 \times 1.41) = \mathbf{338\ V\ pk}\ \text{(neglecting diode drops)}$$

2. Ripple is found for the full-wave circuit,

$$V = \frac{It}{C} = \frac{(40 \times 10^{-3})(8.3 \times 10^{-3})}{50 \times 10^{-6}} = \mathbf{6.65\ V\ p\text{-}p}$$

and for the half-wave circuit,

$$V = \frac{It}{C} = \frac{(40 \times 10^{-3})(16.6 \times 10^{-3})}{100 \times 10^{-6}} = \mathbf{6.65\ V\ p\text{-}p}$$

If the load current is large, the voltage drop on the winding or source resistance will have to be considered. This drop will be about $14I_L r_w$, since both C_1 and C_2 suffer a loss $7I_L r_w$. The drop in dc output voltage due to averaging between the peaks and valleys of the ripple ($\frac{1}{2}V_{rip(p\text{-}p)}$) must also be considered.

3.14 CHOKE-INPUT FILTERS

Choke-input filters are not as common as they were in the days of the vacuum tube, but they may still be encountered in high-voltage power supplies.

The action of the choke on the rectified pulses of dc is quite different from the action of the input capacitor described previously. Whereas the capacitor tends to charge to the *peak* voltage of the input waveform, the choke tends to follow the *average* value as shown in Fig. 3-16. The choke tends to oppose any change in current by producing positive voltages when current decreases *and negative voltages when current increases.* Thus the choke fills in the valleys and brings down the peaks of the rectified waveform. The output voltage can be calculated as

$$V_O \approx 0.9(V_{rms(1/2\ sec)}) - V_D \tag{3-12}$$

since average voltage is approximately 0.9 × rms voltage. V_D is the voltage drop of the diodes.

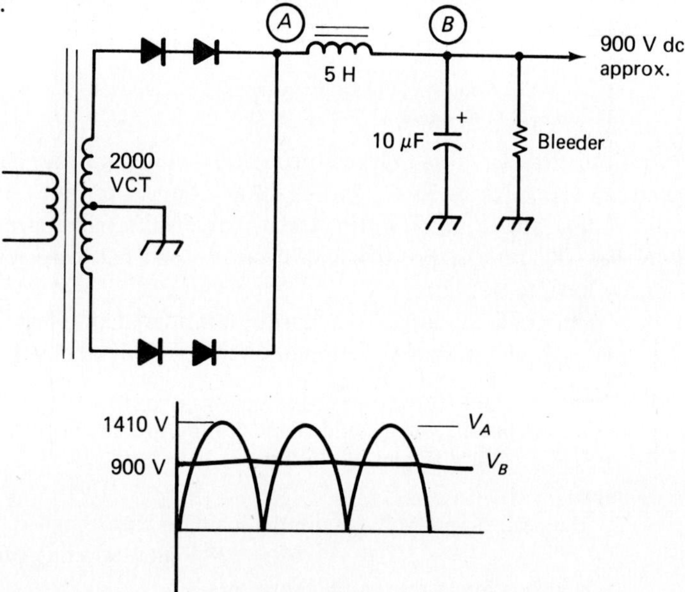

Figure 3-16 Power supply with choke-input filter. The output voltage follows the average value of the rectified waveform.

The output voltage for this type of circuit will tend to soar to the peak value of the rectified ac wave unless a moderate current is maintained through the choke. For this reason a *bleeder* resistor is always connected across the output of the supply to provide for such current flow when the load is removed from the supply. It is good practice to use a bleeder resistor in any high-voltage (above 50 V) supply not only to improve load regulation but to provide a path for discharge of the filter capacitors when the equipment is turned off. If a bleeder is not used, high voltages may remain on the capacitors for a considerable length of time, presenting a real hazard to persons who assume that the circuit is "dead" simply because it is turned off. Bleeder resistance can be calculated by the time-constant formula:

$$R = \frac{\tau}{C} \qquad (3\text{-}13)$$

The time constant, τ, should be a maximum of several seconds. C is the total filter capacitance used in the filter. Much lower values of R_{bleeder} may be necessary to preserve regulation in choke-input filter systems where the load is completely removed.

3.15 CLIPPERS, CLAMPERS, AND AC DETECTORS

A voltage clipper is a circuit that trims off a portion of a signal waveform. In the *shunt peak clipper* of Fig. 3-17(a) this is accomplished by placing diodes across the output to prevent $v_{O(M)}$ from exceeding $\pm V_D$. Since the series resistor is large, V_D

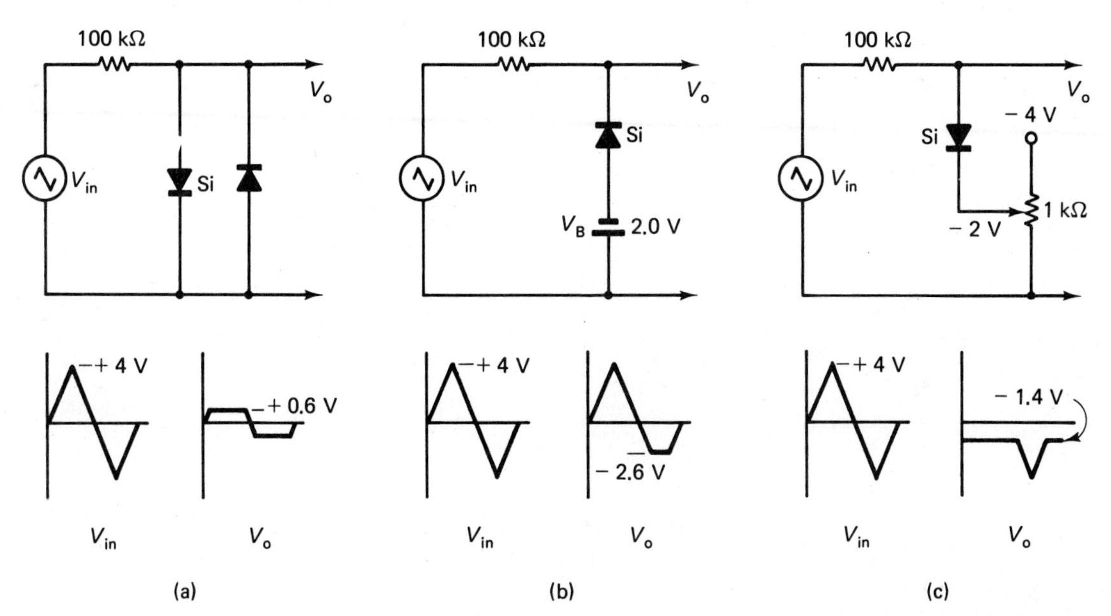

Figure 3-17 (a) Dual-diode clipper limits output to about ± 0.6 V. (b) Biased clipper prevents output from going more negative than $V_B + V_D$. (c) Variable bias clips off outputs above $V_B + V_D$.

will be about 0.6 V for silicon diodes. This circuit is widely used to limit noise spikes and overloads at the input to an amplifier and to form square-wave outputs from sine- or triangle-wave inputs.

Biased clippers are shown in Fig. 3-17(b) and (c). The former will clip off any inputs that attempt to go below (more negative than) −2.6 V. The latter will clip off any input that goes above a variable level, set in the figure to −1.4 V. The load resistance on any of these clippers must be several times higher than the series resistor R_s.

A clamper circuit establishes a new dc reference level for an ac signal without changing the signal waveshape. The operation of a clamper can be understood in terms of two basic facts.

1. A capacitor, if it is large enough, is a short circuit for ac or changing signal waveforms.
2. The forward voltage across a silicon diode cannot exceed approximately 0.6 V in low-current circuits.

Looking at the circuit of Fig. 3-18(a), it can be seen that the ac waveform at B must be identical to that at A, because the capacitor is a short circuit for ac. However, the

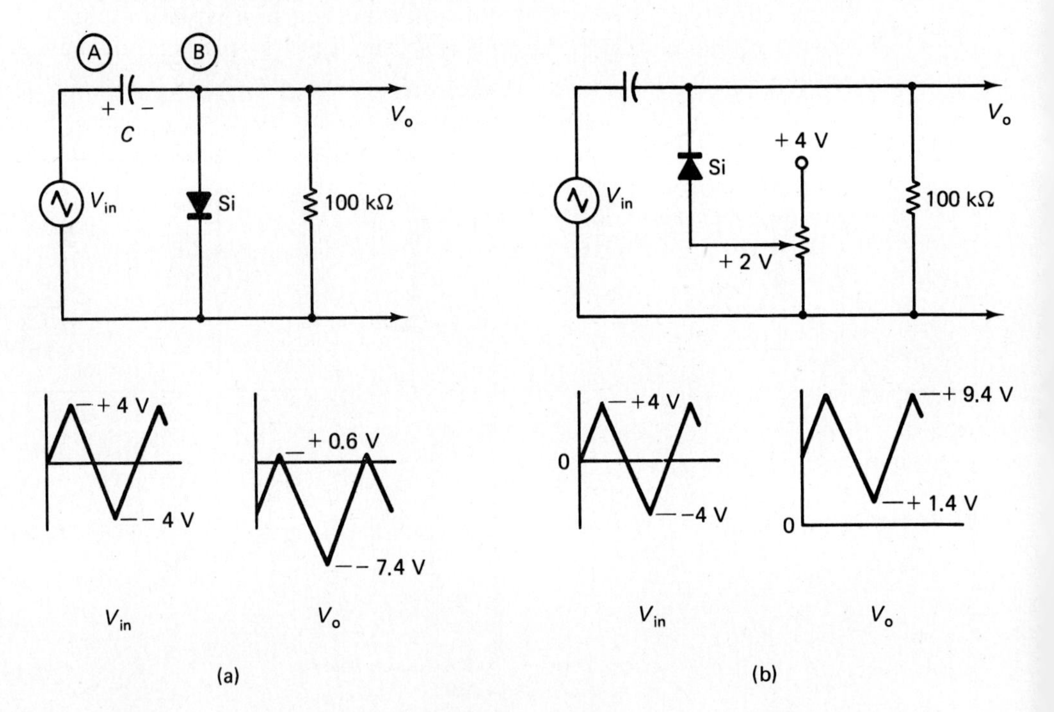

Figure 3-18 (a) Clamper shifts ac waveform so V_o does not go more positive than V_D. (b) Biased clamper sets most-negative point on wave at $V_B - V_D$.

instantaneous voltage at B cannot go more positive than 0.6 V. We say that the waveform is *clamped* at +0.6 V. On the first few cycles from the source the diode conducts heavily, charging C to the peak-positive-source voltage, minus V_D. Thereafter, the diode conducts only slightly at each positive peak to restore charge lost from C to R_L during the previous cycle. To avoid distortion of the output waveform, C must be chosen large enough so that it does not discharge appreciably between input cycles. For 1% discharge

$$C = \frac{1}{0.01fR_L} \tag{3-14}$$

To allow 3% discharge, replace 0.01 with 0.03 in equation (3-14).

A variable-bias clamper is shown in Fig. 3-18(b). In this circuit the output cannot go more negative than a selected dc level, which is +1.4 V in the example.

AC voltage detector: A frequent requirement in electronic systems is to measure the ac voltage at a certain point in a circuit, even though there may be a large amount of dc present at the same point. One example of this is the VU (volume-unit) meter on a tape recorder. In electronic instrumentation systems the need to pick off the ac voltage at a certain point and convert it to dc to operate some metering or control device is also frequently encountered. Of course, a transformer can be used to pick off ac and leave the dc behind, but a capacitor is much smaller, lighter, and less expensive.

If it is desired to convert the ac picked off by the capacitor to dc, a dilemma is encountered. The rectifying diode passes current in only one direction, but the capacitor passes only ac; that is, it passes only currents which flow alternately in each direction. The solution is to connect a second diode the opposite direction to ground, simply to provide a path to discharge the capacitor on alternate half-cycles. This ac voltage detector circuit is shown in Fig. 3-19.

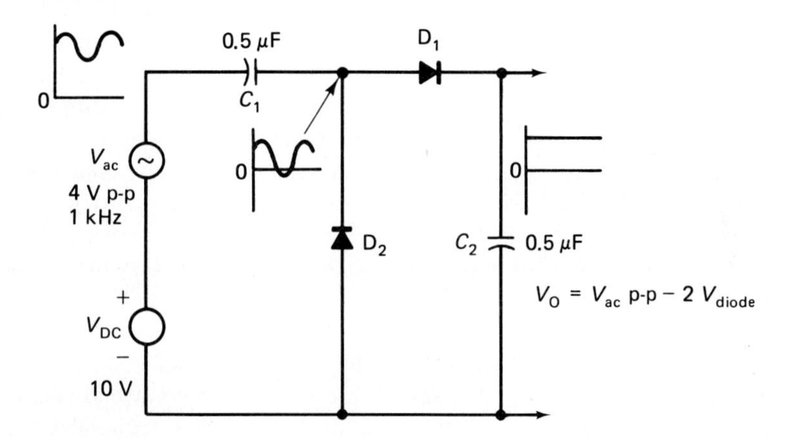

Figure 3-19 Circuit for detecting ac in the presence of dc without using transformers.

3.16 POWER-SUPPLY PRACTICES

This section presents a brief catalog of power-supply applications, shortcuts, and special circuits. Figure 3-20 shows the standard placement of the *ac-off* switch (S_1) in the primary of a power-supply transformer. A slow-blow fuse (F_1) rated at about twice the expected maximum current drain should also be placed in the primary circuit.

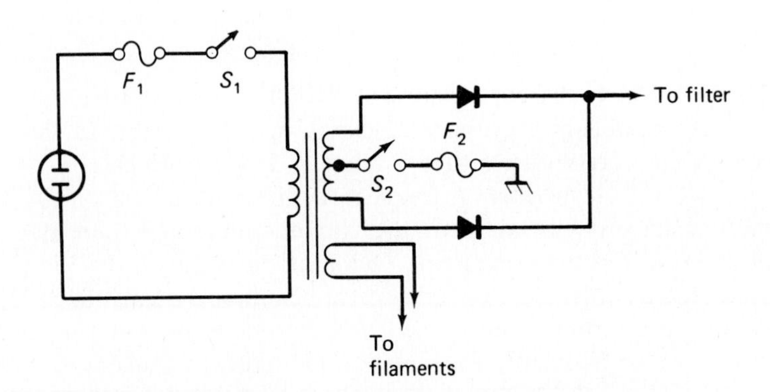

Figure 3-20 Fuse and switch placement in a power supply.

Fuse rating: For electronic equipment the line current is seldom sinusoidal, but rather contains peaks due to the charging of the filter capacitors (see Fig. 3-6). The reading of an ac ammeter can be quite misleading, since most ac meters (including VOMs and DVMs) actually indicate 1.11 times the *average* value of the wave. The fuse, however, responds to the rms value of the line current, which in this case is typically 1.7 times the current indicated on a conventional average-following meter. If a true-rms meter is available, the fuse rating should be 1.5 $I_{\text{line(rms)}}$ to allow a 50% safety margin. If an average-following meter must be used, make the fuse value 2.5 $I_{\text{line(indicated)}}$

Turn-on surge: If the transformer in Fig. 3-20 is very large, its primary resistance will be low (less than 1 Ω). If, by ill luck, S_1 is closed at the zero-voltage point of the line, the transformer core will saturate magnetically, and the primary current (for one cycle or so) will be limited only by the 1-Ω primary resistance. A slow-blowing fuse can stand this, but it does produce horrible noise pulses on the ac line. S_1 is sometimes omitted in favor of S_2, to avoid the turn-on surges encountered with large transformers.

Two-section filter: For critical applications where a very low ripple content is required, two or three filter sections may be cascaded as shown in Fig. 3-21. In analyzing the ripple reduction, L_1 and C_1 are viewed as a voltage divider feeding a second voltage divider L_2–C_2 (see Section 3.10). In high-power applications, L_1 may have specially controlled magnetic saturation characteristics and be termed a *swinging choke*. This means that as higher load current is drawn, the inductance of L_1 decreases.

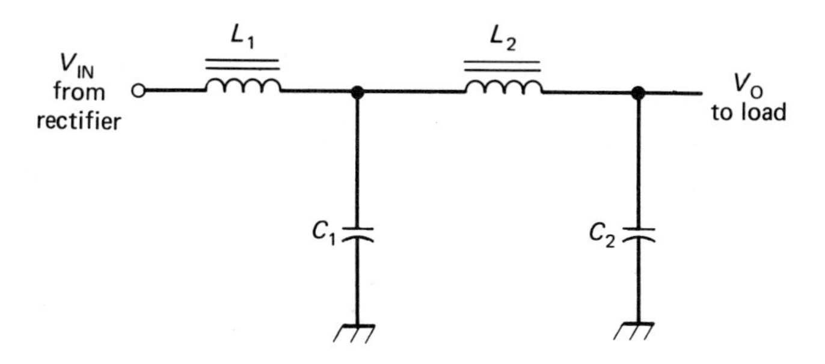

Figure 3-21 A dual-section choke-input filter. L_1 may be a *swinging choke* to improve regulation by saturating at high currents.

Thus L_1 tends to present less opposition to the current pulses from the rectifier, and C_1 charges up to a higher voltage. Selection of a swinging choke with an operating current rating and saturation characteristic to match the rest of the supply can produce a system in which the drop in voltage caused by the resistance of the chokes and transformers is very nearly offset by the swinging choke.

Two output voltages, one equal to half of the other, can be obtained using a single center-tapped transformer and a bridge rectifier circuit, as shown in Fig. 3-22. If the circuit is redrawn omitting D_3 and D_4, it can be seen that D_1 and D_2 appear in a conventional full-wave center-tapped rectifier circuit to produce the low voltage.

Two equal output voltages of opposite polarities can be obtained from a single transformer using the circuit of Fig. 3-23. Often the negative supply is required to deliver only a few milliamperes of current, in which case D_4 can be omitted.

In high-voltage power supplies using a filter choke, it becomes a problem to insulate the choke windings (which may be carrying several thousand volts dc) from the chassis. This problem can be solved by placing the choke in the ground return

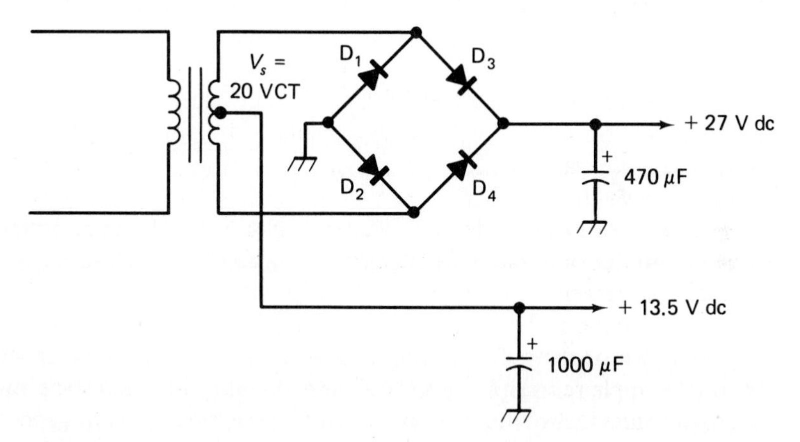

Figure 3-22 A center-tapped transformer and a bridge rectifier provide two output voltages from a single transformer.

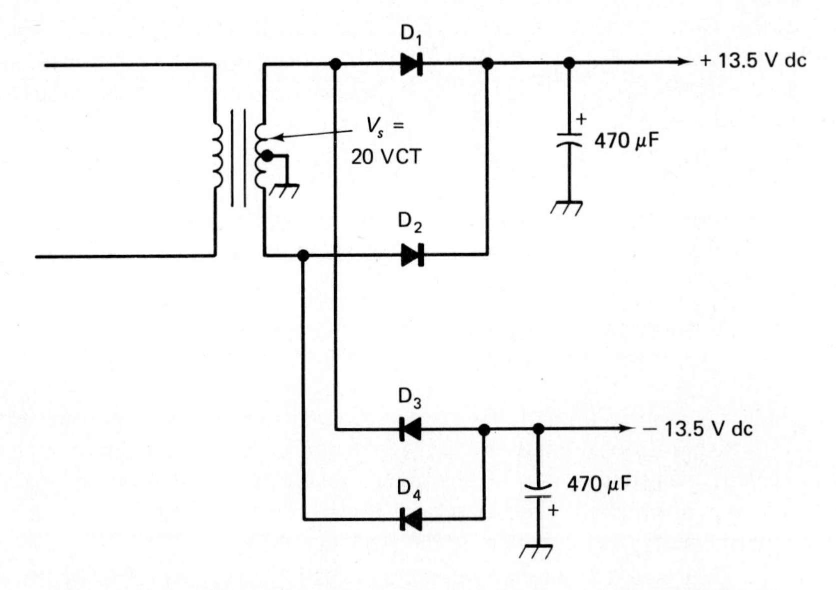

Figure 3-23 Positive and negative rectifiers can be operated from a single transformer using the full-wave center-tapped circuit.

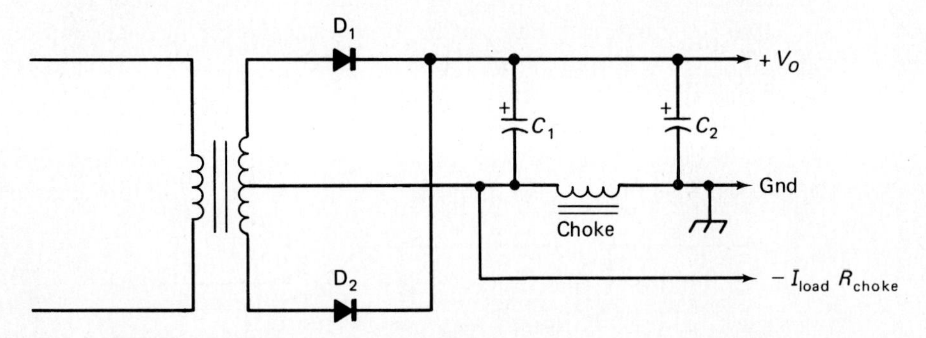

Figure 3-24 Negative-lead filtering reduces choke insulation requirements and provides a small negative bias voltage.

side of the power supply circuit, as shown in Fig. 3-24. The dc resistance of the choke windings produces a small negative dc voltage ($V = I_{LOAD}R_{CHOKE}$) which may be used for bias in vacuum-tube circuits.

Series-strung electrolytics: It is often difficult to find capacitors of a sufficiently high voltage rating for use in kilovolt-level power supply filters. Oil-filled capacitors are available in ratings of several thousand volts, but these are more bulky and expensive than electrolytic capacitors. Electrolytic types are not generally available in voltage ratings much above 500 V, however. Electrolytic capacitors can be placed in series to obtain higher voltage ratings provided that equal-value resistors are shunted across each capacitor in the series string as shown in Fig. 3-25. These *equalizing resistors* are necessary to ensure equal division of the total supply voltage among the several series capacitors, since the leakage resistance of each capacitor will no doubt be different. Leakage resistances may be in the range of a few megohms to several tens of megohms for 450-V electrolytic capacitors, so the equalizing resistors should have a value of 100 kΩ or less to completely swamp out the effects of leakage. The equalizing resistors also serve the function of safety bleeder resistors to discharge the filter capacitors when the supply is turned off.

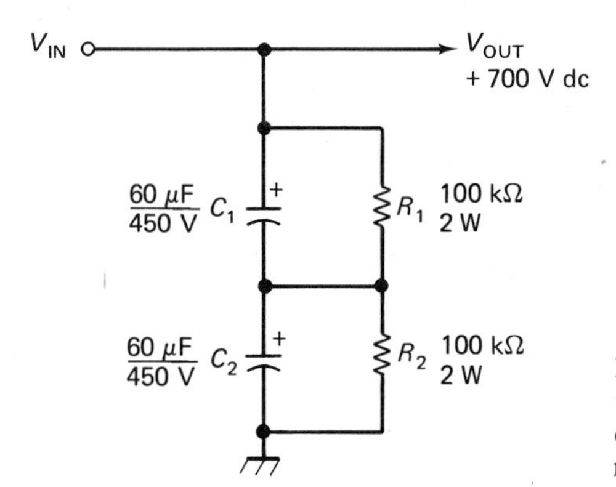

Figure 3-25 Voltage-equalizing resistors must be used when stringing capacitors in series to increase voltage rating.

The wiring of low-ripple power supplies is critical because hookup wire and printed circuit tracks do not have perfectly zero resistance—10 to 50 mΩ is common for 10-cm in-chassis or on-board runs. In a 500-mA supply, charging pulses could reach 5 A, developing 50-mV- to 0.25-V-peak pulses across these conductors. To keep these voltage pulses from appearing in the output, the following precautions should be observed, as illustrated in Fig. 3-26.

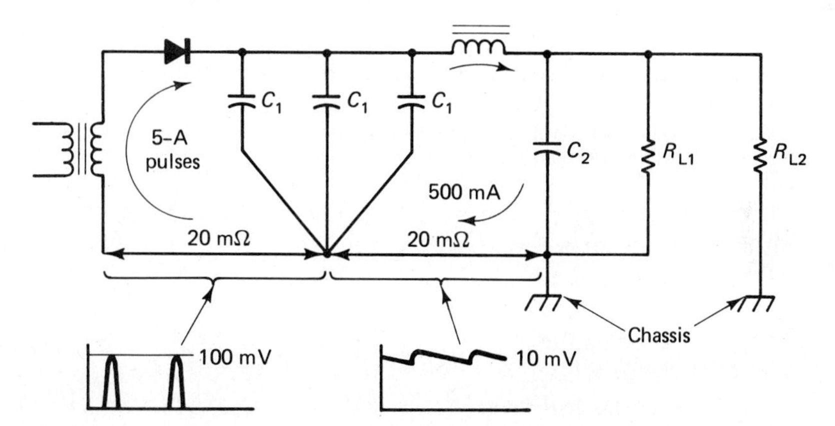

Figure 3-26 The charging path to the first capacitors must not involve other wiring, as fairly large voltage pulses may exist across it.

1. Run the wire from the transformer center tap or "cold" side directly to the first filter capacitor C_1. Do not ground this wire, and do not connect any other wires along its length or to the transfomer cold side.
2. If there are several "C_1" capacitors, either in parallel or from separate supplies, locate them close together and strap their "cold" terminals together with heavy wire or extra-wide circuit-board tracks. If there is a second filter capacitor C_2, run a short wire from the cold side of C_1 to the cold side of C_2 and then branch all other circuit ground connections from this point.
3. Do not chassis-ground the cold side of the transformer at one point and the cold side of C_1 at another. This passes charging pulses through the chassis and causes voltage pulses to appear between two supposedly grounded points.
4. In noise-critical circuits do not use the chassis as a conductor at all—use it as a shield. In this case there will be only one connection from the circuit ground to the chassis, usually at the signal input.

CHAPTER SUMMARY

1. Ripple at the first filter capacitor of a power supply (or the required value of that capacitor) may be calculated from

$$C_1 V_{C1} = I_L t \tag{3-2}$$

where t is the time between charging pulses ($1/f$ for half-wave, $1/(2f)$ for full-wave circuits).

2. Where a choke and second capacitor are used, V_{C1} is found as above and output ripple V_{C2} is calculated by voltage division.

$$\frac{V_{C1(\text{p-p})}}{X_L} = \frac{V_{C2(\text{p-p})}}{X_{C2}} \tag{3-10}$$

where X_L and X_{C2} are calculated at the ripple frequency. In low-current supplies X_L is often replaced by R, a filter resistor.

3. Dc output voltage is found at no load and full load by

$$V_{\text{O(NL)}} = V_{\text{s(pk)}} - V_{\text{D(0.6)}} \tag{3-7}$$

$$V_{\text{O(FL)}} = V_{\text{s(pk)}} - FI_L r_w - V_{\text{D(0.9)}} - \tfrac{1}{2}V_{C1(\text{p-p})} - I_L r_{\text{ch}} \tag{3-11}$$

where

- $V_{\text{s(pk)}}$ is $1.41 V_{\text{s(rms)}}$, the no-load secondary voltage.
- V_{D} is the diode drop (0.6 V NL, 0.9 V FL for silicon).
- F is the ratio of charging-current pulses to load current (typically 7 for half-wave, 4 for full-wave circuits).
- r_w is transformer-winding resistance, estimated as $2r_S$ or calculated from equation (3-5).

- $V_{C1\,(p\text{-}p)}$ is the first-capacitor ripple, calculated in item 1.
- r_{ch} is the dc resistance of the choke or filter resistor, if any.

4. Ripple factor $= \dfrac{V_{o(rms)}}{V_{O(DC)}} = \dfrac{V_{o(p\text{-}p)}}{2.82V_{O(DC)}}$ (3-3)

5. Load regulation $= \dfrac{V_{O(NL)} - V_{O(FL)}}{V_{O(FL)}}$ (3-4)

6. The PIV rating of the rectifier diodes should be at least $4V_{s(pk)}$, which allows a safety factor of 2.

7. In the full-wave center-tapped circuit, use the voltage and winding resistance of one-half the secondary. In the bridge circuit allow two diode drops V_D forward and $2V_{s(pk)}$ for each diode's reverse voltage (PIV).

8. A clipper circuit literally clips off the peaks of a waveform above or below a selected voltage. A clamper accepts an ac waveform of zero average value and shifts it up or down, establishing a new dc reference voltage. An ac detector picks off the ac component of a signal with a capacitor and rectifies it using two diodes.

9. Power-supply fuses must be rated for the rms current they carry (plus a safety factor of about 50%). This is not the current indicated by most average-responding meters.

10. In wiring a power supply, the conductors that carry charging pulses to the first filter capacitor should not be connected to the chassis, nor should any other wires be connected to them, except at the capacitor terminals.

QUESTIONS AND PROBLEMS

3-1. The transformer in Fig. 3-27 has a peak secondary voltage of 5.0 V. Sketch the waveform across R_L, indicating its polarity and peak value.

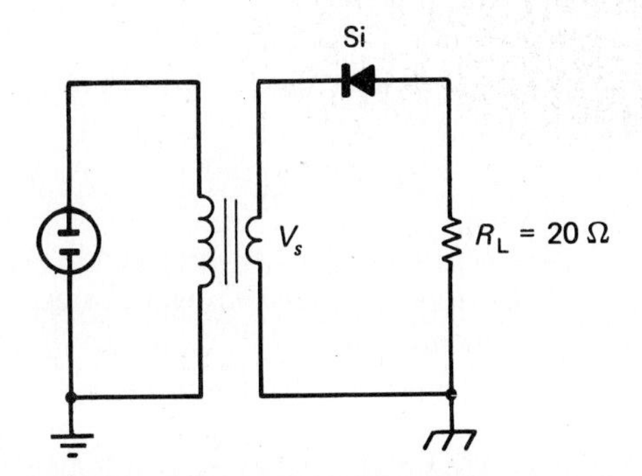

Figure 3-27 Problems 3-1 and 3-2.

3-2. A peak voltage of 6.0 V is required across R_L in the rectifier circuit of Fig. 3-27. What must be the rms secondary voltage?

3-3. Some power transformers have a 1:1 ratio—120-V primary and 120-V secondary. Wouldn't it be cheaper and more efficient to eliminate the transformer and drive the rectifier directly in this case? Explain your answer.

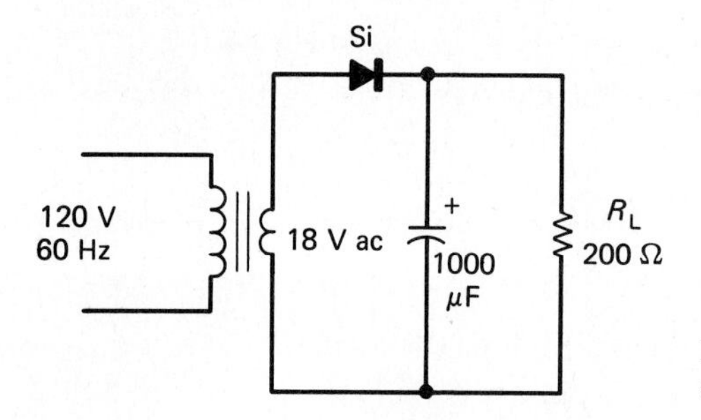

Figure 3-28 Problems 3-4 through 3-8 and Problem 3-11.

For Problems 3-4 through 3-8, refer to Fig. 3-28.

3-4. Find the peak voltage across R_L.

3-5. Find the peak-to-peak ripple voltage across R_L.

3-6. Find the percent ripple in the output voltage.

3-7. Find the dc average voltage across R_L.

3-8. What value of filter capacitor would be required to make the ripple equal 0.5 V p-p?

3-9. A power supply delivers 150 mA at 36 V to its load. When the load is removed the output voltage soars to 45 V. What is the percent load regulation?

3-10. A power supply delivers 850 V at full load and is guaranteed to have a load regulation of 20% or better. How high might the voltage soar at no load?

3-11. What is the approximate ratio of peak transformer-secondary current to dc load current in Fig. 3-28? Why are these two currents not equal?

3-12. A power transformer has a no-load secondary voltage of 240 V with a 120-V primary voltage. The primary winding resistance measures 4 Ω and the secondary measures 20 Ω. What is the effective winding resistance r_w?

3-13. The transformer of Problem 3-12 is placed in the circuit of Fig. 3-6(a). I_L is 150 mA but C is 50 μF. Find the dc average voltage across R_L.

3-14. Find the percent ripple for the power supply of Problem 3-13.

3-15. Find the percent load regulation for the power supply of Problem 3-13.

3-16. What is the surge-current rating required of the diode in Problem 3-13 if no extra surge resistance is added?

3-17. If the diode in Problem 3-13 has a surge-current rating of 6 A, what surge-limiting resistance should be added?

3-18. What is the peak inverse voltage across the diode in Problem 3-13 under normal circumstances? If 1N4006 and 1N4007 diodes (800 V and 1000 V rated PRV, respectively) are available, what would you recommend using?

3-19. Give one advantage and one disadvantage of the full-wave center-tapped rectifier compared to the half-wave rectifier.

For Problems 3-20 through 3-30, refer to Fig. 3-29.

3-20. Find the effective transformer-winding resistance r_w in Fig. 3-29. Remember that only one-half of the secondary is used at a time, although the resistance is given for the entire secondary, as is common industry practice.

3-21. Find the peak voltage at point A.

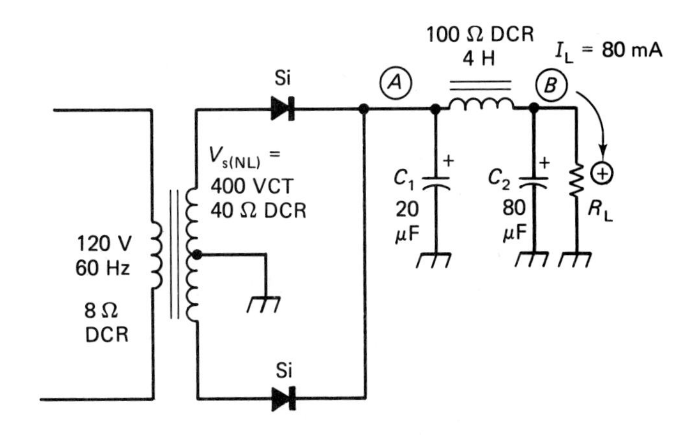

Figure 3-29 Problems 3-20 through 3-30.

3-22. Find the peak-to-peak ripple at point A.

3-23. Find the dc voltage at point B.

3-24. Find the peak-to-peak ripple at point B.

3-25. Find the dc voltage at point B if R_L is removed.

3-26. The diodes have a 10-A surge rating. What value surge-limiting resistor, if any, is needed?

3-27. What is the normal peak reverse voltage across each diode?

3-28. If the first filter capacitor is doubled, what will happen to the full-load output voltage? Use the graph of Fig. 3-6(b) to explain why.

3-29. The first filter capacitor of Fig. 3-29 is to be changed from 20 μF to 100 μF, thus decreasing the drop from peak to average voltage at point A. The 4-H choke is then to be replaced with a resistor of such a value that the extra voltage drop across it matches the reduced drop at C_1. Thus the load regulation will be the same as before, but the expensive choke will be eliminated. Find the required value of the resistor.

3-30. Refer to Problem 3-29 and find the value of C_2 that will restore the output ripple to its original value.

3-31. State two advantages and two disadvantages of the bridge rectifier compared to the full-wave center-tapped circuit.

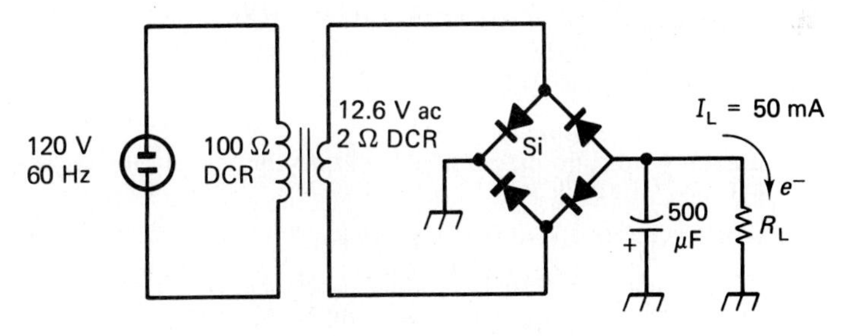

Figure 3-30 Problems 3-32 through 3-35.

For Problems 3-32 through 3-35, refer to Fig. 3-30.

3-32. Find the dc average voltage across R_L in Fig. 3-30.

3-33. Find the no-load dc output voltage.

3-34. Find the peak-to-peak ripple across R_L.

3-35. Find the normal PRV across each diode.

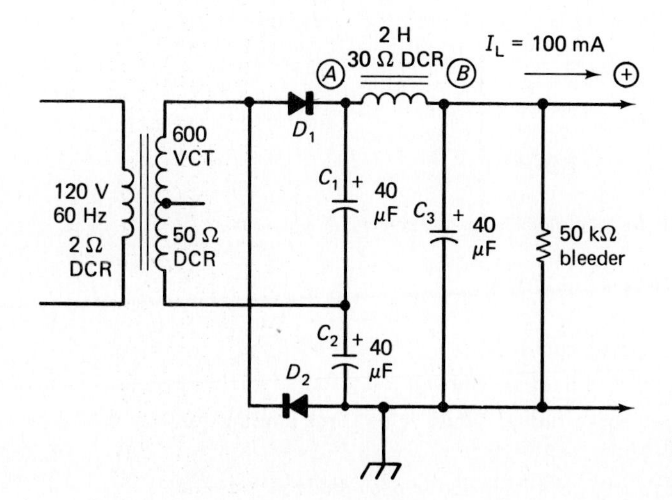

Figure 3-31 Problems 3-36 through 3-41.

For Problems 3-36 through 3-41, refer to Fig. 3-31.

3-36. Draw the transformer secondary and the diode and capacitor that conduct when the top of the secondary is positive. Now draw a second partial schematic showing the secondary and the diode and capacitor that conduct when the top of the transformer is negative. Indicate conventional-current directions and voltage polarities on each diagram.

3-37. Find the peak-to-peak ripple voltage at point *A* in Fig. 3-31.

3-38. Find the rms ripple voltage at point *B*.

3-39. Find the dc output voltage under full load.

3-40. Find the percent load regulation.

3-41. The only diodes available have a PRV rating of 600 V. How many of these should be strung in series at each position, allowing a safety factor of 2 over normal operating PRV?

3-42. Draw a schematic diagram of a biased clipper that will limit the output voltage to the range ±3 V. The input is an 8-V p-p triangle wave. Sketch the output.

3-43. Draw a schematic diagram of a circuit that will input the waveform of Fig. 3-32(a) and output that of Fig. 3-32(b).

3-44. In Fig. 3-19, V_{ac} is 10 V p-p. Find V_O.

3-45. An instrument contains a power supply delivering 500 mA at 5 V. The transformer secondary center tap, the negative of the first filter capacitor, and various parts of the instrument circuitry are all grounded to the chassis, each at the nearest accessible point. What problem is likely to result?

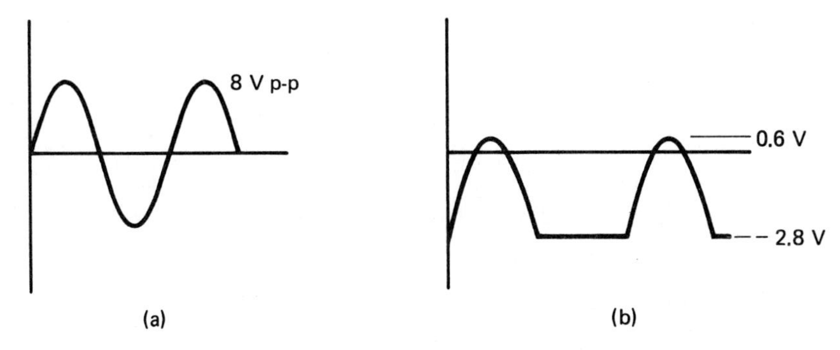

Figure 3-32 Input and output waveforms for Problem 3-43.

4

TRANSISTOR CHARACTERISTICS

4.1 THE TRANSISTOR CONCEPT

The electronic components encountered thus far have all been of a class referred to as *passive* devices. Resistors, capacitors, inductors, and diodes can all *limit* currents in various ways, and transformers can even increase voltage at the expense of current, but none of these devices can produce a higher-powered output signal than the input signal it receives. Devices which can perform this feat are called *active* components. The chief modern-day representative of this class is the transistor.

Of course, transistors are not capable of making high-power signals directly out of low-power signals—the action is better described by saying that a small input current is used to control a large output current. Then it is clear that there must be a relatively large source of current (battery or power supply) to supply the energy of the output signal.

The number of devices that are referred to loosely as *transistors* has increased considerably since the original device was introduced, so it may be advisable to state specifically that we are here concerned with the *bipolar* transistor (also called a bijunction transistor or BJT), as opposed to the newer types which are still less popular. The bipolar transistor is basically a pair of back-to-back diodes which share a common piece of semiconductor material between them. The interaction of these two closely tied diodes causes some remarkable effects which would not be noticed if two separate diodes were used, however. This internal action is not easy to visualize, and a detailed discussion of its exact nature at this point would only tend to obscure the main point of how a transistor behaves as a current-controlling device. Therefore, a superficial (but very functional) description of transistor behavior will be given here, with a more complete explanation of the internal action appearing at the end of the chapter.

The transistor is represented as a pair of back-to-back diodes in Fig. 4-1(a). Diode D_1 is forward-biased by V_{S1}, and diode D_2 is reverse biased by V_{S2}. We might expect that current I_2 would be near zero, but because of the interaction of the two diodes, we find that this is not the case. The small current I_1 flowing into the central region common to the two diodes enables a much larger current I_2 to flow across the reverse-biased diode D_2. The resistor R_1 is necessary in the circuit to limit I_1 to a small value. The current I_2 is produced by battery V_{S2} and is limited by the transistor itself. The value of I_2 is typically 100 times I_1. Thus if R_1 is adjusted for $I_1 = 0.1$ mA, I_2 would be 10 mA. If the input battery V_{S1} were removed so that $I_1 = 0$, then I_2 would drop to zero also.

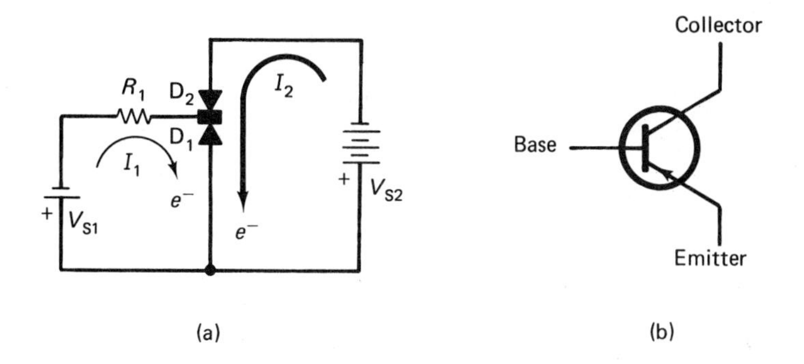

Figure 4-1 (a) Transistor as a pair of interacting diodes. A small forward current (I_1) through D_1 enables a much larger reverse current (I_2) through D_2. (b) Standard transistor schematic symbol. The base is the common element between the collector and emitter diodes. The diode arrow is shown only on the forward-biased (emitter) lead.

The standard symbol for one type of bipolar transistor is shown in Fig. 4-1(b), along with the names of the three leads. These names and the symbol used to be much more descriptive than they are today, since the earliest *point contact* types of bipolar transistors were actually fabricated as two fine wire contacts close together on a base piece of germanium crystal, as depicted by the symbol. Then the symbol was most often drawn on its side with the base at the bottom. The input current was applied at the emitter, and the output was taken at the collector. The base was grounded and served as the common element between the input and output.

This *common-base* circuit is still used to a limited extent, but much more popular today is the *common-emitter* circuit which uses the *base as the input*, the *collector as the output*, and the *emitter as the grounded element*. Most modern transistors are made of silicon, although germanium transistors are still fairly common. In either case, the physical structure of the transistor has undergone a series of evolutions so that the modern transistor looks nothing like the symbol. Many designers draw the symbol without the circle, while others prefer it with the circle to draw attention to the transistor as the focal point of the circuit.

NPN and PNP: Bipolar transistors are available in two distinct types, one designed for use with a positive supply at the collector and the other designed for a negative supply at the collector. The types are called *NPN* (positive supply) and *PNP* (negative supply). The letters refer to the three layers of the transistor structure (collector, base, and emitter) which may carry current by means of free negative electrons or free positive holes. Both *N*- and *P*-type materials are involved in either type of transistor—hence the term *bipolar transistor.*

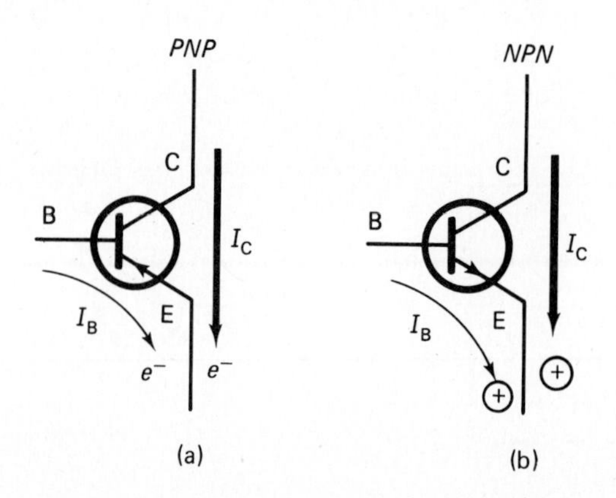

Figure 4-2 The *PNP* transistor generally takes a negative supply (positive ground), while the *NPN* type normally uses a positive supply.

Notice from Fig. 4-2 that the only difference in the symbols for the *PNP* and *NPN* transistors is the direction of the emitter arrow. The middle letter of the name denotes the polarity of the power supply normally used at the collector—negative for *PNP*, positive for *NPN* types. In the most common circuits the collector is connected to the supply side and the emitter to the ground side of the circuit.

Transistors are identified by type numbers which are usually printed on the case. Unfortunately, there is no significance to the type number itself; all the information about the transistor's characteristics must be obtained by looking under the type number in a transistor data book. Several thousands of transistor types are registered with the Joint Electron-Device Council (JEDC) and each may be produced by several manufacturers. Registered types have the prefix 2N- followed by three or four digits. Additional thousands of nonregistered types are given various letter/number symbols by their individual manufacturers.

4.2 BETA

The basic feature of a transistor is that a small current from the base to the emitter will enable a much larger current from the collector to the emitter. *The ratio of the collector current to the base current is the current gain of the transistor, usually referred to as* β *(beta).*

More technically, current gain can be specified as static (dc) beta, denoted as h_{FE}, or as dynamic (ac) beta, denoted as h_{fe}:

$$h_{FE} = \frac{I_C}{I_B} \tag{4-1}$$

$$h_{fe} = \frac{I_c}{I_b} \tag{4-2}$$

For most transistors, static and dynamic beta are equal within 10 or 20%, so the distinction between the two is not often made. In this text we shall treat them all the same, unless the distinction is specifically pointed out. Thus

$$h_{FE} \approx h_{fe} \approx \beta \tag{4-3}$$

The importance of beta in transistor-circuit calculations is best illustrated by an example. The objective will be to predict the output voltage at the collector of the transistor. The base current is easily found by Ohm's law. The collector current is calculated from $I_C = \beta I_B$.

Example 4-1

Find the collector voltage V_C in the circuit of Fig. 4-3.

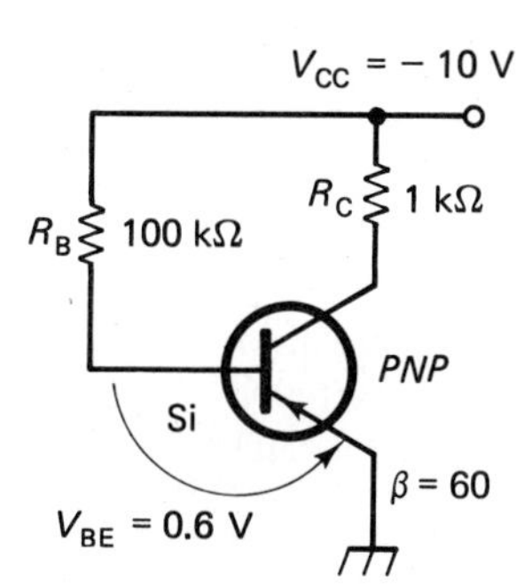

Figure 4-3 Beta is the key to calculating the output (collector) voltage of the transistor (Example 4-1).

Solution The base-emitter junction is a silicon diode forward biased at less than 10 mA, and will have a voltage drop of approximately 0.6 V, as explained in Section 2.4. The resistor R_B then has all the supply voltage except 0.6 V, and I_B can be found:

$$I_B = \frac{V_{R(B)}}{R_B} = \frac{10 - 0.6}{100\ \text{k}\Omega} = 0.094\ \text{mA}$$

The collector current is beta times larger than the base current:

$$I_C = \beta I_B = (60)(0.094) = 5.6\ \text{mA}$$

All the collector current must come through R_C, so the voltage drop $V_{R(C)}$ is determined:

$$V_{R(C)} = I_C R_C = (5.6\ \text{mA})(1\ \text{k}\Omega) = 5.6\ \text{V}$$

The remaining voltage appears from the collector to ground:

$$V_C = V_{CC} - V_{R(C)} = 10 - 5.6 = \mathbf{4.4\ V}$$

4.3 VARIATIONS IN BETA

Since the current gain of a transistor is its most important property, it is indeed unfortunate that the exact value of this gain factor beta cannot be relied upon. Two transistors of the same type number and from the same manufacturer may exhibit betas which are different by a factor of 5. When one is used to resistors with a tolerance of 5 or 10%, and capacitors with a tolerance of 10 or 20%, it is perhaps a little difficult to accept the fact that the most essential parameter of a transistor is so poorly controlled, but this is the nature of the manufacturing processes involved. It is common for manufacturers to specify minimum, typical, and maximum values of h_{fe} for a given transistor type. As an example, the type 2N3569 silicon *NPN* transistor has a minimum h_{fe} of 100, typical of 150, and maximum of 300. The type 2N3638 silicon *PNP* transistor has a minimum h_{fe} of 20, typical of 70, with no maximum limit specified.

Even more distressing in many cases are the variations in beta which occur with changes in operating temperature. A transistor which has a beta of 100 at room temperature, for example, may have a beta of 150 at a temperature of 80°C. Much of the problem of transistor-circuit design is to minimize the effects of changes in current gain (beta) without destroying the amplification effect made possible by high current gain.

The following example will illustrate how changes in beta can affect the collector output voltage of a simple transistor circuit.

Example 4-2

The transistor type used in the circuit of Fig. 4-4 has an h_{FE} which may range from a minimum of 30 to a maximum of 140 at 25°C. In addition, h_{FE} may increase by 30% at 80°C. Predict the minimum and maximum output voltage V_C.

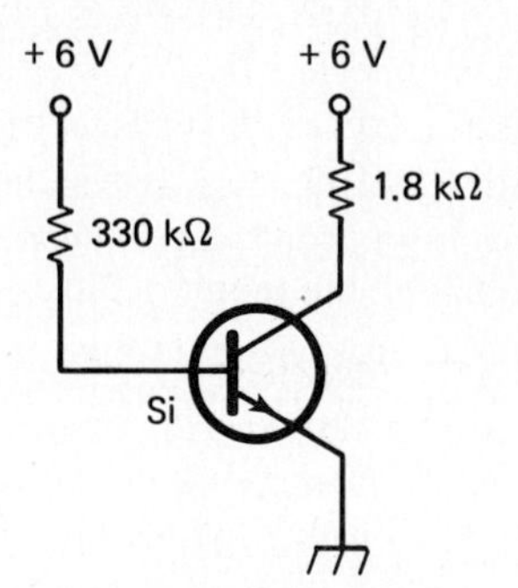

Figure 4-4 Example 4-2.

Solution If beta is 30, the base current is

$$I_B = \frac{V_{CC} - V_{BE}}{R_B} = \frac{6\text{ V} - 0.6\text{ V}}{330\text{ k}\Omega} = \frac{5.4\text{ V}}{330\text{ k}\Omega} = 0.0164\text{ mA}$$

Collector current is beta times base current:

$$I_C = \beta I_B = (30)(0.0164\text{ mA}) = 0.49\text{ mA}$$

Collector voltage can now be found from Ohm's law:

$$V_{R(C)} = I_C R_C = (0.49 \text{ mA})(1.8 \text{ k}\Omega) = 0.88 \text{ V}$$

$$V_C = V_{CC} - V_{R(C)} = 6 - 0.88 = \mathbf{5.12 \text{ V maximum}}$$

The minimum collector voltage will occur when beta is a maximum, since there will be more collector current and more voltage will be lost across R_C.

The maximum beta at 80°C is

$$\beta_{max} = 140 \times 130\% = \mathbf{182}$$

The base current is not changed by the higher beta, but the collector current is much larger:

$$I_C = \beta I_B = (182)(0.0164 \text{ mA}) = 2.99 \text{ mA}$$

$$V_{R(C)} = I_C R_C = (2.99 \text{ mA})(1.8 \text{ k}\Omega) = 5.4 \text{ V}$$

$$V_C = V_{CC} - V_{R(C)} = 6.0 - 5.4 = \mathbf{0.6 \text{ V minimum}}$$

The change in collector voltage due to variation in beta can easily be as severe as demonstrated in the example (V_C changed about 4.6 V out of a total supply voltage of 6 V) unless special circuitry is used to stabilize the transistor.

4.4 LEAKAGE CURRENT

Semiconductor diodes are not perfect; we saw in Chapter 2 that there is a small leakage current when a semiconductor diode is reverse-biased. This problem is also present in the transistor's collector-base junction, but it is more severe because the leakage current flows across the reverse-biased junction into the base where it is amplified by a factor of beta. Thus if a transistor has a collector-base leakage current of 1 μA and a beta of 100, a collector current of 100 μA will appear even though there is no externally applied base-input current. [See Fig. 4-5(a).]

The collector-base diode leakage is specified as I_{CBO} and can be readily determined by the circuit of Fig. 4-5(b). The amplified leakage current from collector to

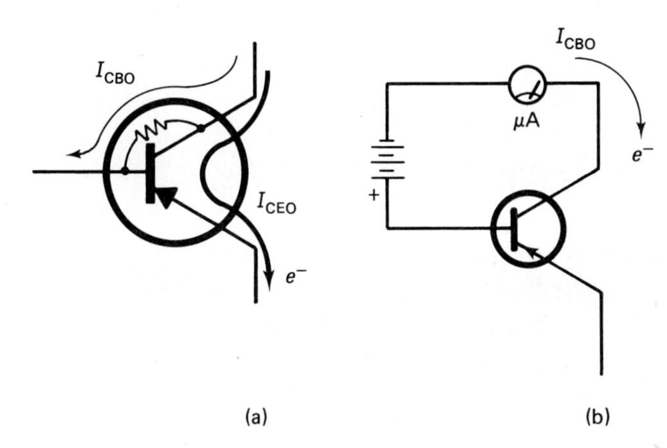

Figure 4-5 (a) The small reverse current which leaks across the collector-base diode (I_{CBO}) is amplified by beta to produce a much larger collector-to-emitter leakage; (b) lab setup to measure I_{CBO}.

emitter can be determined by moving the positive lead of the test battery from the base lead to the emitter lead of the transistor. The amplified leakage is termed I_{CEO} (the O indicates that the unnamed element—the base in this case—is to be left open-circuited for the test).

Manufacturer's data sheets generally specify the maximum I_{CBO} leakage current, although it may be called simply I_{CO} or *collector cutoff current*. Conversion to the more useful I_{CEO} figure can be done quickly if beta is known:

$$I_{CEO} = \beta I_{CBO} \tag{4-4}$$

Thermal runaway: A very real danger from leakage currents is *thermal runaway*, a condition in which increased transistor temperature causes increased leakage currents, which cause transistor temperature to increase still further in a spiral which ends in the destruction of the transistor.

The leakage current problem has been effectively solved by the development of silicon (rather than germanium) transistors. A typical low-power germanium transistor might have a maximum I_{CBO} specification of 5 μA and an operating temperature limit of 85°C, whereas a comparable silicon unit would have a maximum leakage of 50 nA (0.05 μA) and an operating temperature limit of 175°C. The leakage currents for most silicon transistors are low enough to be completely neglected in most circuits, even at fairly high temperatures. For germanium transistors, leakage and thermal runaway present a real problem, and this is probably the main reason germanium is seldom used in new designs.

4.5 SATURATION VOLTAGE

We have seen that the collector current of a transistor is limited by the base current and the beta of the transistor:

$$I_C = \beta I_B \tag{4-5}$$

However, most transistor circuits include a resistance between the collector and the supply, and this resistance also places a limit on collector current:

$$I_{C(sat)} = \frac{V_{CC}}{R_C} \tag{4-6}$$

If the transistor's collector current is limited not by the transistor itself but by the collector resistor, we say that the transistor is *saturated;* that is, it is drawing all the current available in that circuit. The characteristics of a saturated transistor are as follows:

1. Collector-emitter voltage near zero ($V_{CE} \rightarrow 0$).
2. Collector current nearly equals V_{CC}/R_C.
3. Further increase of I_B does not produce any appreciable increase in I_C or decrease in V_{CE}.

The collector-emitter saturation voltage $V_{CE(sat)}$ should ideally be zero to allow the transistor to appear as a perfect short circuit when fully turned on. Of course, this perfection cannot be achieved; typical low-power transistors have saturation voltages of 0.2 V or so, while power transistors handling several amperes may have $V_{CE(sat)}$ of 1 to 2 V or more.

Example 4-3

What input voltage is required to just saturate the transistor in the circuit of Fig. 4-6 under all conditions?

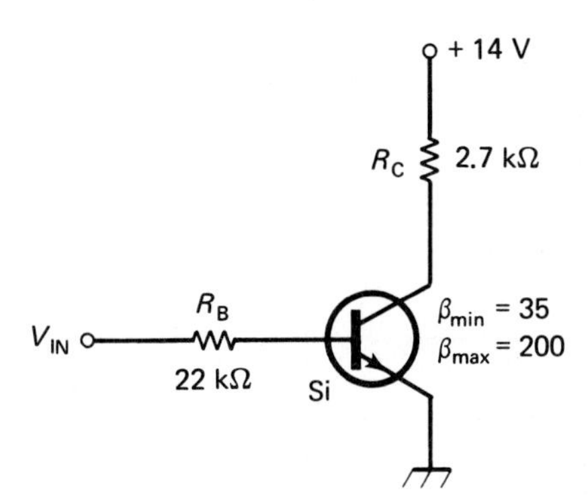

Figure 4-6 Example 4-3.

Solution At saturation the entire supply voltage (with the exception of the few tenths volt $V_{CE(sat)}$) will be dropped across R_C. The collector current is then

$$I_C = \frac{V_{CC}}{R_C} = \frac{14\text{ V}}{2.7\text{ k}\Omega} = 5.2\text{ mA}$$

It is most difficult to saturate a transistor when beta is low, so the minimum beta will be used to find I_B:

$$I_B = \frac{I_C}{\beta} = \frac{5.2\text{ mA}}{35} = 0.148\text{ mA}$$

The voltage drop across R_B which is required to deliver this 0.148 mA to the base is found from Ohm's law:

$$V_{R(B)} = I_B R_B = (0.148\text{ mA})(22\text{ k}\Omega) = 3.3\text{ V}$$

The transistor is silicon, so an additional 0.6-V drop will appear across the base-emitter junction:

$$V_{IN} = V_{R(B)} + V_{BE} = 3.3 + 0.6 = \mathbf{3.9\text{ V}}$$

4.6 CHARACTERISTIC CURVES

A remarkably complete picture of the parameters of any individual transistor can be obtained from its collector characteristic curves. Samples of such curves are given in Fig. 4-7. Manufacturers' data sheets often include a typical set of characteristic curves, but the word *typical* must be emphasized—we have seen that transistor parameters vary drasitcally from one unit to the next, and the characteristic curves of one unit may be quite different from those of another unit of the same type.

As can be seen from the figure, transistor curves are a series of graphs of collector current vs. collector voltage for a number of successive values of base current. Figure 4-7(a) shows that collector current is not so much a function of collector voltage as it is of base current. The static (dc) beta of the transistor can be determined at any point as the ratio of I_C/I_B. At point A in Fig. 4-7(a), this value is 5 mA/100 μA or $h_{FE} = 50$.

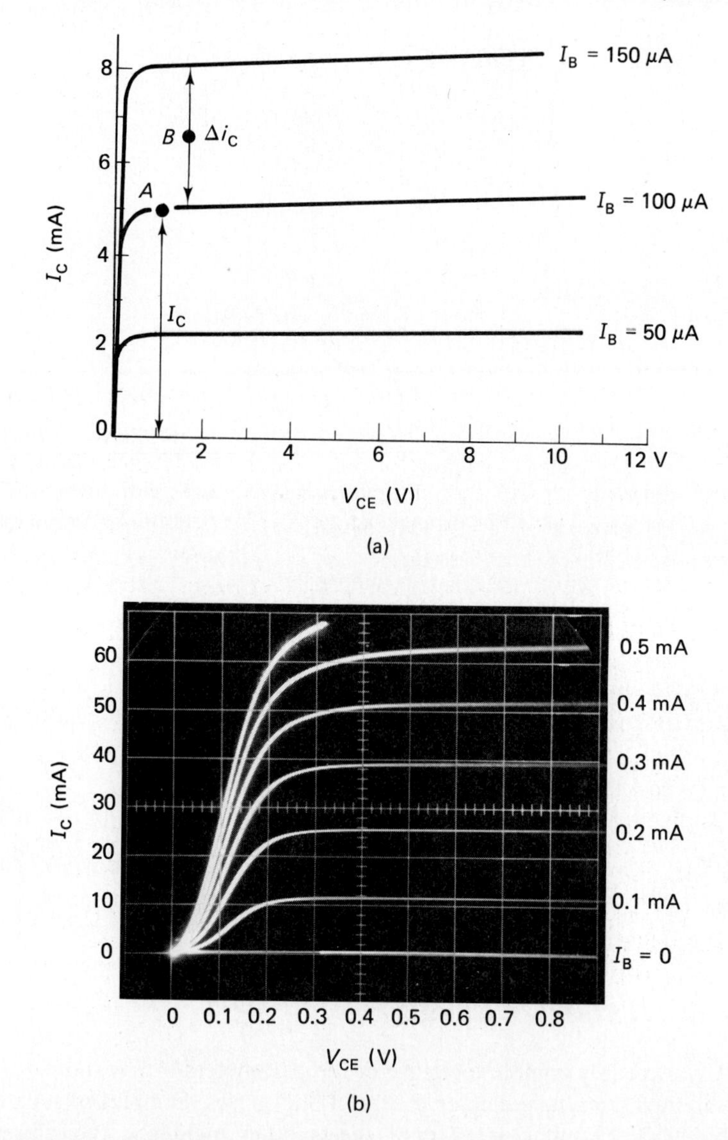

Figure 4-7 (a) Representative transistor characteristic curves. I_C depends primarily on I_B rather than on V_{CE}. (b) Horizontally expanded view of saturation region 2N3569 *NPN* silicon transistor. (c) Characteristic curves for a typical low-power germanium *PNP* transistor (type 2N404).

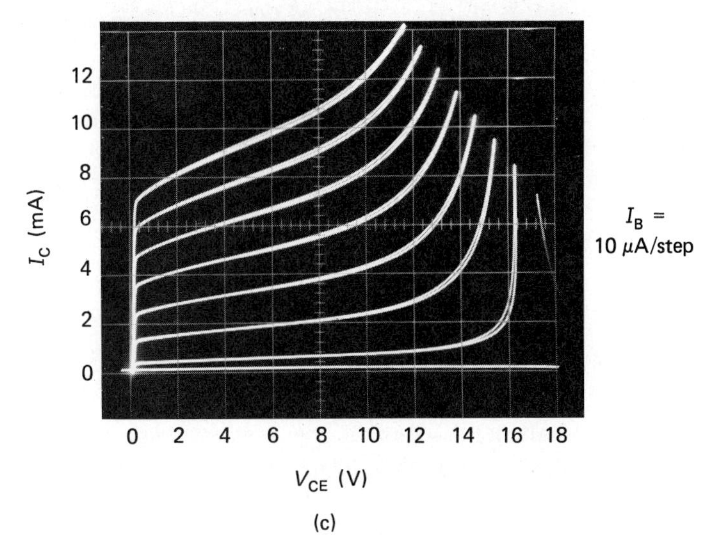

(c)

Figure 4-7 cont.

Dynamic (ac) beta can be determined as the ratio $\Delta i_C/\Delta i_B$. At point *B* in Fig. 4-7(a) this ratio is 3 mA/150-100 μA or $h_{fe} = 60$.

Collector saturation voltage can also be determined from the curves, and a special expanded view of the saturation region [as shown in Fig. 4-7(b)] is often given. As can be seen, the saturation voltage is higher for higher collector currents.

A display of transistor characteristic curves can be obtained on an oscilloscope tube using a curve tracer or adapter. The display can also be constructed one line at a

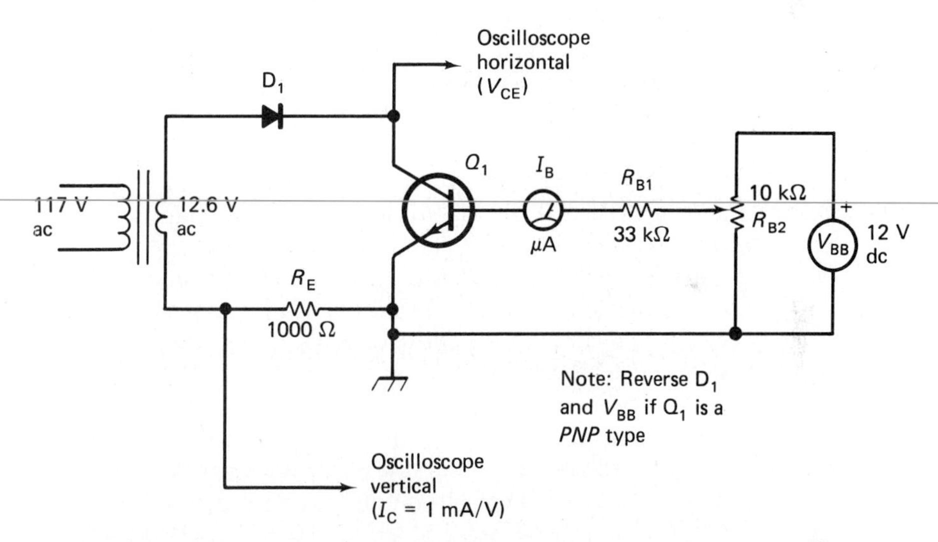

Figure 4-8 Laboratory setup for displaying characteristic curves on an oscilloscope, one line at a time. Q_1 is the transistor under test. R_{B2} sets the base current line.

time using the circuit of Fig. 4-8. If a 'scope camera is available, the entire "family" of curves can be constructed by multiple exposure of the film. In operation the 10-kΩ pot is adjusted for the desired base current reading on the microamp meter. Each volt appearing across R_E then represents 1 mA of collector current. The circuit shown is for *NPN* transistors, but it can be made to work with *PNP* transistors by reversing the polarity of the diode and the base supply.

4.7 POWER AND VOLTAGE RATINGS

Collector voltage: A glance at Fig. 4-7(c) will show that there is an upper limit called the *collector sustaining voltage* above which the collector current increases rapidly. The transistor will not necessarily be destroyed if this region is entered, but the extremely high currents which may be drawn and the nonlinear nature of the curves in the breakdown region make it a "forbidden" region of operation except for a few highly unusual circuits. There are other voltage breakdown effects which are definitely destructive, and these must also be avoided. Manufacturers' data sheets generally specify the maximum voltage from collector to base with the emitter lead *open* (V_{CBO}). Collector-to-emitter specifications V_{CES}, V_{CER}, and V_{CEV} are effectively the same since they *short* the base to emitter (perhaps through a *resistance*), or put a reverse *voltage* on it. All of these ratings are about twice as high as the $V_{CE(sus)}$ which the transistor can sustain in normal common-emitter operation. V_{CEO} and $V_{BR(CEO)}$ are specifications which are only slightly higher than $V_{CE(sus)}$.

Base voltage: The emitter-base diode of the transistor also has a reverse voltage limit beyond which the diode will break down and allow high reverse currents in the base. This voltage is often quite low, typically in the vicinity of 7 V. For some transistor types the base-emitter diode is also quite leaky in the reverse direction so that several microamperes of reverse current will appear at reverse voltages well below the base-emitter voltage limit.

Power limits: Most of the power dissipated in a transistor appears at the collector-base junction as heat. If this heat raises the junction temperature above the limit for the type of transistor involved, permanent damage will result, Typical low-power transistors have power-dissipation limits on the order of 150 to 300 mW. Transistors designed to dissipate 1 W or more will generally have a metal case or contact plate which is connected directly to the collector element. The transistor is then mounted on a heat sink to keep the collector element cool.

It is an elementary electrical concept that power is the product of voltage and current together. Thus a transistor can have a high voltage from collector to emitter and dissipate very little power if it draws very little current. Similarly, it can draw a very high current if it is near saturation (V_{CE} near zero). The power-dissipation limit of a transistor can be graphed on the characteristic curves, as shown in Fig. 4-9.

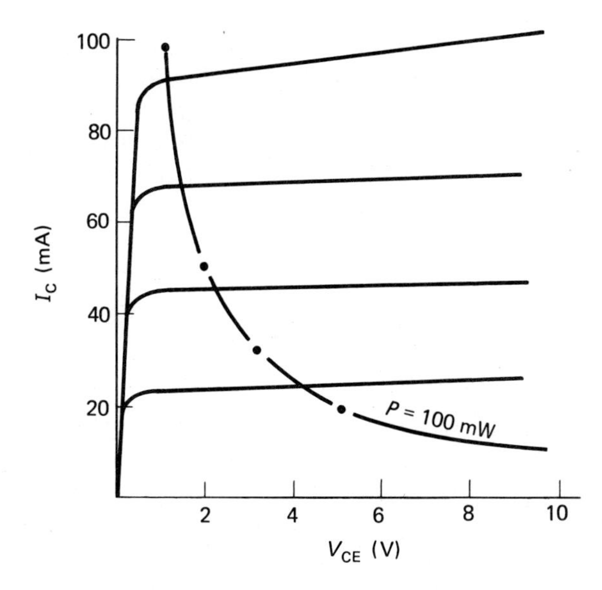

Figure 4-9 Constant-power-dissipation line on a family of characteristic curves.

4.8 TRANSISTOR TESTS AND MEASUREMENTS

Transistors are generally more vulnerable than their related circuit components; although they do not wear out with age, an accidental short circuit or misapplied voltage for just a fraction of a second can easily destroy them. Furthermore, a blown transistor will not normally smoke, crackle, or show any physical signs of failure. Sometimes a transistor will feel suspiciously hot just after failure, but even this symptom cannot be relied upon. Some means of testing transistors quickly is therefore of great importance in troubleshooting.

Transistor curve tracers and curve tracer adapters for conventional oscilloscopes are commercially available, and these provide the most complete and accurate check of a transistor. Many commerical meter-type transistor checkers are also available. Some of these are *in-circuit* testers which greatly facilitate checking soldered-in transistors. The instrument manual on these types should be studied closely, however, since some of the test functions may not operate and the accuracy of the basic beta readings may be greatly impaired for in-circuit tests. For a quick go–no-go check of a transistor when no special test equipment is available, a standard volt-ohmmeter can be used, as illustrated in Figs. 4-10 and 4-11. At least two leads of the transistor must be disconnected from the circuit when performing these tests. Many modern multimeters have a *low ohms* function which deliberately uses a test voltage too low to cause a junction to conduct. Where available, the *high ohms* function should be used.

Junction test: It has been noted that a transistor is basically a pair of interacting diodes. Thus it is possible to check a transistor with an ohmmeter by simply looking for a low forward and high reverse resistance between the emitter and base leads, and then between the collector and base leads, as shown in Fig. 4-10. If either of these junctions shows high resistance in both directions, it is an open junction. A shorted junction will show low resistance in both directions. The majority of transistor failures will show up in this simple check of the two junctions. The transistor should also be checked for high resistance both ways between the collector and emitter, since it is possible for these two elements to short without destroying the two diode junctions.

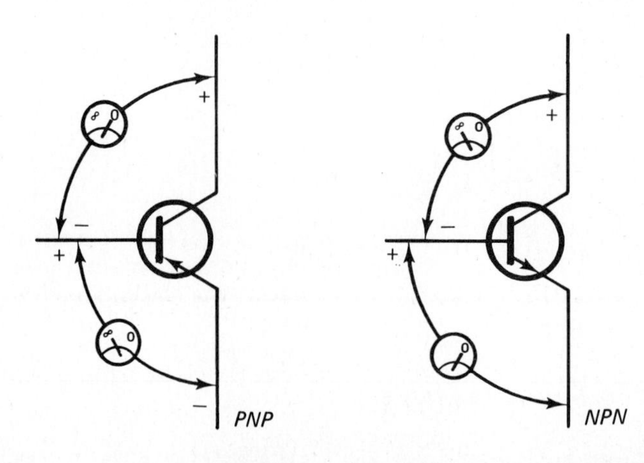

Figure 4-10 An ohmmeter can be used to check the base-emitter and collector-base diode junctions of a transistor.

The highest ohms scale of the VOM should not be used for these tests since most 20 000-Ω/V meters use a 9- to 30-V internal battery on this range, and this voltage may be enough to cause reverse voltage breakdown of the transistor junctions. For power transistors the lowest ($R \times 1$) scale of the meter should be used, since power transistors may have a rather high reverse leakage current which would show up as a rather large meter deflection on the higher ohms scales.

The base lead of a transistor with nonstandard pin connections can also be identified by the test of Fig. 4-10. Assuming that the transistor is good, two leads will be found which show a high resistance between them for either direction of ohmmeter connection. These will be the emitter and collector leads, and the remaining lead must then be the base. Distinguishing between the emitter and collector leads may be a more difficult problem, since for some transistors the major difference between them is the power dissipation capability of the collector. Once the polarity of the test voltage at the ohmmeter probes has been determined, the type of transistor (*PNP* or *NPN*) can also be determined by noting whether the diode junctions conduct with negative (*PNP*) or positive (*NPN*) base voltage.

Beta test: A rough estimate of the beta and leakage current of a transistor can be obtained very quickly with an ohmmeter using the test setup shown in Fig. 4-11. For small-signal transistors the ohmmeter is set to the $R \times 100$ or $R \times 1$ kΩ scale and the probes are connected between the collector and emitter, observing proper polarity. The lead of the ohmmeter which provides a positive test voltage on the ohms scale is connected to the collector for *NPN* transistors or to the emitter for *PNP* types. The meter deflection with the base lead open will give some indication of the I_{CEO} leakage of the transistor. (Be certain that the fingers are not touching the test leads or a false leakage due to body resistance will appear.) Small-signal silicon transistors generally show no deflection at all for this test. Germanium units will show progressively more deflection for higher leakage currents.

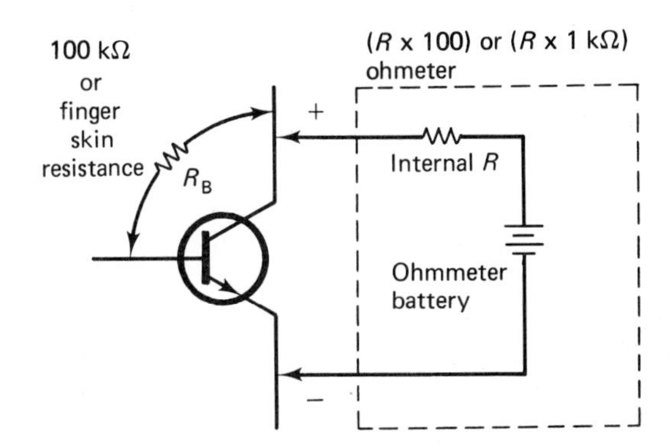

Figure 4-11 An ohmmeter and 100-kΩ resistor are all that is required to make a rough check on the beta and leakage of a transistor.

To check for beta, a 100-kΩ resistor is clipped between the collector and base leads, providing a base turn-on current for the transistor. This will cause a meter deflection much greater (a lower resistance reading) than the value of the 100-kΩ resistor itself, because of the transistor's current gain. Larger meter deflections indicate higher beta transistors. No deflection at all indicates an open transistor. Slight deflection indicates no current gain, possibly as the result of an emitter-base short. Large deflection, which does not change when the 100-kΩ resistor is removed, indicates a collector-base or collector-emitter short.

For a rough check on whether a transistor has any current gain at all, the collector and base leads can simply be held with the fingers, letting body resistance do for R_B. Once some practice has been gained with a particular meter and transistor type, defective units, and even units whose beta or leakage are abnormal, can be spotted with this test.

4.9 TRANSISTOR SUBSTITUTIONS

Although there are many thousands of different registered transistor types on the market, and more thousands of unregistered in-house types, finding a substitute for a particular transistor presents no difficulty in most cases, because transistors of the same type number may have widely varying characteristics, and most circuits are designed to minimize the effects of changes in these parameters. Although it should be emphasized that there are circuits which require matched transistors, or transistors with specially controlled characterstics, the following guidelines for transistor substitution will prove sufficient for the majority of small-signal applications:

1. Replace a *PNP* type with a *PNP* type, and an *NPN* type with an *NPN* type. This is a must for all cases.
2. Replace a silicon type with silicon, and a germanuim with germanium. Although some circuits will function with either type, many will not. In cases where both types seem to perform equally well, silicon offers greater reliability.
3. Be sure the transistor is capable of dissipating the power required by the circuit. Small-signal transistors in plastic or TO-5 round metal cans will typically handle 100 to 300 mW.
4. Be sure the transistor can handle the voltages applied by the circuit. Small-signal transistors typically have collector breakdown voltages in the range of 15-40 V. although transistors are available which can withstand several hundred volts.
5. Be sure the transistor has adequate frequency response (or switching speed in the case of pulse circuits). Frequency specifications on transistors can be misleading. For example, a transistor with a specification of $f_{T(min)} = 40$ MHz will likely have a serious reduction in gain at frequencies of 1 MHz and above. Chapter 9 deals with this problem in detail.

Four transistor types (a silicon *PNP* and an *NPN* at 300 mW and 5 W) will be found adequate to cover the majority of low-frequency replacement needs. Additional types will be required for replacing high-power, high-voltage, or high-frequency transistors.

4.10 TWO-JUNCTION DEVICE BEHAVIOR

In this chapter we have treated the transistor from a completely functional point of view, without regard for the actual internal action of the device. In this section we shall present a highly simplified account of internal transistor action. It should be emphasized that this is done mostly to satisfy the curiosity of the reader, since most technicians find an analysis in terms of internal device physics to be more confusing than enlightening. For practical technician's work the functional approach of the previous sections is recommended.

Common-base action: If a sandwich of three alternately doped layers of semiconductor material is made (*PNP* or *NPN*), and if the middle layer is kept extremely thin (a few thousandths of an inch), a bipolar transistor is formed. A first step toward understanding the action of this device is best taken by viewing it in the common-base connection, as in Fig. 4-12.

The emitter-base junction is forward biased by battery B_1, so that there is a large current across the first junction. However, because of the extreme thinness of the base region, most of the electrons simply keep right on going across the second junction (the collector-base junction). Only about 1 to 4% of the electrons injected at the emitter are able to "find" the base lead and flow out of the base. The other 96 to 99% flow from the emitter directly across to the collector.

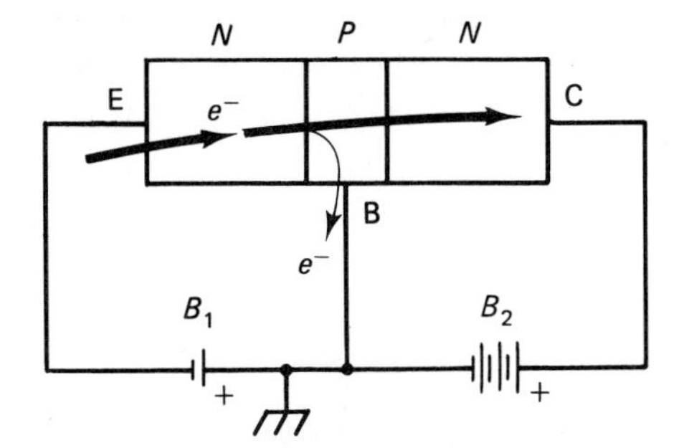

Figure 4-12 Basic transistor action: Nearly all the emitter current becomes collector current, but because the collector voltage is higher than the base voltage, the device has a total power gain.

The collector junction is reverse biased by battery B_2, and the full potential of B_2 appears across that junction. The current across this junction is not due to forward bias but to charge carriers injected through the emitter. Notice that the collector-base voltage is much higher than the emitter-base voltage, because the collector junction is reverse biased while the emitter junction is forward biased. An emitter current change will cause an almost equal collector current change, but the collector voltage is higher than the emitter voltage. Therefore the common-base circuit has a current gain which is slightly less than 1, but it has a high voltage gain and hence a total power gain.

Common-emitter action: In the early years of transistor technology, the common-base circuit was most widely used, and the foregoing analysis was quite relevant. Today we most often think of the transistor in the common-emitter circuit, and a few additional considerations are necessary. First, notice from Fig. 4-12 that the base current is quite small compared to the main-line emitter-to-collector current. Then consider that the flow of emitter current depends on whether the emitter-base junction voltage can be affected by altering either the emitter or the base voltage. Since the base current is so low, it turns out that much less power is required to alter the base voltage than the emitter voltage. This is the idea behind the common emitter circuit.

Figure 4-13 shows an *NPN* transistor connected in a common-emitter configuration. The two batteries of the previous circuit have been combined into a single battery,

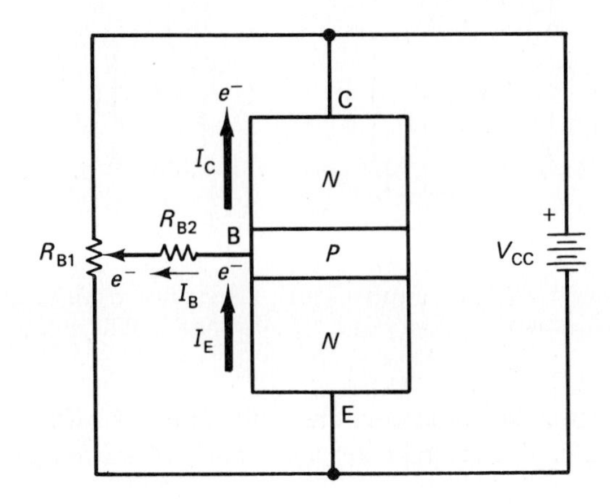

Figure 4-13 Basic common-emitter circuit action: Controlling the base voltage provides a low-power, low-current means of turning the base-emitter diode on and off.

V_{CC}. The base voltage (which must be somewhere between the collector and emitter potentials, but rather closer to the emitter) is now provided by a voltage-dividing potentiometer. This is quite practical, since the base current is small, as has been noted.

With R_{B1} at the bottom of its range, the emitter-base junction has no voltage across it, and there is no current across either junction. If R_{B1} is rotated toward the top of its range, the emitter-base diode will be forward biased when V_{BE} reaches 0.6 V (assuming a silicon device), and current will begin to flow across the emitter-base junction.

A few percent of this emitter junction current will find its way out the base and through R_{B2}. The resulting voltage drop across R_{B2} will tend to lower the emitter-base voltage and limit the emitter-base junction current. Thus R_{B2} provides a self-balancing effect on this current. If the junction current becomes too heavy bacause of a too-high voltage from the potentiometer, there is a greater current through R_{B2} and some of the input voltage is dropped across R_{B2}. Without R_{B2} in the circuit, the emitter-junction current would increase almost without limit as soon as V_{BE} reached 0.6 V. The result would be the destruction of the transistor.

4.11 TRANSISTOR FABRICATION

The development of the transistor is traced in Figs. 4-14 and 4-15. The original transistor was a *point contact* device, fabricated as two fine wires making contact close together on a germanium crystal [Fig. 4-14(a)]. One wire was the emitter, the other the collector, and the crystal itself was the base. The large base area of this device allowed too many of the emitter charge carriers to flow out the base, and betas were consequently low.

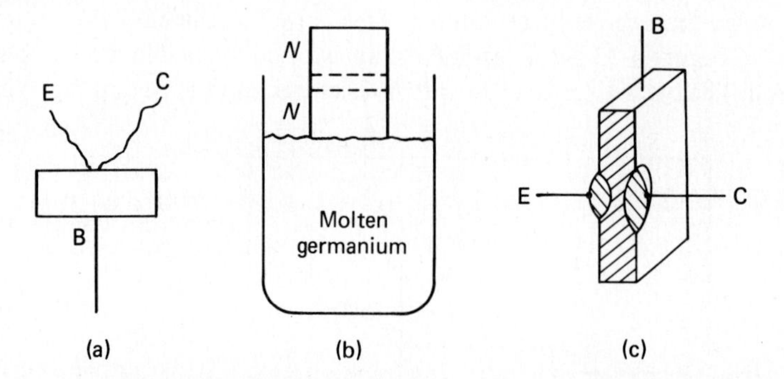

Figure 4-14 Early transistor fabrication techniques: (a) the first crude point contact devices; (b) rate-grown transistors by "crystal pulling"; (c) alloy junction transistor.

A second method of transistor fabrication, illustrated in Fig. 4-14(b) and known as *rate growing*, involved "pulling" a crystal from a container of melted germanium.

The impurity content of the molten germanium was changed from predominantly *P* type to predominantly *N* type and then back to *P* type as the solidified crystal was slowly extracted.

Alloy junction transistors were fabricated by using heat to bond-separate *P*-type emitter and collector pellets into opposite sides of a central slab of *N*-type base material, as shown in Fig. 4-14(c). The depth of penetration of the emitter and collector pellets determines the thickness of the base region and hence the efficiency of the device as an amplifier.

The silicon planar process, illustrated step by step in Fig. 4-15 is currently the most popular transistor fabrication technique. and provides the basis for the monolithic integrated-circuit manufacturing technique. The process begins with a very heavily doped (N+) silicon *substrate,* as shown in Fig. 4-15(a). A gas containing silicon with a light concentration of *N*-type impurities is then passed over the substrate at high temperature and a thin *N*-type silicon *epitaxial layer* is vapor-deposited on the substrate. This epitaxial layer consists of a crystal-lattice structure which is

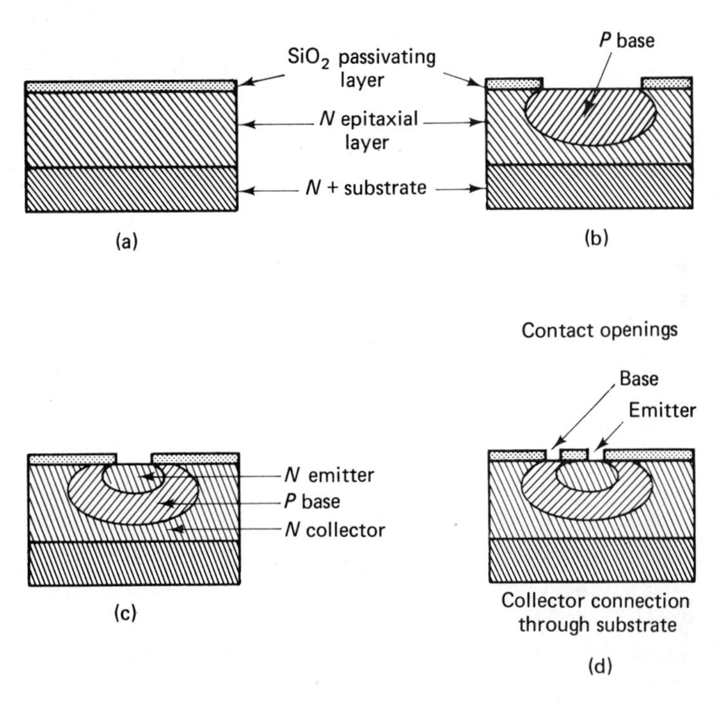

Figure 4-15 Silicon-planar process of transistor fabrication. (a) An *N*-doped epitaxial layer of silicon with a highly regular crystal structure is grown on top of a *P*-type substrate and covered with a protective layer of silicon dioxide. (b) A window is etched in the protective layer and *P*-type impurities are vapor-diffused into the *N*-type expitaxial layer. (c) A second smaller window is etched an an *N*-type region is diffused into the *P* region. (d) Openings are etched into the protective layer, and metal is vapor-deposited to contact the collector, base, and emitter regions of the transistor.

much more perfect than can be produced by refining bulk silicon. The substrate will serve as the collector contact and heat sinking path for the transistor; the epitaxial layer will become the collector itself. A silicon dioxide *passivating* layer is deposited over the epitaxial layer as a protective cover.

A photographic negative *mask* is then used to fix an acid-resistant coating over the passivating layer everwhere except in the center where the base is to be formed. [See Fig. 4-15(b).] An acid solution etches away the protective silicon dioxide layer in this central region. Another vapor containing *P*-type impurities is then passed over the device, and the impurities diffuse into the epitaxial layer through the opening in the passivating layer.

A second passivating layer, photomasking, and etching leave a smaller opening for the diffusion of *N*-type impurities to form an emitter region nested in the base region, as shown in Fig. 4-15(c). A final passivating layer, masking, and etching are necessary to expose the areas where contacts to the emitter and base regions will be made. [See Fig. 4-15(d),] Aluminum contact points are vapor-deposited through the openings in a final *metallization* process.

The above processes are generally carried out on a group of sliver-dollar-sized disks, each containing several thousand invididual transistors. When the process is completed the units can be tested, marked good or bad, sawed into single chips, and mounted in plastic or metal packages.

CHAPTER SUMMARY

1. A bipolar transistor uses a small base-emitter current to control a large collector-emitter current from a battery or dc supply.
2. The ratio of collector current to base current is β (beta), also called h_{FE} (dc currents) and h_{fe} (ac currents).

$$\beta = \frac{I_C}{I_B} = h_{FE} \approx h_{fe} \qquad (4\text{-}3)$$

Beta varies considerably from one transistor to the next, even among transistors of the same type number.
3. Bipolar transistors are available in two opposite-polarity types. With the emitter grounded, *NPN* types require positive voltages at the base and collector, whereas *PNP* types require negative supply voltages.
4. A transistor is *saturated* when it is fully turned on and collector current is limited by the collector supply and external coilector resistance. At saturation V_{CE} is small and increasing I_B no longer increases I_C.
5. Collector characteristic curves show I_C (vertically) vs. V_{CE} (horizontally) for several fixed steps of I_B. They can be used to read h_{FE}, h_{fe}, $V_{CE(sat)}$, and maximum collector voltage.
6. Bipolar transistors can be tested by using an ohmmeter to check for low forward

and high reverse resistance of the B–E and B–C junctions, and high resistance both ways from C to E.

7. Transistor operation depends upon the thinness of the base region. Most of the charge carriers crossing the emitter-base junction continue right on to the collector, even though the base-collector junction is reverse biased. Base current is therefore much less than main-line emitter–collector current.

QUESTIONS AND PROBLEMS

4-1. Fill in the blanks with *positive* or *negative*: In an *NPN* transistor the base voltage is more ________ than the emitter, and the collector voltage is more ________ than the emitter. In a *PNP* transistor the base is more ________ than the emitter and the collector is more ________ than the emitter.

4-2. Fill in the blanks with *emitter*, *base*, or *collector*: In a *PNP* transistor a small electron flow from ________ to ________ controls a larger electron flow from ________ to ________. In an *NPN* transistor a large conventional current from ________ to ________ is controlled by a small conventional current from ________ to ________.

4-3. A transistor has a base current of 150 μA and a collector current of 12 mA. What is its β?

4-4. What base current is required to produce a collector current of 1.5 A if β is 60?

4-5. What is the collector current in a transistor with $\beta = 180$ if $I_B = 40\ \mu$A?

4-6. In Fig. 4-3, V_{CC} is changed to -3 V and the transistor β is now 75. Find V_C.

4-7. The transistor in Fig. 4-4 has $\beta_{min} = 45$ and $\beta_{max} = 250$. R_C is changed from 1.8 kΩ to 1.0 kΩ. Find $V_{C(max)}$ and $V_{C(min)}$.

4-8. What β is required to just saturate the transistor in Fig. 4-4? All circuit values are as given in the figure.

4-9. Read the curves of Fig. 4-7(b) for dc beta (h_{FE}) at $V_{CE} = 0.6$ V and $I_C = 25$ mA.

4-10. Read the curves of Fig. 4-7(b) for ac beta (h_{fe}) at $V_{CE} = 0.6$ V and $I_C = 25$ mA.

4-11. A transistor has leads 1, 2, and 3. With the positive ohmmeter probe on 1 and the negative on 2, a low resistance is observed (1 – 2, LO). Other ohmmeter tests show (1 – 3, LO), (2 – 3, HI), (2 – 1, HI), (3 – 1, HI), and (3 – 2, HI). Tell whether the transistor is *NPN* or *PNP* and which lead is the base.

4-12. In a certain transistor 98.3% of the emitter current reaches the collector. What is the β of this transistor?

4-13. Why is the common-emitter circuit generally preferred over the common-base circuit?

4-14. Define the following terms: substrate; epitaxial; passivation; diffusion.

5

TRANSISTOR SWITCHING CIRCUITS

5.1 TRANSISTOR SWITCH

The low collector-to-emitter saturation voltage of a transistor which is turned on by a high base current gives the transistor many of the properties of a closed switch. The low current passed from collector to emitter when no base current is applied gives the transistor many of the properties of an open switch. Transistors are, in fact, very extensively used as electrically operated, no-moving-part switches.

On-off switching of electric currents in a highly organized fashion is the basic internal function of the digital computer. A few of the earliest computers used relays to perform this function, but even the fastest relays require a millisecond or more to switch from one state to the other. By contrast, early slow-speed transistors were capable of switching in less than a microsecond, and devices are currently available with switching times in the range of 10 ns (10×10^{-9} s). Incredible as these speeds may seem, the call is for even faster devices, and continuing research is sure to bring them.

An elementary example of a transistor used as a switch to control an instrument-panel indicating light is shown in Fig. 5-1. As simple as it is, this circuit finds wide application in electronic instruments. To pick a few examples, it may be used to indicate when an automatic test has been completed, or when an overvoltage is present. It may be driven by an ac voltage detector like the one in Fig. 3-19 to indicate the presence of an ac "call" tone on an intercom line.

Example 5-1

1. Find the value of R_C that will make $I_{LED(max)} = 15$ mA in Fig. 5-1. Use $V_{CE(sat)} = 0.1$ V and $V_{LED} = 1.4$ V.

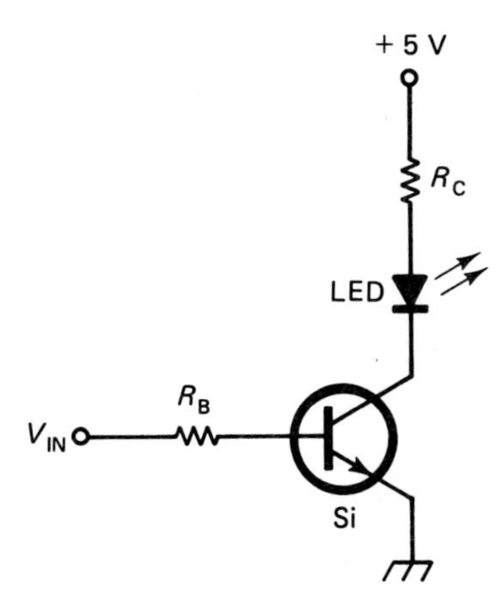

Figure 5-1 Transistor LED driver. R_C limits the LED current to the desired value.

2. Find R_B such that $V_{IN} = 2.4$ V will saturate the transistor if $\beta_{min} = 30$.

Solution

1.
$$V_{R(C)} = V_{CC} - V_{LED} - V_{CE(sat)}$$
$$= 5 - 1.4 - 0.1 = 3.5 \text{ V}$$
$$R_C = V_{R(C)}/I_{LED} = 3.5/15 \text{ mA} = 233\ \Omega$$

2.
$$V_{R(B)} = V_{IN} - V_{BE} = 2.4 - 0.6 = 1.8 \text{ V}$$
$$I_{R(B)} = I_B = I_C/\beta_{min} = 15 \text{ mA}/30 = 0.5 \text{ mA}$$
$$R_B = V_{R(B)}/I_{R(B)} = 1.8 \text{ V}/0.5 \text{ mA} = 3.6 \text{ k}\Omega$$

Relay driver: Transistors that can switch 10 A dc are readily available, but sometimes the load current is ac, or the $V_{CE(sat)}$ loss cannot be tolerated, or the load must be isolated from the control-circuit ground. In these cases the transistor switch may be used to energize the coil of a relay whose contacts then control the load. Figure 5-2 shows such a circuit.

The diode shown across the relay coil does not conduct when the relay is energized. However, when V_{IN} is suddenly removed I_C and I_{COIL} drop to zero, and the coil's magnetic field collapses, creating a self-induced "kickback" voltage. This voltage could be large enough to damage the transistor if a discharge path through the diode were not provided. Turn-off of the relay is delayed by approximately one time constant ($\tau = L_{coil}/r_{coil}$) as the inductance discharges.

Example 5-2

1. In Fig. 5-2, what is the lowest β that will be certain to turn the relay coil fully on? Use $V_{CE(sat)} = 0$.
2. What is the turn-off delay time caused by L_{coil}?
3. What is P_{max} of the transistor if V_{IN} can assume any value?

Solution

1.
$$I_C = V_{CC}/r_{coil} = 12 \text{ V}/60\ \Omega = 200 \text{ mA}$$
$$I_B = (V_{IN} - V_{BE})/R_B = (6 - 0.6)/470 = 11.5 \text{ mA}$$
$$\beta = I_C/I_B = 200/11.5 = \mathbf{17.4}$$

2.
$$\tau = L_{coil}/r_{coil} = 0.5/60 = \mathbf{8.3 \text{ ms}}$$

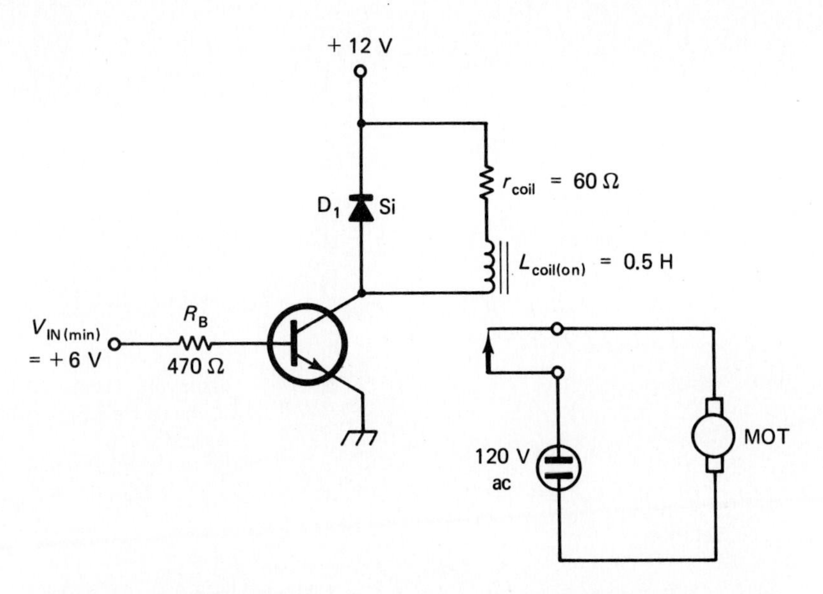

Figure 5-2 Transistor relay driver. D_1 suppresses inductive kickback voltage from the coil's inductance.

3. Power is maximum when transistor dc resistance equals collector-line resistance r_{coil}, and $V_{CE} = V_{COIL} = 6$ V.

$$I_C = V_{COIL}/r_{coil} = 6/60 = 0.1 \text{ A}$$

$$P_C = IV = (0.1)(6) = \mathbf{0.6\ W}$$

5.2 TRANSISTOR INVERTER (NOT GATE)

Many times it is desired to *reverse the sense* of a control signal. For example, a straightforward lamp-driver circuit could be designed to turn on a light when a photoresistor is turned on by sunlight. However, the reverse is generally what is required, i.e., to turn on a light when sunlight is *not* present. There are several ways to reverse the sense of the signal in such an application. One is to use a normally closed set of relay contacts to control the lamp, so turning on the transistor will open the contacts. This method may be quite practical in a simple photo-control application, but in large control systems relays would be too slow and too expensive.

The *transistor inverter*, consisting of R_2, R_3, and Q, in Fig. 5-3(a), is generally used to reverse the sense of a control signal. It is sometimes called a NOT gate, because it produces an output voltage when it does NOT have an input voltage.

Example 5-3

The emergency exit lamp in Fig. 5-3(b) requires 1 A of current. The transistors are silicon. Find the resistances of the photoresistive cell that will turn the lamp full-on (1) and full-off (2). Both transistors have $\beta = 50$.

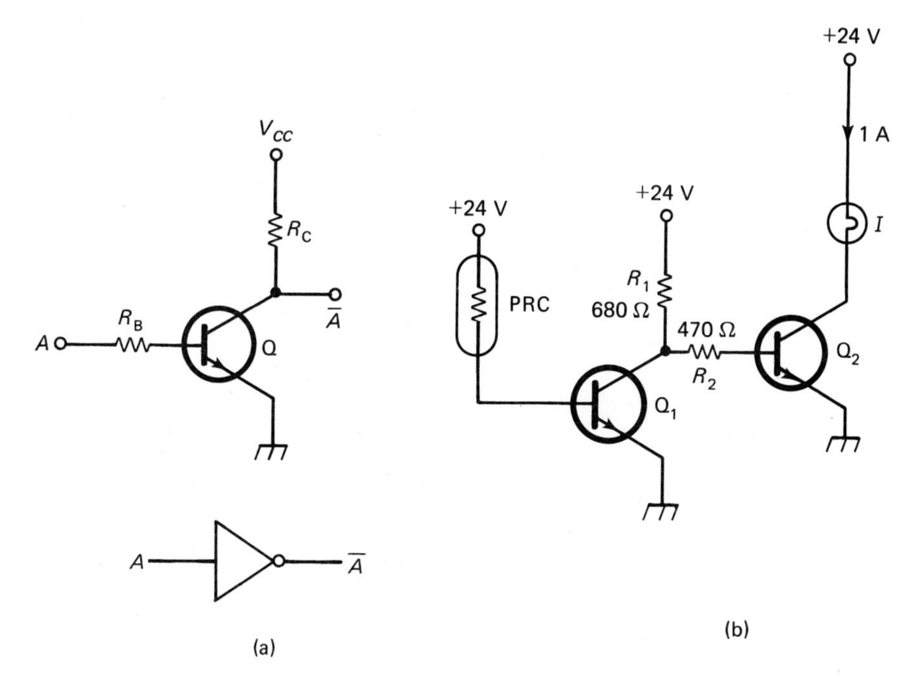

Figure 5-3 (a) Transistor inverter schematic diagram and logic symbol; (b) design example—inverter Q_1 and lamp driver Q_2.

Solution

1. For $I_I = 1$ A:

$$I_{R2} = \frac{I_I}{\beta} = \frac{1\text{ A}}{50} = 20\text{ mA}$$
$$V_{R2} = I_{R2}R_2 = (20\text{ mA})(470\ \Omega) = 9.4\text{ V}$$
$$V_{1C} = V_{R2} + V_{2B} = 9.4 + 0.6 = 10.0\text{ V}$$
$$V_{R1} = V_{CC} - V_{1C} = 24 - 10.0 = 14.0\text{ V}$$
$$I_{R1} = \frac{V_{R1}}{R_1} = \frac{14.0}{680} = 20.6\text{ mA}$$
$$I_{PRC} = \frac{I_{R1}}{\beta} = \frac{20.6}{50} = 0.412\text{ mA}$$
$$R_{PRC} = \frac{V}{I} = \frac{24 - 0.6}{0.412\text{ mA}} = \mathbf{56.8\ k\Omega}$$

2. For $I_I = 0$:

$$V_{1C} = V_{2B(\text{threshold})} = 0.6\text{ V}$$
$$V_{R1} = V_{CC} - V_{1C} = 24 - 0.6 = 23.4\text{ V}$$
$$I_{R1} = \frac{V_{R1}}{R_1} = \frac{23.4}{680} = 34.4\text{ mA}$$
$$I_{PRC} = \frac{I_{R1}}{\beta} = \frac{34.4\text{ mA}}{50} = 0.69\text{ mA}$$
$$R_{PRC} = \frac{V}{I} = \frac{24 - 0.6}{0.69} = \mathbf{33.9\ k\Omega}$$

Notice that the inverter, Q_1, provides current gain as well as inverting the sense of the input signal. A single transistor could not have operated the lamp from the photocell, since the maximum available input current from the PRC would be insufficient to saturate transistor Q_2. Notice also that *decreasing* the current to the base of Q_1 causes V_{1C} to *increase*. This causes I_{2B} to increase and V_{2C} to decrease. This fact—that *increasing* base input voltage causes *decreasing* collector output voltage—is very important, both here and in many circuits to follow.

5.3 SCHMITT TRIGGER

It is often desirable to provide a toggling or snap action in a control function. For example, in the case of the light-operated emergency-exit lamp, it may be noticed that the lamp "mushes in" as the light intensity slowly increases. It is therefore good practice to place a trigger circuit between the photocell and the lamp driver. The trigger circuit should ignore all signals up to a predetermined voltage but turn completely on for any signal even slightly above this triggering level. If the input signal drops below a predetermined level, the trigger should automatically switch completely off without delay.

The circuit that performs this function is called a *Schmitt trigger*. The difference between the turn-on and turn-off levels is called *hysteresis*. For example, a Schmitt trigger which turns on at +6 V but does not turn back off until the voltage drops to +4 V is said to have a 2-V hysteresis zone. Excessive hysteresis is generally undesirable, as demonstrated by the case of the thermostat control system which turns on at 65°F but does not turn back off until the temperature is up to 80°F.

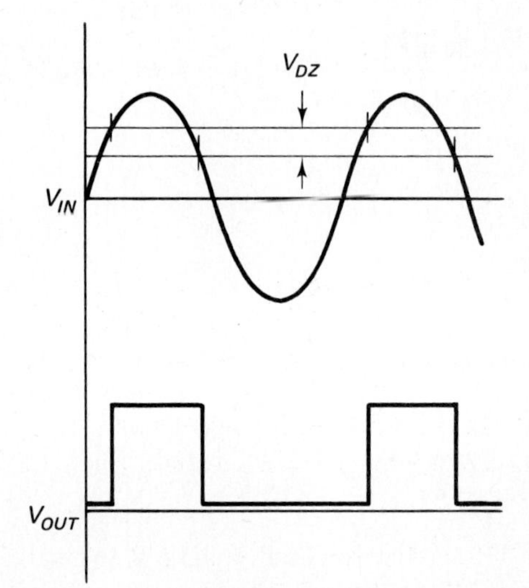

Figure 5-4 Schmitt-trigger input and output waveforms showing hysteresis effect.

The output waveform produced by a Schmitt trigger with a sine-wave input is shown in Fig. 5-4. Note the effect of hysteresis, as evidenced by the relative input voltages at which the output switches on and off. The Schmitt trigger is sometimes referred to as a squaring circuit because it takes "rounded" input signals and shapes them into square-cornered output signals. This term may be confusing, however, since there are also circuits which are designed to square an input signal (i.e., multiply the input signal by itself).

Figure 5-5 shows a transistor Schmitt-trigger circuit. Notice the similarity to the inverter and lamp driver of Fig. 5-3. Increasing V_{IN} causes decreasing V_{1C}, which causes increasing V_O. Here, however, there is a *feedback* resistor R_F so that a high V_O reinforces the high V_{IN} which caused it. R_F therefore provides the hysteresis effect, tending to make Q_2 stick in either the high- or low-voltage output condition. R_s and R_v are used to bias the Q_1 input. Negative voltage from R_v means that positive V_{IN} is required to turn Q_1 *on*; positive voltage from R_v means that negative V_{IN} will switch Q_1 *off*.

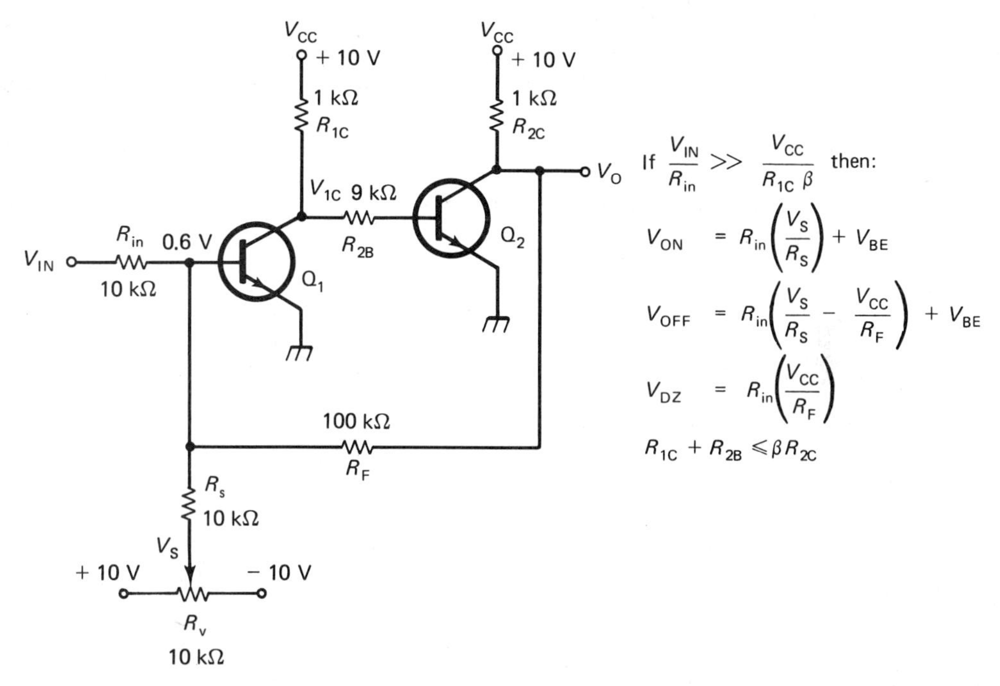

Figure 5-5 Schmitt-trigger circuit and design equations.

Circuit action: Let's watch the Schmitt trigger in action. V_S is −2.4 V, so Q_1 is turned *off* by the negative base voltage while V_{IN} is zero. V_{1C} is +9 V, so Q_2 is turned on and V_O is zero.

- V_{IN} starts to go positive. When it reaches +3.6 V there will be 3.0 V across R_{in}, attempting to turn Q_1 *on* and −3.0 V across R_s, attempting to turn it *off*.
- When V_{IN} goes above 3.6 V it starts to turn Q_1 *on* and V_{1C} drops below 9 V. Soon there is not enough voltage across R_{2B} to keep Q_2 turned on.

- As soon as Q_2 starts to turn *off*, V_O rises and additional turn-on current is sent through R_F to the base of Q_1.
- Q_1 now turns on harder, lowering V_{1C} to zero, turning Q_2 *off* completely. V_O rises to nearly +10 V, adding about 10 V/100 kΩ or 0.1 mA to the Q_1 turn-on current.

The Schmitt trigger will not turn *off* again until the input current from R_{in} is reduced by this 0.1-mA margin. This means a V_{IN} reduction of (0.1 mA) (10 kΩ) = 1 V. The turn-off point is therefore 3.6 − 1.0 = 2.6 V.

Some notes on the selection of component values may be helpful.

- R_{IN} is made high enough to avoid loading the driving source. Lower values produce narrower hysteresis zones. A too-high value would make it impossible to turn Q_1 fully on.
- R_s is chosen to set the range of trigger voltages. If $R_s = R_{in}$, the range is approximately $\pm V_{CC}$. If R_{in} is smaller than R_s the range is proportionately smaller.
- R_v is generally chosen about equal to R_s.
- R_F is chosen to set the hysteresis zone. Lower values produce wider dead zones.
- R_{2C} is chosen to provide the load current required at V_O. It should be much lower than R_F.

b R_{1C} and R_{2B} may be made equal and should be sized to provide adequate turn-on current for the base of Q_2 (i.e., $R_{1C} + R_{2B} \leq \beta_{min} R_{2C}$). Making these values excessively low will widen the hysteresis zone and require lower values of R_{in}.

Example 5-4

Figure 5-6 shows the emergency-exit-lamp switch of Example 5-3 with a feedback

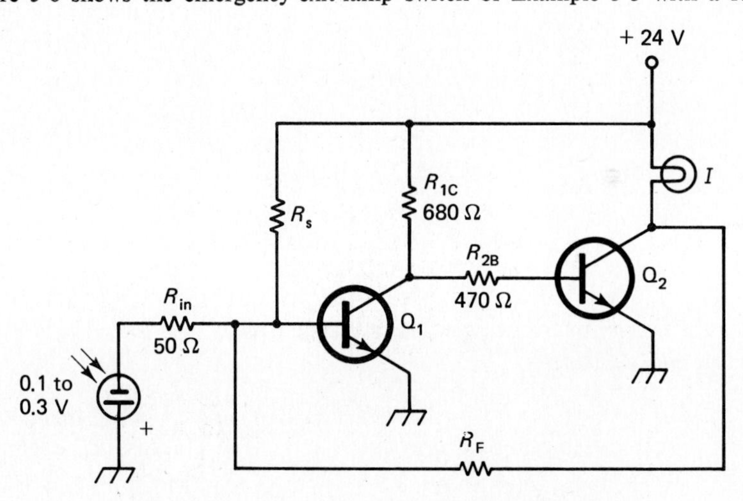

Figure 5-6 Snap-on exit-lamp switch of Example 5-3.

resistor to provide Schmitt-trigger snap action. The input has been changed to a photovoltaic cell which has an internal resistance of 50 Ω. Its no-load output is 0.3 V at normal room light and 0.1 V with lights out. Find values of R_F and R_s to trigger the lamp *off* and *on* as these levels are crossed.

Solution To turn Q_1 on at 0.1 V from the PVC:

$$I_{R(\text{in})} = I_{R(s)}$$

$$\frac{V_{1B} - V_{IN}}{R_{in}} = \frac{V_{CC} - V_{1B}}{R_s}$$

$$\frac{0.6 - (-0.1)}{50} = \frac{24 - 0.6}{R_s}$$

$$R_S = \mathbf{1670\ \Omega}$$

For a dead-zone of 0.2 V, R_F is

$$R_F = \frac{R_{in} V_{CC}}{V_{DZ}} = \frac{(50)(24)}{0.3 - 0.1} = \mathbf{6.0\ k\Omega}$$

5.4 LOGIC GATES

OR gate: Many decision-making operations which are highly structured in the sense that they do not require the making of value judgments can be taken care of automatically by electronic logic circuits. An elementary example is provided by an elevator for a three-story building. If there is a call from floor 1, OR from floor 2, OR from floor 3, the elevator motor is signaled to run. A circuit which implements this directive is shown in Fig. 5-7.

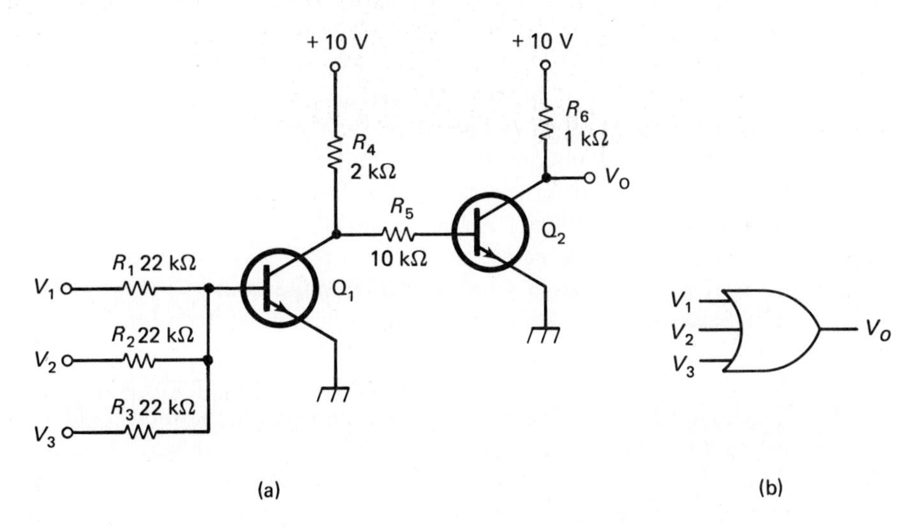

Figure 5-7 Transistor OR gate and logic symbol.

If V_1, or V_2, or V_3 is positive 10 V, Q_1 is turned on, and the voltage at the collector of Q_1 goes to zero. If V_{1C} is zero, Q_2 is *not* turned on, there is no current through the collector of Q_2, and no voltage is dropped across R_6. Hence the collector voltage of Q_2, which is V_O, will be $+10$ V if any or all of the inputs are $+10$ V. This circuit is called a three-input OR gate, and any system which functions in this manner is indicated by the symbol in Fig. 5-7(b).

Nor gate: It may be noticed that R_5, R_6, and Q_2 comprise an inverter, as discussed in section 5.2. The circuitry remaining when these three components are removed is called a NOR or NOT-OR gate, since the output of Q_1 will go NOT positive (i.e., zero, or ground) when any of the inputs go positive.

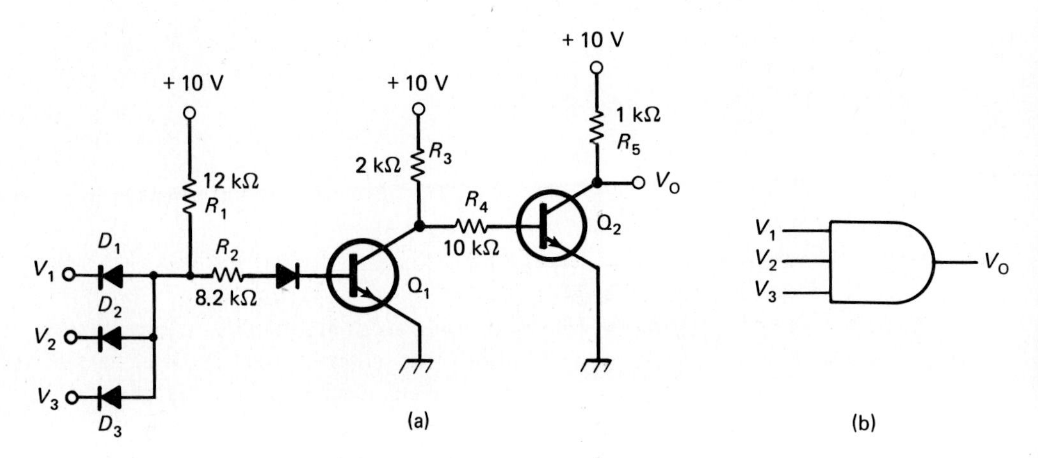

Figure 5-8 Transistor AND gate and logic symbol.

AND gate: Another type of logic gate, called the AND gate, is shown in Fig. 5-8. In this circuit, V_A and V_B and V_C must all be positive in order to let Q_1 turn on. If any one of the inputs is grounded, the base current will be drained away from Q_1 and it will turn off. (It should be noted than an open-circuited input is forbidden in logic circuitry—all inputs must be either at $+10$ V or ground.) As in the previous case, Q_2 may be left out, in which case the circuit is called a NAND gate, and the output (of Q_1) is *not* positive if V_A *and* V_B *and* V_C are all positive. The input circuitry of either of the two basic gates may be expanded with any number of resistors or diodes to provide the desired number of input conditions.

Combinational logic: It is possible to design logic systems for rather complicated decision-making schemes. One final example will be given to demonstrate the combination of logic gates. Recalling the elevator problem for which the circuit of Fig. 5-7 was designed, it may be realized that the elevator motor should *not* be commanded to run for a call at the second floor if the car is already at the second floor. To exclude this kind of possibility, a more complex logic system must be used. This system is diagrammed in symbolic form (called a logic diagram) in Fig. 5-9.

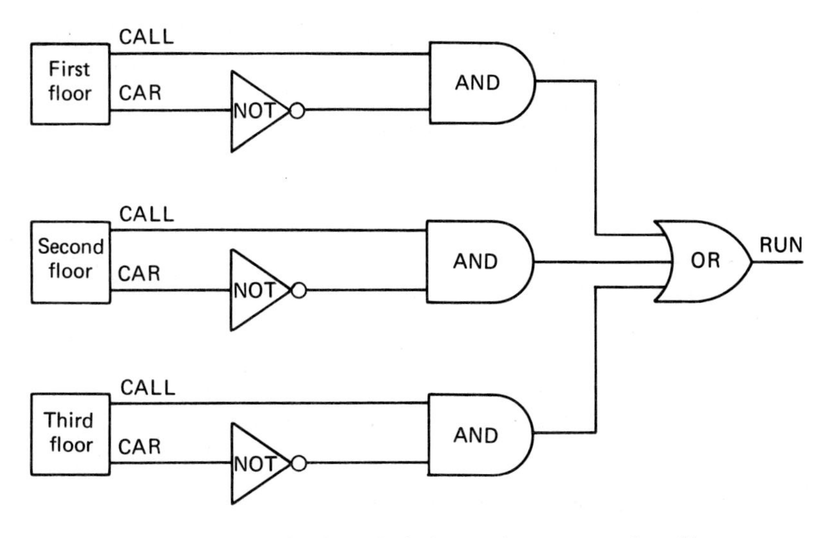

Figure 5-9 Combinational logic for an elevator control problem.

A verbal expression for the function of this system is as follows: If a call is at 1 AND the car is NOT at 1, OR if a call is at 2 AND the car is NOT at 2, OR if a call is at 3 AND the car is NOT at 3, then RUN.

5.5 THE FLIP-FLOP CIRCUIT

Electronic circuits are often required to store or "remember" certain bits of information. As an example, let us consider the elevator problem from the previous section. We would like to push the floor-call button only momentarily and have the control system *remember* that a call has been placed, so we do not have to hold the push button down until the car arrives. A circuit to fill this function is shown in Fig. 5-10. It is variously called an Eccles–Jordan circuit or a bistable multivibrator, but the most common and certainly the most descriptive term is *flip-flop*.

If the circuit is examined closely, it may be realized that it is identical with the Schmitt-trigger circuit of Fig. 5-5. However the feedback resistor (R_{1B} in this circuit) has so low a value that once a high Q_2 output turns Q_1 on, Q_1 remains on, even though the input (at S in this circuit) may be removed. Two resistors have been supplied to connect the S (set) and R (reset) signals to the inputs of the transistors. When a positive voltage is present at S (assuming no voltage at R), transistor Q_1 will turn on and its collector voltage will go to zero. Thus there will be no voltage across R_{2B}, and Q_2 will be turned off. The collector of Q_2 (which is marked X, indicating the output of the flip-flop) is then at +10 V, and we say that the flip-flop has been *set*. If the voltage at S is removed, Q_1 will remain on because the voltage at the collector of Q_2 will supply it with base current through R_{1B}.

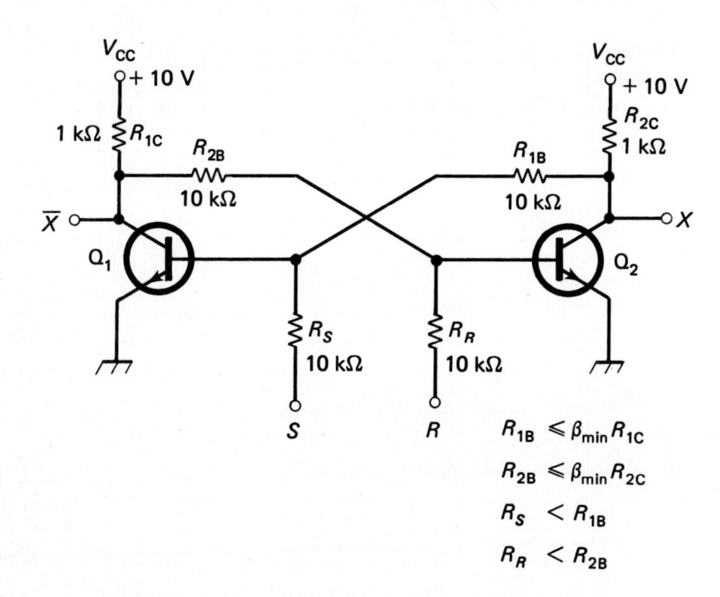

Figure 5-10 Transistor bistable multivibrator or flip-flop.

Input modes: Once the flip-flop is set, it will remain set until Q_1 is turned off. This can be done in a number of ways:

1. Ground the base of Q_1, or feed it a negative spike through a resistor, diode, or capacitor. This will rob Q_1 of its base bias and turn it off.
2. Open the emitter-ground connection of Q_1 momentarily. This will make the Q_1 collector go positive since there will be no path from collector to ground.
3. Short point X to ground. This will remove the voltage from R_{1B}, turning Q_1 off.
4. Feed a positive current through R_R to the base of Q_2. This will turn on Q_2, bringing point X to ground and removing the base bias from Q_1. This RESET signal may be fed to the base of Q_2 through a resistor, a diode, or, if it is a short pulse-type signal, through a capacitor.

Types of flip-flops: The flip-flop shown in the figure is of a particular type called an *R-S flip-flop*. The letters S and R refer to the fact that it can be set and reset by dc voltage at inputs S and R, respectively. There are a number of advanced flip-flop circuits which contain input circuitry to determine under what conditions the flip-flop will be set or reset. In all these circuits, however, the primary intent is to store information, i.e., to remember which of two signals was received last.

Notice that the output signal, X, and its complement $\bar{X}$ (read "X-NOT") are both available from the flip-flop. Thus whenever the X output is $+10$ V, the $\bar{X}$ output is 0 V, and whenever $X = 0$, $\bar{X} = +10$ V. This eliminates the need for an inverter to produce the complement, or NOT, of a signal. Notice also that the flip-flop has only two possible output voltages, 0 and +10 V (approximate). As long as the base-resistor values are low enough to saturate the transistors, this will always be the case.

5.6 ONE-SHOT MULTIVIBRATOR

A few modifications of the basic flip-flop multivibrator can produce a circuit which will remember an input pulse for a given period of time, and then *automatically* "forget" it. The output waveform of this circuit is a voltage pulse which begins in immediate response to a positive-going input pulse and ends a predetermined time, *t*, later. As might be expected, the time, *t*, is determined by the time-constant charging rate of a capacitor. The circuit is sometimes called a monostable multivibrator because it has only one stable state, in contrast to the two stable states of the bistable circuit just discussed. The name *one-shot* is perhaps more descriptive, since it indicates that the circuit produces a single pulse of a predetermined length each time its input is triggered.

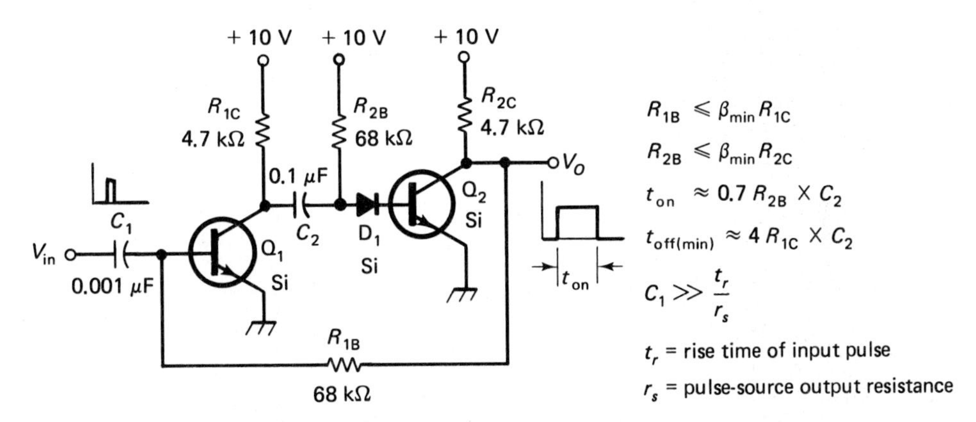

Figure 5-11 Monostable multivibrator or one-shot circuit.

The operation of the one-shot circuit (referring to Fig. 5-11, with $V_{CC} = +10$ V) is as follows: The positive-going edge of the input pulse is coupled through C_1 to the base of Q_1, turning it on momentarily. The collector of Q_1 therefore drops quickly from +10 to 0 V, grounding the left side of C_2. C_2 has been previously charged to $(V_{CC} - V_{D1} - V_{2B})$ or 8.8 V—positive on the left, negative on the right. The turn-on of Q_1 thus places the right side of C_2 at −8.8 V. D_1 prevents this reverse voltage from breaking down the base-emitter junction of Q_2 (see Section 4.7).

With its base voltage negative, Q_2 turns off and its collector voltage rises to nearly +10 V in immediate response to the input pulse. R_{2C} and R_{1B} now supply base current to Q_1, holding it on even after the positive input pulse from C_1 is gone.

As soon as Q_1 is turned on by the input pulse and V_{2B} jumps to a negative voltage, R_{2B} begins to charge C_2 toward the +10-V supply level. C_2 is never allowed to reach +10 V, however, since at +1.2 V, D_1 and the base of Q_2 turn on. The current from R_{2B} is then diverted to Q_2, and the charging process stops. The rise from −8.8 to +1.2 V represents approximately one-half of the potential rise from −8.8 to +10 V. Referral to a time constant chart shows that a rise to 50% of the full charging potential requires a time of approximately 0.7 RC. Thus the length of the output pulse of the one-shot can be calculated as $t = 0.7\ R_{2B}C_2$.

When the base of Q_2 reaches 0.6 V and Q_2 turns on, the collector of Q_2 drops to 0 V, and R_{1B} no longer keeps Q_1 on. The collector voltage of Q_1 then begins to rise, but since C_2 is also connected to the collector of Q_1, this rise follows the time constant charging curve of $R_{1C}C_2$. Until C_2 is completely recharged through R_{1C}, a new output pulse cannot be properly initiated. After four time constants, the capacitor will be charged to 98% of its full value, so the minimum recovery time for the one-shot is calculated as $4\ R_{1C}\ C_2$. Attempts to initiate an output pulse before the circuit is completely recovered will result in an output pulse shorter than the calculated time. This problem does not commonly occur because R_{1C} is generally 10 to 30 times smaller than R_{2B}, making the recovery time quite short in comparison to the output pulse time.

To summarize the operation of the one-shot,

1. Input pulse turns Q_1 on.
2. Collector of Q_1 goes from +10 to 0 V.
3. Bottom of R_{2B} goes from +1.2 to − 8.8 V. Q_2 turns off, and V_{2C} holds Q_1 turned on.
4. R_{2B} charges C_2 toward +10 V but is clamped at +1.2 V by D_1 and base of Q_2.
5. Q_2 turns on and V_{2C} goes to 0 V.
6. Q_1 turns off and R_{1C} recharges C_2, bringing V_{1C} back to +10 V.

Waveforms for various points in the one-shot circuit are given in Fig. 5-12.

Example 5-5

A certain electronic instrument has been blowing transistors at a prodigious rate, and it is suspected that the problem is being caused by high positive voltage spikes getting into the power supply line. It is desired to connect a solenoid-operated counter to record the number of occurrences of spikes above a set level over a period of time. The counter requires 12 V at 250 mA for a minimum of 100 ms to operate, but the pulses may be less than a microsecond in duration. Design a circuit to detect and count the pulses.

Solution A circuit to detect the pulses is shown in Fig. 5-13. D_1 is used to prevent negative pulses from getting to the base of Q_1 and turning it off prematurely. D_2 is therefore required to provide a discharge path for C_1. Q_1 and Q_2 comprise the one-shot, which will be timed for 200 ms to ensure counter operation. Since Q_2 is normally on and turns off during the output pulse, an inverter consisting of R_6 and Q_3 is used to activate the solenoid coil during the output pulse. Q_3 also provides added current gain, which is important since the solenoid requires a fairly heavy current. D_3 is required to suppress inductive kickback voltages from the solenoid coil.

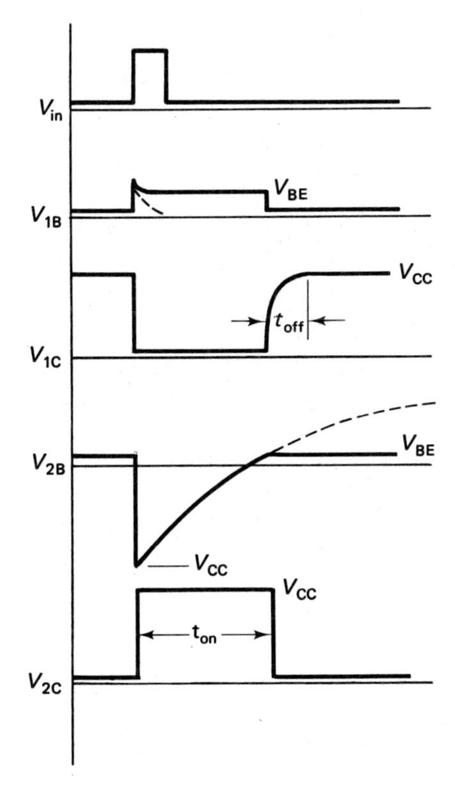

Figure 5-12 Voltage waveforms for the one-shot of Fig. 5-11.

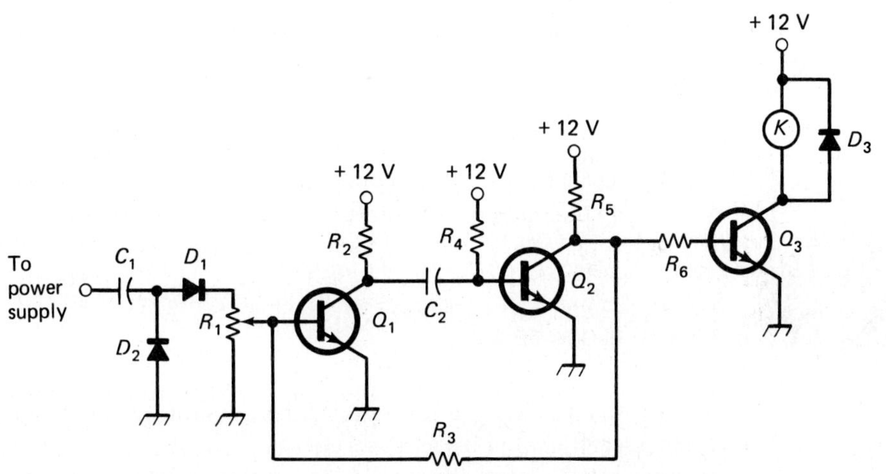

Figure 5-13 One-shot and relay driver design (Example 5-6).

Assuming a $\beta_{\min}$ of 25 for all transistors,

$$I_{3B} = \frac{I_K}{\beta} = \frac{250\text{ mA}}{25} = 10\text{ mA}$$

$$R_5 + R_6 = \frac{V_{CC} - V_{BE}}{I_{3B}} = \frac{11.4\text{ V}}{10\text{ mA}} = \mathbf{1.14\ k\Omega}$$

This calculation ignores the loading effect of R_3, which will be small if R_3 is large. However, it will be wise to make $R_5 + R_6$ a little smaller than 1.14 kΩ to ensure enough turn-on current. R_5 is chosen as 470 Ω, and R_6 as 560 Ω.

$$R_{4(\max)} = \beta R_5 = (25)(470) = \mathbf{12\ k\Omega}$$

$$C_2 = \frac{t_{on}}{0.7R_4} = \frac{200 \times 10^{-3}}{0.7(12 \times 10^3)} = 24 \times 10^{-6} = \mathbf{24\ \mu F}$$

A 25-μF capacitor would be used, and in this range anything except an electrolytic capacitor would be physically quite large. A look at Fig. 5-12 shows that V_{1C} is always more positive than V_{2B}, so the positive side of C_2 would be at the left. The recovery time, t_{off}, is not critical and is set arbitrarily at 100 ms.

$$R_2 = \frac{t_{off}}{4C_2} = \frac{100 \times 10^{-3}}{4(25 \times 10^{-6})} = 1 \times 10^3 = \mathbf{1\ k\Omega}$$

$$R_3 = \beta R_2 = (25)(1\text{ k}\Omega) = \mathbf{25\ k\Omega}$$

R_2 would be **1 kΩ**, and R_3 would be **22 kΩ**.

R_1 sets the pulse voltage level at which the one-shot will trigger. It must be low enough in value to provide enough turn-on current for Q_1, but it should not be chosen so low as to cause undue circuit loading. A value of 5 kΩ would be reasonable.

The input triggering capacitor C_1 is calculated next:

$$C_1 \gg \frac{t_r}{R_s}$$

We are looking for pulses with a rise time of approximately 1 μs, and the source resistance (which now includes at least a part of R_1) may be approximately 5 kΩ. Therefore

$$C_1 \gg \frac{10^{-6}}{5 \times 10^3} = \mathbf{0.0002 \times 10^{-6}}$$

This is a minimum value and should be made larger as long as the time constant of $R_S C_1$ does not reach the range of the pulse output time, t_{on}. A value of $\mathbf{C_1 = 0.002\ \mu F}$ is chosen for the design.

CHAPTER SUMMARY

1. A transistor may be switched from saturation to cutoff to control an LED, relay coil, or incandescent lamp in its collector circuit.

2. A common-emitter circuit *inverts* the sense of the signal from input to output; that is, a more positive voltage at the base input resistor produces a less positive voltage at the collector output.

3. A Schmitt trigger is formed by two inverters with a high-value resistor feeding

back from the output of the second to the input of the first. The output of this circuit *toggles* or snaps from high to low voltage as the input voltage crosses the upper and lower limits of the *hysteresis* zone.
4. A flip-flop is formed by two inverters with a low-value feedback resistor. The flip-flop will remain in the *set* state with no input signal, and change states in response to a voltage at the *reset* input.
5. A logic OR gate can be made by connecting several input resistors to the base of one common-emitter transistor, followed by an inverter. The output is *true* (high) if any one input is *true*.
6. A logic AND gate can be made by connecting several input diodes and a turn-on resistor to the base of a common-emitter transistor, followed by an inverter. The output is *true* (high) only if all inputs are *true*.
7. NOR and NAND gates operate as noted in 5 and 6 above, except that the outputs are *false* (low) in response to *true* OR or AND inputs. They are formed by omitting the inverter stages.
8. A one-shot circuit produces a high-level output of a predetermined length (of time) in response to a single short input pulse. The circuit is formed of two inverters with a capacitor and base turn-on resistor coupling the output of the first to the input of the second. Output-pulse length is determined by the time constant of this *R* and *C*. A low-value resistor feeds back from the output of inverter 1 to the input of inverter 2.

QUESTIONS AND PROBLEMS

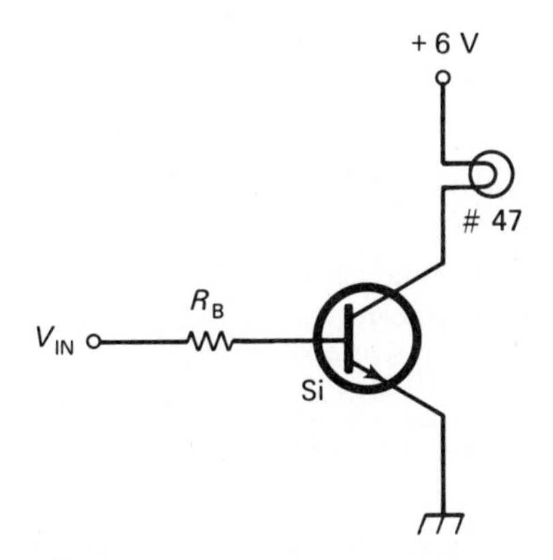

Figure 5-14 Problem 5-1.

5-1. In the circuit of Fig. 5-14, what value of base resistor should be used to ensure full turn-on of the 150-mA lamp with an input voltage of 3 V dc? Assume that $\beta_{min} = 25$.

The following specifications apply to the circuit components in Fig. 5-15: Transistor: $\beta_{min} = 30$, $\beta_{max} = 180$. Relay: Coil rated at 60 mA to pick—guaranteed to drop at 10 mA or higher. Coil resistance $= 200\ \Omega$.

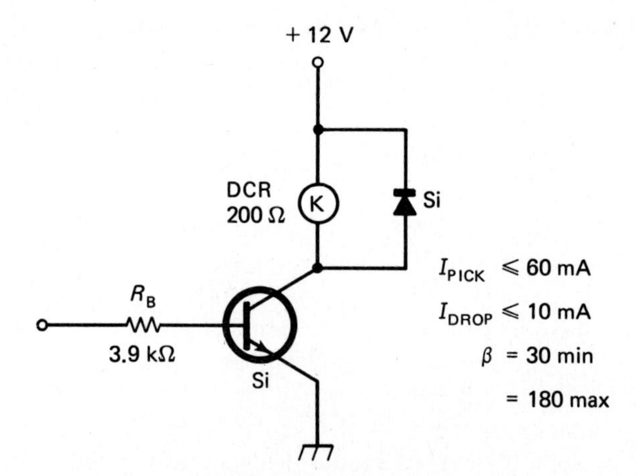

Figure 5-15 Problems 5-2 through 5-4.

5-2. What is the lowest value of V_{IN} which will ensure that the relay in Fig. 5-15 will pick under all conditions?

5-3. What is the highest value of V_{IN} which will ensure that the relay in Fig. 5-15 will drop under all conditions?

5-4. The relay coil in Fig. 5-15 has an inductance of 100 mH. When the coil is deenergized, inductive kickback will cause a current through the coil and the diode which will delay the "drop" of the relay contacts. What is the maximum delay to be expected between transistor turn-off and relay drop? *Hint:* Use the inductive time-constant formula $\tau = L/R$ and the universal time-constant chart in Appendix D.

5-5. Redraw the inverter-and-lamp-driver circuit of Fig. 5-3(b), changing V_{CC} to +6 V, R_1 to 270 Ω, and R_2 to 150 Ω. Assume that both transistors have $\beta = 35$ and calculate the photocell resistances for $I_{LAMP} = 180$ mA and $I_{LAMP} = 0$.

5-6. For the Schmitt trigger of Fig. 5-5 V_S is set to −0.8 V. Find the turn-on and turn-off values of V_{IN} and the dead zone between them.

5-7. In the circuit of Fig. 5-7, what is the minimum value of β for Q_1 which will turn on Q_1 completely with a single input of 5 V? Use no safety factor.

5-8. In the circuit of Fig. 5-7, assuming that Q_2 is turned off, how many 22-kΩ input resistors to ground can be connected at V_O before the output voltage is loaded down to 5 V? This is the number of similar gates that can be driven by the first gate and is termed the *fan out.*

5-9. In the circuit of Fig. 5-8, what is the lowest value of V_1 which will keep Q_1 completely turned on? Assume that $\beta_{min} = 25$ and that $\beta_{max} = 150$.

5-10. In the circuit of Fig. 5-16, what is the *actual* output voltage at X when the flip-flop is set? Do not neglect V_{BE}.

5-11. What is the lowest positive voltage at R that will be certain to reset the flip-flop of Fig. 5-16? *Hint:* First find the lowest voltage at X that will keep Q_1 turned on. Then find the voltage V_R which will produce that value of V_X. Use $\beta_{max} = 150$ for Q_1 and $\beta_{min} = 25$ for Q_2.

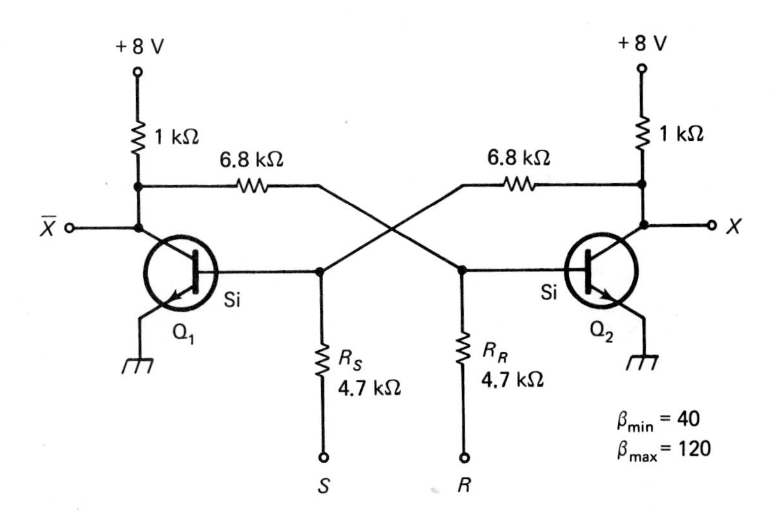

Figure 5-16 Problems 5-10 and 5-11.

In the circuit of Fig. 5-11, R_{2C} is 1 kΩ and R_{2B} is 15 kΩ. $\beta_{min} = 30$ and $\beta_{max} = 150$ for both transistors. Problems 5–12 through 5–15 refer to this circuit.

5-12. Use the data above to calculate a value of C_2 which will produce 1-ms output pulses.

5-13. It is expected that input pulses to the circuit above may occur as fast as 600 pulses per second. Find the largest permissible value of R_{1C} to ensure complete recovery between pulses.

5-14. If R_{1C} is chosen to be 1 kΩ, find the largest permissible value of R_{1B} in the circuit above.

5-15. Pulses having a rise time of 0.1 μs are fed to the circuit above from a square-wave generator having an output impedance of 600 Ω. What is the lowest possible value for C_1?

6

THE COMMON-EMITTER AMPLIFIER

6.1 AN ELEMENTARY AMPLIFIER

Electronic amplification was unquestionably the feat that gave birth to the art of electronics. It is the indispensible element in electronic communications, computation, and control systems. The first device capable or performing this feat was the vacuum tube, which dominated electronics from the 1920s to the 1960s. Today's amplifiers most often rely on the transistor.

An amplifier circuit uses a small input voltage or current waveform as a sort of pattern to reproduce a higher-power copy of that waveform at the output. The larger voltages and currents required are obtained from a battery or dc power supply.

Most transistors operate effectively only for one direction of input current because of the diode junctions involved in their structure. However, most signals requiring amplification are inherently ac. Elementary amplifiers solve this dilemma by establishing a zero-signal dc current, called *bias*, through the input and output circuits. The ac input signal is combined with the dc bias, producing a fluctuating dc input. The transistor uses this small input current to control the current from the dc supply, thus producing a fluctuating dc output current. Coupling capacitors are used at the input and output to let the ac signals pass without permitting the dc bias to reach the source and load elements.

The simplest possible transistor amplifier is shown in Fig. 6-1. Although this circuit is not reliable enough to be used in practice, the fact that it illustrates amplifier principles in an easy-to-understand circuit makes it well worth studying. A look back at the transistor test circuit of Fig. 4-4 and at the lamp driver of Fig. 5-1 will show that the only thing new about this circuit is the capacitors. C_{in} is used to couple the ac input signal to the base, where it alternately adds to and subtracts from the base bias current delivered by R_B.

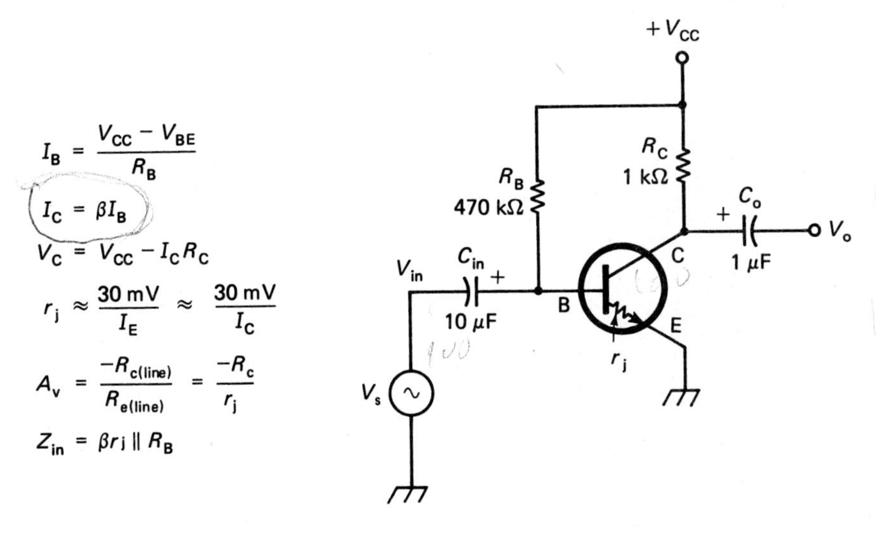

Figure 6-1 Simple but rather unreliable transistor amplifier. Collector current is completely dependent on β.

As the ac swing increases the base current, the collector current increases, but by a much larger (β times) amount. This increased collector current causes an increased drop in voltage across R_C, so v_C drops as v_{IN} rises. On the negative half-cycle the input signal i_B drops, i_C decreases, and v_C rises.

Capacitor C_o couples these ac swings at the collector to the output but blocks the dc collector voltage. An ideal bias point for this amplifier would have V_C halfway between V_{CC} and ground, so that the output would be free to swing equally up and down without running into either limit.

6.2 THE TRANSISTOR BASE-EMITTER DIODE

To understand the behavior of the transistor amplifier shown in Fig. 6-1, you must first understand the behavior of a diode that is forward biased with dc but has an ac signal riding on that dc level. Figure 6-2 shows the V vs. I curve for such a diode—the base-emitter junction of a silicon transistor. The diode is forward biased at 3 mA dc and has a V_D of 0.7 V dc. However, there is an ac signal riding on the dc level. This swings the current from 4 mA to 2 mA at the peaks, which run from 0.71 to 0.69 V. The resistance of the diode junction *to the ac signal* is then

$$r_j = \frac{\Delta V}{\Delta I} = \frac{(0.71 - 0.69)\ \text{V}}{(4 - 2)\ \text{mA}} = 10\ \Omega \tag{6-1}$$

It is easy to visualize that, if the diode were biased at a higher dc current, the steeper slope of the curve would produce a lower ratio of $\Delta V/\Delta I$, that is, a lower ac junction resistance r_j. Similarly, lower dc bias currents produce higher junction resistances.

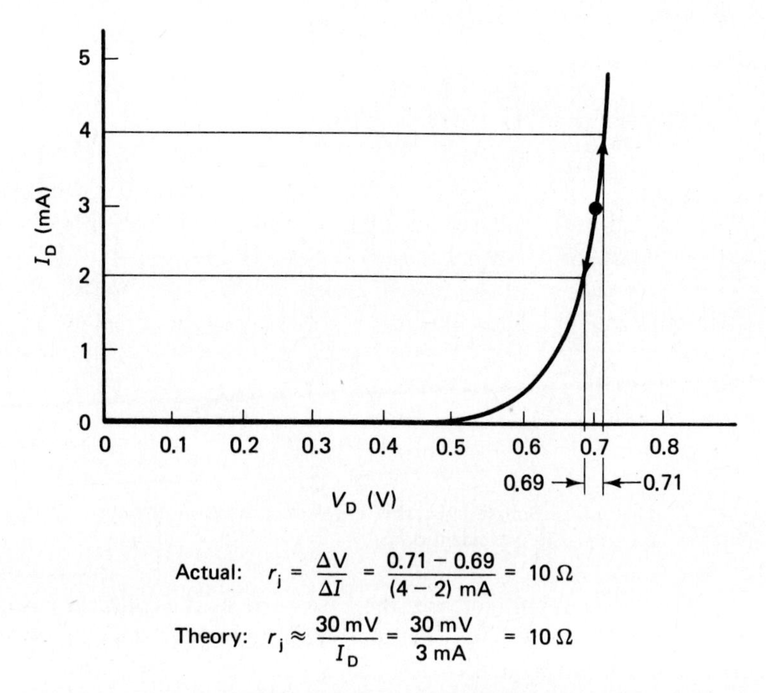

Figure 6-2 Resistance of a junction diode to ac-signal swings riding on a dc-bias level.

Dynamic resistance: These resistances are called *dynamic* because of the active nature of the ac signal swings. By contrast, the *static* or dc resistance at the bias point in Fig. 6-2 is simply

$$R_D = \frac{V}{I} = \frac{0.7\text{ V}}{3\text{ mA}} = 233\ \Omega$$

The idea that the same device can have a dynamic (ac) resistance that is so different from its static (dc) resistance may be a little difficult to accept at first. However, it is a commonplace with nonlinear devices, and it will soon seem quite natural to analyze amplifier circuits twice—first for the dc bias levels, and then for the ac signals. It is helpful to think of the *slope* of the line on the V vs. I curve as representing dynamic resistance. The steeper the slope, the lower the ratio of $\Delta V/\Delta I$ and the lower the dynamic resistance.

Calculating r_j: The dynamic resistance of silicon and germanium junction diodes can be estimated without the aid of a V vs. I graph by the following formula:

$$r_j \approx \frac{30\text{ mV}}{I_D} \tag{6-2}$$

Shockley, one of the inventors of the transistor, determined a theoretical value of 26 mV for the numerator of this important equation. Manufacturing imperfections cause experimental values to range from 25 to 50 mV, with 30 mV being a typical value. Although its accuracy is limited, the utility of equation (6-2) will soon become apparent.

6.3 AMPLIFIER VOLTAGE GAIN

In Fig. 6-1 the ac input signal is connected directly to the base of the transistor, since C_{in} is chosen to have a reactance that is near zero at the signal frequency. Thus V_{in} appears across the base-emitter junction resistance r_j.

The output signal V_o appears across the collector resistor R_C, since C_o presents essentially zero reactance and the V_{CC} supply is an ac ground. (Look back at the power-supply circuits of Chapter 3—they all have a large capacitor across the output providing an ac short circuit from the supply line to ground.)

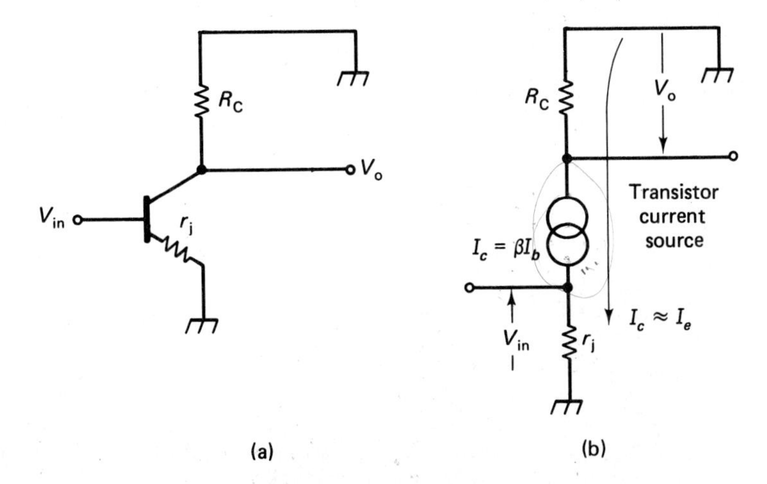

Figure 6-3 The basic transistor amplifier (a) appears to ac as a voltage divider (b) with the input voltage across r_j and the output voltage across R_C.

The emitter current is equal to the collector current (with the negligible addition of the base current). This is true for the ac current swings as well as for the dc bias current. R_C and r_j are therefore somewhat similar to a voltage divider with V_{in} across r_j and V_o across R_C, and the same current $I_c \approx I_e$ passing through them. This is illustrated in Fig. 6-3. If we define $R_{c(\text{line})}$ as the resistance to ac from the collector to ground, and $R_{e(\text{line})}$ similarly, the voltage gain for this amplifier is

$$A_v = \frac{V_o}{V_{in}} = \frac{-I_c R_{c(\text{line})}}{I_e R_{e(\text{line})}} \tag{6-3}$$

$$A_v = \frac{-R_{c(\text{line})}}{R_{e(\text{line})}} \tag{6-4}$$

$$A_v = \frac{-R_c}{r_j} \tag{6-5}$$

Equation (6-3) is the definition of voltage gain. Equation (6-4) is a general form for any *common-emitter* amplifier, that is, one where the input is applied at the base and the output is taken from the collector. The minus sign simply indicates that rising input-signal swings produce falling output-signal swings. This signal inversion is also termed a 180° phase shift, since for a sine wave, the effect is the same. Equation (6-5) is a specific form, applicable to the circuit of Fig. 6-1 only.

6.4 AMPLIFIER INPUT IMPEDANCE

What does the ac signal V_{in} "see" when it looks into the amplifier of Fig. 6-1? Another way of asking this is: What is the ratio V_{in}/I_{in}? We call this ratio the *input impedance* Z_{in}. (Even though it may be purely resistive, common practice is to call it Z_{in} rather than r_{in}.)

Z_{in} is important because low values of Z_{in} will cause loading of the source voltage. For example, if r_s is 10 kΩ and Z_{in} is 10 kΩ resistive, half of V_s will be dropped across r_s and half will appear as V_{in} across the amplifier Z_{in}. However, if Z_{in} drops to 1 kΩ, only $\frac{1}{11}$ of V_s will appear at V_{in}, the other $\frac{10}{11}$ being dropped across r_s.

To determine Z_{in} in Fig. 6-1 we must know the following facts:

- C_{in} is chosen to have essentially zero reactance at the frequency of the signal V_s.
- R_B is connected (as far as the ac signal is concerned) from the input to ground. This is so because the supply V_{CC} terminal is effectively at ground for ac.
- The transistor collector-base junction is reverse biased, and so presents an extremely high impedance to the ac signal at the base. There is current in the collector, of course, but as illustrated in Fig. 4-12, this is current injected from the emitter. For reverse voltages from base to collector this junction remains an open circuit.
- The transistor base-emitter junction is a forward-biased diode with a dynamic resistance approximated by 30 mV/I_E.
- The base current is less than the emitter current by approximately a factor of β. This is true for the ac signal current as well as for the dc bias current.

Base-input resistance r_b: From the foregoing, and an examination of Fig. 6-1, it should be clear that there are two parallel paths for ac current from V_{in} to ground. The first is simply through R_B to V_{CC}, which is ac ground. The second is from the transistor base to its emitter through dynamic resistance r_j. However, since the base current is less than the base-emitter junction current I_j by an approximate factor of β, the dynamic resistance from the base to ground (r_b) is greater than r_j by a factor of beta.

$$\Delta V_{be} = \Delta I_j r_j$$

$$\Delta I_b = \frac{\Delta I_j}{\beta}$$

$$r_b = \frac{\Delta V_{be}}{\Delta I_b} = \beta r_j \qquad (6\text{-}6)$$

The input impedance of the amplifier is the parallel combination of the bias resistor and the base input resistance

$$Z_{in} = R_B \| r_b$$
$$= R_B \| \beta r_j$$
$$\approx R_B \left\| \frac{30 \text{ mV} \times \beta}{I_E} \right. \qquad (6\text{-}7)$$

The vertical bars are a shorthand representation of the parallel-resistance formula. Thus $R_1 \| R_2$ is calculated as $1/(1/R_1 + 1/R_2)$.

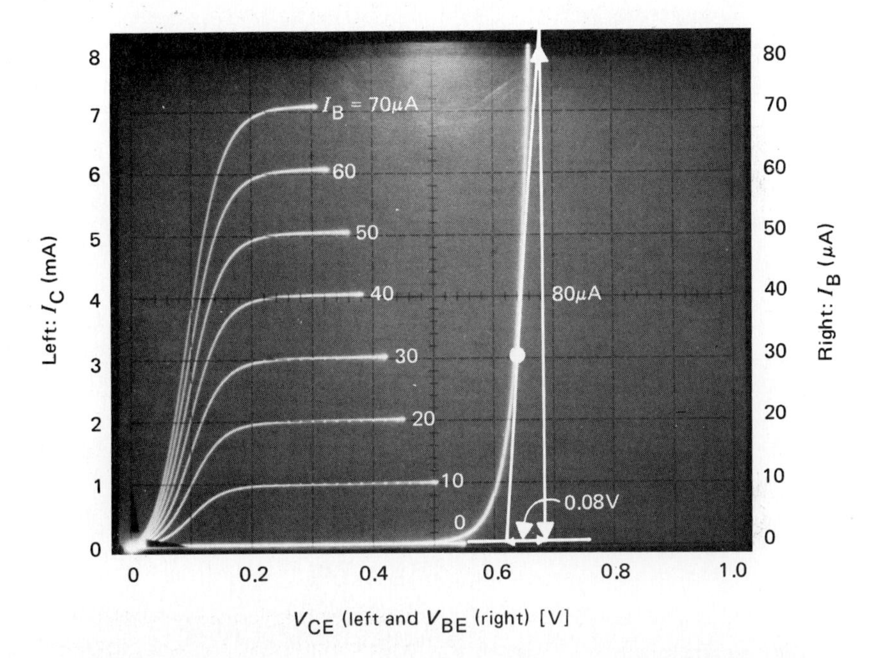

Figure 6-4 Right-hand curve shows V_{BE} vs. I_B for a silicon transistor. Slope $\Delta V/\Delta I$ is 1000 Ω at 30 μA. Left-hand curves show the transistor has $\beta = 100$, so $I_E \approx 100 \times 30\ \mu$A $= 3$ mA at this point. Predicted r_b is $\beta r_j = 100 \times 30$ mV/3 mA $= 1000\ \Omega$.

These equations are applicable to the circuit of Fig. 6-1 only. The "approximately equal to" sign in the last form recalls that the number 30 mV is an average and varies unpredictably from one transistor to the next.

Figure 6-4 shows the actual input V_B-vs.-I_B curves for a silicon transistor biased at $I_E = 3$ mA. The transistor's β is 100. We expect the base input resistance to be

$$r_b = \frac{0.03\beta}{I_E} = \frac{(0.03)(100)}{0.003} = 1\ \text{k}\Omega \tag{6-8}$$

Laying a tangent line to the actual curve at $I_B = I_E/\beta = 3\ \text{mA}/100 = 30\ \mu\text{A}$, we find that

$$r_b = \frac{\Delta V}{\Delta I} = \frac{0.1\ \text{V}}{0.1\ \text{mA}} = 1\ \text{k}\Omega \tag{6-9}$$

Example 6-1

Select R_B for $V_C = 5$ V, and find Z_{in} and A_v for the circuit of Fig. 6-5.

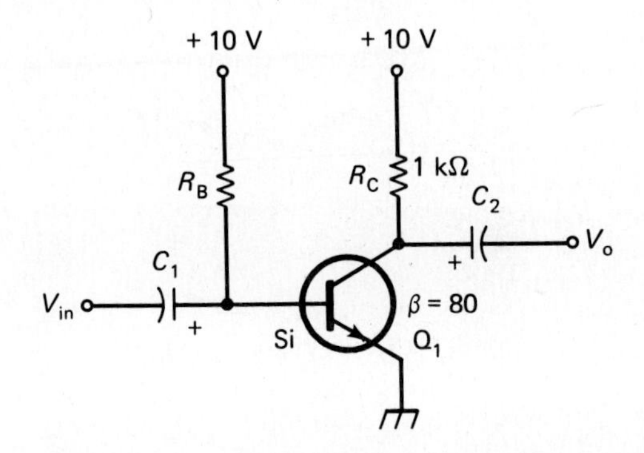

Figure 6-5 Simple common-emitter amplifier (Example 6-1).

Solution

$$V_{R(C)} = V_{CC} - V_C = 10 - 5 = 5\ \text{V}$$

$$I_C = \frac{V_{R(C)}}{R_C} = \frac{5\ \text{V}}{1\ \text{k}\Omega} = 5\ \text{mA}$$

$$I_B = \frac{I_C}{h_{FE}} = \frac{5\ \text{mA}}{80} = 0.063\ \text{mA}$$

$$V_{R(B)} = V_{CC} - V_{BE} = 10 - 0.6 = 9.4\ \text{V}$$

$$R_B = \frac{V_{R(B)}}{I_B} = \frac{9.4\ \text{V}}{0.063\ \text{mA}} = \mathbf{150\ k\Omega}$$

$$r_j = \frac{0.03}{I_C} = \frac{0.03}{0.005} = 6\ \Omega$$

$$r_b = h_{fe}\, r_j = (80)(6) = 480\ \Omega$$

$$Z_{in} = r_b \,||\, R_B = 480\ \Omega \,||\, 150\ \text{k}\Omega = \mathbf{478\ \Omega}$$

$$A_v = \frac{-R_C}{r_j} = \frac{-1000}{6} = \mathbf{-167}$$

6.5 AMPLIFIER OUTPUT IMPEDANCE

Generally, an amplifier output is expected to deliver current to some kind of load. The load may be a loudspeaker, the input impedance of a following amplifier, or even a servomotor. Whatever the load, we usually represent it as a resistance. Adding a load resistance to an amplifier causes its output voltage to be loaded down, so the voltage gain of an amplifier with load is lower than its gain without load.

If we know the output impedance Z_o of an amplifier, we can tell how much its V_o and A_v will drop when a given load is connected. For example, an amplifier with $Z_o = 1\ \text{k}\Omega$ and $A_v = 100$ will have A_v reduced to 50 if a load $R_L = 1\ \text{k}\Omega$ is connected.

Output impedance is the resistance seen by the load looking back into the amplifier's output terminals. Note that Z_o does not include the load R_L. For the simple circuit of Fig. 6-5, Z_o is approximately equal to R_C. The only other path from V_o to ground is through the transistor's collector, which is a very high dynamic resistance.

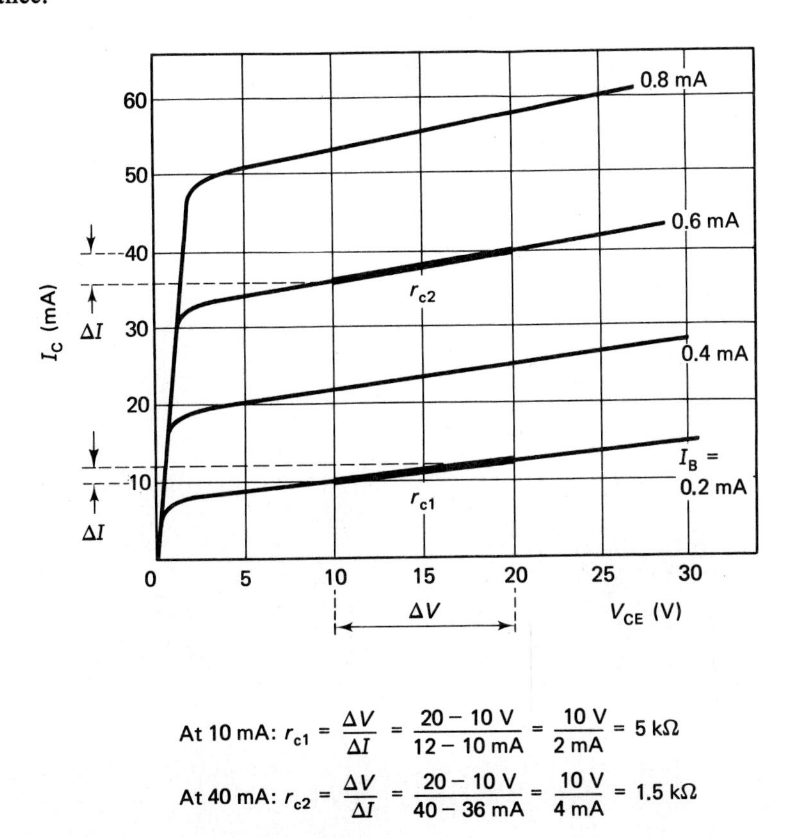

$$\text{At 10 mA: } r_{c1} = \frac{\Delta V}{\Delta I} = \frac{20-10\text{ V}}{12-10\text{ mA}} = \frac{10\text{ V}}{2\text{ mA}} = 5\ \text{k}\Omega$$

$$\text{At 40 mA: } r_{c2} = \frac{\Delta V}{\Delta I} = \frac{20-10\text{ V}}{40-36\text{ mA}} = \frac{10\text{ V}}{4\text{ mA}} = 1.5\ \text{k}\Omega$$

Figure 6-6 A steeper slope of the characteristic curve line indicates a lower dynamic collector resistance.

Collector ac resistance shunts output: In some cases where R_C is very high, or where Q_1 is a power transistor, the dynamic collector resistance r_c may be low enough to become important. In this case, Z_o is R_C in parallel with r_c.

$$Z_o = R_C \| r_c \tag{6-10}$$

The slope of the characteristic-curve lines can be used to determine r_c for any transistor as shown in Fig. 6-6. Note that r_c becomes lower as I_C becomes higher.

6.6 AMPLIFIER WITH LOAD RESISTANCE

Figure 6-7 shows a simple common-emitter amplifier with a load resistor. In this circuit the collector signal current flows through R_C and R_L, which are in parallel for ac (C_o connects the bottom of R_C to the top of R_L, and the power supply output capacitor connects the top of R_C to ground). The voltage-gain formula then becomes

$$A_v = \frac{-R_{c(\text{line})}}{R_{e(\text{line})}} = \frac{-R_C \| R_L}{r_j} \tag{6-11}$$

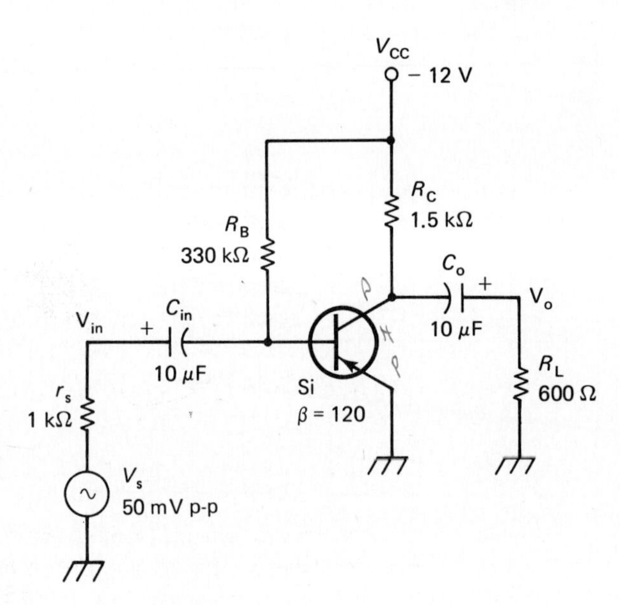

Figure 6-7 R_L appears in parallel with R_C for ac, lowering amplifier gain.

Example 6-2

Find V_C, Z_{in}, V_{in}, Z_o, and V_o for the circuit of Fig. 6-7.

Solution

$$I_B = \frac{V_{CC} - V_{BE}}{R_B} = \frac{12 - 0.6}{330\ \text{k}\Omega} = 34.5\ \mu\text{A}$$

$$I_C = \beta I_B = (120)(34.5\ \mu\text{A}) = 4.14\ \text{mA}$$

$$V_{R(C)} = I_C R_C = (4.14\ \text{mA})(1.5\ \text{k}\Omega) = 6.21\ \text{V}$$

$$V_C = V_{CC} - V_{R(C)} = 12 - 6.21 = \mathbf{5.79\ V}$$

$$r_j = \frac{30\ \text{mV}}{I_E} = \frac{30\ \text{mV}}{4.14\ \text{mA}} = 7.25\ \Omega$$

$$Z_{in} = R_B \| r_b = R_B \| \beta r_j$$
$$= (330\ \text{k}\Omega) \| (120)(7.25\ \Omega) = \mathbf{868\ \Omega}$$

$$V_{in} = V_s \frac{Z_{in}}{r_s + Z_{in}} = (50\ \text{mV}) \frac{868}{1000 + 868} = \mathbf{23.2\ mV\ p\text{-}p}$$

$$Z_o = R_C = 1.5\ \text{k}\Omega$$

$$A_v = \frac{R_C \| R_L}{r_j} = \frac{1500 \| 600}{7.25} = 59$$

$$V_o = V_{in} A_v = (23.2\ \text{mV})(59) = \mathbf{1.37\ V\ p\text{-}p}$$

6.7 LOAD-LINE ANALYSIS

An analysis of a transistor amplifier can be made using the transistor's characteristic curves if these are available. Such a graphical analysis is not often made in practice because transistor characteristic curves vary so radically from one unit to the next, making it difficult to know when a given set of curves truly represents a particular transistor. However, the graphical analysis will provide a clearer visualization of the operation and limitations of the amplifier.

We shall use the example from Fig. 6-5 as a basis for the graphical analysis. Notice that the curves of Fig. 6-8 show a collector current of 8 mA for a base current of 100 μA, indicating an h_{FE} of 80, as in Example 6-1.

The dc collector-current path for this circuit consists of the transistor in series with the 1-kΩ collector resistor across the 10-V supply. Since the transistor's current equals the resistor's current, and the transistor's voltage is the supply minus the resistor's voltage, the range of possible transistor voltages and currents is given by a linear equation:

$$I_C = \frac{V_{R(C)}}{R_C} = \frac{V_{CC} - V_C}{R_C} \tag{6-12}$$

This indicates that there is a straight line on the V vs. I graph which represents all possible combinations of voltage and current for the transistor. As long as the circuit has $V_{CC} = 10$ V and $R_C = 1$ kΩ, voltage and current combinations which do not lie on the line are not possible.

Drawing the load line: This line, called the *dc* or *static load line*, can be readily positioned by its end points, which represent the extreme conditions of the transistor's conduction, i.e., fully on (saturated) or fully off (cutoff).

If the transistor's base current is zero, the collector current is likewise zero (assuming no leakage), and no voltage will be dropped across the collector resistor. Hence, the transistor's collector voltage will equal the supply voltage when I_C is zero. This determines the *cutoff* end of the load line, as shown in Fig. 6-8.

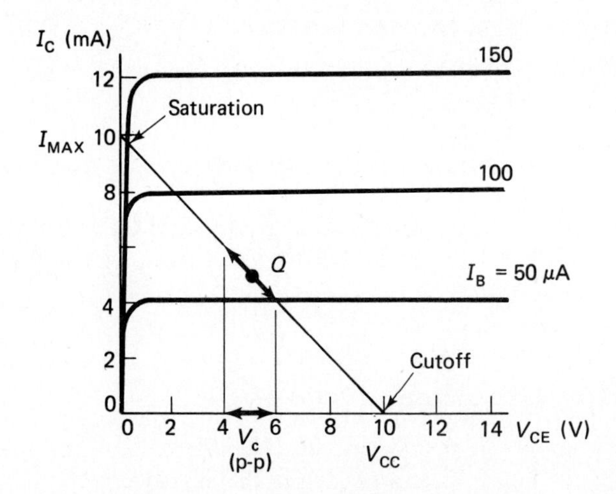

Figure 6-8 Transistor characteristic curves with dc load line and Q point.

If the transistor is fed enough base current to cause saturation, only a few tenths of a volt appear from collector to emitter, and almost the full supply voltage appears across R_C. The transistor acts very much like a short circuit in this case, and the collector current is found by

$$I_{C(sat)} = \frac{V_{CC}}{R_C} \tag{6-13}$$

This *saturated* collector current with near-zero collector voltage determines the second end point of the static load line. All possible circuit conditions lie on the straight line from saturation to cutoff.

Signal swings: Movement up and down the load line is caused by changes in transistor base current. Notice that as higher input voltages cause larger base current, the collector output voltage decreases. The amplifier, therefore, gives an output signal which is inverted from the input signal. This inversion is often termed a 180° phase shift, since for a symmetrical signal like a sine, square, or triangle wave, the effect is the same.

***Q* point**: Notice that in Fig. 6-8, the bias point of the example problem ($I_B = 63\ \mu A$, $I_C = 5$ mA, $V_C = 5$ V) is marked with a Q. This refers to the *quiescent* or rest point, with no signal present. Input signals are interpreted as swings up and down the load line, equal distances around the Q point. In Fig. 6-8, the base current, initially 63 μA, swings down to 50 μA and up to approximately 75 μA in response to a 25-μA p-p input signal. This causes an output voltage swing from 6 to 4 V, or 2 V p-p.

If it is desired to confirm the voltage gain of 167 as determined in Example 6-1, we can compute the input voltage which causes the 25-μA p-p input current swing:

$$\begin{aligned} V_{in} &= I_b r_b = I_b Z_{in} \\ &= (25 \times 10^{-6})(480) \\ &= 12 \text{ mV p-p} \end{aligned}$$

Now A_v can be determined from this input voltage and the output voltage of $6 - 4$ V or 2 V p-p:

$$A_v = \frac{V_o}{V_{in}} = \frac{2}{12 \times 10^{-3}} = 167$$

It should be emphasized again that the analysis method presented in the preceding sections is less cumbersome and more practical than the load line method where transistors are concerned, but it is helpful sometimes to sketch a crude set of curves and a load line for a particular circuit to visualize the effect of changing the supply voltage, bias point, etc.

6.8 DYNAMIC LOAD LINE

Load-line analysis can provide some helpful insight into the effects of adding a load resistance to an amplifier. Assuming a 1-kΩ load added to the example circuit of Fig. 6-5, the resistance seen by an ac signal looking out of the collector is now $R_C \| R_L$, or 500 Ω, in place of the previous 1000 Ω. The load line must reflect this lower dynamic resistance by taking on a more steep vertical slope. The dc bias conditions are unchanged, however, since the load resistor is coupled to the circuit through a capacitor, which is an open circuit to dc. Therefore, two load lines are necessary—one having the slope of R_C, which is used to establish the Q point, and a second having the slope of $R_C \| R_L$, which is used to determine the ac output voltage and current. These two are termed the static (dc) and *dynamic (ac) load lines* (SLL and DLL), respectively.

The criteria for the dynamic load line are that it must pass through the Q point and that it must have the slope of R_C in parallel with R_L. The first condition is easily satisfied; the second is met with the help of a third line, called the slope line. This line has no function other than to assist in finding the proper slope for the dynamic load line. The procedure for laying out the slope line is as follows:

1. Choose a "test" voltage and mark a point at this voltage and zero current on the curves. The test voltage can be any value, but it is usually most convenient to make it one-half to one-fifth of the supply voltage V_{CC}. In Fig. 6-9, a test voltage of 2 V was chosen.
2. Calculate the parallel combination of R_C and R_L. For the present example this is 1 kΩ || 1 kΩ or 500 Ω.
3. Calculate the current that the assumed test voltage would push through the parallel resistance. For this case

$$I = \frac{V}{R} = \frac{2}{500} = 0.004 = 4 \text{ mA}$$

Mark this current at zero voltage on the curves.
4. Connect the zero-voltage and zero-current points. This line has the slope of $R_C \| R_L$.

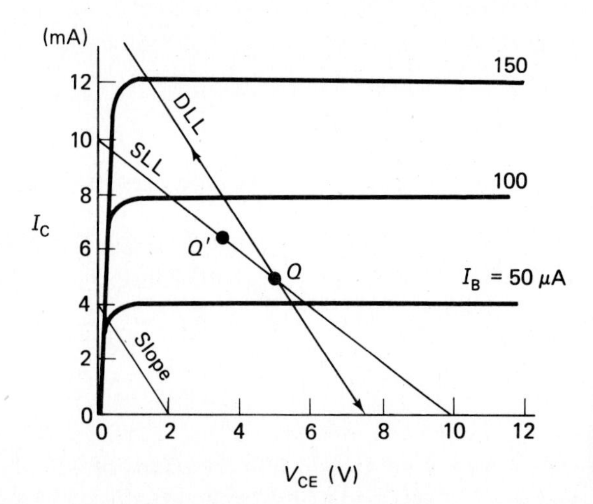

Figure 6-9 Transistor curves with dc (static) load line, slope line, and ac (dynamic) load line parallel to the slope line but running through the Q point.

A triangle and straightedge can then be used to transfer this slope to a line though the Q point. When an ac signal is applied, the operating point is now shifted up and down along the dynamic load line. Note that the lower total resistance caused by adding R_L causes the DLL to be steeper than the SLL, resulting in less output voltage swing and lower voltage gain. Lower values of R_C or R_L will cause even steeper load lines and even lower A_v.

Maximizing output voltage: It is evident from the static load line that where no load resistance is used, the collector should be biased midway between its saturation and cutoff voltages ($V_C = 5$ V in the example of Fig. 6-8). This Q point will allow the maximum swing of output signal up and down the load line. If the transistor were biased at $V_C = 8$ V, for example, the upward voltage swing would be limited to 2 V above the 8-V Q point, since 8 V + 2 V or 10 V is the maximum possible peak output voltage. The total output voltage is thus limited to 4 V p-p (2-V swing up and an equal 2-V swing down). An attempt to produce a larger output wave would result in distortion—the clipping off of the top peaks of the wave. If the collector bias point is set too low, say 2 V dc, this premature clipping will occur on the bottom of the waveform. With the bias point in the center of the load line, maximum swing-up and -down is possible without distortion.

Notice from Fig. 6-9 that when a load resistance is added, the maximum possible output voltage swing is considerably reduced, since the DLL has a steeper slope than the SLL. The degree of this reduction in V_o can be found from the formula

$$V_{o(FL)} = V_{o(NL)} \frac{R_C \| R_L}{R_C} \tag{6-14}$$

where FL and NL refer to full-load and no-load output voltages. Thus for the example in which $R_C = 1$ kΩ, $R_L = 1$ kΩ, and $V_{CC} = 10$ V, the maximum full-load output voltage with the collector biased at 5 V is found as

$$V_{o(FL)} = 10 \text{ V p-p}\left(\frac{1 \text{ k}\Omega \| 1 \text{ k}\Omega}{1 \text{ k}\Omega}\right)$$
$$= 10 \text{ V p-p}(0.5)$$
$$= \mathbf{5 \text{ V p-p}}$$

A close examination of Fig. 6-9 will show that, although the Q point is midway between saturation and cutoff, the voltage swing along the DLL meets the zero *current* clipping point (at about 7.5 V) well before the zero *voltage* clipping point is reached by an equal swing in the opposite direction. This illustrates the fact that the optimum bias point for a circuit changes when a load resistance is added.

If a new bias point is established at Q' in Fig. 6-9, and a new DLL of the same slope is drawn through Q', it will be seen that a larger peak-to-peak swing will be made possible without waveform clipping. The exact optimum point is best determined experimentally by lowering the value of the circuit base resistance to raise the collector current while increasing the input signal level to observe the waveform clipping. At the optimum point, the output waveform will begin to show clipping at the top and bottom simultaneously as the input signal is increased.

6.9 VOLTAGE-DIVIDER BIAS

The simple circuit of the preceding sections is not used in commercial practice because it allows collector current to vary directly with transistor beta. For typical small-signal transistors beta may vary from 40 to 200 (a factor of 5). Thus it would be impossible to position the Q point reliably near the center of the load line. One transistor may have such a high beta that it is biased into saturation (V_C near zero) while another may have such a low beta that it is biased near the V_{CC} supply. In either case, large output signal swings will be clipped off. In addition, as β changes, I_E changes proportionally and r_j varies inversely. A_v and Z_{in} are then also at the mercy of changing beta.

A circuit which avoids all these problems is shown in Fig. 6-10. The bias technique employed here tends to keep I_C constant, regardless of changes in transistor beta.

- R_{B1} and R_{B2} form a voltage divider which fixes the voltage at the transistor base.

$$V_B \approx V_{CC}\frac{R_{B2}}{R_{B1} + R_{B2}} \tag{6-15}$$

- With V_B fixed, V_E is fixed at 0.6 V lower than V_B for a silicon transistor (0.2 V lower for germanium).

$$V_E = V_B - V_{BE} \tag{6-16}$$

- With V_E fixed, I_E is fixed by the emitter resistor.

$$I_E = \frac{V_E}{R_E} \tag{6-17}$$

- Collector current is nearly identical to emitter current so I_C, and hence the voltage across R_C, are fixed.

$$V_{R(C)} = I_C R_C \approx I_E R_C \tag{6-18}$$

- Collector voltage is the supply voltage minus the drop across R_C.

$$V_C = V_{CC} - V_{R(C)} \tag{6-19}$$

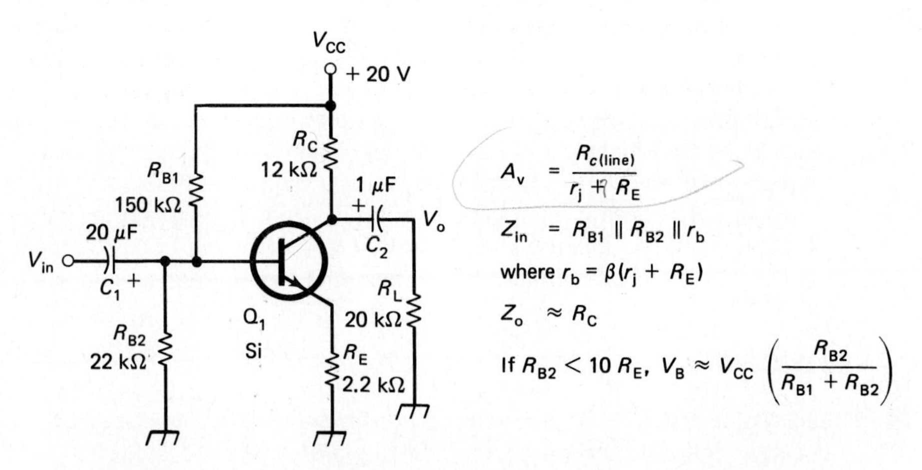

Figure 6-10 Voltage-divider bias in a common-emitter amplifier provides bias stability, moderate voltage gain, and high input impedance.

6.10 BIAS STABILITY

The success of the bias circuit of Fig. 6-10 rests on two assumptions: negligible V_{BE} change and negligible I_B.

Changes in base-emitter voltage V_{BE} cause changes in bias current I_E because even with V_B fixed, V_E changes when V_{BE} changes. V_{BE} is typically 0.6 V for silicon transistors, but it may easily change by ± 0.1 V from one transistor to the next, or with temperature changes. To swamp out the changes in V_{BE}, we must make sure that $V_{R(E)}$ is very much larger than the anticipated ± 0.1-V changes in V_{BE}. Making $V_{R(E)} = 2.0$ V would limit bias-current shift due to changes in V_{BE} to about $\pm 0.1/2.0$, or $\pm 5\%$, and this value of $V_{R(E)}$ is recommended as a minimum for general-purpose low-power amplifiers.

The voltage across R_E is subtracted from the supply V_{CC} when doing load-line analysis and determining output signal swing. Where V_{CC} is only a few volts to begin, a 2-V loss may be objectionable, and $V_{R(E)} = 1.0$ V may be used if a bias shift of $\pm 10\%$ can be tolerated.

Changes in transistor base current will occur to keep I_E constant as beta changes in the circuit of Fig. 6-10. The voltage divider R_{B1}-R_{B2} must be able to maintain a constant voltage regardless of these changes in base current. This means that I_B must be much lower than the main-line current through R_{B1} and R_{B2}.

Let us assume that I_E is to be 10 mA and $\beta = 100$. I_B must then be 0.1 mA. If β drops to 50, perhaps as a result of transistor replacement, then I_B would have to increase to 0.2 mA. To ensure that this current would not load the voltage divider, we would design the main-line current to be many times greater than 0.2 mA (say, 2 mA, if β_{min} were 50).

An easy-to-apply rule of thumb is to make R_{B2} *not larger* than $10R_E$. If β_{min} is 40 (a reasonable number for modern transistors), this will limit the range of possible emitter-bias currents to a maximum value as given by equations (6-15) through (6-17) (at very high beta) and a minimum that is 20% less than this value. If it is necessary to limit the loading to 10%, the rule of thumb becomes $R_{B2} \leq 5R_E$.

Example 6-3

Calculate the collector voltage for the circuit of Fig. 6-10 using the component values given and assuming a very high beta. Estimate the actual V_C if β drops to a minimum of 40.

Solution

$$V_B \approx V_{CC}\frac{R_{B2}}{R_{B1} + R_{B2}} = (20)\frac{22}{150 + 22} = 2.56 \text{ V}$$

$$V_E = V_B - V_{BE} = 2.56 - 0.6 = 1.96 \text{ V}$$

$$I_E = \frac{V_E}{R_E} = \frac{1.96 \text{ V}}{2.2 \text{ k}\Omega} = 0.89 \text{ mA}$$

$$V_{R(C)} = I_C R_C \approx I_E R_C = (0.89 \text{ mA})(12 \text{ k}\Omega) = 10.7 \text{ V}$$

$$V_C = V_{CC} - V_{R(C)} = 20 - 10.7 = \mathbf{9.3\ V}$$

If β were as low as 40, we would expect I_E, and consequently $V_{R(C)}$, to drop by about 20%, since $R_{B1} = 10\ R_E$.

$$V_{R(C)\min\beta} \approx 80\% \times 10.7 \text{ V} = 8.6 \text{ V}$$

$$V_{C(\min\beta)} \approx V_{CC} - V_{R(C)} = 20 - 8.6 = \mathbf{11.4\ V}$$

Bias-stability factor: For a more exacting comparison of bias circuits we may define a bias-stability factor S as the ratio of emitter current at minimum β to emitter current at maximum β:

$$S = \frac{I_{E(min)}}{I_{E(max)}} \tag{6-20}$$

Since many transistors have no maximum β specified, we will let β_{max} go to infinity. Then the stability-factor equation for the circuit of Fig. 6-10 (derived in Appendix E) is

$$S = \frac{\beta_{min}R_E}{\beta_{min}R_E + (R_{B1} \| R_{B2})} \tag{6-21}$$

The ideal stability factor would be unity, indicating no bias change from minimum to infinite beta. The unstabilized circuit of Fig. 6-1 has $S = 0$, indicating total dependency of I_E on beta. $S = 0.9$ means that I_E at minimum β would be 90% of its value if β were near infinity.

Example 6-4

Determine the stability factor for the circuit of Fig. 6-10 and the actual value of V_C if β drops to 40. Compare with the estimated V_C from Example 6-3.

Solution

$$S = \frac{\beta_{min} R_E}{\beta_{min} R_E + (R_{B1} \| R_{B2})} \tag{6-21}$$

$$= \frac{(40)(2.2)}{(40)(2.2) + (150 \| 22)} = 0.82$$

$$I_{C(min\ \beta)} = (0.82)(0.89\ \text{mA}) = 0.73\ \text{mA}$$

$$V_{R(C)min\ \beta} = I_C R_C = (0.73)(12) = 8.8\ \text{V}$$

$$V_{C(min\ \beta)} = V_{CC} - V_{R(C)} = 20 - 8.8 = \mathbf{11.2\ V}$$

The estimated value of V_C from Example 6-3 was **11.4 V**.

6.11 AC BEHAVIOR OF THE STABILIZED AMPLIFIER

The behavior of the stabilized amplifier of Fig. 6-10 for ac signals can be predicted using the same techniques that were used for the simple circuit of Fig. 6-1.

Ac voltage gain A_v is the ratio of the resistance through which the ac collector current flows divided by the resistance through which the ac emitter current flows. V_{in} is applied across the transistor's emitter junction resistance r_j in series with R_E. V_o is taken across R_C and R_L, which appear in parallel to the ac signal. For the circuit of Fig. 6-10,

$$A_v = \frac{-R_{c(line)}}{R_{e(line)}} = \frac{-R_C \| R_L}{r_j + R_E} \tag{6-22}$$

Input impedance Z_{in} is the parallel combination of three resistance paths to ground: R_{B1} (the supply is ac ground), R_{B2}, and the transistor base input resistance r_b. As before, r_b is the emitter-line resistance *multiplied by* β, because I_b is less than I_e by a factor of β. For the circuit of Fig. 6-10,

$$Z_{in} = R_{B1} \| R_{B2} \| r_b \tag{6-23}$$

where

$$r_b = \beta(r_j + R_E) \tag{6-24}$$

and

$$r_j \approx \frac{30\ \text{mV}}{I_E} \tag{6-2}$$

Output impedance is simply R_C in parallel with the transistor's collector resistance r_c, although r_c is usually high enough to be neglected in comparison with R_C.

$$Z_o = R_C \| r_c \tag{6-10}$$

Example 6-5

Determine A_v, Z_{in}, and Z_o for the bias-stabilized circuit of Fig. 6-10. Use $I_C = 0.73$ mA from Example 6-4, and $\beta = 40$.

Solution

$$r_j \approx \frac{30 \text{ mV}}{I_E} = \frac{30 \text{ mV}}{0.73 \text{ mA}} = 41\ \Omega \tag{6-2}$$

$$A_v = \frac{-R_C \| R_L}{r_j + R_E} = \frac{-12 \| 20}{0.041 + 2.2} = \mathbf{-3.35} \tag{6-22}$$

$$Z_{in} = R_{B1} \| R_{B2} \| \beta(r_j + R_E) \tag{6-23}$$

$$= 150 \| 22 \| 40(0.041 + 2.2) = 15.8\ \text{k}\Omega$$

$$Z_o = R_C = \mathbf{12\ k\Omega}$$

Notice that the voltage gain of the example circuit is very much less than that of the unstabilized circuit, and that the input impedance is quite a bit higher. Low voltage gain is one of the less desirable features of the circuit of Fig. 6-10.

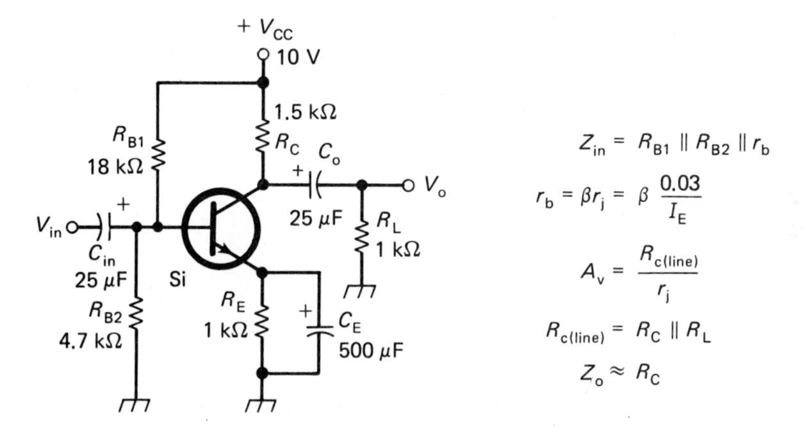

Figure 6-11 Bypassing the emitter resistor in a stabilized common-emitter amplifier provides maximum voltage gain but lower Z_{in}.

6.12 EMITTER BYPASSING

Bias stability and high ac-signal gain can be achieved together in a common-emitter amplifier by simply bypassing the emitter resistor R_E with a large capacitor as shown in Fig. 6-11. The ac signal sees C_E as a short circuit, leaving r_j alone in the emitter line. Voltage gain and Z_{in} for Fig. 6-11 are therefore given by

$$A_v = \frac{R_C \| R_L}{r_j} \tag{6-11}$$

$$Z_{in} = R_{B1} \| R_{B2} \| r_b \tag{6-23}$$

where

$$r_b = \beta r_j \tag{6-6}$$

Note that these are the same formulas that apply to the unstabilized circuit of Fig. 6-1, with the single addition of R_{B2}. The voltage gain is high, and the input impedance is relatively low.

Amplifier linearity is a desirable feature wherein the output-signal waveshape is an exact reproduction of the input-signal waveshape. Linearity is also called *fidelity* and it is generally specified by a *percent harmonic distortion* figure. In the circuit of Fig. 6-11, V_o is directly dependent on A_v, which varies inversely with r_j, which varies inversely with i_E. If the amplifier handles a large signal, i_E may change considerably from the bias point to the peaks of the signal swing. Thus r_j and hence A_v will be modified by the signal. The net effect will be to increase A_v on the positive input-signal swing (for *NPN* transistors). This corresponds to the *negative* output signal swing, and this will be overemphasized in the output waveform, as shown in Fig. 6-12(a).

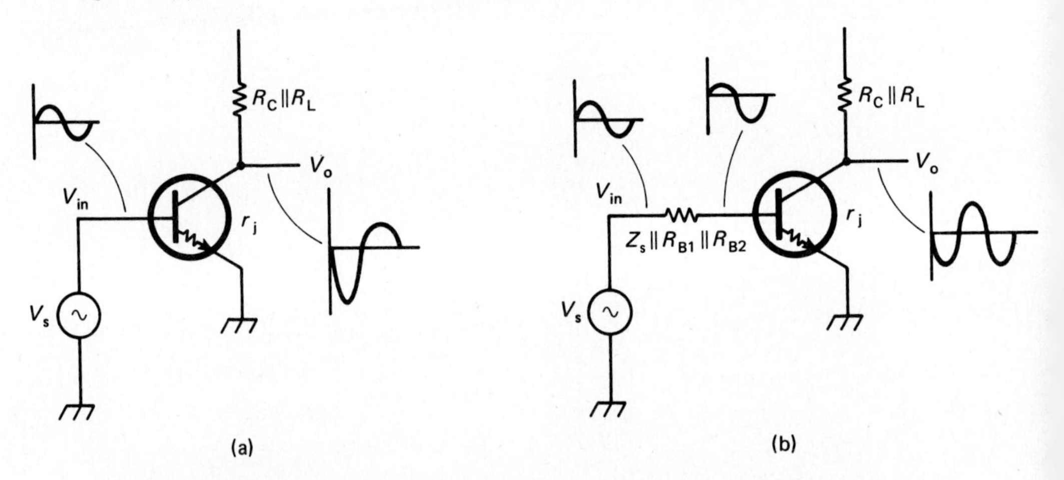

Figure 6-12 (a) If ac collector-current swings are nearly as large as dc collector-bias current, r_j changes with signal swings and V_o is distorted. (b) This effect is swamped out if the source impedance driving the base is much higher than r_b.

Remember, this nonlinearity becomes a problem only when the ac signal current becomes an appreciable fraction of the dc bias current. This can be recognized when $V_{o(ac)}$ becomes an appreciable fraction of $V_{R(C)}$. Also, the problem does not appear when Z_s of the signal is much higher than Z_{in} of the amplifier, as in Fig. 6-12(b). In this case Z_s swamps out the effects of changing r_j, and V_o is undistorted, although V_{in} appears distorted by the changing Z_{in} of the amp.

Partial bypassing: To raise the input impedance and improve the large-signal linearity of the amplifier of Fig. 6-11, it is common to leave a portion of the R_E unbypassed, as shown in Fig. 6-13. We will call the unbypassed portion R_{E1}. The dc bias current sees $R_{E1} + R_{E2}$, and the voltage drop across them is made 2 V or more for bias stability. The ac signal sees $r_j + R_{E1}$, and these determine A_v and Z_{in}. For the circuit of Fig. 6-13,

$$A_v = \frac{R_C \| R_L}{r_j + R_{E1}} \tag{6-25}$$

$$Z_{in} = R_{B1} \| R_{B2} \| r_b \tag{6-23}$$

where

$$r_b = \beta(r_j + R_{E1}) \tag{6-26}$$

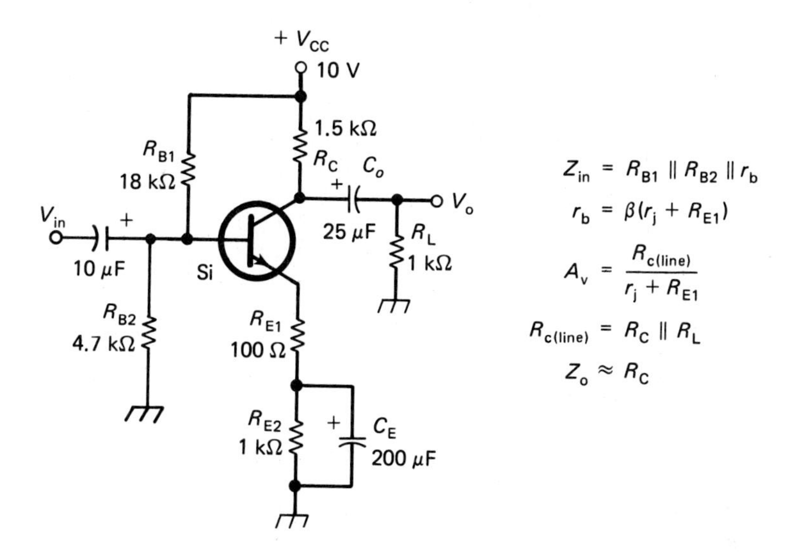

Figure 6-13 Bypassing only a part of the total R_E permits a compromise between high A_v and high Z_{in}.

If R_{E1} is made equal to r_j, the following changes will be noted from the fully bypassed amplifier of Fig. 6-11.

- A_v is reduced to one-half
- r_b is doubled, increasing Z_{in}, but by something less than double, because of the swamping effect of R_{B1} and R_{B2}
- Distortion due to nonlinearity of r_j is reduced by one-half

If R_{E1} is made $5r_j$, the factors of 2 above become factors of 5.

Example 6-6

Analyze the circuit of Fig. 6-14 for V_C, Z_{in}, Z_o, A_v, and S.

Solution

$$\frac{V_B}{6.8\text{ k}\Omega} = \frac{-9\text{ V}}{15\text{ k}\Omega + 6.8\text{ k}\Omega}; \qquad V_B = 2.8\text{ V}$$

$$V_E = V_B - V_{BE} = 2.8 - 0.2 = 2.6\text{ V}$$

$$\frac{V_{R(C)}}{R_C} = \frac{V_{R(E)}}{R_E}; \qquad \frac{V_{R(C)}}{3.9\ \text{k}\Omega} = \frac{2.6\ \text{V}}{2.7\ \text{k}\Omega + 0.1\ \text{k}\Omega}; \qquad V_{R(C)} = 3.6\ \text{V}$$

$$V_C = V_{CC} - V_{R(C)} = 9 - 3.6 = \mathbf{5.4\ V}$$

$$I_C = \frac{V_{R(C)}}{R_C} = \frac{3.6\ \text{V}}{3.9\ \text{k}\Omega} = 0.92\ \text{mA}$$

$$r_j = \frac{0.03}{I_E} = \frac{0.03}{0.00092} = 33\ \Omega \qquad (6\text{-}2)$$

$$r_b = \beta(r_j + R_{E1}) = 35(33 + 100) = 4.6\ \text{k}\Omega \qquad (6\text{-}24)$$

$$Z_{in} = R_{B1} \,\|\, R_{B2} \,\|\, r_b = 15\ \text{k}\Omega \,\|\, 6.8\ \text{k}\Omega \,\|\, 4.6\ \text{k}\Omega = \mathbf{2.3\ k\Omega}$$

$$Z_o = \mathbf{3.9\ k\Omega}$$

$$R_{c(\text{line})} = R_C \,\|\, R_L = 3.9\ \text{k}\Omega \,\|\, 5\ \text{k}\Omega = 2.19\ \text{k}\Omega$$

$$A_v = \frac{R_{c(\text{line})}}{r_j + R_{E1}} = \frac{2190}{33 + 100} = \mathbf{16.5} \qquad (6\text{-}22)$$

$$S = \frac{\beta_{\min} R_E}{\beta_{\min} R_E + R_{B1} \,\|\, R_{B2}} = \frac{35 \times 2.8}{(35)(2.8) + (15 \,\|\, 6.8)} = \mathbf{0.95} \qquad (6\text{-}21)$$

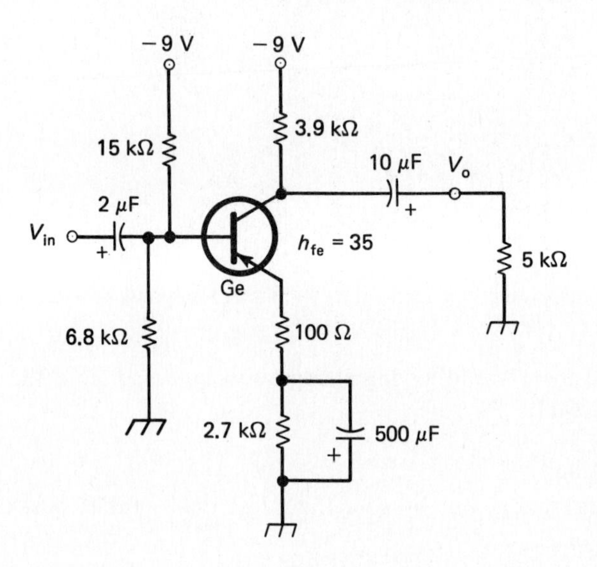

Figure 6-14 Example 6-6.

6.13 MAXIMUM-OUTPUT-VOLTAGE DETERMINATION

The load-line techniques of Sections 6.7 and 6.8 may be used to predict the maximum V_o of the stabilized amplifiers of Figs. 6-10, 6-11, and 6-13 if two points are kept in mind.

1. Any voltage across a bypassed emitter resistor must be subtracted from V_{CC} before laying down the maximum-voltage zero-current point.

2. The load-line swing is the total swing across the collector-line resistance plus any unbypassed emitter resistance ($R_C \| R_L + R_{E1}$). The actual output is found by

$$V_o = V_{swing}\left(\frac{R_C \| R_L}{R_C \| R_L + R_{E1}}\right) \tag{6-27}$$

Often R_{E1} will be negligible.

A mathematical determination of the maximum peak undistorted output of an amplifier can be made by the following steps.

1. The maximum peak swing down toward ground is the collector voltage change from the zero-signal Q point to the full-on point with the transistor V_{CE} at saturation.

$$V_{o(pk)down} = V_{C(Q)} - V_{C(on)} \tag{6-28}$$

The former ($V_{C(Q)}$) is calculated in a routine bias analysis, and the latter ($V_{C(on)}$) is easily found by analyzing a series circuit consisting of R_C, the transistor saturated at $V_{CE(sat)}$, and any unbypassed emitter resistance.

2. The maximum peak swing up toward V_{CC} is the voltage developed across the collector-line resistance $R_C \| R_L$ by a complete cutoff of collector current, from I_C to zero.

$$V_{o(pk)up} = I_C(R_C \| R_L) \tag{6-29}$$

The bypassed emitter resistance has fixed voltage across it for purposes of ac-signal swings. We might conceive of C_E as a battery with a voltage $V_{R(E)2}$ which subtracts from V_{CC}.

An amplifier is biased for maximum V_o when the voltage limit of step 1 equals the current limit of step 2. If the current limit is lower than the voltage limit, a higher bias current I_C will increase maximum V_o. If the current limit is higher than the voltage limit, a lower I_C will increase $V_{o(max)}$.

Example 6-7

Find the maximum $V_{o(p\text{-}p)}$ for the circuit of the previous example, Fig. 6-14. Let $V_{CE(sat)} = 0.2$ V.

Solution

1. $V_{C(Q)}$ was determined in Example 6-6 to be 5. 4 V. Figure 6-15(a) shows the determination of $V_{C(sat)}$.

$$V_{R(C)sat} = (V_{CC} - V_{CE(sat)} - V_{E2})\frac{R_C}{R_C + R_{E1}}$$

$$V_{E2} = V_E\frac{R_{E2}}{R_E} = (2.6)\frac{2.7}{2.7 + 0.1} = 2.5 \text{ V}$$

$$V_{R(C)sat} = (9 - 0.2 - 2.5)\frac{3.9}{3.9 + 0.1} = 6.1 \text{ V}$$

$$V_{C(sat)} = V_{CC} - V_{R(C)sat} = 9 - 6.1 = 2.9 \text{ V}$$

$$V_{o(pk)down} = V_{C(Q)} - V_{C(sat)} = 5.4 - 2.9 = \mathbf{2.5\ V}$$

2. The collector current at the Q point was found in Example 6-6 to be 0.92 mA. Figure 6-15(b) shows the determination of $V_{o(pk)up}$:

$$V_{o(pk)up} = I_C(R_C \parallel R_L) = (0.92\text{ mA})(3.9\text{ k}\Omega \parallel 5\text{ k}\Omega) = \mathbf{2.0\ V\ pk}$$

The current limitation governs, and $V_{o(p\text{-}p)max}$ is twice the peak determination, or 4.0 V p-p. Slightly greater output could be achieved by increasing the bias current I_C.

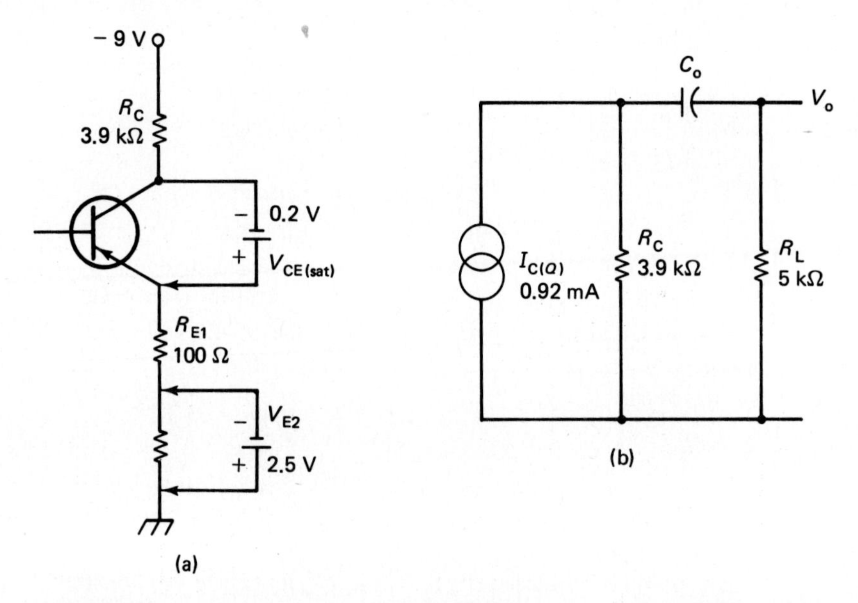

Figure 6-15 Maximum V_o determination in Example 6-7: (a) voltage limitation when Q turns on; (b) current limitation when Q turns off.

6.14 AMPLIFIER TESTS AND MEASUREMENTS

The preceding sections of Chapter 6 have dealt primarily with predicting the bias point, voltage gain, input impedance, and output impedance of various types of amplifiers in the audio-frequency range. It is now appropriate to discuss some of the laboratory techniques for verifying these predictions.

Bias point can, of course, be determined by measuring V_C with a dc-coupled 'scope or a voltmeter. This measurement should be taken with no ac input signal present, since some voltmeters respond to the peak signal rather than the dc average value.

Voltage gain can be determined by measuring the ac input voltage and the ac output voltage of the amplifier with a calibrated oscilloscope, and calculating

$$A_v = \frac{V_o}{V_{in}} \tag{6-3}$$

Care must be taken to ensure that the input signal is large enough to be measured on the scope but not so large as to cause distortion of the output-signal waveform. In the event that the scope does not have enough sensitivity to measure the input signal, a 100-to-1 voltage divider can be used between the signal generator and the amplifier, as shown in Fig. 6-16. The signal at point A will be large enough to be measured easily, and the actual amplifier input signal at point B will be $\frac{1}{100}$ of V_A.

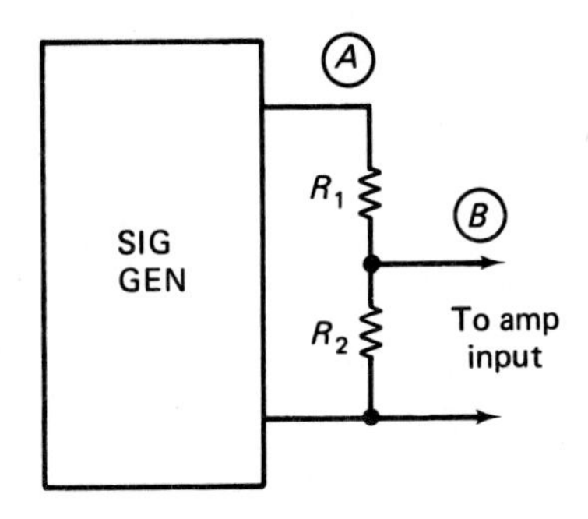

Figure 6-16 A 100 : 1 input attenuator for measuring V_{in} of high-gain amplifiers. Voltage is measured at point A. Input signal at point B is approximately $\frac{1}{100}$ V_A.

The input impedance of an amplifier can be measured by inserting a variable test resistance in series with the amplifier input, as shown in Fig. 6-17. With R_{TEST} set to zero resistance, the signal generator is set to produce an even value of output voltage (say 2 V p-p) as measured on the 'scope. The value of R_{TEST} is then increased until the measured V_o drops to one-half of its original value (in this case, 1 V p-p). Variable resistor R_{TEST} is then removed and its resistance is measured with an ohm-meter. The value of R_{TEST} will equal Z_{in} of the amplifier, since half of the signal generator voltage was dropped across each of them.

For this technique to be successful, *the source impedance of the signal generator must be much less then the input impedance of the amplifier.* As a rule of thumb, the

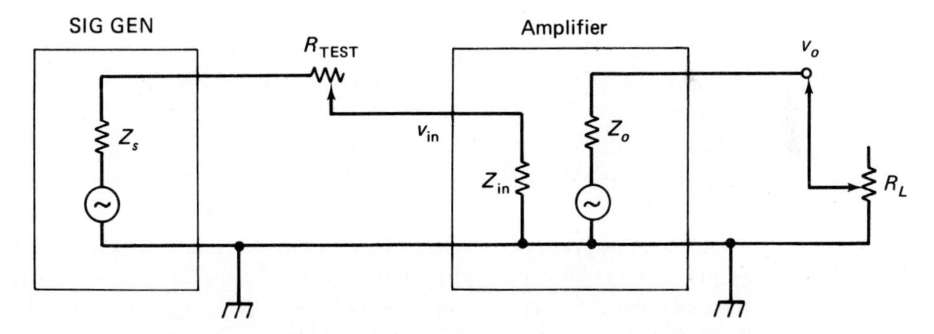

Figure 6-17 Laboratory setup for determining Z_{in} and Z_o of an amplifier. It may be necessary to use the voltage divider of Fig. 6-16 to keep Z_s less than $\frac{1}{20}$ Z_{in} to ensure accuracy of the Z_{in} determination.

maximum Z_s should be 50 Ω for small-signal amplifiers, and 1 Ω for power amplifiers. This low signal-source impedance can be realized by using the input voltage divider of Fig. 6-16 with a conventional signal generator. If noise pickup becomes a problem, it may be necessary to use a number of small fixed resistors in place of the variable reistor for R_{TEST}.

Output impedance can be determined by means of a variable load resistance R_L, as shown in Fig. 6-17. R_{TEST} is shorted out, and R_L is temporarily disconnected. The signal generator is set to produce an even-number ac output voltage (say $V_o =$ 0.4 V p-p). R_L is then connected and adjusted until V_o drops to one-half of its original value ($V_o = 0.2$ V p-p in this case). Finally R_L is removed and measured with an ohmmeter. Since half of the original 0.4 V p-p appeared across the load and the other half across the internal Z_o of the amplifier, $Z_o = R_L$. If adding R_L causes distortion of the output waveform the input-signal level must be reduced for this test.

A typical test frequency for audio amplifiers is 1 kHz, although 440 Hz will produce a more pleasant test tone if the amplifier is actually driving a loudspeaker. In any case, extremely low or extremely high test frequencies should be avoided, because of the capacitive effects that would be encountered.

CHAPTER SUMMARY

Bias Conditions:

1. Dc voltages and currents (called bias) must be applied to a transistor to allow it to respond properly to ac signals. In simple transistor amplifiers a common objective is to bias V_{CE} midway between saturation and cutoff.

2. A common-emitter amplifier can be biased with a single resistor from the V_{CC} supply to the base, but this method is seldom used because bias point changes drastically with transistor-beta changes.

3. A stable bias circuit uses voltage-divider resistors R_{B1} and R_{B2} to fix the base voltage V_B. V_E is less than V_B by one diode drop (≈ 0.6 V for silicon), and I_E is thus held stable at V_E/R_E. This technique requires $V_E \geq 2$ V and $R_{B2} \leq 10R_E$ to hold bias variations to about $\pm 20\%$ using transistors with $\beta_{min} = 40$.

4. Collector-resistor voltage is $V_{R(C)} = I_C R_C \approx I_E R_C$. Collector voltage is $V_C = V_{CC} - V_{R(C)}$.

Signal Analysis:

5. Voltage gain A_v of the common-emitter amplifier may be calculated as the ratio of ac resistance in the collector circuit to ac resistance in the emitter circuit, since collector and emitter currents are nearly equal. Collector-line resistance includes R_C, R_L (if present), and transistor output resistance r_c in parallel. For silicon transistors in elementary circuits r_c is usually high enough to be neglected. Emitter-line resistance includes base-emitter junction resistance $r_j \approx 0.03/I_E$ in series with any unbypassed resistance R_{E1} from the emitter to ground.

$$A_v = \frac{R_C \| R_L \| r_c}{r_j + R_{E1}} \tag{6-25}$$

6. Input impedance Z_{in} of a common-emitter amplifier may be calculated as the parallel combination of the base resistor(s) and the base-input resistance r_b. Because I_b is less than I_e by a factor of β, r_b is larger than the emitter-line resistance by a factor of β.

$$r_b = \beta(r_j + R_{E1}) \tag{6-24}$$

7. All or part of the emitter resistance may be bypassed for ac with a capacitor. The effect is to increase A_v and decrease Z_{in}.
8. Load lines may be drawn on a set of transistor characteristic curves to visualize large-signal operation and optimize bias (Q point). The static (dc) load line represents the no-load case (R_C only). The dynamic load line represents the case with R_L added. Adding a load resistor causes $V_{o(max)}$ and A_v to decrease.
9. Maximum output may be calculated from the lower of these two:

$$V_{o(pk)down} = V_{C(Q)} - V_{C(on)} \tag{6-28}$$

$$V_{o(pk)up} = I_C(R_C \| R_L) \tag{6-29}$$

10. Amplifier output impedance Z_o may be calculated as $Z_o = R_C \| r_c$. Z_o can be measured by removing any R_L, noting V_o, and adding a variable R_L such that V_o drops to one-half its original value. Then $Z_o = R_L$.
11. Amplifier input impedance can be measured by placing a variable resistor R_{TEST} in series with the input signal and adjusting it until V_o drops to one-half its original value. Then $Z_{in} = R_{TEST}$. Generator source resistance r_g must be much smaller than Z_{in} for this test.

QUESTIONS AND PROBLEMS

6-1. Define these terms: bias; static resistance; dynamic resistance.

6-2. A diode curve (similar to Fig. 6-2) shows 0.5 mA at 0.60 V and 1 mA at 0.65 V. What is the dynamic resistance of the diode in this range?

6-3. Estimate the dynamic resistance of a junction diode carrying a dc bias current of 1.5 mA using equation (6-2).

6-4. In the circuit of Fig. 6-1, R_C is 1500 Ω and I_E is 3 mA. Find r_j and A_v.

6-5. In Fig. 6-1, V_{CC} is +12 V, R_C is 680 Ω, R_B is 120 kΩ, and β is 50. Find in order: I_B, I_C, and A_v.

6-6. What is base-input resistance r_b compared to emitter junction resistance r_j? Why is r_b not equal to r_j?

6-7. What is the ac base-input resistance r_b for a transistor whose emitter current is 250 μA and whose β is 120?

6-8. What is the input impedance of the amplifier in Problem 6-4?

6-9. Read the output resistance of the transistor whose curves appear in Fig. 4-7(c). Take the reading at $V_{CE} = 6$ V and $I_C = 2$ mA. Note that this is a germanium type.

6-10. Read the output resistance of the silicon transistor in Fig. 4-7(b) at 0.6 V and 50 mA.

6-11. In Example 6-2 (Fig. 6-7) the transistor is replaced with one having $\beta = 50$. What are the new A_v and Z_{in}?

6-12. What is the Q point in transistor load-line terminology?

6-13. Trace the characteristic curves of Fig. 6-6 on your own paper. Draw the dc load line for $V_{CC} = 25$ V and $R_C = 820\ \Omega$.

6-14. On the curves of Problem 6-13 draw a slope line representing R_C of 820 Ω in parallel with R_L of 390 Ω. Use 5 V as the assumed voltage. Adjust the position of the ac load line for maximum output signal. Draw this line and give the Q-point current.

6-15. Why is the circuit of Fig. 6-1 not used in practice? How does the circuit of Fig. 6-10 overcome this problem?

6-16. Draw the stabilized amplifier circuit of Fig. 6-10, substituting the following values for the ones given: $R_{B1} = 8.2$ kΩ, $R_{B2} = 1$ kΩ, $R_E = 150\ \Omega$, $R_C = 1$ kΩ, $R_L = 680\ \Omega$, and $V_{CC} = 18$ V. Find the following for the circuit you have drawn:
(a) V_C (voltage, collector to ground)
(b) A_v (do not neglect to find r_j)
(c) Z_{in} (assume C_{in} and C_o have very low reactances)
(d) S, the bias-stability factor. $\beta_{min} = 30$, $\beta_{max} \longrightarrow \infty$
(e) Z_o, the output impedance

6-17. Add a large-value bypass capacitor from emitter to ground in the amplifier of Problem 6-16. Calculate the new A_v and Z_{in}.

6-18. What is amplifier linearity? Is it better with or without an emitter-bypass capacitor? Explain why.

6-19. In the circuit of Fig. 6-11, transistor β doubles. Which of the following are substantially affected, and in which direction? A_v; Z_{in}; V_C.

6-20. Calculate A_v for the bypassed amplifier of Fig. 6-11. Draw the static and dynamic load lines for the circuit (the actual transistor curves will not be needed). For an output-signal swing of 4 V p-p, find $i_{E(max)}$ and $i_{E(min)}$, $r_{j(min)}$ and $r_{j(max)}$, and $A_{v(max)}$ and $A_{v(min)}$.

6-21. For the partially bypassed CE amplifier of Fig. 6-13, calculate (a) V_C; (b) Z_{in}; (c) A_v; (d) S.

6-22. An *NPN* common-emitter amplifier produces waves which are clipped off at the top (positive) peaks. The collector bias current should be (increased or decreased?). Sketch a load line supporting your answer.

6-23. Find the maximum $V_{o(p\text{-}p)}$ for the amplifier of Fig. 6-13, (a) using the load-line technique; (b) using the mathematical technique.

6-24. V_{in} to an amplifier is 24 mV p-p. V_o is 3.7 V p-p. Calculate A_v.

6-25. A signal generator uses a voltage divider (Fig. 6-16) to feed an amplifier. $V_A = 8$ mV p-p, $R_1 = 1000\ \Omega$, $R_2 = 10\ \Omega$, and $V_o = 870$ mV p-p. What is A_v?

6-26. A signal generator has $Z_s = 1500\ \Omega$ and a no-load output of 50 mV. What voltage does it deliver when connected to an amplifier with $Z_{in} = 800\ \Omega$ resistive?

6-27. A signal generator with $Z_s = 50\ \Omega$ feeds an amplifier. $V_o = 4.0$ V p-p. A variable resistance is inserted in series between the generator output and the amplifier input, and is adjusted until $V_o = 2.0$ V p-p. The resistance is then removed and measured as 3.8 kΩ. Find Z_{in}.

6-28. The test of Problem 6-27 is repeated on a different amplifier. This time the variable resistance reads 85 Ω. Find Z_{in}. *Hint:* A voltage divider (Fig. 6-16) should have been used, but Z_{in} can still be calculated by circuit analysis. Z_s is not negligible.

6-29. An amplifier outputs 300 mV p-p with no load. Adding a 600-Ω load causes V_o to drop to 150 mV p-p. What is the Z_o of the amplifier?

6-30. Some power amplifiers can be damaged by operating them with no load or too heavy a load. One such amplifier delivered 7.5 V p-p to a single 16-Ω loudspeaker and 6.4 V p-p to a pair of 16-Ω loudspeakers in parallel. Calculate the Z_o of the amplifier.

7

SPECIAL AMPLIFIER TYPES AND CIRCUITS

7.1 THE EMITTER FOLLOWER

The bipolar transistor's characteristically low input impedance presents frequent problems of matching a high-impedance signal source to the amplifier input. For example, if the circuit of Fig. 6-14 were to be driven with a typical high-impedance microphone having an impedance of 50 kΩ and an output level of 10 mV, only about 0.4 mV would appear across the 2.3 kΩ input impedance of the amplifier; the other 9.6 mV would be lost across the internal impedance of the microphone. One solution to this problem would be to use a step-down transformer to provide a proper impedance match, but transformers are heavy and expensive and have limited frequency response. A circuit that solves the problem with none of these disadvantages is shown in Fig. 7-1.

This circuit is called a *common collector* amplifier, since the input is applied to the base and the output is taken from the emitter, leaving the collector as the common element between them. It is more frequently referred to as an *emitter follower*, however, because the emitter voltage tends to follow the base-signal waveform, the only difference between the two being the 0.6- or 0.2-V diode junction drop. The advantage of this amplifier comes from the fact that the emitter (output) current is greater than the base (input) current by an approximate factor of beta. Therefore a high-impedance source can deliver a relatively small input current to the base of the amplifier, and a much larger current will be delivered to the load, which may have a relatively low impedance.

The voltage gain of the emitter follower is typically slightly less than 1, and there is no inversion of the signal waveform as there is in the case of the common-emitter amplifier. The voltage gain will drop appreciably below unity only if the resistance seen looking out of the emitter ($R_{e(line)}$) is so small that the emitter junction resistance (r_j) drops an appreciable portion of the voltage.

$$A_v = \frac{R_E \| R_L}{r_j + R_E \| R_L} \tag{7-1}$$

Since the base current is smaller than the emitter current by an approximate factor of beta, the resistance seen looking into the base is larger than the resistance in the emitter line by the same factor:

$$r_b = \beta(r_j + R_{e(line)}) \tag{7-2}$$

The resistance seen looking out of the emitter ($R_{e(line)}$) is usually the parallel combination of R_E and R_L. To get the total input impedance, the shunting effect of R_{B1} and R_{B2} must also be taken into account.

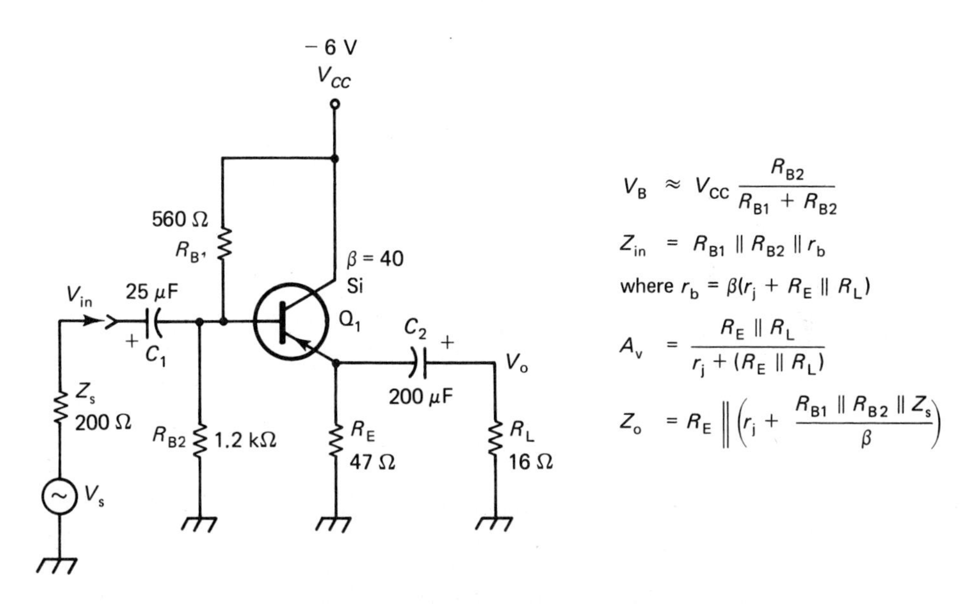

Figure 7-1 Emitter-follower amplifier.

The output impedance of the emitter follower depends primarily on the impedance of the signal source and the beta of the transistor. Although R_E, R_B, and r_j also affect the output impedance to a lesser extent, the emitter follower is often conceived as an impedance transforming device whose output impedance is approximately Z_s/β.

Output impedance is the resistance seen by an ac signal at the output terminals looking back into the amplifier. There are two parallel paths to ground for ac. The first consists solely of R_E. The second begins with r_j and continues through the emitter to the base side of the transistor. Since the current on the emitter side is beta times the current on the base side, the resistance seen looking out of the base will appear, when viewed from the emitter side, to be smaller by a factor of beta. There are three paths to ac ground looking out of the transistor base: R_{B1}, R_{B2}, and Z_s.

All of the above is summarized by the formula for the output impedance of an emitter follower:

$$Z_o = R_E \bigg\| \left(r_j + \frac{R_{B1} \| R_{B2} \| Z_s}{\beta} \right) \tag{7-3}$$

Maximum output voltage is determined as described in Section 6.13, except that the *saturation* or voltage limitation is in the *upward* direction (toward V_{CC}) as the transistor turns on. The *cutoff* or current limit is in the downward direction (toward ground) and equals $I_C(R_E \| R_L)$.

Example 7-1

Determine V_E, A_v, Z_{in}, and Z_o for the circuit of Fig. 7-1.

Solution

$$V_B = V_{CC}\frac{R_{B2}}{R_{B1} + R_{B2}} = (6)\frac{1200}{560 + 1200} = 4.1 \text{ V}$$

$$V_E = V_B - V_{BE} = 4.1 - 0.6 = \mathbf{3.5 \text{ V}}$$

$$I_E = \frac{V_E}{R_E} = \frac{3.5}{47} = 74 \text{ mA}$$

$$r_j = \frac{0.03}{0.074} = 0.41\ \Omega$$

$$A_v = \frac{R_E \| R_L}{r_j + R_E \| R_L} = \frac{47 \| 16}{0.41 + 47 \| 16} = \mathbf{0.97} \tag{7-1}$$

$$Z_{in} = R_{B1} \| R_{B2} \| r_b$$

$$r_b = \beta(r_j + R_E \| R_L) = 40(0.41 + 47 \| 16) = 494\ \Omega \tag{7-2}$$

$$Z_{in} = 560 \| 1200 \| 494 = \mathbf{215\ \Omega}$$

$$Z_o = R_E \bigg\| \left(r_j + \frac{R_{B1} \| R_{B2} \| Z_s}{\beta} \right) \tag{7-3}$$

$$= 47 \bigg\| \left(0.41 + \frac{560 \| 1200 \| 200}{40} \right) = \mathbf{3.4\ \Omega}$$

Some comments on the foregoing example may make the resultant numbers more meaningful. A_v is only 3% less than unity, making V_o nearly equal to V_{in}. The 215-Ω Z_{in} will load the 200-Ω source down to nearly one-half, so V_{in} will be just slightly more than $\frac{1}{2}V_s$.

The input impedance may seem unusually low, especially since the emitter follower was presented as a high-Z_{in} circuit, but this is a much higher powered circuit than the others we have seen so far. Relative to the load impedance, Z_{in} is quite high. If the source were to drive R_L directly, V_s would be loaded down by the factor 16/(200 + 16) or 0.074. The Z_o figure indicates that a 3.4-Ω load would cause V_o to drop to one-half of its no-load value. It should be realized that with a higher-resistance load and appropriate bias resistors an emitter follower can have Z_{in} values in the range 0.1 to 1 MΩ.

Example 7-2

Determine the stability factor S and the maximum V_o for the circuit of Fig. 7-1. Use $\beta_{min} = 40$.

Solution

$$S = \frac{\beta_{min} R_E}{\beta_{min} R_E + R_{B1} \| R_{B2}} = \frac{(40)(47)}{(40)(47) + 560 \| 1200} = \mathbf{0.83} \qquad (6\text{-}18)$$

This indicates that a beta change from infinity to 40 would cause I_C to decrease by a factor of 0.83.

$$V_{o(pk)up} = V_{E(Q)} - (V_{CC} - V_{CE(sat)})$$
$$= 3.5 - (6 - 0.2) = 2.3 \text{ V pk}$$
$$V_{o(pk)down} = I_{E(Q)}(R_E \| R_L) = (0.074)(47 \| 16) = 0.88 \text{ V pk} \qquad (6\text{-}25)$$

The current limit governs, and $V_{o(max)p\text{-}p} = 2(0.88)$ or **1.76 V p-p.**

7.2 BOOTSTRAPPING FOR HIGH Z_{in}

Bootstrapping is a technique for obtaining high input impedance and reasonable voltage gain in a common-emitter amplifier. The circuit is shown in Fig. 7-2. The idea is to feed a signal V_e that is very nearly equal to V_{in} back to point B through capacitor C_B. The voltage across R_{B3} is therefore nearly zero, and the ac signal current through it is nearly zero. The base-bias resistors are therefore removed from the Z_{in} path, and Z_{in} consists (ideally) of base-input resistance r_b only. If the transistor is selected for high beta, Z_{in} can be remarkably high. In order for the fed-back voltage to be nearly equal to V_{in}, R_{E1} must be much larger than r_j, dictating much less than maximum voltage gain. The input impedance can be predicted by the equation

$$Z_{in} = R_{B3}\left(\frac{r_j + R_{E1}}{r_j}\right) \Big\| \beta(r_j + R_{E1}) \qquad (7\text{-}4)$$

provided that

$$R_{B3} \gg R_{B1} \| R_{B2} \qquad \text{and} \qquad R_{E1} \ll R_{B1} \| R_{B2}$$

The term to the left of the parallel sign in equation (7-4) represents the effective resistance of R_{B3}, while the term to the right is the base input resistance r_b.

The addition of the bootstrap resistor R_{B3} degrades the bias stability of the circuit. The new formula for stability factor S is

$$S = \frac{\beta_{min} R_E}{\beta_{min} R_E + (R_{B1} \| R_{B2} + R_{B3})} \qquad (7\text{-}5)$$

Bias conditions, A_v, Z_o, and $V_{o(max)}$ for the bootstrap circuit are calculated as in Sections 6.12 and 6.13.

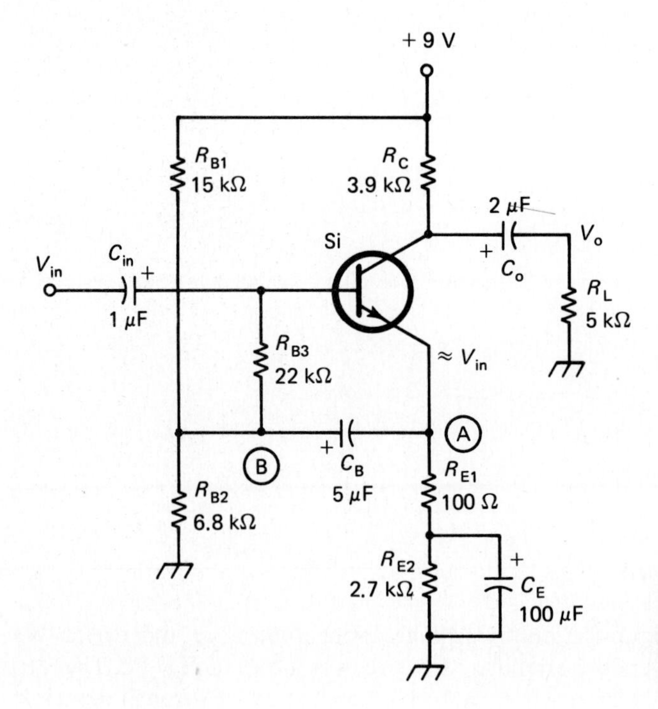

Figure 7-2 Bootstrapping increases input impedance by practically eliminating the shunting effect of the base-bias resistors.

Example 7-3

Find the input impedance and stability factor for the circuit of Fig. 7-2 using the values given and $\beta = 200$. The circuit is the same as Fig. 6-14 (Example 6-6) except for the addition of R_{B3} and C_B.

Solution

$$Z_{in} = R_{B3}\left(\frac{r_j + R_{E1}}{r_j}\right) \bigg\| \beta(r_j + R_{E1}) \tag{7-4}$$

$$= (22\,000)\left(\frac{33 + 100}{33}\right) \bigg\| (200)(33 + 100)$$

$$= \mathbf{20.5\ k\Omega}$$

The value for the nonbootstrap circuit with $\beta = 35$ was **2.3 kΩ**. With $\beta = 200$ it would have been 4.0 kΩ.

$$S = \frac{\beta_{min} R_E}{\beta_{min} R_E + (R_{B1} \| R_{B2} + R_{B3})} \tag{7-5}$$

$$= \frac{(35)(2.8)}{(35)(2.8) + (15 \| 6.8 + 22)} = \mathbf{0.79}$$

The value for the nonbootstrap circuit with $\beta_{min} = 35$ was **0.95**.

The bootstrap circuit may be applied to the emitter-follower amplifier of Fig. 7-1 for even higher input impedances.

7.3 COLLECTOR SELF-BIAS

A variation of the voltage-divider bias scheme that is particularly useful with the bootstrap circuit is shown in Fig. 7-3. The only change is that R_{B1} is taken from the collector end of R_C rather than from V_{CC}. This necessitates lowering the value of R_{B1}, typically to about one-half of its original value.

Bias stability is improved by this circuit change because it introduces *negative feedback* from the dc output (collector) to the dc input (base). Negative feedback is often referred to as a process which tends to "offset the upset."

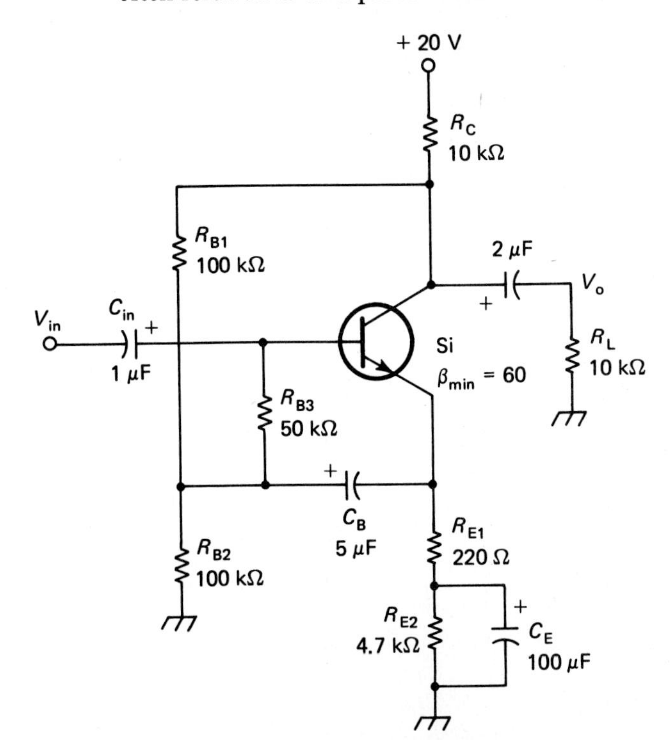

Figure 7-3 Collector self-bias: taking R_{B1} from the collector rather than from V_{CC} automatically lowers the base bias if I_C increases and V_C drops.

Consider, for example, that the beta of the transistor in Fig. 7-3 decreases from 80 to 40. The resultant demand for more base current from R_{B1}-R_{B2}-R_{B3} will cause V_B to drop and $V_{R(E)}$ will drop by a like amount. Thus I_C is reduced and the $V_{R(C)}$ drop is reduced, leaving a higher voltage at V_C. With the old connection of R_{B1} to V_{CC} this shift in the Q point can easily cause serious reductions in the maximum output-voltage swing. However, with R_{B1} connected to the collector, a rise in V_C causes an increase in the voltage-divider output to the base. This tends to *increase* V_B, $V_{R(E)}$, and

I_C back toward their original values. To summarize:

- β drops, requiring I_B to increase.
- Loading of voltage divider drops V_B, V_E, and I_E.
- Lower I_E means lower I_C, less drop across R_C, hence higher V_C.
- But higher V_C means higher V_B, hence higher V_E, and so on.

Of course, it must not be imagined that when β drops, V_C at first rises and then suddenly falls back down as the circuit "belatedly gets the message" from the negative-feedback loop. Neither can the feedback "offset" be so great as to entirely overcome the original "upset." The compensation provided by negative feedback is immediate and it never fully restores the original condition—it only limits the amount of change.

The stability factor for the circuit of Fig. 7-3 is typically two to three times better than given in equation (7-5) for the circuit of Fig. 7-2. In this context "three times better" than $S = 70\%$ would be $S' = 90\%$, that is, a 10% drop in I_C as opposed to a 30% drop. An exact calculation of S' is quite complicated and will not be attempted here. The ideal collector voltage with a very high beta transistor in the collector-self-bias circuit is shown in Appendix E to be

$$V_C = \frac{V_{CC} R_E + V_{BE} R_C}{\dfrac{R_{B2} R_C}{R_{B1} + R_{B2}} + R_E} \tag{7-6}$$

Example 7-4

Find V_C and S for the circuit of Fig. 7-3.

Solution

$$V_C = \frac{V_{CC} R_E + V_{BE} R_C}{\dfrac{R_{B2} R_C}{R_{B1} + R_{B2}} + R_E} \tag{7-6}$$

$$= \frac{(20)(4.9) + (0.6)(10)}{\dfrac{(100)(10)}{200} + 4.9} = \mathbf{10.5\ V}$$

$$S = \frac{\beta_{min} R_E}{\beta_{min} R_E + (R_{B1} \,||\, R_{B2} + R_{B3})} \tag{7-5}$$

$$= \frac{(60)(4.9)}{(60)(4.9) + (100 \,||\, 100 + 50)} = \mathbf{75\%}$$

This is 25% *instability*. S' would have approximately $\frac{25}{2}$ or 12% instability, corresponding to $S' = \mathbf{88\%}$.

7.4 MILLER EFFECT

Why did we save the discussion of collector self-bias until after examining the bootstrap amplifier? Since it improves bias stability and seems to cause no other problems, why not use collector self-bias all the time? The answer is that *connecting an impedance from the output to the input of an inverting amplifier causes a serious lowering of input impedance.* This phenomenon is called the Miller effect, and it would be encountered in a circuit such as the one shown in Fig. 7-4.

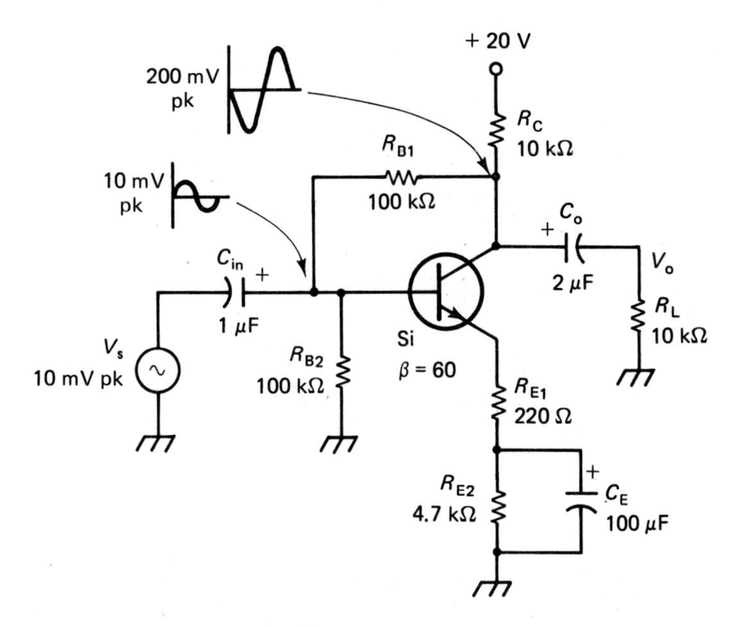

Figure 7-4 Without the bootstrap circuit the Miller effect seriously lowers Z_{in} by making R_{B1} appear to have a low resistance.

The gain of this amplifier is about −20. Thus a 10-mV positive peak at V_{in} (the left end of R_{B1}) will cause a 200-mV negative peak at V_o (the right end of R_{B1}). The total voltage across R_{B1} is then $V_{in}(A_v + 1)$, in this case $21V_{in}$. The ac current in R_{B1} is therefore 21 times the current that would appear if R_{B1} were connected to an ac ground. All of this current comes from the source. Another way of saying this is that the source delivers a current to R_{B1} as if there were a resistor $R_{B1}/(1 + A_v)$ connected from the input to ground. The total input impedance for the circuit of Fig. 7-4 is

$$Z_{in} = \frac{R_{B1}}{A_v + 1} \Big\| R_{B2} \Big\| \beta(r_j + R_{E1}) \tag{7-7}$$

Of course, when bootstrapping is used, R_{B1} and R_{B2} are isolated from the input and do not appear in the equation for Z_{in}, so the lowering of R_{B1} is of no consequence.

Applications of the Miller-effect concept appear in all sorts of amplifier circuits, so the circuit of Fig. 7-4 should be understood as just one example of a general phenomenon that is likely to be encountered again in many other situations. Often the fedback impedance is not a resistance, but a few picofarads of stray capacitance, which the Miller effect makes appear to be 100 pF or more.

Example 7-5

Calculate and compare Z_{in} for the bootstrap circuit of Fig. 7-3 and the Miller-effect circuit of Fig. 7-4.

Solution

$$I_E \approx I_C = \frac{V_{CC} - V_C}{R_C} = \frac{20 - 10.5}{10} = 0.95 \text{ mA}$$

$$r_j \approx \frac{30 \text{ mV}}{I_E} = \frac{30 \text{ mV}}{0.95 \text{ mA}} = 32 \ \Omega \qquad (6\text{-}2)$$

$$A_v = \frac{-R_C \| R_L}{r_j + R_{E1}} = \frac{-10 \| 10}{0.032 + 0.22} = -19.8 \qquad (6\text{-}22)$$

For Fig. 7-3,

$$Z_{in} = R_{B3}\frac{r_j + R_{E1}}{r_j} \Big\| \beta(r_j + R_{E1}) \qquad (7\text{-}4)$$

$$= (50)\frac{0.032 + 0.22}{0.032} \Big\| (60)(0.032 + 0.22)$$

$$= \mathbf{14.6\ k\Omega}$$

For Fig. 7-4,

$$Z_{in} = \frac{R_{B1}}{A_v + 1} \Big\| R_{B2} \Big\| \beta(r_j + R_{E1}) \qquad (7\text{-}7)$$

$$= \frac{100}{19.8 + 1} \Big\| (100) \Big\| (60)(0.032 + 0.22)$$

$$= \mathbf{3.5\ k\Omega}$$

7.5 SPECIAL COMMON-EMITTER BIAS CIRCUITS

The two-supply bias system of Fig. 7-5 can be recognized as a common-emitter amplifier, since V_{in} is at the base and V_o is at the collector. The emitter is bypassed to ground for ac, so the input signal sees

$$Z_{in} = R_B \| \beta r_j$$

The two-supply bias system creates a problem in determining the bias point. It should be understood that there are *two separate supplies* for this circuit—one 10 V negative with respect to ground (its positive terminal grounded) and the other 10 V positive to ground (its negative terminal grounded). Thus the total circuit voltage is 20 V from V_{CC} to V_{EE}.

This bias arrangement can be analyzed by recognizing that, from the base side of the transistor, R_E appears to be beta times larger than its actual value, even to the dc current. The base-emitter diode drop of 0.2V must also be considered. The base-current path then appears as shown in Fig. 7-5(b). The total voltage for this path is $V_{EE} - V_{BE}$, or 9.8 V. Thus

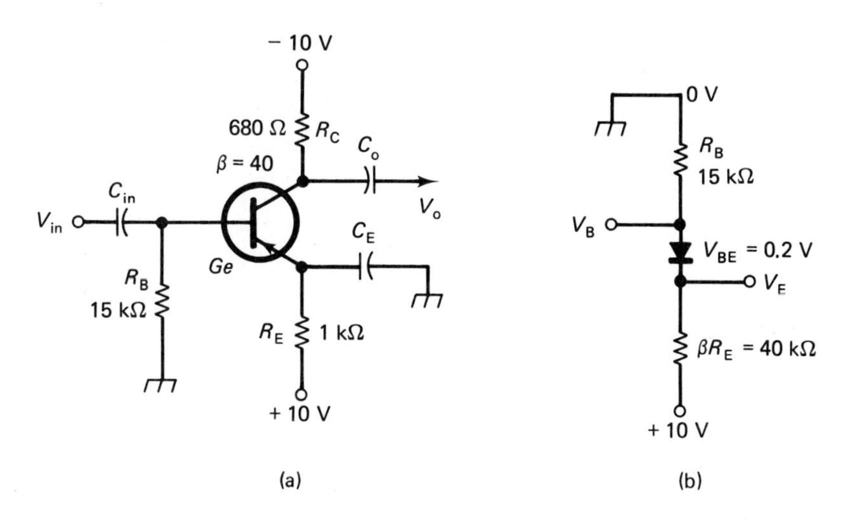

Figure 7-5 (a) Two-supply biasing of a common emitter amplifier; (b) base-current path for two-supply circuit.

$$I_B = \frac{V}{R} = \frac{9.8}{15\text{k}\ \Omega + 40\ \text{k}\Omega} = 0.18\ \text{mA}$$

$$V_{R(E)} = (I_B)(\beta R_E) = (0.18\ \text{mA})(40\ \text{k}\Omega) = 7.1\ \text{V}$$

$$V_E = V_{EE} - V_{R(E)} = 10 - 7.1 = \mathbf{2.9\ V}$$

The remainder of the analysis is now quite conventional:

$$I_E = \frac{V_{R(E)}}{R_E} = \frac{7.1}{1\ \text{k}\Omega} = 7.1\ \text{mA}$$

$$r_j = \frac{0.03}{0.0071} = 4.2\ \Omega$$

$$A_v = \frac{R_C}{r_j} = \frac{680}{4.2} = \mathbf{162}$$

$$V_{R(C)} = I_E R_C = (7.1)(680) = 4.8\ \text{V}$$

$$V_C = V_{CC} - V_{R(C)} = -10 + 4.8 = \mathbf{-5.2\ V}$$

Inverted-ground biasing: The transistor circuits shown so far have invariably used *NPN* types with positive power supplies and *PNP* types with negative supplies. It is possible (and quite common) to reverse this practice so that a *PNP* transistor can be used with a positive supply.

Figure 7-6 shows several ways that such an *inverted-ground* circuit might appear in a schematic. The important thing to recognize is that, although the emitter dc voltage is relatively "hot" (near the supply voltage), the emitter is at signal ground because of the bypass capacitor C_E.

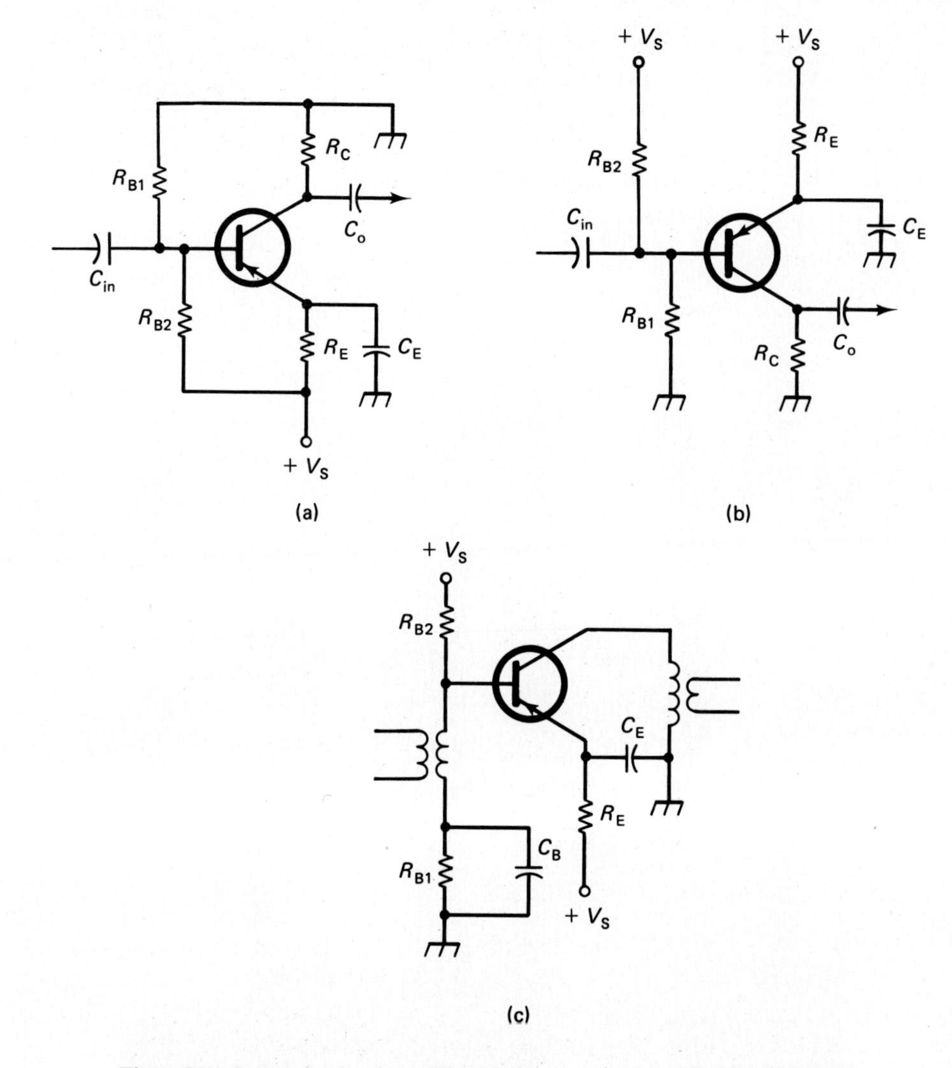

Figure 7-6 Inverted-ground-amplifier biasing circuits: (a) conventional voltage-divider bias-stabilized common-emitter amplifier with ground at negative side of V_S supply; (b) the same circuit redrawn to place $+V_S$ at the top and ground at the bottom; (c) transformer-coupled common-emitter amplifier with inverted ground using *PNP* transistor with positive supply.

7.6 THE DARLINGTON COMPOUND

A circuit connection that is frequently used where a very high-beta transistor is required is the Darlington-pair configuration, shown in Fig. 7-7. The emitter output of Q_1 is fed directly to the base of Q_2. The total current gain from base of Q_1 to collector (or emitter) of Q_2 is approximately $\beta_1\beta_2$, or β^2, assuming that Q_1 and Q_2 have equal betas.

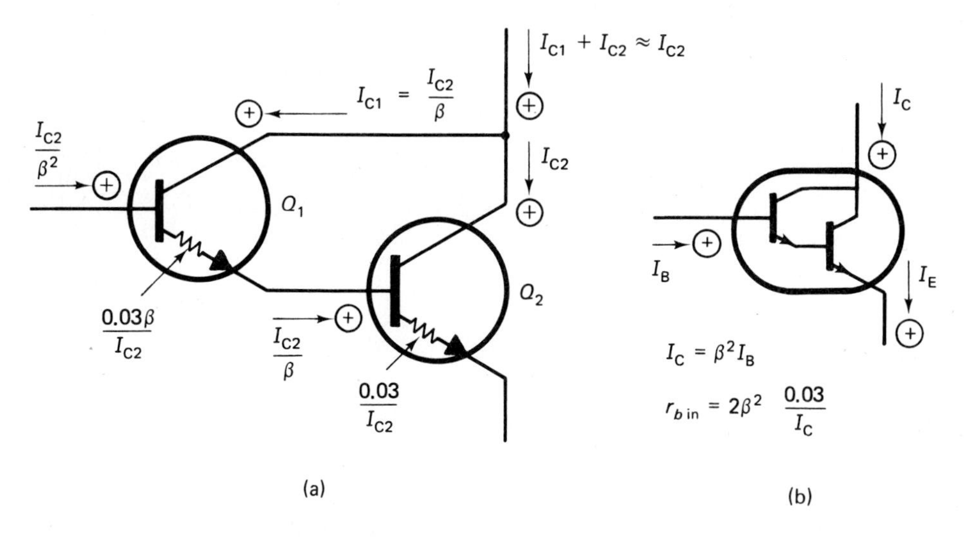

Figure 7-7 Darlington configuration for very high current gain: (a) current relationships in the individual transistors; (b) the Darlington pair considered as a single transistor.

The input resistance of the Darlington pair is also much higher than for a single bipolar transistor. There are two reasons for this: First, the input resistance of the second transistor ($r_{b2} = 0.03/I_{C2}$) is multiplied by the beta of the first transistor; second, the value of emitter junction resistance of the first transistor ($r_{j1} = 0.03/I_{E1}$) is unusually high since I_{E1} is equal to the base current of Q_2, which is quite low.

The Darlington pair can be treated as if it were a single transistor with a current gain of β^2. Indeed, Darlington pairs are available in a single package with three leads, as shown in Fig. 7-7(b). If separate transistors are used in a Darlington connection, they should be silicon types. The I_{CEO} leakage of the first transistor may be enough to completely saturate the second transistor if germanium types are used.

For bias stability a good rule of thumb is to make $R_{B2} \leq 300R_E$ when using a Darlington configuration, as shown in Fig. 7-8. Since there are now two V_{BE} junction voltages to contend with, the variation may be as high as ± 0.2 V. It is a good idea to keep $V_E \geq 3$ V in Darlington amplifier circuits to swamp out this variation.

Example 7-6

For the circuit of Fig. 7-8, find V_C, Z_{in}, and A_v.

Solution Voltage divider $R_{B1} - R_{B2}$ is analyzed first. I_{B1} is neglected in the face of the much larger $I_{R(B)2}$:

$$\frac{V_S}{R_{B1} + R_{B2}} = \frac{V_{B1}}{R_{B2}}; \qquad \frac{12\text{ V}}{860\text{ k}\Omega} = \frac{V_{B1}}{300\text{ k}\Omega}$$

$$V_{B1} = 4.2\text{ V}$$

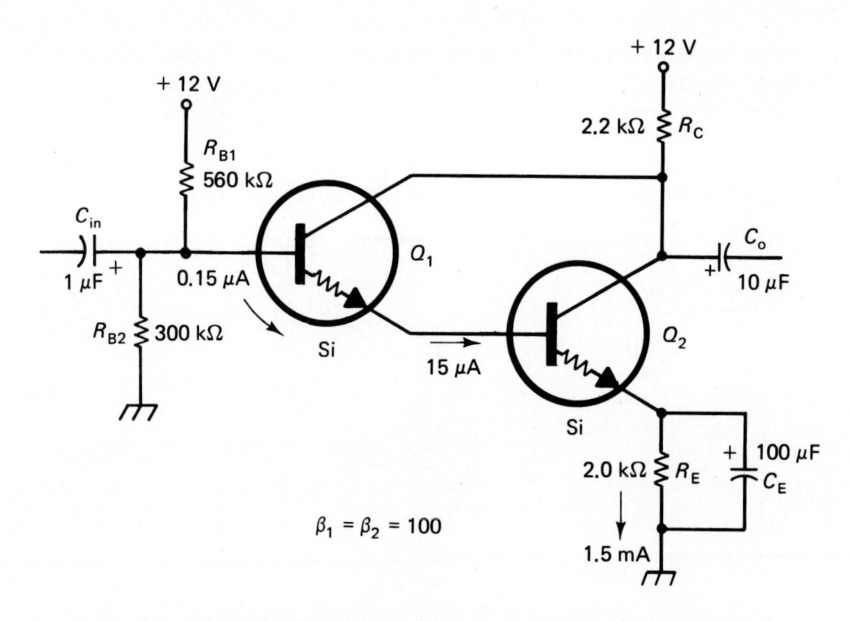

Figure 7-8 Darlington pair in a common-emitter amplifier (Example 7-6).

There are two silicon base-emitter drops from V_{B1} to V_{E2}:

$$V_{E2} = V_{R(E)} = V_{B1} - 2V_{BE} = 4.2 - 1.2 = 3.0 \text{ V}$$

$$I_{R(E)} = \frac{V}{R_E} = \frac{3.0 \text{ V}}{2 \text{ k}\Omega} = 1.5 \text{ mA}$$

The collector current nearly equals the emitter current, so V_C can be found:

$$V_{R(C)} = IR = (1.5 \text{ mA})(2.2 \text{ k}\Omega) = 3.3 \text{ V}$$

$$V_C = V_{CC} - V_{R(C)} = 12 - 3.3 = \mathbf{8.7\ V}$$

Using the formula given in Fig. 7-7,

$$r_b = 2\beta^2 \frac{0.03}{I_C} = (20000)\left(\frac{0.03}{0.0015}\right) = 400 \text{ k}\Omega \tag{7-8}$$

$$Z_{in} = R_{B1} \,||\, R_{B2} \,||\, r_b = 560 \,||\, 300 \,||\, 400 = \mathbf{131\ k\Omega}$$

The voltage gain of the pair is dependent on the emitter junction resistances of the transistors:

$$r_{j2} = \frac{0.03}{I_{E2}} = \frac{0.03}{0.0015} = 20 \ \Omega$$

$$I_{E1} = I_{B2} = \frac{I_{E2}}{\beta_2} = \frac{1.5 \text{ mA}}{100} = 0.015 \text{ mA}$$

$$r_{j1} = \frac{0.03}{0.015 \times 10^{-3}} = 2 \times 10^3 = 2 \text{ k}\Omega$$

$$r_{b2} = \beta r_{j2} = (100)(20) = 2 \text{ k}\Omega$$

Notice that $r_{j1} = r_{b2}$. This means that only half the voltage at the base of Q_1 appears at the base of Q_2. The overall gain of the pair is therefore half of the gain of Q_2 alone:

$$A_v = \frac{1}{2}\frac{R_C}{r_{j2}} = \frac{2200}{(2)(20)} = 55 \tag{7-9}$$

7.7 THE PHASE SPLITTER

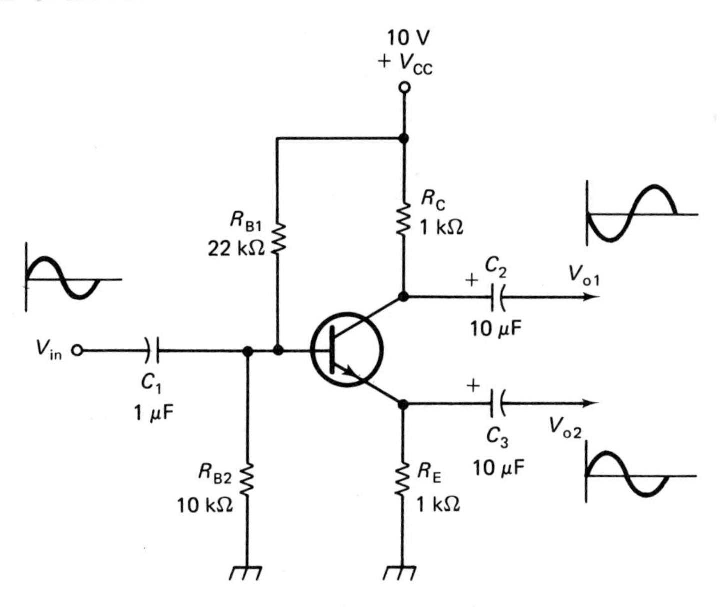

Figure 7-9 Phase-splitter amplifier. $R_C = R_E$, so the two outputs are equal but inverted from one another.

Audio and instrumentation systems often find need for a circuit that will provide two output signals, identical except that one will be inverted from the other. Such a circuit is the *phase splitter*, also called a *phase inverter* or *paraphase amplifier*, shown in Fig. 7-9. It is really nothing more than an emitter follower combined with a gain-of-1 common-collector amplifier. To ensure that $V_{o1} = V_{o2}$ it is necessary that

$$R_C \,\|\, R_{L1} = R_E \,\|\, R_{L2}$$

The collector signal V_{o1} is inverted from the input, but the emitter signal is non-inverted.

7.8 THE COMMON-BASE CIRCUIT

In the early years of transistor technology the common-base circuit was more popular than the common-emitter or common-collector configurations, perhaps because it is inherently bias stable and will function quite well with transistor betas as low as 5 or 10. The input impedance of the circuit is extremely low, however, making it impractical to cascade common-base amplifiers, with the output of one driving the input of the next. Now that high-beta transistors with low leakage currents are readily available, the common-emitter circuit is generally preferred over the common-base circuit, because of its more reasonable input impedance.

The common-base circuit does have a few characteristics which encourage its application in modern designs:

1. It provides voltage gain without the waveform inversion produced by the common-emitter circuit.
2. It has an ability to retain its full gain at higher frequencies than the common-emitter circuit. This leads to its frequent application in radio-frequency amplifiers. Since amplifiers in the range 10 to 100 MHz are normally coupled by tuned transformers anyway, the problem of matching the low Z_{in} of the common-base amplifier at these frequencies is easily taken care of by a step-down transformer.
3. The collector breakdown voltage in the common-base circuit is V_{CBO}, which is typically 150 to 200% of the common-emitter limit V_{CEO}. A transistor which handles a 200-V supply in common emitter could, therefore, handle 300 or perhaps 400 V in common base.
4. Collector dynamic impedance r_c is higher in the common-base circuit than it is in the common-emitter circuit by a factor of β. The voltage gain in both circuits is $(r_c \| R_C \| R_L)/R_{e(line)}$. If R_C and R_L are high, the shunting effect of r_c may cause gain changes in the common-emitter circuit. The common-base circuit can be made more gain stable in this regard. Also, truly fantastic voltage gains (above 10 000) can be achieved in common base if R_L is eliminated and R_C is replaced by a high-impedance choke.

Although the capacitor-coupled circuit of Fig. 7-10 has few applications, it will provide a basis for understanding the transformer-coupled common-base circuits.

Alpha: Notice that the equations given with Fig. 7-10 do not even include the term beta. The relevant term in the common-base circuit is alpha (α), which is defined as the foward current gain from emitter to collector:

$$\alpha = \frac{I_C}{I_E} \tag{7-10}$$

The term h_{FB} is generally used for alpha in engineering literature. Alpha and beta are directly related quantities:

$$\alpha = \frac{\beta}{1+\beta} \quad \text{or} \quad \beta = \frac{\alpha}{1-\alpha} \tag{7-11}$$

For any value of beta above 10, alpha is nearly unity; alpha drops significantly below 1 only for extremely low values of beta. Thus a typical range of beta for a given transistor type might be from 250 to 40, but this would be reflected as an alpha change

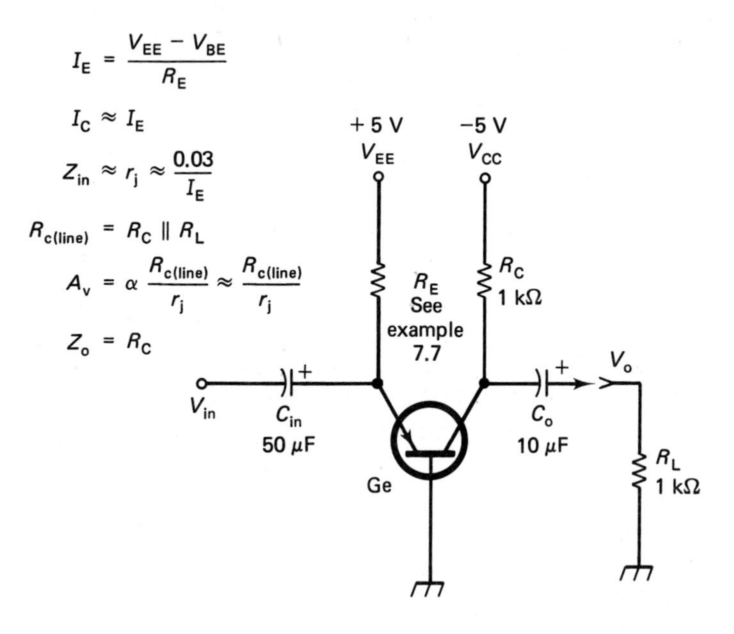

Figure 7-10 Elementary common-base amplifier with bias, gain, and impedance formulas.

from 0.996 to 0.975. This fact explains the relative bias stability of the common-base circuit. The correspondence between alpha and beta is listed for several values as a convenient reference:

β	α
100	0.99
50	0.98
20	0.95
10	0.91
5	0.83
2	0.67

The voltage gain of the common-base circuit does depend on alpha, but since modern transistors all have betas much higher than 10, we can assume alpha to be unity in most cases.

Since the ac input signal looks directly into the emitter of the transistor in this circuit, the input impedance is simply r_j in parallel with R_E. In most cases, R_E is high enough to be neglected.

Example 7-7

In Fig. 7-10, $R_C = 1\ \text{k}\Omega$, $R_L = 1\ \text{k}\Omega$, and the supplies are plus and minus 5 V. The transistor is germanium with $\beta = 40$. Find a suitable value of R_E, and also A_v and Z_{in}.

Solution If V_C is biased at -2.5 V,

$$I_C = \frac{V_C}{R_C} = \frac{2.5\ \text{V}}{1\ \text{k}\Omega} = 2.5\ \text{mA}$$

Since I_E is nearly equal to I_C,

$$R_E = \frac{V_{EE} - V_{BE}}{I_E} = \frac{5\ \text{V} - 0.2\ \text{V}}{2.5\ \text{mA}} = \mathbf{1.92\ k\Omega}$$

$$Z_{in} = \frac{0.03}{I_E} = \frac{0.03}{2.5\ \text{mA}} = 0.012\ \text{k}\Omega = \mathbf{12\ \Omega}$$

$$A_v = \frac{R_{c(\text{line})}}{r_j} = \frac{R_C \parallel R_L}{r_j} = \frac{500}{12} = \mathbf{42}$$

7.9 TRANSFORMER-COUPLED INPUT

The common-base amplifier provides an excellent and not-too-difficult opportunity to see how transformer coupling is used in amplifiers. The circuit is shown in Fig. 7-11. The bias is obtained with the standard voltage-divider circuit first presented in the common-emitter amplifier of Fig. 6-10. However C_B grounds the base for ac and the transformer secondary applies the input signal to the emitter, making the circuit a common-base amplifier.

The transformer is used to step up the otherwise low input impedance of the amplifier. According to basic transformer theory the impedance seen looking into

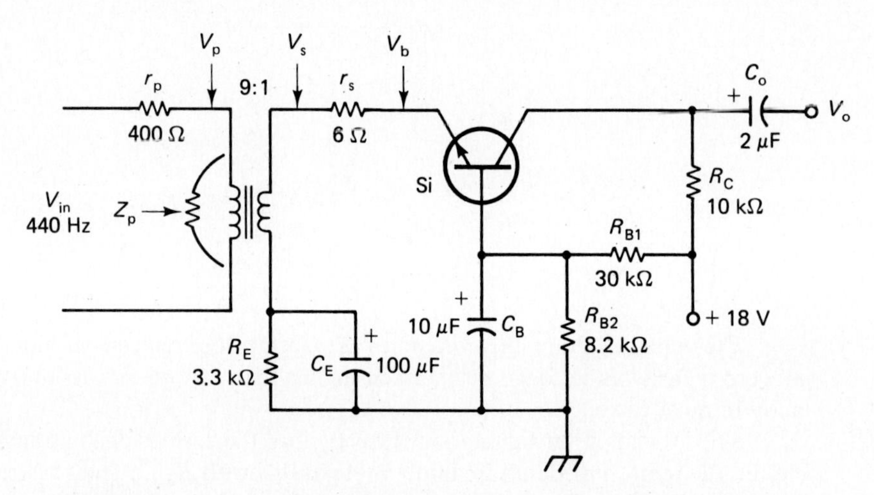

Figure 7-11 A transformer can be used to make a very low resistance look more reasonable, but winding-resistance losses must be considered.

the primary is given by

$$Z_P = \frac{R_o}{n^2} \tag{7-12}$$

where R_o is the resistance seen looking out of the secondary and n is the turns ratio N_S/N_P. The Z_P appears as a reflected resistance across the transformer primary, and the winding resistance r_P forms a voltage divider with it. The transformer must be selected so that at the lowest signal frequency X_L of the secondary is at least three times the emitter-input resistance. Lower values of transformer inductance will load the input signal seriously.

The transformer-winding resistances are not negligible in most practical cases, so we do not neglect them here. A dc-ohmmeter measurement of the winding resistances is generally accurate up to about 10 kHz for transformers in the 1-pound and under category. At higher frequencies skin-effect resistance comes into play and the true ac resistance may be many times the dc resistance.

Example 7-8

Analyze the circuit of Fig. 7-11 for V_C, A_v, and Z_{in}.

Solution

$$V_B = V_{CC}\frac{R_{B2}}{R_{B1} + R_{B2}} = (18)\frac{8.2}{30 + 8.2} = 3.9 \text{ V}$$

$$V_E = V_B - V_{BE} = 3.9 - 0.6 = 3.3 \text{ V}$$

$$I_E = \frac{V_E}{R_E} = \frac{3.3}{3.3} = 1.0 \text{ mA}$$

$$V_C = V_{CC} - I_C R_C = 18 - (1.0)(10) = 8 \text{ V}$$

$$r_j \approx \frac{30 \text{ mV}}{I_E} = \frac{30 \text{ mV}}{1 \text{ mA}} = 30\ \Omega$$

$$Z_{(\text{at S})} = r_S + r_j = 6 + 30 = 36\ \Omega$$

$$Z_{(\text{at P})} = Z_S\left(\frac{N_P}{N_S}\right)^2 = (36)\left(\frac{9}{1}\right)^2 = 2.9 \text{ k}\Omega \tag{7-12}$$

$$Z_{in} = r_P + Z_P = 0.4 + 2.9 = \mathbf{3.3\ k\Omega}$$

$$A_{v(\text{S to out})} = \frac{R_C}{r_S + r_j} = \frac{10\,000}{6 + 30} = 278$$

$$A_{v(\text{P to out})} = A_{v(\text{S to out})}\frac{N_S}{N_P} = (278)\left(\frac{1}{9}\right) = 31$$

$$A_{v(\text{in to out})} = A_{v(\text{P to out})}\frac{Z_P}{r_P + Z_P} = (31)\frac{2.9}{0.4 + 2.9} = \mathbf{27.2}$$

7.10 THE *h*-PARAMETER TECHNIQUE

Advanced circuit analysis long ago figured out a way to completely describe the behavior of any *linear two-port* network by means of four *hybrid* parameters. A *linear* network is one composed of V_s, I_s, R, L, and C components whose values do not change with voltage or current. A *two-port* network is one having two pairs of terminals—input and output in this case. The term *hybrid* indicates that the four parameters are dissimilar quantities—one has units of ohms, one has units of siemens, and two are unitless ratios.

A great deal was written about amplifier analysis with h parameters in the early years of transistor technology. Apparently, there were many who hoped that transistors would soon be developed with parameters stable enough to make the h-parameter approach practical. Unfortunately, such transistors have not been developed. Parameter changes by a factor of 2 or more from unit to unit and from minimum to maximum specified temperature are still common, and developments do not seem to be moving toward markedly better stability. Furthermore, transistors are not linear devices as required by h-parameter theory. They are approximately linear only for ac signals which are much smaller than the dc bias levels, and only if the dc bias levels remain fixed. Thus h-parameter analysis is limited to small-signal ac analysis where the parameters of an individual transistor have been measured at the specific temperature and bias point where the device is to be used.

Modern transistor circuit design tends to rely on swamping and feedback techniques to stabilize the behavior of devices which are inherently unstable, rather than on attempts to stabilize the devices themselves. Thus h-parameter analysis is somewhat less popular than it once was. Nevertheless, we have inherited a number of terms from the h-parameter theory, and it would be well to give them some attention.

The *h* parameters for a transistor in the common-emitter connection are:

h_{ie} = input impedance V_b/I_b; output shorted for ac
h_{fe} = forward current transfer ratio I_c/I_b; output shorted for ac
h_{re} = reverse voltage transfer ratio V_b/V_c; input open for ac
h_{oe} = output admittance I_c/V_c; input open for ac

A rough correspondence exists between three of these parameters and three that we have been using in our approximation analyses:

$$h_{ie} \approx r_{be} = \beta r_j = \beta \frac{30 \text{ mV}}{I_E} \tag{7-13}$$

$$h_{fe} \approx \beta = \frac{I_C}{I_B} \tag{7-14}$$

$$h_{oe} \approx \frac{1}{r_c} = \frac{\Delta I_C}{\Delta V_C} \quad \text{from the characteristic curves} \tag{7-15}$$

We have no approximate parameter corresponding to h_{re} because the ratio of V_c fed back to V_b in real transistors is so small as to be negligible.

Similar sets of four h parameters exist for the common-base and common-collector connections, and these are sometimes specified in manufacturers' literature. The following equations will permit conversion to the more useful common-emitter parameters.

$$h_{ie} = h_{ic} \approx \frac{h_{ib}}{1 + h_{fb}} \tag{7-16}$$

$$h_{fe} = -h_{fc} - 1 \approx \frac{-h_{fb}}{1 + h_{fb}} \qquad (7\text{-}17)$$

$$h_{oe} = h_{oc} \approx \frac{h_{ob}}{1 + h_{fb}} \qquad (7\text{-}18)$$

$$h_{re} = 1 - h_{rc} \approx \frac{h_{ib}h_{ob}}{1 + h_{fb}} - h_{rb} \qquad (7\text{-}19)$$

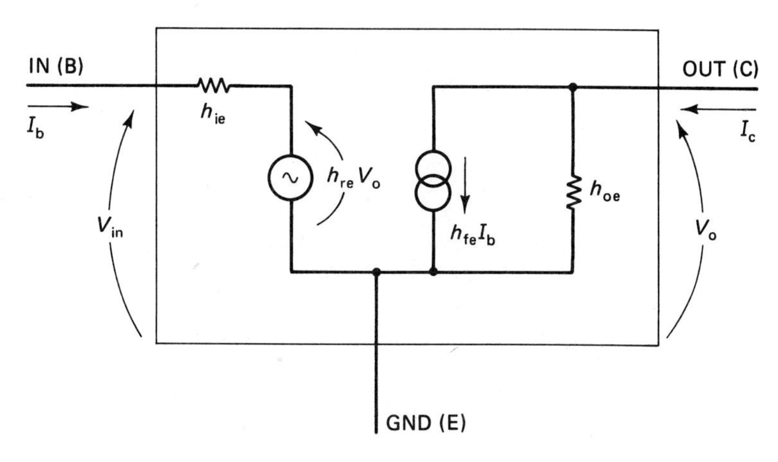

Figure 7-12 The h-parameter ac-equivalent circuit for a transistor in the common-emitter connection.

The h-parameter equivalent circuit for a transistor in the common-emitter configuration is shown in Fig. 7-12. The equivalent circuit for the common-collector configuration is similar except that IN, GND, and OUT are B, C, and E, respectively, and the parameters are h_{ic}, h_{rc},, h_{fc} and h_{oc}. In common-base IN, GND, and OUT are E, B, and C and the parameters are h_{ib}, h_{rb}, h_{fb}, and h_{ob}. The input resistance (IN to GND), voltage gain (IN to OUT), and output resistance (OUT to GND) are given by the equations below. Z_s is the source resistance driving the IN terminal of the transistor and r_L is the ac resistance seen looking from the OUT to the GND terminal.

$$r_{in} = h_i - \frac{h_f h_r r_L}{1 + h_o r_L} \qquad (7\text{-}20)$$

$$A_v = \frac{-h_f r_L}{h_i + (h_i h_o - h_f h_r) r_L} \qquad (7\text{-}21)$$

$$r_o = \frac{1}{h_o - \dfrac{h_f h_r}{h_i + Z_s}} \qquad (7\text{-}22)$$

Circuit calculations using h parameters must take into account biasing components which are external to the transistor, as well as the transistor characteristics given by the equations above. An example will serve to illustrate the technique.

Example 7-9

Use the h-parameter technique to find Z_{in}, A_v, and Z_o for the common-emitter amplifier of Fig. 7-13(a). All the capacitors have virtually zero reactance. The transistor parameters at the bias point are: $h_{ie} = 2000\ \Omega$, $h_{re} = 5 \times 10^{-4}$, $h_{fe} = 80$, and $h_{oe} = 30\ \mu S$.

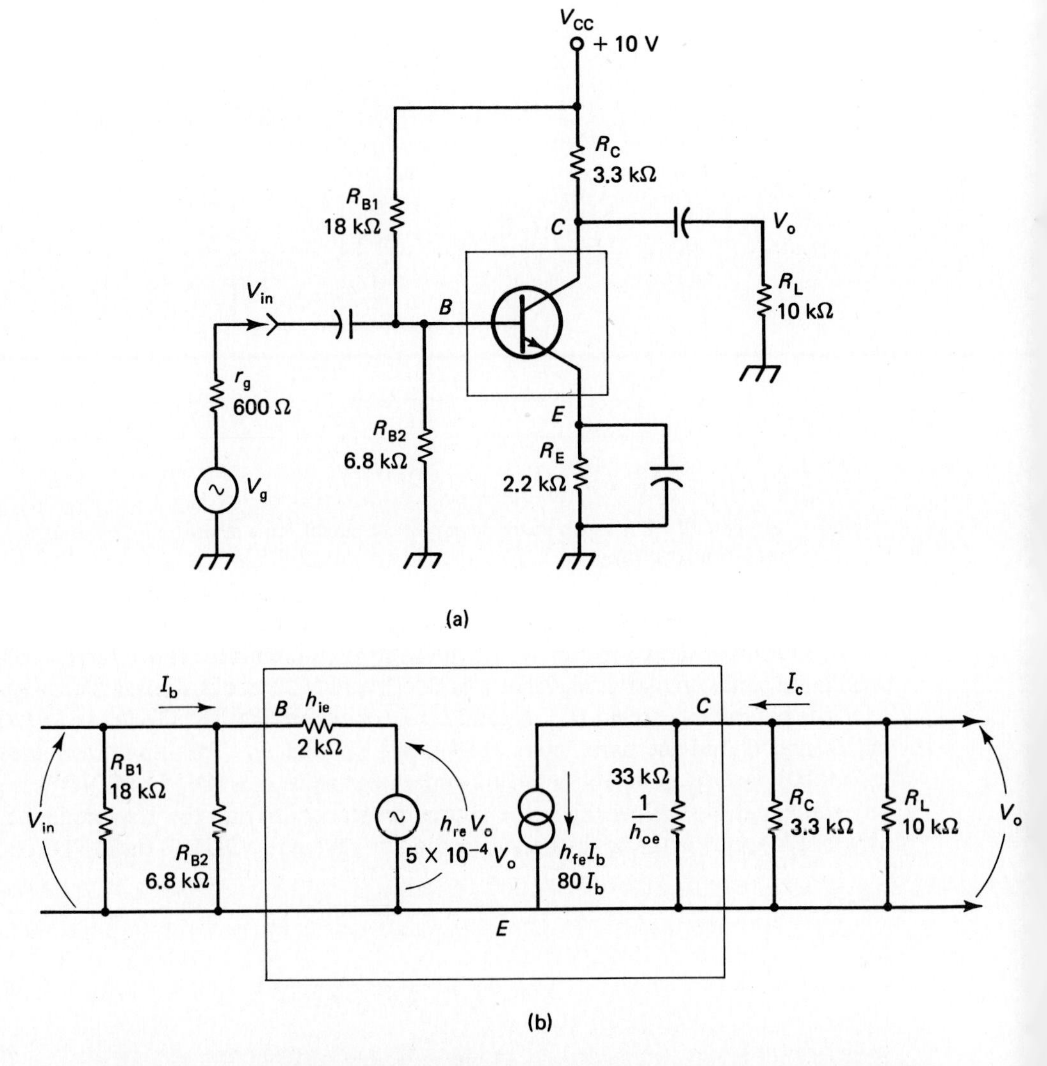

Figure 7-13 (a) Amplifier to be analyzed by the h-parameter technique; (b) h-parameter ac-equivalent circuit for the entire amplifier.

Solution Figure 7-13(b) shows the h-parameter ac-equivalent circuit for the complete amplifier.

$$r_L = R_C \,\|\, R_L = 3300 \,\|\, 10\,000 = 2481\ \Omega$$

$$r_{in} = h_{ie} - \frac{h_{fe} h_{re} r_L}{1 + h_{oe} r_L} \tag{7-20}$$

$$= 2000 - \frac{(80)(5 \times 10^{-4})(2481)}{1 + (30 \times 10^{-6})(2481)} = \mathbf{1908\ \Omega}$$

$$Z_{in} = R_{B1} \,||\, R_{B2} \,||\, r_{in} = 18\,000 \,||\, 6800 \,||\, 1908 = 1376\ \Omega$$

$$A_v = \frac{-h_{fe} r_L}{h_{ie} + (h_{ie} h_{oe} - h_{fe} h_{re}) r_L} \tag{7-21}$$

$$= \frac{-(80)(2481)}{(2000) + [(2000)(30 \times 10^{-6}) - (80)(5 \times 10^{-4})](2481)}$$

$$= \mathbf{-96.8}$$

$$Z_s = r_g \,||\, R_{B1} \,||\, R_{B2} = 600 \,||\, 18\,000 \,||\, 6800 = 535\ \Omega$$

$$r_o = \frac{1}{h_{oe} - \dfrac{h_{fe} h_{re}}{h_{ie} + Z_s}} = \frac{1}{(30 \times 10^{-6}) - \dfrac{(80)(5 \times 10^{-4})}{2000 + 535}}$$

$$= 70\,320\ \Omega$$

$$Z_o = r_o \,||\, R_C = 70\,320 \,||\, 3300 = \mathbf{3152\ \Omega}$$

CHAPTER SUMMARY

1. The emitter follower (or common-collector) amplifier has the input at the base and the output at the emitter. It is used to drive a low-impedance load while presenting a relatively higher impedance to the source. Voltage gain is slightly less than one, noninverting. Bias is obtained from base voltage dividers and an emitter stabilizing resistor, as in the common-emitter amplifier. There is generally no collector resistor.
2. Voltage gain, input impedance, and output impedance for the emitter follower may be calculated from

$$A_v = \frac{R_E \,||\, R_L}{r_j + R_E \,||\, R_L} \tag{7-1}$$

$$Z_{in} = r_b \,||\, R_{B1} \,||\, R_{B2}$$

where

$$r_b = \beta(r_j + R_E \,||\, R_L)$$

$$Z_o = R_E \left|\right| \left(r_j + \frac{R_{B1} \,||\, R_{B2} \,||\, Z_s}{\beta}\right) \tag{7-3}$$

3. Bootstrapping can be used to increase the input impedance of a common-emitter or emitter-follower amplifier by nearly eliminating the shunting effect of the base-bias resistors. Bias stability is degraded slightly by the addition of bootstrap resistor R_{B3}. Input impedance may be calculated as

$$Z_{in} = R_{B3}\left(\frac{r_j + R_{e(line)}}{r_j}\right) \left|\right| \beta(r_j + R_{e(line)}) \tag{7-4}$$

where $R_{e(line)}$ is the ac resistance seen looking out of the emitter terminal to ground. For a common-emitter amplifier $R_{e(line)} = R_{E1}$, the unbypassed emitter resistor. For an emitter follower, $R_{e(line)} = R_E \| R_L$.

4. Collector self-bias improves bias stability of a common-emitter amplifier by connecting the top end of base resistor R_{B1} to the collector rather than to V_{CC}.

5. The Miller effect makes any impedance from output to input of an inverting amplifier appear to be lower by a factor of $A_v + 1$ when viewed from the input. A collector-self-bias resistor R_{B1} from collector to base is such an impedance.

6. Inverted ground biasing has the supply at the emitter side and ground at the collector side. It permits use of *PNP* transistors with positive supplies and *NPN* transistors with negative supplies. Dual supplies (one positive ground, the other negative ground) are often employed in larger systems. A bypass capacitor may be used to establish ac-signal ground in these circuits.

7. The Darlington pair produces effectvely a single transistor with a gain of β^2. It is formed by feeding the emitter of Q_1 to the base of Q_2, and connecting the collectors together.

8. A phase splitter (paraphase amplifier) is a combination emitter follower and gain-of-1 common-emitter amplifier. It provides two equal outputs, one inverted from the other.

9. The common-base amplifier has an extremely low input impedance but it may permit higher frequency response, higher supply voltages, higher voltage gain, and better bias stability than other circuits. The input signal is applied at the emitter and the output is taken from the collector. The base is grounded, at least for ac. Input impedance and voltage gain are found from

$$Z_{in} \approx r_j$$

$$A_v = \frac{R_C \| R_L}{r_j}$$

10. Beta changes have much less effect on the common-base circuit than on other circuits. The relevant factor is alpha (α), which is directly obtainable from β.

$$\alpha = \frac{I_C}{I_E} = \frac{\beta}{1 + \beta} \tag{7-13}$$

11. Transformer coupling is often used to increase Z_{in} of a common-base amplifier. The impedance seen looking into the primary is

$$Z_{in} = r_P + r_{refl}$$

where

$$r_{refl} = \frac{r_S + r_o}{n^2}$$

Turns ratio $n = N_S/N_P$, r_{refl} is the resistance reflected to the primary, r_o is resistance seen looking out of the secondary (into the amplifier proper), and r_P and r_S are the primary and secondary winding resistances. The transformer reduces voltage gain by

$$\frac{r_{refl}}{r_P + r_{refl}} (n) \frac{r_P}{r_S + r_o}$$

where the first factor represents primary-resistance loss, the second is the step-down ratio, and the third represents secondary-resistance loss.

12. A set of four h (hybrid) parameters completely specifies the behavior of a transistor. Unfortunately, these parameters vary widely from unit to unit and with bias-point changes. Many manufacturers continue to specify h parameters, for example: h_{fe} for β, h_{oe} for collector resistance r_c, and h_{ie} for base-input resistance r_{be}.

QUESTIONS AND PROBLEMS

7-1. Draw a schematic diagram of an emitter follower (as in Fig. 7-1) with the following values: $Z_s = 600\ \Omega$, $R_{B2} = 10\ \text{k}\Omega$, $R_E = 2.2\ \text{k}\Omega$, $R_L = 1\ \text{k}\Omega$, $V_{CC} = +12$ V, and Q_1 is a silicon *NPN* with $\beta = 90$. V_E is to be 8.0 V. Find the following: (a) R_{B1}; (b) A_v; (c) Z_{in}; (d) Z_o; (e) S; (f) $V_{o(max)p\text{-}p}$.

7-2. List two advantages and one disadvantage of the emitter follower compared to the common-source amplifier.

7-3. Draw a schematic diagram of an emitter follower (Fig. 7-1) with bootstrap components added. Use a *PNP* transistor and indicate the polarity on the supply and on all capacitors.

7-4. For the circuit drawn in Problem 7-3, mark $R_{B3} = 330\ \Omega$ and $C_B = 100\ \mu\text{F}$. Calculate the new Z_{in} and S. Don't forget that R_L is in parallel with R_E for ac. Therefore, R_{E1} in equation (7-4) must be replaced with $R_E \| R_L$.

7-5. What is the Miller effect? What causes it?

7-6. Explain how collector self-bias improves bias stability in a common-emitter amplifier.

7-7. In the inverted-ground circuit of Fig. 7-6(b), $V_{CC} = +9$ V, $R_{B1} = 10\ \text{k}\Omega$, $R_{B2} = 10\ \text{k}\Omega$, $R_E = 2.2\ \text{k}\Omega$, $R_C = 1\ \text{k}\Omega$, and $\beta = 75$. Find A_v and Z_{in}.

7-8. Redraw the emitter follower of Fig. 7-1 replacing Q_1 with a Darlington compound Q_1–Q_2, both silicon with $\beta = 40$. Find the new V_E, A_v, Z_{in}, and Z_o. Compare the results with those of Example 7-1.

7-9. Find V_E and V_C for the phase splitter of Fig. 7-9.

7-10. List four advantages and one big disadvantage of the common-base amplifier circuit.

7-11. What is α (alpha) for a transistor with $\beta = 30$?

7-12. For the common-base circuit of Fig. 7-10, $V_{EE} = -12$ V, $V_{CC} = +12$ V, $R_E = 68\ \text{k}\Omega$, $R_C = 33\ \text{k}\Omega$, and $R_L = 50\ \text{k}\Omega$. Find Z_{in} and A_v.

7-13. Draw the circuit of a 2:1 step-down transformer with $r_P = 10\ \Omega$ and $r_S = 5\ \Omega$. Add a 1-V no-load source with a 50-Ω internal resistance to the primary and a 15-Ω R_L to the secondary. Calculate $V_{R(L)}$

7-14. In a common-base circuit similar to Fig. 7-11 the transformer is as specified in Problem 7-13, $V_{CC} = +12$ V, $R_E = 270\ \Omega$, $R_C = 820\ \Omega$, $R_{B1} = 2.2\ \text{k}\Omega$, and $R_{B2} = 560\ \Omega$. Find A_v and Z_{in}.

7-15. A transistor data sheet lists the following parameters at $V_{CE} = 5$ V and $I_C = 1$ mA: $h_{ib} = 40\ \Omega$, $h_{fb} = -0.98$, $h_{ob} = 0.6\ \mu\text{s}$, and $h_{rb} = 1 \times 10^{-6}$. Find β, base-input resistance r_{be}, and collector output resistance r_c in the common-emitter configuration.

8

FIELD-EFFECT TRANSISTORS

8.1 JUNCTION FETS

As early as 1939, William Shockley attempted to build a device that would use the electrostatic *field* around a metal wire screen to control the current through it. The attempt failed, as did a number of other attempts along the same lines throughout the 1940s. In fact, it was an experiment designed to investigate the failure of one of these attempts to build a *field-effect* transistor that led to the development of the bipolar transistor in 1947, Shockley finally laid the foundation for a successful field-effect transistor in 1952, but production difficulties and preoccupation with the bipolar transistor kept the field-effect transistor pretty much in the background until the mid 1960s.

Construction: The basic field-effect transistor (FET) is fabricated as a sheet of *N*-type silicon with a connection terminal at each end, as shown in Fig. 8-1(a). Current is free to flow in either direction through this sheet, the dc resistance being typically several hundreds of ohms. A deposit of *P*-type impurity along the top of the sheet has the ability to limit the current through the sheet if the resulting *P-N* junction is reverse-biased.

Operation: As explained in Section 2.8, the negative voltage on the *P* side of the junction and the relatively more positive voltage on the *N* side will cause the charge carriers to be drawn away from the vicinity of the junction, leaving a region depleted of charge carriers which will not pass current.

As the negative voltage on this *gate* element is increased, the electron charge carriers are drawn farther away from the junction, further constricting the channel

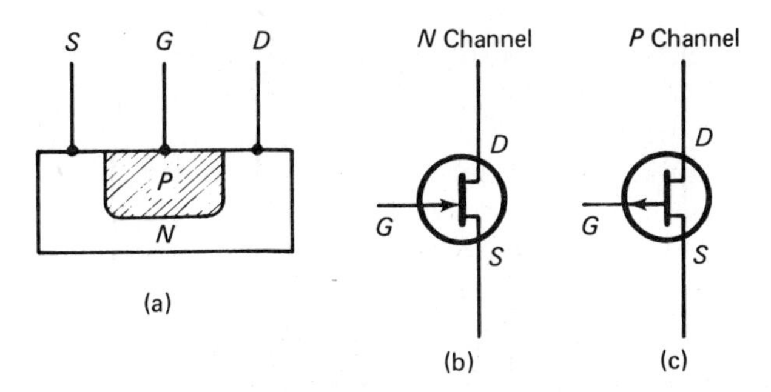

Figure 8-1 Junction field-effect transistors: (a) basic structure; (b) *N*-channel schematic symbol; (c) *P*-channel symbol.

width which remains conductive, until eventually a *pinch-off* voltage is reached at which essentially no current is able to flow through the channel.

The schematic symbol for an *N*-channel FET is shown in Fig. 8-1(b). The end of the channel which injects the negative charge carriers is called the *source* and the end which collects the charge carriers is termed the *drain.*

Since all the current through the FET is confined to a single polarity of charge carrier (*N* type in this case), the FET is referred to as a *unipolar* transistor. This is in contrast to the conventional *bipolar* transistor in which current flows through both *N*- and *P*-type materials as it passes through the emitter, base, and collector regions.

Although the *N*-channel FET is more common, it is possible to construct a *P*-channel FET with an *N*-doped gate as shown in Fig. 8-1(c).

The characteristic curves of a typical *N*-channel junction FET are shown in Fig. 8-2. Notice that for drain-to-source voltages below 2 or 3 V, the curves show a steep slope, indicating low dynamic resistance and low voltage gain. This is in contrast to the 0.2- to 0.3-V saturation regions typical of bipolar transistors. Above 4 or 5 V the FET curves are horizontal, indicating a high dynamic drain resistance.

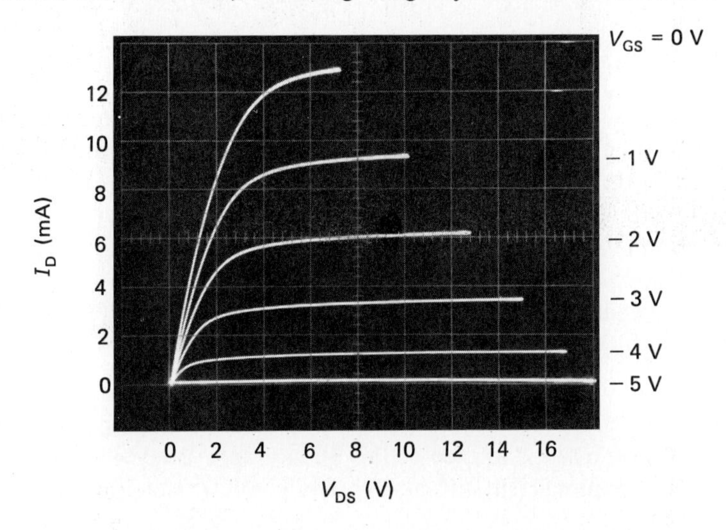

Figure 8-2 Characteristic curves for a 2N3819 *N*-channel junction FET.

FET parameters: As with most solid-state devices, the characteristic curves for a particular type of FET must be regarded as only typical for that type, since the actual characteristics may vary considerably from unit to unit. FET specifications more commonly give the maximum and minimum values for a number of device parameters, from which design calculations can be made. The most important of these parameters from an ac signal point of view is *transconductance*, g_{fs}, which is also called *mutual conductance*, g_m, and *forward transfer admittance*, y_{fs}. This parameter specifies how much output current change will be produced by a given input voltage change:

$$y_{fs} = g_{fs} = \frac{\Delta I_D}{\Delta V_{GS}}\bigg|\, V_{DS} = \text{constant} \tag{8-1}$$

The subscript letters f and s indicate that the parameter specifies a *forward* comparison (i.e., output over input) in the common *source* circuit configuration.

It must be emphasized that y_{fs} is an ac property of the FET and that it is not useful in calculating dc bias conditions. In this respect it differs markedly from beta of the bipolar transistor, which is essentially the same for ac or dc. For bias calculations with the FET, two parameters are important. The first of these is the gate to source pinch-off voltage $V_{GS(OFF)}$ (sometimes called V_P) at which the FET draws essentially zero drain current. This is often specified as V_{GS} with $I_D = 100\ \mu A$ or some other small drain current. The second dc parameter is the drain current with zero gate bias voltage, I_{DSS}. Both of these parameters vary widely for most FET types and will have maximum and minimum specifications.

Parameter relationships: The relation between the controlling gate voltage and the controlled drain current is expressed by the equation

$$I_D = I_{DSS}\left(1 - \frac{V_{GS}}{V_{GS(off)}}\right)^2 \tag{8-2}$$

This *square-law* relation is in marked contrast to the simple (though approximate) linear relation $I_C = \beta I_B$ of the bipolar transistor. Figure 8-2 shows how I_D is more greatly affected by V_{GS} when V_{GS} is near zero, as predicted by equation (8-2). In using this equation it must be understood that maximum values of I_{DSS} occur with maximum values of $V_{GS(off)}$, and likewise for minimums.

Equation (8-2) may be solved for V_{GS} with the help of the quadratic formula to yield an equation giving gate bias required for a specified drain current.

$$V_{GS} = V_{GS(off)}\left(1 - \sqrt{\frac{I_D}{I_{DSS}}}\right) \tag{8-3}$$

The transconductance of the FET at any gate voltage can be predicted if the zero-bias drain current and pinch-off gate voltage are known.

$$y_{fs} = \frac{2I_{DSS}}{V_{GS(off)}}\left(1 - \frac{V_{GS}}{V_{GS(off)}}\right) \tag{8-4}$$

8.2 MOS DEVICES

Further development of the junction field-effect transistor has resulted in an FET which does not draw gate current for either positive or negative input voltages. This device has variously been called an IGFET (insulated-gate field-effect transistor), MOSFET (metal-oxide silicon field-effect transistor), MOST (metal-oxide silicon transistor), and simply MOS (metal-oxide semiconductor). The references to metal oxide indicate that the gate is separated from the source-drain channel by a very thin layer of silicon dioxide, which is an electrical insulator.

For an *N*-channel MOS, negative voltages applied across the insulating layer are still capable of pinching off current flow in the thin drain-source channel by forcing the negative charge carriers out of the neck of the channel. In addition to this *depletion* mode of operation, it is also possible to attract additional *N*-charge carriers to the channel region by placing a positive voltage on the gate, thus *enhancing* the conductivity of the MOS. Junction FETs are not operated in the enhancement mode because this would forward-bias the gate diode and result in a large input current. The insulated gate of the MOS, however, permits either polarity of gate voltage to be applied with no input current resulting.

An MOS designed to be operated strictly in the depletion mode is termed a type *A* FET. An FET designed to be operated in either the depletion or enhancement mode is known as type *B*. These types are given the schematic symbol shown in Fig. 8-3(a). Type *A* and *B* FETs are also known as "normally on" FETs since they pass considerable drain current with zero gate-source voltage.

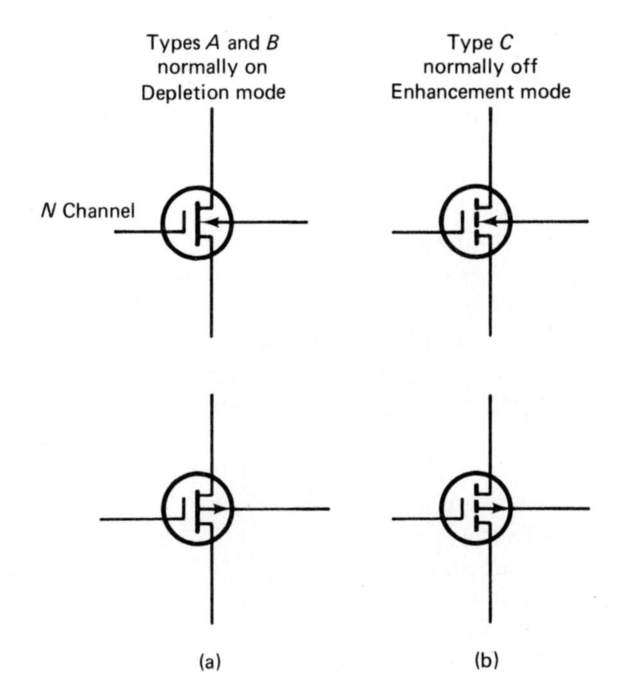

Figure 8-3 MOS field-effect transistors: (a) depletion *N*-channel and *P*-channel types; (b) enhancement types.

An MOS which passes negligible drain current with zero gate-source voltage and must be biased forward into conduction is termed an enhancement or type *C* FET and is given the schematic symbol shown in Fig. 8-3(b). Enhancement-mode MOSFETS begin to turn on at a V_{GS} specified as V_{TH}, the threshold voltage. The $V_{GS(off)}$ and I_{DSS} specifications of depletion-mode devices are not applicable. Drain current for a type *C* FET is given by the square-law equation

$$I_D = k(V_{GS} - V_{TH})^2 \tag{8-5}$$

where k is a constant which must be determined for each FET by substitution of known values of the other three terms. V_{TH} ranges from 0.1 to 6 V for commonly available type *C* FETs.

Incidentally, you should be wary of the natural tendency to call depletion MOS devices "DMOS" and enhancement MOS devices "EMOS." "DMOS" is a term used by manufacturers to designate a *double diffusion* fabrication process—a DMOS device may be enhancement mode.

The fourth lead shown in the symbols is the substrate upon which the device is constructed. Normally, this substrate lead is connected together with the source of the FET.

Static protection: One precaution must be stressed when working with or even just handling MOS devices. The oxide layer insulating the gate from the channel is extremely thin and can be punctured by voltages on the order of 100 V. Such destructive potentials can be found on soldering pencils with two-wire ungrounded power cords and can even appear from electrostatic buildup on clothing if the humidity is low enough to produce static shocks. Many manufacturers are supplying MOS devices with clips around the base pins to prevent insulation breakdown. These clips should be left in place until after the MOS has been soldered in the circuit. Some MOSFETs have internal *zener diodes* (see Section 15.1) across the terminals to protect against damage from electrostatic discharge, but these diodes increase leakage and deteriorate high-frequency response.

VMOS devices: For many years the FET was strictly a low-power device, valued for its nearly infinite input resistance and, in certain applications, for its square-law transfer characteristic. The development of the VMOS transistor has given us high-input-impedance devices with current specifications of several amperes and voltage limitations in the 100-V range. The "V" in the acronym stands for "vertical," indicating that the current path within the device is vertical, rather than horizontal as it is in low-power FETs.

VMOS devices are generally enhancement-mode (type *C*) FETs with a turn-on threshold V_{TH} of 1 to 5 V. Saturation voltage at $I_D = 1$ A is typically 2 to 4 V, somewhat higher than bipolars of comparable power rating. Forward transconductance is surprisingly high—0.2 to 2 S typical, as opposed to 5 to 10 mS for low-power FETs. One drawback is input capacitance, which may be 50 to 1000 pF—enough to make the device difficult to drive at high frequencies and fast switching rates.

8.3 COMMON-SOURCE FET BIASING

The basic bias circuit for a depletion-mode FET (such as the common *N*-channel JFET) is similar to that for the bipolar transistor, with the following reservations.

1. The FET requires no input current of the gate-bias resistors, so loading is not a problem and the bias resistors can have values in the megohm range.
2. A silicon *NPN* bipolar has a voltage drop of 0.5 to 0.7 V from base to emitter. An *N*-channel JFET has a voltage *rise* of 0.5 to 8 V from gate to source. By choosing a slightly more expensive type this 7.5-V range may be reduced to 2 or 3 V, but the 0.2-V range of the bipolar is so far only a dream.

A word of warning about FET biasing is in order here. With bipolars it may be possible to "fudge" a bias circuit together which will work for most transistors. Worst-case calculations are recommended, of course, but there are certainly "designers" who have done this and been lucky enough to get away with it. No such luck can be hoped for with FETs. FET parameters vary widely and there is no chance that a circuit fudged together for one FET will work with all FETs of that type.

A common-source bias circuit (analogous to the common-emitter circuit of Fig. 6-10) is shown in Fig. 8-4. The supply voltage is quite high and a scandalous amount of it is wasted across R_S, but this is necessary if any bias stability at all is to be expected. Figure 8-4(b) shows an equivalent bias circuit that is especially convenient where dual-polarity supplies are available.

Elaborate equations and charts can be developed to relate bias voltage V_G to I_D for the circuits fo Fig. 8-4, but the accuracy of the results obtainable thereby do not justify such efforts. The following technique is entirely adequate and far simpler.

1. Make certain that $I_{DSS(min)}$ for the FET is larger than the drain current required by the circuit. A factor of 2 or more is recommended.
2. Estimate the maximum $V_{R(S)}$ as $V_G + V_{GS(off)max}$. Estimate the maximum $V_{R(D)}$ as $V_{R(S)}(R_D/R_S)$. Make certain that $V_{DD} - V_{R(D)} - V_{R(S)}$ leaves enough voltage V_{DS} to keep the FET in the active region. Minimum is 2 V, even for small signals.
3. Calculate maximum and minimum values of V_{GS} from equation (8-3) using maximum, then minimum values of $V_{GS(off)}$ and I_{DSS}. I_D may be an estimated value obtained from reasonable expectations of $V_{R(D)}$ or $V_{R(S)}$, or it may be a design target value.
4. Analyze the circuit for maximum and minimum I_D, $V_{R(D)}$, $V_{R(S)}$, and V_{DS}. If I_D turns out to be more than 50% different from the estimated I_D, return to step 3 with a new estimate.

Example 8-1

The FET of Fig. 8-4(a) has the following specifications:

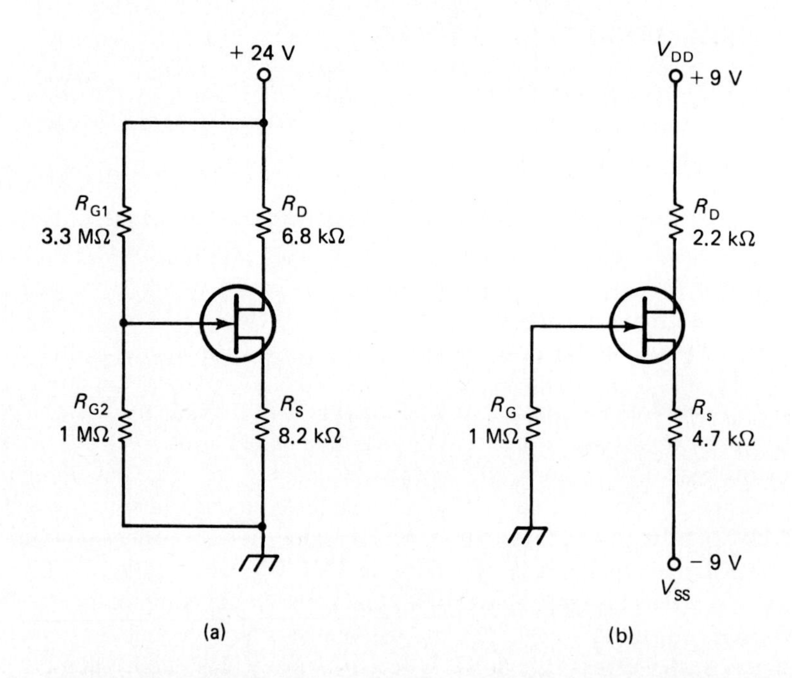

Figure 8-4 FET-bias circuits can be similar to bipolar-bias circuits, except that there is a voltage *rise* from G to S, which is larger and much less predictable than the V_{BE} drop. (a) Single-supply; (b) dual-supply circuit.

	minimum	*maximum*
$V_{GS(off)}$	1.0 V	7.0 V
I_{DSS}	3.0 mA	12 mA

Find the minimum $V_{R(D)}$ (at minimum I_{DSS}) and minimum V_{DS} (at maximum I_{DSS}).

Solution

$$V_G = V_{CC}\frac{R_{G2}}{R_{G1} + R_{G2}} = 24\left(\frac{1}{3.3 + 1}\right) = 5.6 \text{ V}$$

Estimating V_{GS} as 3 V average yields an estimate of I_D:

$$V_{R(S)} = V_G + V_{GS} \approx 3 + 5.6 = 8.6 \text{ V}$$

$$I_D = I_S = \frac{V_{R(S)}}{R_S} \approx \frac{8.6}{8.2} = 1.05 \text{ mA}$$

At minimum I_{DSS},

$$V_{GS} = V_{GS(off)}\left(1 - \sqrt{\frac{I_D}{I_{DSS}}}\right) = 1\left(1 - \sqrt{\frac{1.05}{3}}\right) = 0.41 \text{ V} \qquad \text{(8-3)}$$

$$V_{R(S)} = V_G + V_{GS} = 5.6 + 0.41 = 6.0 \text{ V}$$

$$I_D = I_S = \frac{V_S}{R_S} = \frac{6.0}{8.2} = 0.73 \text{ mA}$$

$$V_{R(D)} = I_D R_D = (0.73)(6.8) = \mathbf{5.0 \text{ V minimum}}$$

At maximum I_{DSS}:

$$V_{GS} = V_{GS(off)}\left(1 - \sqrt{\frac{I_D}{I_{DSS}}}\right) = 7\left(1 - \sqrt{\frac{1.05}{12}}\right) = 4.9\text{ V} \tag{8-3}$$

$$V_{R(S)} = V_G + V_{GS} = 5.6 + 4.9 = 10.5\text{ V}$$

$$I_D = I_S = \frac{V_S}{R_S} = \frac{10.5}{8.2} = 1.28\text{ mA}$$

$$V_{R(D)} = I_D R_D = (1.28)(6.8) = 8.7\text{ V maximum}$$

$$V_{DS} = V_{CC} - V_{R(D)} - V_{R(S)} = 24 - 8.7 - 10.5 = \mathbf{4.8\text{ V minimum}}$$

The FET is not in danger of cutoff ($V_{R(D)} = 0$) or saturation ($V_{DS} < 3$ V). These results could not easily be obtained experimentally because FETs at each limit of the specifications are not easily obtainable in the sample batches available at the product-design stage. At the production stage, however, the worst-case limits are sure to appear.

A bias stability factor for FETs may be defined, similar to that given for bipolars in equation (6-17). For Fig. 8-4 this becomes

$$S = \frac{I_{D(min)}}{I_{D(max)}} = \frac{V_G + V_{GS(min)}}{V_G + V_{GS(max)}} \tag{8-5}$$

where V_G is the gate bias voltage with respect to the bottom end of R_S. $V_{GS(min)}$ and $V_{GS(max)}$ should ideally be the voltages at the target drain current. These can be calculated from equation (8-2), but it is usually adequate in small-signal amplifiers to use the minimum- and maximum-specified values of $V_{GS(off)}$ in place of V_{GS} values.

Example 8-2

Determine the bias stability factor for the circuit of Fig. 8-4(a) using (1) the $I_{D(min)}$ and $I_{D(max)}$ values from Example 8-1, (2) equation (8-5) and the V_{GS} values calculated in Example 8-1, and (3) equation (8-5) and the values of $V_{GS(off)}$ specified in Example 8-1.

Solution

1. $S = \dfrac{I_{D(min)}}{I_{D(max)}} = \dfrac{0.73\text{ mA}}{1.28\text{ mA}} = \mathbf{57\%}$

2. $S = \dfrac{V_G + V_{GS(min)}}{V_G + V_{GS(max)}} = \dfrac{5.6 + 0.41}{5.6 + 4.9} = \mathbf{57\%}$

3. $S \approx \dfrac{V_G + V_{GS(off)min}}{V_G + V_{GS(off)max}} = \dfrac{5.6 + 1}{5.6 + 7} = \mathbf{52\%}$

8.4 THE COMMON-SOURCE JFET AMPLIFIER

Input impedance: A complete FET amplifier is shown in Fig. 8-5. The input impedance is simply the parallel combination of the two gate resistors, since $r_g \longrightarrow \infty$.

$$Z_{in} = R_{G1} \,||\, R_{G2} \tag{8-6}$$

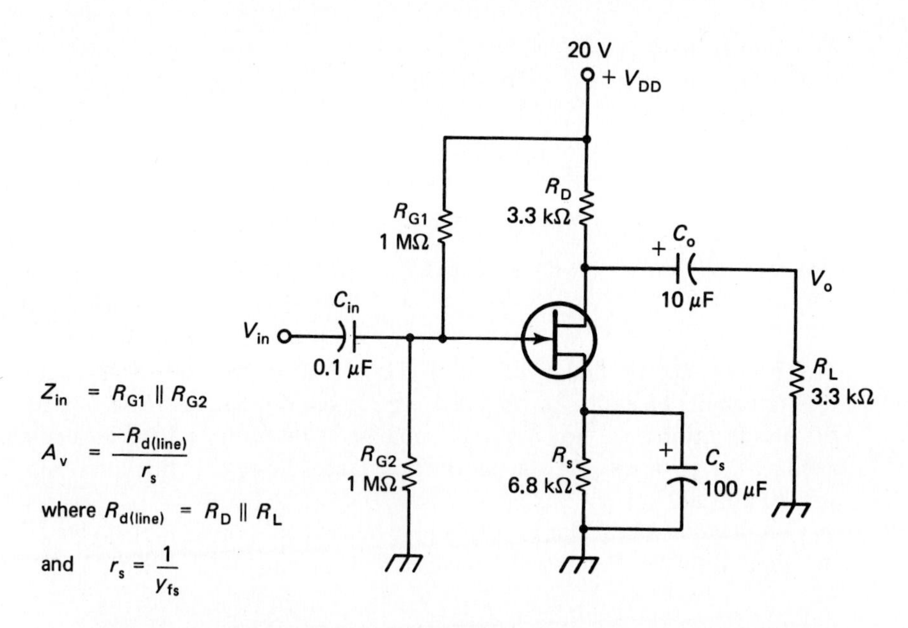

Figure 8-5 Stabilized common-source FET amplifier. Imput impedance is much higher than comparable bipolar circuits, but voltage gain and bias stability are poorer.

Voltage gain: With R_S bypassed, all of V_{in} appears as V_{gs}. Remembering that $y_{fs} = I_d/V_{gs}$, it is obvious that $I_d = y_{fs}V_{in}$. The ac output current I_d flows through the drain-line resistance $R_D \| R_L \| r_d$, where r_d is the dynamic resistance of the FET drain. The specification y_{os} (output admittance) given by the manufacturers can be translated into r_d by the simple equation

$$r_d = \frac{1}{y_{os}} \tag{8-7}$$

or r_d can be calculated from the slope of the characteristic-curve lines, $\Delta V_D/\Delta I_D$. Typical r_d values for small-signal FETs range from 5 kΩ at low V_{DS} and high I_D to 100 kΩ at high V_{DS} and low I_D. Often r_d is negligible in comparison to lower R_D and R_L values, but this is less likely to be so than in the case of r_c of the bipolar transistor. Combining the information above yields

$$A_v = \frac{V_o}{V_{in}} = \frac{-I_d R_{d(line)}}{V_{gs}} = -y_{fs}(R_D \| R_L \| r_d) \tag{8-8}$$

Source resistance r_s: To keep the FET analysis similar to the bipolar analysis, and to deal with partially bypassed source resistance and source-follower amplifiers, it will be helpful to define a quantity r_s analogous to the ac emitter-junction resistance r_j of the bipolar transistor.

$$r_s = \frac{1}{y_{fs}} \tag{8-9}$$

It should be understood that r_s is a purely fictitious resistance, even though it is very useful in visualizing FET ac-signal behavior. The equation for voltage gain in Fig. 8-5 using the r_s concept becomes

$$A_v = \frac{-(R_D \| R_L \| r_d)}{r_s} \tag{8-10}$$

Large-signal operation of the circuit of Fig. 8-5 will result in changing y_{fs} as the signal swings over wide ranges of I_D. (This is assuming that large-signal swings can be accommodated under all bias conditions.) Changing y_{fs} values will produce changes in A_v—higher for more positive input signal swings in an N-channel device. The positive input (negative output) peaks will therefore be emphasized over the opposite-polarity swings. This effect can be minimized by leaving part of R_S unbypassed, as shown in Fig. 8-6. The voltage-gain equation then becomes

$$A_v = \frac{-R_D \| R_L \| r_d}{r_s + R_{S1}} \tag{8-11}$$

Often the inclusion of significant R_{S1} reduces A_v to a value far below 10.

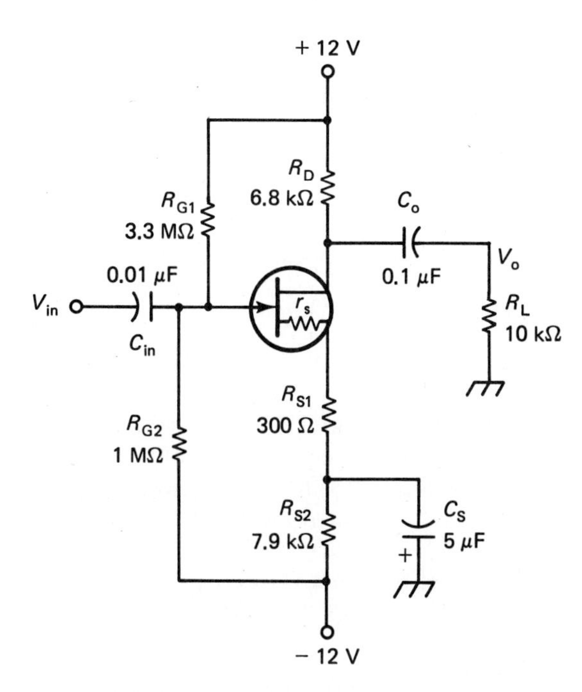

Figure 8-6 Adding an unbypassed source resistor will minimize distortion of large signals. R_{S1} must be larger than r_s, and an already-low gain will become even lower.

Example 8-3

The bias conditions for the circuit of Fig. 8-6 are the same as those for Fig. 8-4(a), except that the ground reference has been shifted to −12 V. Find minimum A_v using the data and V_{GS} value from Example 8-1, and $y_{os(max)} = 0.1$ mS.

Solution

$$y_{fs(min)} = \frac{2I_{DSS(min)}}{V_{GS(off)min}}\left(1 - \frac{V_{GS(min)}}{V_{GS(off)min}}\right) \quad (8\text{-}4)$$

$$= \frac{(2)(3 \times 10^{-3})}{1}\left(1 - \frac{0.41}{1}\right) = 3.5 \text{ mS}$$

$$r_{s(max)} = \frac{1}{y_{fs(min)}} = \frac{1}{3.5 \text{ mS}} = 286\ \Omega \quad (8\text{-}9)$$

$$r_{d(min)} = \frac{1}{y_{os(max)}} = \frac{1}{0.1 \text{ mS}} = 10 \text{ k}\Omega \quad (8\text{-}7)$$

$$A_v = \frac{-R_D \| R_L \| r_d}{r_s + R_{S1}} = \left(\frac{-6.8 \| 10 \| 10}{0.286 + 0.30}\right) = \mathbf{-4.9} \quad (8\text{-}11)$$

8.5 IMPROVING FET BIAS STABILITY

By now it must be apparent that biasing the FET presents a considerable problem. We have seen that a typical low-cost FET just barely avoids saturation or cutoff under worst-case conditions with a total supply voltage of 24 V—and we have not even considered temperature effects on bias. Better stability can be achieved by dropping more voltage across R_S, but this would require even higher supply voltages, and many systems operate with a single supply of 12 or 9 V.

The collector-self-bias approach (Section 7.3) which was so successful with bipolar transistors is of little help with FETs because the unit-to-unit V_{GS} variations which we are trying to offset are generally larger than the tolerable variations in V_D which would be fed back to offset them. Three approaches are available to solve the bias problem:

1. Select FETs with a narrow range of $V_{GS(off)}$. The 2N4393, for example, has $V_{GS(off)}$ specifications of 0.5 V minimum and 3.0 V maximum. Relative prices at this writing are about $0.50 for the low-cost types vs. $1.50 for the 2N4393.
2. Abandon the objective of achieving voltage gain in an FET amplifier. Use a source follower with a gain of 1 (Section 8.6) for high Z_{in} and build up A_v with a bipolar stage.
3. Use current-source bias.

Current-source bias is shown in Fig. 8-7. If an FET with $V_{GS(off)}$ of 3 V is used, it permits the operation of reliable common-source amplifiers on supply lines as low as 8 V. If the supply is raised to the range 12 to 15 V, FETs with $V_{GS(off)}$ to 7 V can be accommodated.

The key to bias stability is the bipolar transistor Q_2, which is biased exactly as described in Section 6.9. The collector current of Q_2, which is held stable by voltage divider R_{B1}–R_{B2} and resistor R_E, is also the drain current of the FET. The voltage at the Q_2 collector is free to vary as required by the V_{GS} of Q_1. The current I_C, and hence

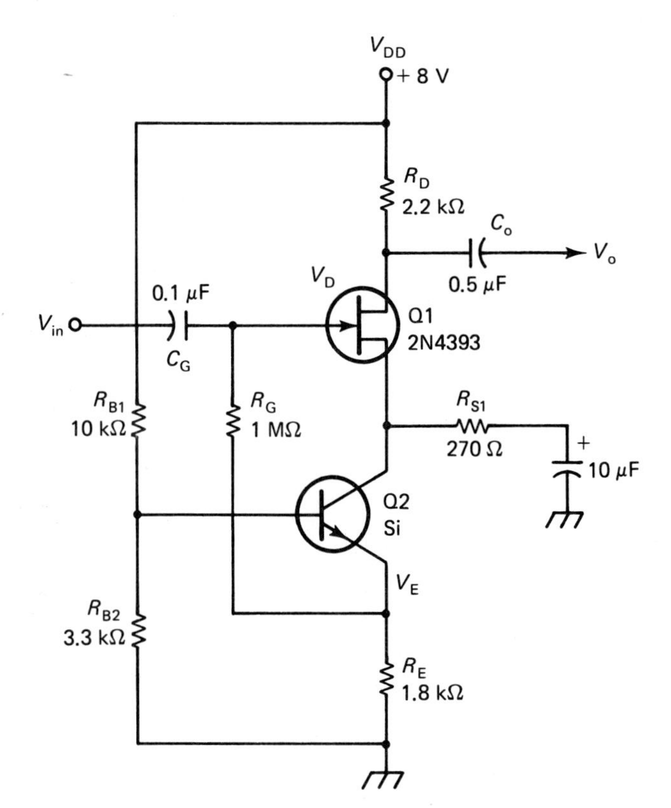

Figure 8-7 A bipolar-transistor current source stabilizes the bias current of the unruly FET.

I_D, $V_{R(D)}$, and V_D are all fixed, provided only that I_D is less than I_{DSS} of the FET. A series path from V_{DD} to ground includes voltage drops $V_{R(D)}$, V_{DS}, V_{SG}, and $V_{R(E)}$, and it is necessary that the sum of these four equal V_{DD}. Minimum values are $V_{R(D)}$: 1 V small signal; V_{DS}: 2 V to avoid saturation; V_{SG}; 3V for selected FETs, 7 V for low-cost types; $V_{R(E)}$: 2 V for high stability, 1 V minimum.

Capacitor C_S is required since the Q_2 collector acts like a nearly infinite ac resistance in the FET source. If it is desired to limit A_v or to improve linearity by swamping out changes in r_s, R_{S1} may be placed in series with C_S. Note that in this position R_{S1} does not add to the voltage drop in the dc bias circuit.

Example 8-4

Find $V_{R(D)}$, V_{DS}, and A_v for the circuit of Fig. 8-7. Assume that $V_{GS(off)} = 3.0$ V and $I_{DSS} = 30$ mA.

Solution

$$V_B = V_{DD}\frac{R_{B2}}{R_{B1} + R_{B2}} = (8)\frac{3.3}{10 + 3.3} = 1.98 \text{ V}$$

$$V_E = V_B - V_{BE} = 1.98 - 0.6 = 1.38 \text{ V}$$

$$I_E = I_D = \frac{V_E}{R_E} = \frac{1.38}{1.8} = 0.77 \text{ mA}$$

$$V_{R(D)} = I_D R_D = (0.77)(2.2) = \mathbf{1.7\ V}$$

$$V_{GS} = V_{GS(off)}\left(1 - \sqrt{\frac{I_D}{I_{DSS}}}\right) = 3\left(1 - \sqrt{\frac{0.77}{30}}\right) = 2.52 \text{ V} \qquad (8\text{-}3)$$

$$V_{DS} = V_{DD} - V_{R(D)} - V_{GS} - V_{R(E)}$$

$$= 8 - 1.7 - 2.52 - 1.38 = \mathbf{2.4\ V}$$

$$y_{fs} = \frac{2I_{DSS}}{V_{GS(off)}}\left(1 - \frac{V_{GS}}{V_{GS(off)}}\right) \qquad (8\text{-}4)$$

$$= \frac{(2)(30 \times 10^{-3})}{3}\left(1 - \frac{2.52}{3}\right) = 3.2 \text{ mS}$$

$$r_S = \frac{1}{y_{fs}} = \frac{1}{3.2 \text{ mS}} = 312\ \Omega$$

$$A_v = \frac{-R_D}{r_s + R_{S1}} = \frac{-2200}{312 + 270} = \mathbf{-3.8}$$

8.6 THE SOURCE-FOLLOWER AMPLIFIER

The disadvantages of the FET common-source amplifier are many:

- Bias instability
- Supply requirements above 8 V, even for special FET types
- Voltage-gain instability
- Distortion, except on very small signals
- Low voltage gain

All of these problems except the last are greatly alleviated in the common-drain configuration, more often called a source follower. This circuit, shown in several bias arrangements in Fig. 8-8, is the FET counterpart of the emitter follower discussed in Section 7.1.

The simple circuit of Fig. 8-8(a) requires a V_{DD} supply of only $V_{GS(max)} + V_{DS}$, which amounts to 3 + 2 or 5 V if low-V_{GS} FETs are used. This means that it can be operated from the +5-V supplies which have become standard since the advent of digital integrated circuits. Its bias current I_{DS} is entirely at the mercy of the FET parameters, however. This is not as serious a problem for a source follower as it is for the common-source amplifier since saturation will not occur unless V_{GS} nears V_{DD}, and cutoff will not occur at all. It is shown in Appendix E that for Fig. 8-8(a),

$$I_D = \frac{I_{DSS}}{\left(0.5 + \sqrt{0.25 + \dfrac{I_{DSS}R_S}{V_{GS(off)}}}\right)^2} \qquad (8\text{-}12)$$

which, for large R_S simplifies to $I_D = V_{GS(off)}/R_S$. It is recommended that R_S be chosen to limit $I_{D(max)}$ to a few milliamperes at $V_{GS(off)max}$.

$$R_S = \frac{V_{GS(off)max}}{I_{D(max)}} \qquad (8\text{-}13)$$

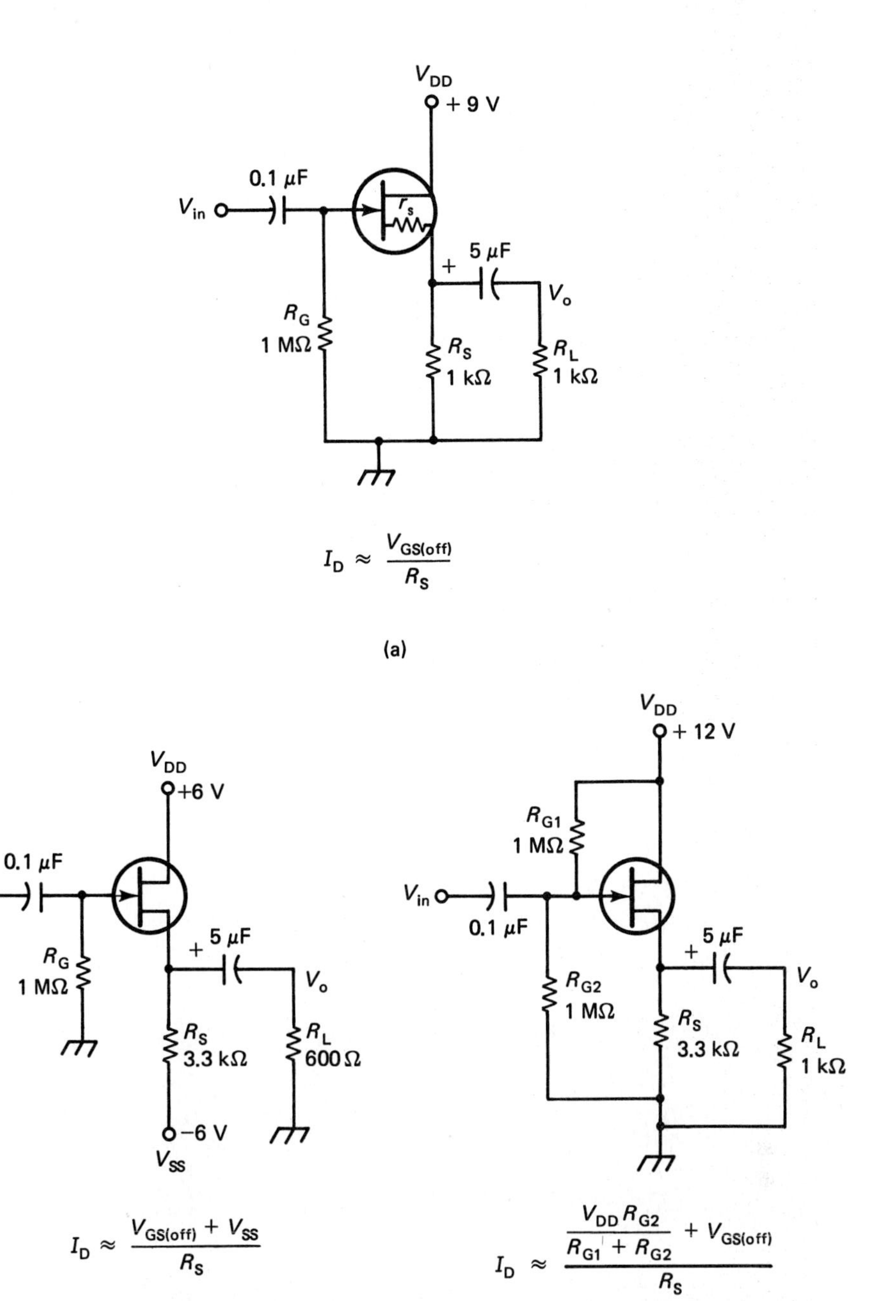

Figure 8-8 The source follower is much more tolerant of bias instability than the common-source amplifier. (a) Simple bias system: V_S and I_S may vary widely. (b) Two-supply system: bias is more stable. (c) More-stable single-supply system: higher V_{DD} supply is required.

The voltage gain for all the circuits of Fig. 8-8 is noninverting and less than 1.

$$A_v = \frac{R_S \| R_L}{r_s + (R_S \| R_L)} \tag{8-14}$$

If $V_{in(p\text{-}p)}$ is not many times smaller than the dc bias $V_{R(S)}$, the input signal will cause wide variations in i_S which will change r_s and hence A_v. The circuits of Fig. 8-8(b) and (c) minimize changes in i_S and r_s (while at the same time improving bias stability) by dropping a voltage much larger than V_{GS} across R_S.

The output impedance of the source followers is of interest since they are usually employed to permit a high-impedance source to drive a low-impedance load. The development of Z_o here is similar to that given with the emitter follower for equation (7-3), except that we conceive of the FET as a transistor with infinite beta ($I_S/I_G \longrightarrow \infty$). Thus any impedance at the gate, no matter how high, appears as Z_g/∞ or zero when viewed looking back into the source. For all the circuits (Fig. 8-8)

$$Z_o = R_S \| r_s \tag{8-15}$$

A super source follower with a gain very near unity can be built by keeping all the impedances in the source line many times greater than r_s, as in Fig. 8-9. Note that R_S is replaced with a current-source transistor Q_2 and R_L is replaced with the 40-kΩ r_b of emitter follower Q_3. Even if r_s is 400 Ω (a rather high value), A_v will be 0.99 for the FET. Assuming 6 V for V_E, $r_j = 0.03/5\text{ mA} = 6\ \Omega$ and A_v for the emitter follower is 0.98, giving an overall gain of 0.97.

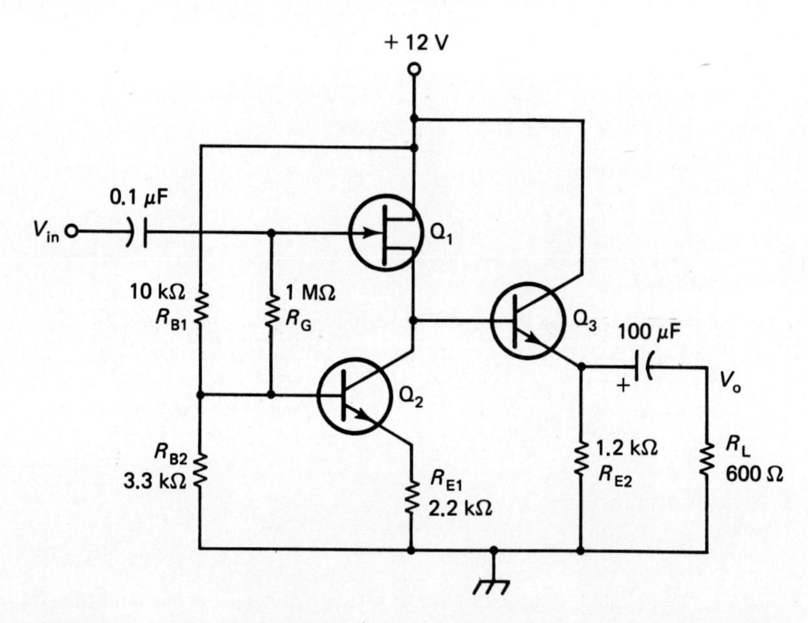

Figure 8-9 Super source follower with current-source bias and emitter-follower output.

Example 8-5

Find V_{DS}, A_v, and Z_o for the circuit of Fig. 8-8(b). The supplies are ± 6 V, $R_S = 3.3$ kΩ, $V_{GS(off)} = 5$ V, and $I_{DSS} = 20$ mA.

Solution Estimating

$$I_D \approx \frac{V_{GS(off)} + V_{SS}}{R_S} = \frac{5+6}{3.3} = 3.3 \text{ mA}$$

we have

$$V_{GS} = V_{GS(off)}\left(1 - \sqrt{\frac{I_D}{I_{DSS}}}\right) = 5\left(1 - \sqrt{\frac{3.3}{20}}\right) = 3.0 \text{ V} \tag{8-3}$$

$$V_{DS} = V_{DD} - V_S = 6 - 3 = \mathbf{3\ V}$$

$$y_{fs} = \frac{2I_{DSS}}{V_{GS(off)}}\left(1 - \frac{V_{GS}}{V_{GS(off)}}\right) \tag{8-4}$$

$$= \frac{(2)(20 \times 10^{-3})}{5}\left(1 - \frac{3}{5}\right) = 3.2 \text{ mS}$$

$$r_s = 312\ \Omega$$

$$Z_o = R_S \| r_s = 3300 \| 312 = \mathbf{285\ \Omega}$$

$$A_v = \frac{R_S \| R_L}{r_s + R_S \| R_L} = \frac{3300 \| 600}{285 + 3300 \| 600} = \mathbf{0.64}$$

8.7 MOSFET BIAS CIRCUITS

Type *A* and *B* MOSFETs can be operated in the circuits given previously for junction FETs. Type-*C* MOSFETs (enhancement mode) require forward gate-bias voltage and use differing bias circuits. Typical values of the turn-on gate voltage V_{TH} range

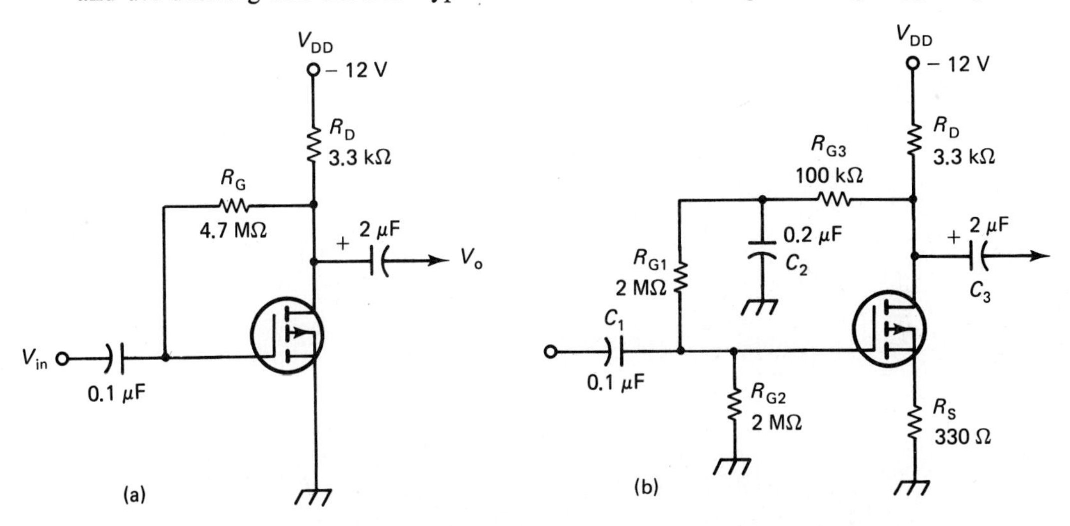

Figure 8-10 Amplifiers using enhancement-mode insulated-gate FETs: (a) simple circuit for small-signal amplifiers; (b) bias-stable circuit with decoupling capacitor C_2 to nullify Miller effect.

from 0.5 V minimum to 1.5 V maximum for some type *C* MOSFETs, and 3 V minimum to 6 V maximum for others. For small-signal amplifiers the high-V_{TH} FETs may be used in the simple circuit of Fig. 8-10(a). The Miller effect will lower Z_{in} to $R_{G1}/(A_v + 1)$.

For larger signals and lower-V_{TH} FETs the circuit of Fig. 8-10(b) is more bias stable. R_S has been added to stabilize A_v against changes in y_{fs}. R_{G3} and C_2 decouple R_{G1} from the output signal, so Z_{in} is simply $R_{G1} \| R_{G2}$ for this circuit.

CHAPTER SUMMARY

1. The principal advantage of the FET is its almost infinite input resistance.

2. The following table compares the silicon bipolar transistor and the junction FET in a common-emitter (source) circuit:

	Bipolar transistor	JFET
Input element	Base	Gate
Common element	Emitter	Source
Output element	Collector	Drain
Positive-supply type	*NPN*	*N*-channel
Negative-supply type	*PNP*	*P*-channel
Zero-bias condition	$I_B = 0$, $I_C = 0$	$V_{GS} = 0$, $I_D = \text{max}$
Bias voltage, *NPN* and *N*-channel devices	V_{BE}, +0.5 to +0.7 V	V_{GS}, −0.5 to −7 V

3. The dc-bias characteristics of an FET are specified by two parameters: I_{DSS}, which is the drain current with $V_{GS} = 0$, and $V_{GS(off)}$, which is the gate-source voltage needed to reduce I_D to a specified value near zero. The gate-source voltage needed to produce a given drain current is then given by

$$V_{GS} = V_{gs(off)}\left(1 - \sqrt{\frac{I_D}{I_{DSS}}}\right) \qquad (8\text{-}3)$$

4. The ac-signal gain of an FET is measured by its forward transfer admittance y_{fs}, which is the ratio of ac drain current to ac gate voltage.

$$y_{fs} = \frac{\Delta I_D}{\Delta V_{GS}} \qquad (8\text{-}1)$$

This value is largest for low values of V_{GS} and high values of I_D. Specifically,

$$y_{fs} = \frac{2I_{DSS}}{V_{GS(off)}}\left(1 - \frac{V_{GS}}{V_{GS(off)}}\right) \qquad (8\text{-}4)$$

5. MOS transistors have the gate insulated from the channel by a metal-oxide-of-silicon layer. They draw even less gate current than junction FETs. MOSFETs are available in three types:

Type	With $V_{GS} = 0$, $I_D =$	I_D increased by:	I_D decreased by:
A	Full value	Not applicable	Reverse V_{GS}
B	Mid-value	Forward V_{GS}	Reverse V_{GS}
C	Zero	Forward V_{GS}	Not applicable

Type *C* FETs have a turn-on threshold specification V_{TH} in lieu of $V_{GS(off)}$. VMOS FETs are type *C* devices capable of operation at high voltages and currents.

6. Bias stability is more difficult to achieve with FETs than with bipolar transistors because V_{GS} varies much more from unit to unit and with temperature than does V_{BE}. This problem is minimized by:

(a) Dropping large voltages across R_S so that changes in V_{GS} are small by comparison (Fig. 8-4),

(b) Using a bipolar transistor as a current source in the FET source line (Fig. 8-7), or

(c) Using the FET in the source-follower circuit (Fig. 8-8) and obtaining voltage gain in a following bipolar stage.

7. Voltage gain for the common-source amplifier is

$$A_v = \frac{-(R_D \| R_L \| r_d)}{r_s} \tag{8-10}$$

where r_d is the dynamic drain resistance of the FET ($\Delta V_{DS}/\Delta I_D$, obtained from the characteristic curves), and r_s is a fictitious source resistance equal to $1/y_{fs}$.

8. Voltage gain for the source follower is

$$A_v = \frac{R_D \| R_L}{r_s + (R_D \| R_L)} \tag{8-14}$$

The output impedance of this amplifier is

$$Z_o = R_s \| r_s \tag{8-15}$$

QUESTIONS AND PROBLEMS

8-1. An *N*-channel junction FET has its source terminal grounded. The normal bias voltage on its drain is __________ and on its gate is __________. To increase drain current, gate voltage must be made more __________. (Answer positive or negative in each blank.)

8-2. Answer Question 8-1 for (a) a *P*-channel junction FET; (b) an *N*-channel enhancement-mode MOSFET; (c) a *P*-channel enhancement-mode MOSFET.

8-3. Read y_{fs} from the curves of Fig. 8-2 at $V_{DS} = 8$ V and $I_D \approx 6$ mA.

8-4. Read $V_{GS(off)}$ and I_{DSS} at $V_{DS} = 8$ V from Fig. 8-2.

8-5. Calculate y_{fs} at $I_D = 6$ mA using equation (8-4) and the data from Problem 8-4. Compare to y_{fs} determined in Problem 8-3.

8-6. Use equation (8-2) and the data from Problem 8-4 to predict I_D at $V_{GS} = -2.0$ V. Read I_D at $V_{GS} = -2.0$ V from the curves of Fig. 8-2 ($V_{DS} = 8$ V) and compare with the predicted value.

8-7. Match the following FET terms into pairs: type *A*, type *B*, type *C*; enhancement mode, depletion mode, depletion/enhancement mode.

8-8. What FET characteristic is indicated by the term "DMOS"? What is indicated by "VMOS"?

8-9. Compare typical V_{BE} for silicon *NPN* junction transistors with typical V_{GS} for *N*-channel junction FETs.

8-10. An FET with the following parameters is placed in the circuit of Fig. 8-4(b).

	Minimum	Maximum
$V_{GS(off)}$	0.5 V	4.0 V
I_{DSS}	4.4 mA	20 mA

Find the minimum and maximum $V_{R(D)}$.

8-11. Compare typical A_v and Z_{in} values for the common-source circuit of Fig. 8-5 and the common-emitter circuit of Fig. 6-11.

8-12. Calculate Z_{in} and A_v for the common-source amplifier of Fig. 8-5. The FET curves are given in Fig. 8-2.

8-13. In Example 8-4 (Fig. 8-7) the FET is replaced with another having $V_{GS(off)} = -0.8$ V and $I_{DSS} = 6$ mA. Find $V_{R(D)}$ and A_v.

8-14. The FET in the source-follower circuit of Fig. 8-8(a) has $V_{GS(off)} = 2.5$ V and $I_{DSS} = 15$ mA. (a) Find I_D using equation (8-12). (b) Find A_v.

8-15. Repeat Problem 8-13 for Fig. 8-8(b).

8-16. Repeat Problem 8-13 for Fig. 8-8(c).

9

DECIBELS AND FREQUENCY LIMITATIONS

9.1 THE DECIBEL SYSTEM

The decibel (dB) system of measure is widely used in the audio, radio, TV, and instrument industries for comparing two signal levels. There are two main points to remember in applying this system:

- A decibel measure is a ratio, not an amount. It tells how many times greater or how many times less a signal is compared to some reference.
- Decibel measure is nonlinear. 20 dB is not twice as much power, voltage, or current as 10 dB.

The original unit "bel" was defined as the logarithm (to base 10) of the power ratio of two signals. This unit proved too large, and was split into tenths, or decibels:

$$\alpha_{\text{dB}} = 10 \log \left(\frac{P_2}{P_1}\right) \qquad \text{dB} \tag{9-1}$$

where P_1 is the reference power (the input signal level, or perhaps an industry-wide standard-reference level) and P_2 is the signal in question (the output signal). A little reflection, or a few moments with a calculator will reveal that α_{dB} is always *positive* for a signal greater than the reference level, and always negative for a signal less than the reference level. Zero dB means equal levels—no gain, no loss. For reference, the transposition of equation 9-1 is given:

$$\frac{P_2}{P_1} = \log^{-1}\left(\frac{\alpha_{\text{dB}}}{10}\right) = 10^{\alpha_{\text{dB}}/10} \tag{9-2}$$

Choosing a log-based system allows a tremendous range of power ratios to be encompassed using only two-digit numbers, without sacrificing discernment at the low end of the scale:

$$1 \text{ dB} = 1.26:1 \text{ power ratio}$$

$$50 \text{ dB} = 100{,}000:1 \text{ power ratio}$$

The graphs of Fig. 9-1 illustrate this point. Both graphs show the same data.

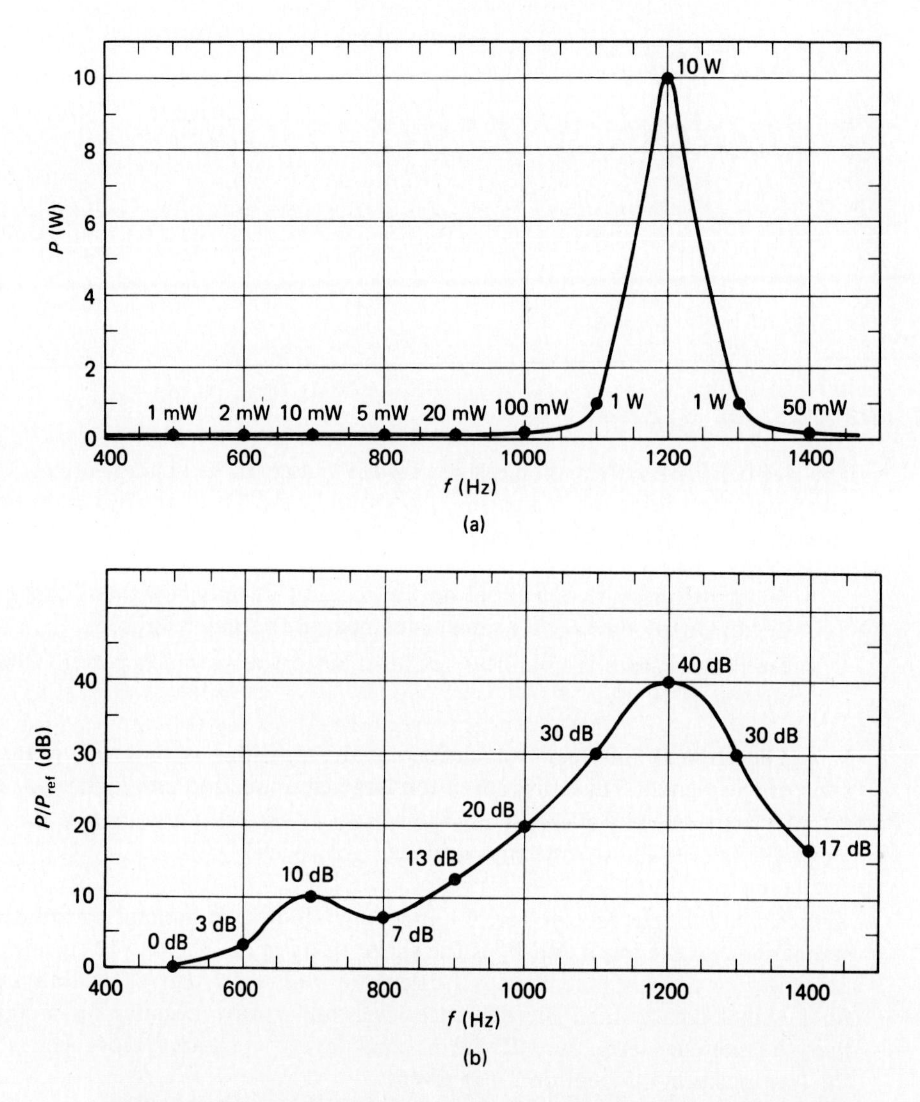

Figure 9-1 (a) A graph of power output versus frequency for a tuned amplifier so compresses the power levels below 100 mW that they cannot be seen. (b) The same data, converted to decibels with 1 mW–0 dB, shows all power levels clearly.

Voltage and decibels: Voltage is much easier to measure than power, so it would be convenient to apply the decibel system to voltage measure. Since $P = V^2/R$, and squaring a number is accomplished by multiplying its logarithm by 2, we can write equations (9-1) and (9-2) as

$$\alpha_{dB} = 20 \log\left(\frac{V_2}{V_1}\right) \tag{9-3}$$

$$\frac{V_2}{V_1} = \log^{-1}\left(\frac{\alpha_{dB}}{20}\right) = 10^{\alpha_{dB}/20} \tag{9-4}$$

These equations rest on the assumption that R_1 for the reference voltage V_1 is the same as R_2 for the measured voltage V_2. This restriction is often ignored in practice, but the results obtained thereby are not correct. If the impedance driven by V_1 is not known to be equal to the impedance driven by V_2, the impedances must be measured, the two power levels calculated, and equation (9-1) applied. If the values of R_1 and R_2 cannot be determined, the decibel system cannot properly be applied.

Example 9-1

An amplifier has a 45-dB gain and a 2.8-mV input signal. The input impedance and load impedance are both 300 Ω. What is the output voltage?

Solution

$$\frac{V_2}{V_1} = \log^{-1}\left(\frac{\alpha_{dB}}{20}\right) = \log^{-1}\left(\frac{45}{20}\right) = 178 \tag{9-4}$$

$$V_2 = 178V_1 = (178)(2.8) = \mathbf{500\ mV}$$

Adding decibels: Signal transmission in a large system involves multiplying by some factor A_v each time an amplifier is encountered and dividing by a loss factor F_v each time a signal splitter, attenuator, or length of cable is encountered. However, if all the system-component voltage ratios (V_o/V_{in}) were specified as decibels (positive for gain, negative for loss), we could determine the total system gain by algebraically *adding* the dB contribution of each component. This is because adding logs of numbers effectively multiplies the numbers, and subtracting logs effectively divides. This is often done in practice, because additions and subtractions are much easier to keep track of than multiplications and divisions.

Example 9-2

An audio system has a 12-dB gain in a preamplifier, an 8-dB loss in a mixer, and a 32-dB gain in an amplifier. What is the overall gain?

Solution

$$\alpha_{dB} = \alpha_1 + \alpha_2 + \alpha_3 = 12 - 8 + 32 = \mathbf{36\ dB}$$

Fast decibel conversions: People who work frequently with decibels soon commit the following table to memory. With it, any voltage ratio can be converted approximately to dB (or the reverse) without using a pencil or calculator.

α (dB)	V_2/V_1	α (dB)	V_2/V_1
0	1	20	10
1	1.12	30	31.6
3	1.41	40	100
6	2.0	50	316
10	3.16	60	1000

9.2 COUPLING CAPACITORS

All the amplifiers discussed in Chapters 6 through 8 were of the ac-coupled type; that is, they used capacitors to couple the signals in and out while blocking the dc bias voltages from the input and output terminals. We have been analyzing the performance of these amplifiers at audio frequencies (typically 440 or 1000 Hz), and we have assumed that the reactance of the coupling capacitors was so small that they could be regarded simply as short circuits to ac and open circuits to dc.

At low audio frequencies the reactance of the coupling capacitors increases, thus limiting the input current and lowering the gain of the amplifier. It is often necessary to know the lowest frequency of the input signal in order to determine the required values of these capacitors.

The low-frequency cutoff point of an amplifier is defined as the frequency at which the output power is reduced to one-half of its maximum or *midband* value for a constant input signal. Midband for an audio amplifier can generally be taken as 1000 Hz.

This *half-power point* is more easily identified by the voltage drop than by the power drop because voltage is more easily measured than power. Since power varies as the square of voltage for a given load resistance, voltage varies as the square root of power. Thus if power is divided by 2, voltage must have been divided by $\sqrt{2}$, which is equivalent to multiplying by 0.707. The lower cutoff point of an amplifier is identified as the frequency at which the output voltage drops to 0.707 of its midband value (assuming a constant input voltage level and a constant load resistance).

In the decibel system of measure, half-power is expressed by -3 dB. Therefore, the cutoff frequency is also called the *3-dB down point* or *the minus 3-dB point.*

Vector circuit analysis can demonstrate that the voltage across any resistor in a series-resistive ac circuit is reduced to 0.707 by the insertion of a series capacitor whose reactance is equal to the total resistance. This is illustrated in Fig. 9-2(a) and (b), which represent the input-signal circuit of an amplifier at midband and at the lower -3-dB point, respectively.

Bode plot: An approximate curve of the attenuation of V_{in} at frequencies below f_{low} is shown in Fig. 9-2(c). This curve is called a Bode plot, and it is remarkably easy to draw. The curve is horizontal above f_{low} and falls off at a rate of 20 dB/decade below f_{low}. Note that 20 dB/decade is a voltage factor of 10 per frequency factor of 10. Also note that the straight-line approximate curve is -3 dB in error of f_{low}.

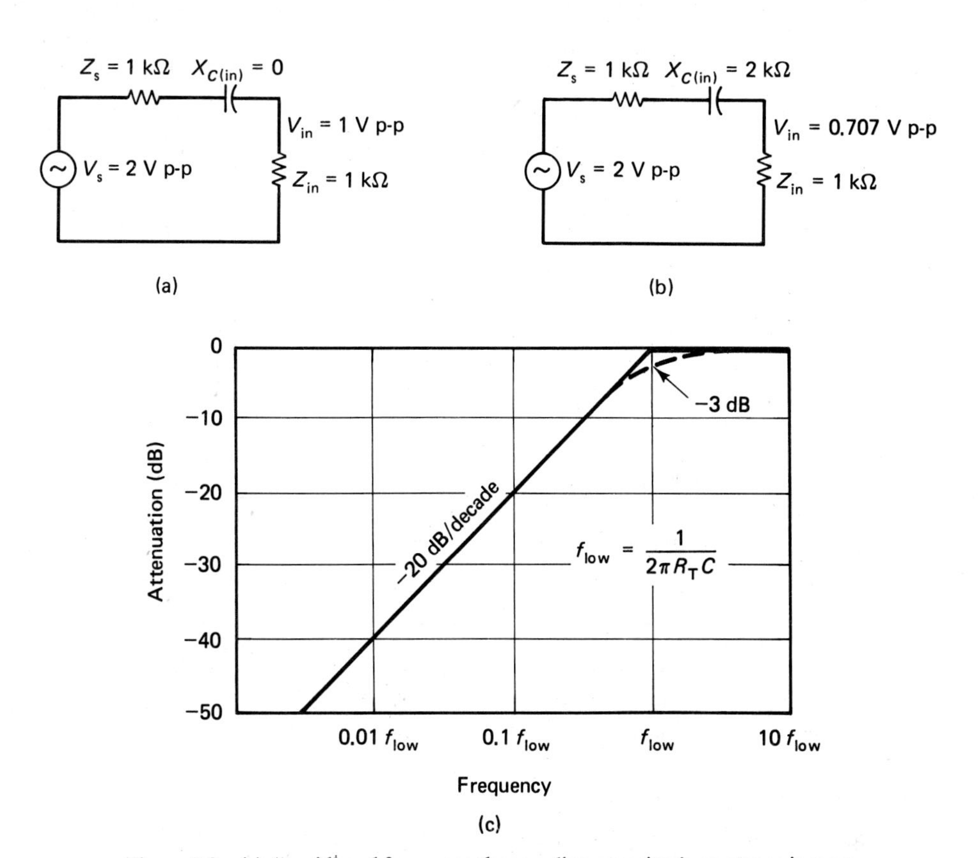

Figure 9-2 (a) At midband frequency the coupling capacitor's reactance is essentially zero. (b) At the lower cutoff frequency, X_C reduces the input voltage to 0.707 of its midband value. (c) The Bode (say BO-dee) plot gives the attenuation at frequencies below f_{low}.

Multiple cutoffs: When $X_{C(\text{in})} = Z_{\text{in}} + Z_s$, the input voltage is reduced to 0.707. A similar situation exists for the output-signal circuit. When $X_{C(\text{o})} = Z_o + R_L$, the output signal is reduced to 0.707 of its value at midband where $X_{C(\text{o})}$ is near zero. If the input and output coupling capacitors have cutoff frequencies which are quite different, the overall amplifier cutoff point is determined by the *higher* of the two frequencies. However, if both input and output coupling capacitors have the same cutoff frequency, the overall amplifier gain will, of course, be 0.707×0.707, or 0.5 times the midband gain at that frequency. The overall −3-dB point will then be slightly higher in frequency than it would be if one capacitor's cutoff was considerably lower than the other's. Therefore, it is a good idea to choose capacitors which are slightly larger than would be calculated by $X_{C(\text{in})} = Z_{\text{in}} + Z_s$, or $X_{C(\text{o})} = Z_o + R_L$. One way to ensure this and to simplify things in the process is to *ignore Z_s and R_L*, since these are not always known at the time of amplifier design. The *rule-of-thumb design formulas* then become

$$\left.\begin{array}{l} X_{C(\text{in})} \approx Z_{\text{in}} \quad (9\text{-}5) \\ X_{C(\text{o})} \approx Z_o \quad (9\text{-}6) \end{array}\right\} \text{ at the lower cutoff frequency}$$

The capacitor values can be calculated from the formula

$$X_C = \frac{1}{2\pi f C} \tag{9-7}$$

where f is the low-frequency half-power point. However, it is usually much simpler to determine the approximate capacitor value from the reactance chart which appears in Appendix C.

9.3 BYPASS CAPACITORS

Some of the most popular biasing circuits utilize voltage-dropping resistors in the emitter or source, as shown in Fig. 9-3. These resistors must generally be bypassed with a comparatively large capacitor to keep ac voltage from appearing across them and reducing the circuit gain. Approximate formulas for the bypass capacitors are given for each circuit in Fig. 9-3. The reactances in the formulas are for the capacitors at the low cutoff frequency of the amplifier.

In the bipolar transistor amplifier [Fig. 9-3(a)] it is necessary that the reactance of the *emitter bypass cappacitor be small compared to the unbypassed resistance in the emitter line*. Otherwise, the effective impedance in the emitter line will be increased, and it will be recalled from Chapter 6 that voltage gain depends on emitter-line impedance:

$$A_v = \frac{Z(\text{looking out of collector})}{Z(\text{in emitter line})} \tag{9-8}$$

For circuits where R_{E1} is omitted and the entire emitter resistance is bypassed, $X_{C(E)}$ must equal r_j, which sometimes results in a rather large bypass capacitor. For example, if $r_j = 6\ \Omega$ and the low-frequency cutoff is to be 20 Hz, the value of C_E must be 1500 μF. Fortunately, the voltage rating required is usually low.

For the FET circuit $X_{C(S)}$ must equal the fictitious source resistance r_s (which equals $1/y_{fs}$) at the low cutoff frequency.

In choosing C_{in} for an FET amplifier it is wise to consider the problem of 60-Hz noise pickup. Consider that $R_G = 1\ \text{M}\Omega$, $Z_s = 50\ \text{k}\Omega$, and $f_{low} = 120$ Hz. If C_{in} were chosen to make $X_{C(in)} = R_G + Z_s = 1.05\ \text{M}\Omega$ at 120 Hz, the reactance of C_{in} would be about 2 MΩ at 60 Hz. Line noise picked up on the gate wiring would see an impedance of nearly 1 MΩ to ground and could be troublesome, especially in prototype or breadboard circuits where long unshielded wires are common. Choosing C_{in} for a reactance of 50 kΩ at 60 Hz would short most of this noise out through Z_s.

Calculating gain reduction: It may sometimes be desired to know just how much reduced the gain of an amplifier will be at some particular frequency below the low cutoff. A very simple rule of thumb is that *for every factor of 2 that the frequency in question is below the cutoff frequency, the gain will be down by a factor of 2*, and this will happen *for each capacitor* in the circuit which produces a low cutoff point. Thus if an amplifier having an input, output, and emitter (or source)-bypass capacitor is designed to have a low cutoff of 180 Hz on each capacitor, the 60-Hz line noise picked

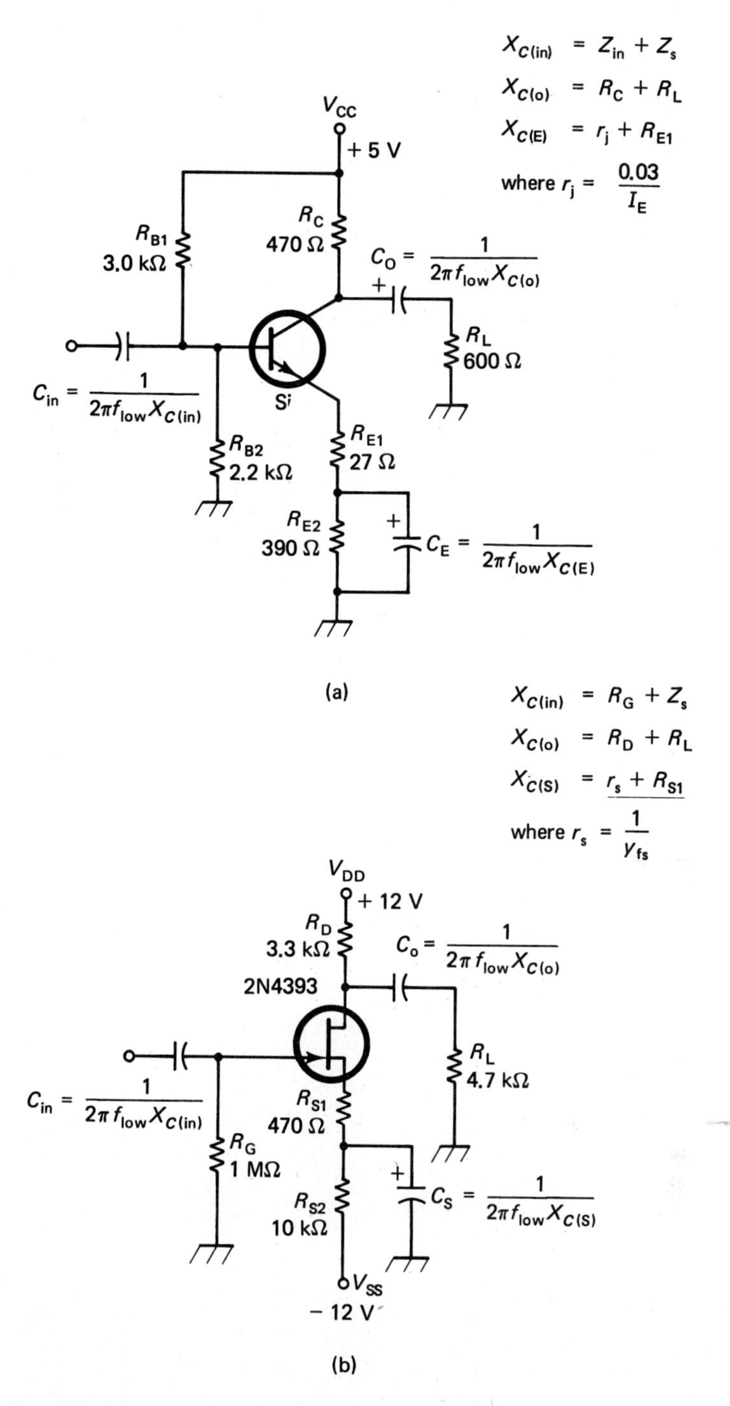

Figure 9-3 Design formulas for input, output, and bypass capacitors for (a) transistor and (b) amplifiers. Formulas give capacitor reactance X_C at the lower cutoff frequency. Capacitor value is then calculated from $C = \frac{1}{2}\pi f_{lo} X_C$.

up by the microphone will be down by $(\frac{180}{60})(\frac{180}{60})(\frac{180}{60}) = (3)(3)(3) =$ **27 times** from the midband voice and music signals.

If the three capacitors produce different −3-dB frequencies (i.e., 30, 120, and 240 Hz), the total attenuation at 60 Hz would be

$$(1)(\tfrac{120}{60})(\tfrac{240}{60}) = (1)(2)(4) = \textbf{8 times}$$

Note that the first cutoff frequency of 30 Hz causes no attenuation of the 60-Hz signal—hence the factor of 1. The cutoff frequency of the total amplifier stage in this example would be 240 Hz, determined entirely by the highest-frequency capacitive cutoff.

In designing coupling- and bypass-capacitor values it will be found that, if two capacitors cut off at the same frequency, choosing $X_C = 0.65R$ will provide a cumulative attenuation of −3 dB at f_{low}. If three capacitors cut off at the same frequency, X_C should be chosen as $0.5R$.

Example 9-3

In the circuit of Fig. 9-4, determine A_v at midband, A_v at 60 Hz, and the low cutoff frequency.

Solution First it is necessary to find the bias point so that r_j, Z_{in}, and A_v at midband can be determined:

$$\frac{V_B}{4.7\text{ k}\Omega} = \frac{12\text{ V}}{10\text{ k}\Omega + 4.7\text{ k}\Omega}; \qquad V_B = 3.8\text{ V}$$

$$V_E = V_B - V_{BE} = 3.8 - 0.2 = 3.6\text{ V}$$

$$I_E = \frac{V_E}{R_E} = \frac{3.6\text{ V}}{1\text{ k}\Omega} = 3.6\text{ mA}$$

$$r_j = \frac{0.03}{I_E} = \frac{0.03}{0.0036} = 8.3\ \Omega$$

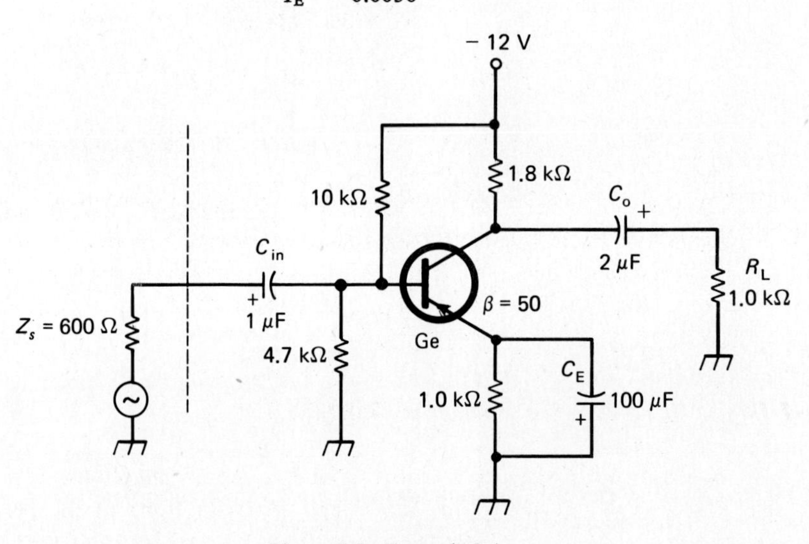

Figure 9-4 Example 9-1.

$$R_{c(\text{line})} = 1.8\ \text{k}\Omega \,||\, 1.0\ \text{k}\Omega = 640\ \Omega$$

$$A_v\ (\text{midband}) = \frac{R_{c(\text{line})}}{r_j} = \frac{640}{8.3} = \mathbf{77}$$

$$r_b = \beta r_j = (50)(8.3) = 410\ \Omega$$

$$Z_{in} = R_{B_1} \,||\, R_{B_2} \,||\, r_b = 10\ \text{k}\Omega \,||\, 4.7\ \text{k}\Omega \,||\, 0.41\ \text{k}\Omega = \mathbf{360\ \Omega}$$

Now the low cutoff frequency for each of the capacitors can be determined:

$$\text{for } C_{in}\text{:}\quad f = \frac{1}{2\pi XC}, \qquad \text{where } X = Z_s + Z_{in}$$

$$f = \frac{0.159}{(600 + 360)(1 \times 10^{-6})} = \mathbf{166\ Hz}$$

$$\text{for } C_o\text{:}\quad f = \frac{1}{2\pi XC}, \qquad \text{where } X = Z_o + R_L$$

$$f = \frac{0.159}{(1800 + 1000)(2 \times 10^{-6})} = \mathbf{28\ Hz}$$

$$\text{for } C_E\text{:}\quad f = \frac{1}{2\pi XC}, \qquad \text{where } X = r_j, \text{ since } R_{E1} = 0$$

$$f = \frac{0.159}{(8.3)(100 \times 10^{-6})} = \mathbf{191\ Hz}$$

The low cutoff frequency for the amplifier is 191 Hz, determined by C_E, which produces the highest cutoff of the three capacitors. C_o has a cutoff considerably below 60 Hz and causes little attenuation at that frequency.

The approximate attenuation of C_{in} and C_E is given by the ratio of their cutoff frequencies to 60 Hz:

$$\text{attenuation factor} = \left(\frac{166}{60}\right)\left(\frac{191}{60}\right) = \mathbf{8.8}$$

The A_v at 60 Hz is the midband A_v divided by the attenuation:

$$A_{v(60\,\text{Hz})} = \frac{77}{8.8} = \mathbf{8.7}$$

In calculating low-frequency response by these methods, it is wise to remember that the technique is only approximate and that the tolerance of the electrolytic capacitors involved may be as broad as -20 to $+75\%$. Where low-frequency response is required, it is therefore wise to double the values of capacitance calculated by the above methods, and where low-frequency response must be minimized, the calculated capacitor values should be halved, to provide a margin of safety.

9.4 INTERELECTRODE CAPACITANCES

Bipolar and field-effect transistors have a certain small capacitance (typically a few picofarads) between their electrodes. At frequencies in the megahertz range, the reactance of these capacitances becomes low enough to affect the operation of the device. The usual result is that this *interelectrode capacitance* places an upper limit on the frequency response of the amplifier.

Figure 9-5 shows the primary interelectrode capacitances in the bipolar transistor and JFET. In high-frequency-amplifier circuit analysis, we are usually not so much interested in the direct interelectrode capacitances as we are in the input, output, and feedback (output-to-input) capacitances.

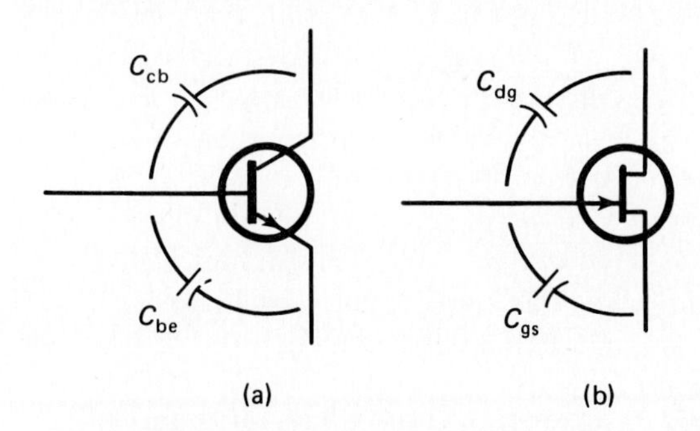

Figure 9-5 Interelectrode capacitances in (a) transistors, and (b) FETs.

The output capacitance of a transistor in either common-emitter or common-base configuration is essentially equal to C_{cb}. If this capacitance presents a reactance equal to collector load r_L, the dynamic load impedance seen by the collector will be lowered sufficiently to reduce the voltage gain of the amplifier by 0.707, and the upper cutoff frequency will have been reached. Above this upper 3-dB point the voltage gain of the amplifier will drop progressively lower. This collector output capacitance is typically 5 to 20 pF for small-signal transistors and may run into hundreds of picofarads for power transistors. Stray wiring capacitance can add greatly to this collector-ground capacitance if the proper precautions are not taken. Shielded wire, for example, will present typically 5 pF/in. of capacitance from center-conductor to ground, and hookup wire if dressed flat down against a metal chassis can present 1 pF/in. or more of capacitance. The relatively large metal bulk of some output coupling capacitors can add another 5 pF or so to ground if mounted close to a chassis or other grounded surface.

Example 9-4

The transistor in the circuit of Fig. 9-4 has a C_{cb} specification of 12 pF. The "load resistance" is actually an ultrasonic loudspeaker mounted nearly 1 ft from the transistor in the instrument, and it is estimated that the connecting wiring adds another 10 pF to the collector-ground capacitance. What is the upper frequency cutoff of the amplifier?

Solution The total capacitance from collector to ground is 12 + 10 or 22 pF. The resistive load impedance is 1.8 kΩ || 1.0 kΩ or 640 Ω. The upper cutoff occurs at the frequency where X_C of 22 pF equals 640 Ω:

$$f = \frac{1}{2\pi XC} = \frac{0.159}{(640)(22 \times 10^{-12})} = 11.3 \text{ MHz}$$

This does not guarantee that the amplifier will perform up to 11.3 MHz; rather it indicates that output capacitance restricts the amplifier to frequencies below 11.3

MHz. There are at least two other factors which place limits on the high-frequency performance of the amplifier. The lowest of these three high-frequency limits determines the actual upper -3-dB point of the total amplifier.

The Miller effect was discussed in Section 7.4 in connection with the collector-base feedback resistor in a bias-stabilized transistor amplifier. The conclusion was that when a resistance is connected from output to input of an inverting amplifier, the ac voltage across that resistor is approximately $V_{in}A_v$, and the current through it is A_v times larger than it would be if the resistor were connected directly to ground. Hence the resistance, when seen from the input, has the appearance of a lower-value resistance connected from the input to ground. The approximate value of this *Miller resistance* is R/A_v.

In the common-emitter and common-source amplifiers there is a capacitance of typically several picofarads between the input and output elements, and when multiplied by an A_v of 10 or more, this Miller input capacitance C_m may lower the amplifiers's input impedance at high frequencies, making it difficult to drive through the impedance of the signal source.

Although Miller input capacitance does not decrease the voltage gain of an amplifier, it does decrease the power gain because it causes a lower Z_{in}, and hence a larger input current requirement for a given output power. In practice, the usual manifestation of Miller input capacitance is a drop in input voltage caused by loading of the signal source. Of course, this results in a drop in V_o at high frequencies. The critical -3-dB drop will occur when the reactance of the input capacitance C_m equals the parallel combination of the driving source Z_s and the amplifier's resistive input impedance, Z_{in}:

$$\text{at } f_{hi}, \qquad X_{C(m)} = Z_s \,||\, Z_{in} \tag{9-9}$$

Therefore,

$$f_{hi} = \frac{1}{2\pi(Z_s \,||\, Z_{in})C_m} \tag{9-10}$$

Example 9-5

In the circuit of Example 9-4 (Fig. 9-4) the transistor's input capacitance (C_{be}) is 18 pF and its output feedback capacitance (C_{cb}) is 6 pF. Find the upper cutoff frequency.

Solution The A_v has already been determined as 77.

$$C_m = C_{be} + A_v C_{cb} = 18 + (77 \times 6) = \mathbf{480\ pF}$$

$$f = \frac{1}{2\pi XC}, \qquad \text{where } X = Z_s \,||\, Z_{in} = 600 \,||\, 360 = 225\ \Omega$$

$$= \frac{0.159}{(225)(480 \times 10^{-12})} = \mathbf{1.5\ MHz}$$

Solving the input-capacitance problem: As illustrated by the examples in the preceding two sections, input capacitance generally curtails amplifier operation at a lower frequency than output capacitance. However, if the amplifier can be driven by a very *low-impedance signal source*, the input-capacitance problem can be practically eliminated, since a low Z_s would be capable of supplying the extra current demanded by the input capacitance without loading down the input voltage. The emitter follower (or source follower) presents a low output impedance and can be inserted between the actual signal source and the voltage-amplifier stage to prevent the input loading which would normally appear at high frequencies.

Amplifiers of the voltage-follower type (i.e., emitter follower or source follower) exhibit no Miller effect since there is no voltage increase from input to output. Since there is no phase inversion, the voltage across the gate-source capacitance (to take the FET source follower as an example) is only a small fraction of the input voltage, and C_{gs} appears to be much smaller than it actually is. The drain is at ac ground in the source follower, so C_{gd} appears from the input to ground. Thus for the FET source follower input stray capacitance (gate to ground) is

$$C_g = C_{gd} + C_{gs}(1 - A_v) \tag{9-11}$$

Since A_v is usually quite close to 1, the second term is considerably the smaller.

In an emitter-follower amplifier the input current I_b actually comprises a small part of the output current I_e. This means that any output capacitive load will be reflected back to the base input but will appear smaller by a factor of beta. For the emitter follower, input capacitance C_b is then comprised of the collector-base capacitance plus a fraction of the capacitance seen looking out of the emitter:

$$C_b = C_{cb} + \frac{C_{e(out)}}{\beta} \tag{9-12}$$

FETs are voltage-operated, not current-operated, and therefore output capacitance is not reflected back to the input in these devices as it is in the bipolar transistor.

9.5 TRANSISTOR CUTOFF FREQUENCIES

The ability of the bipolar transistor to act as a current amplifier begins to drop off at high frequencies, and this is manifested as a drop in ac beta, h_{fe}. Transistors are commonly specified by the *terminal frequency* f_T, *at which* h_{fe} *drops to unity*, even though the transistor must usually be operated well below this frequency. The parameter f_T is often referred to as the *gain-bandwidth product*, because of the relationship between high-frequency beta (h'_{fe}) and operating frequency (f):

$$f_T = h'_{fe}f, \qquad \text{provided } h'_{fe} \leq h_{fe} \tag{9-13}$$

where h'_{fe} is the high-frequency beta and h_{fe} is the low-frequency beta, usually measured at 1 kHz.

This equation states that if h'_{fe} is 1 at 100 MHz, it will be 2 at 50 MHz, 10 at 10 MHz, 50 at 2 MHz, and so on, provided that h'_{fe} can never become greater than the low-frequency beta, h_{fe}. This is illustrated in Fig. 9-6(a).

The frequency at which beta begins to drop seriously (−3 dB) is termed f_β or $f_{h(fe)}$, and is given by

$$f_{h(fe)} = \frac{f_T}{h_{fe}} \tag{9-14}$$

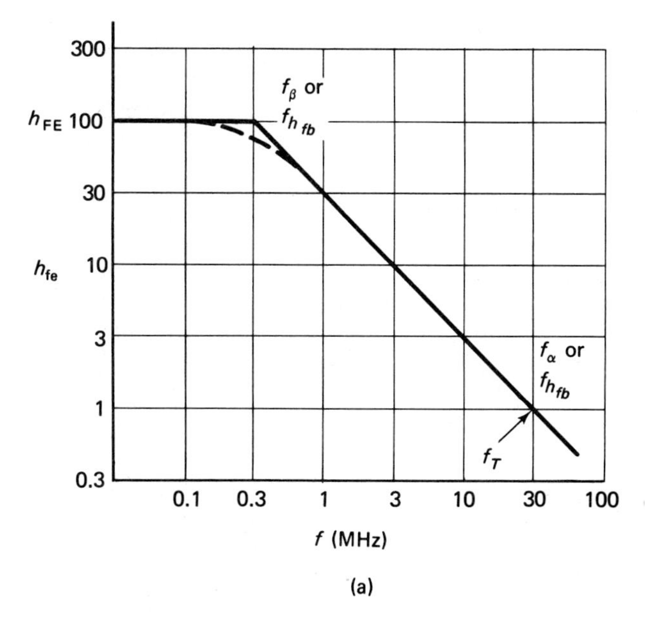

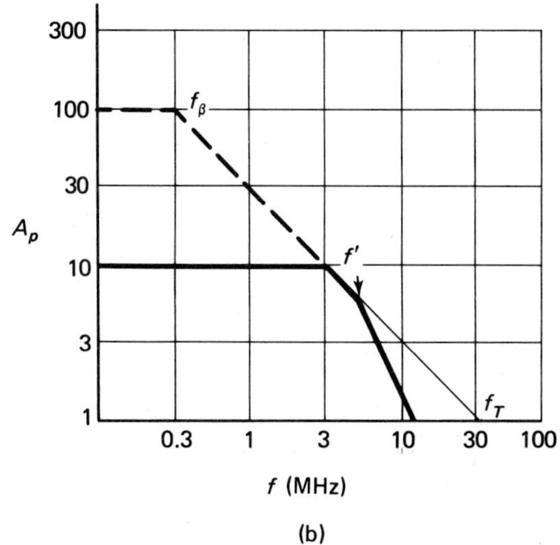

Figure 9-6 (a) Beta drops to unity at f_T. Beta begins to drop off at $f_\beta = f_T/h_{fe}$. (b) The power gain of a $\times 100$ amplifier using a 30-MHz f_T transistor begins to drop off at 0.3 MHz, but the same transistor in a $\times 10$ amplifier holds up to 3 MHz. Note the second break in the gain curve at f' (about 5 MHz) resulting from output capacitance.

Thus if a certain transistor has $f_T = 100$ MHz and low-frequency beta $h_{fe} = 125$, we may begin to expect problems resulting from beta drop at a frequency

$$f = \frac{f_T}{h_{fe}} = \frac{100\text{ MHz}}{125} = 0.8\text{ MHz} = 800\text{ kHz}$$

In the usual common-emitter circuit, the problem will take the form of lowered Z_{in} long before an actual decrease in A_v is noticed. If beta drops as low as 3 or 4, A_v may be directly reduced by -3 dB. However, a beta drop from 125 to 25 will reduce the amplifier's Z_{in} by a factor of 5 in unstabilized circuits and will cause considerable reduction in Z_{in} even in stabilized circuits where the input impedance has been swamped by low-value base voltage-divider resistors. As with the Miller-input capacitance problem, an emitter-follower stage inserted between the signal source and the voltage amplifier will go a long way toward solving the problem of reduced output resulting from lowered beta.

As a rule of thumb, transistors should be chosen which have an f_T that is 100 times the highest operating frequency for high-gain or high-impedance amplifiers. For low-gain amplifiers driven by low-Z sources or preceded by an emitter follower, f_T should be about 10 times the highest operating frequency for good results. The value of high-frequency beta (β') which produces a $\times 0.7$ drop in input voltage is shown in Appendix E to be

$$\beta' = \frac{0.7}{(0.3r_E/R_B) + (0.35r_E/Z_s) + (1/\beta)} \tag{9-15}$$

where r_E is the unbypassed emitter resistance including r_j, R_B is the parallel combination of R_{B1} and R_{B2}, and Z_s is source impedance.

Example 9-6

It has already been determined that the circuit of Fig. 9-4 has an upper cutoff frequency of 1.5 MHz because of Miller input capacitance. If the transistor has $f_T = 25$ MHz, what is the frequency at which the output begins to drop because of decreased beta?

Solution

$$\beta' = \frac{0.7}{(0.3r_E/R_B) + (0.3r_E/Z_s) + (1/\beta)} \tag{9-15}$$

$$\beta' = \frac{0.7}{[(0.3 \times 8.3)/3200] + [(0.3 \times 8.3)/600] + (1/50)} = \mathbf{27}$$

The frequency at which the transistor's beta is reduced to this value can be found from the gain-bandwidth product formula:

$$f_T = f h'_{fe} \tag{9-16}$$

$$f = \frac{f_T}{h'_{fe}} = \frac{25 \text{ MHz}}{27} = \mathbf{0.9 \text{ MHz}}$$

This is lower than the cutoff frequencies resulting from either output capacitance or input capacitance and is therefore the chief determiner of the upper frequency limit of the total amplifier of Fig. 9-4.

9.6 HIGH-FREQUENCY CIRCUITS

There are a number of techniques which can be used to extend the high-frequency limit of transistor amplifiers. The most obvious include selecting transistors with high f_T and low C_{be} and C_{cb} specifications and keeping lead lengths short and suspended away from other parts of the circuit. Beyond this, there are certain circuit designs that are better suited to high-frequency operation. A number of these are discussed below.

Keep circuit impedance low. This will make the frequency-dependent reactances of the input and output capacitances appear high in comparison to the amplifier Z_{in} and Z_o. Low-value base voltage-divider resistors will swamp the Z_{in} of the amplifier so that changes caused by Miller capacitance and lowered beta will amount to only a small part of the total amplifier Z_{in}. Low values of R_C will swamp out the effects of collector output capacitance. Emitter followers before each stage allow amplifiers to be designed with lower Z_{in}.

Use low gain in each stage and build up gain with multiple stages, rather than trying to achieve high gain in a single stage. Miller input capacitance decreases with decreased stage gain, thus increasing the cutoff frequency. An unbypassed emitter resistor is excellent for this purpose since it reduces input capacitance by decreasing A_v and also keeps r_b high when the transistor's beta begins to drop a high frequencies.

Use tuned circuits at the input and output if only a single frequency or a narrow band of high frequencies is to be amplified. Figure 9-7(a) shows a tuned circuit in the collector of a transistor amplifier. Note that the output capacitance C_{cb}, of the transistor simply becomes a part of the tuning capacitance in parallel with C_t. The total capacitive reactance is canceled by the inductive reactance of the coil at the resonant frequency, so the output capacitance C_{cb} is no longer a problem.

Use common-base rather than common-emitter circuitry. As pointed out in Section 7-8, the gain of a common-base amplifier depends on alpha, not beta, and beta can drop to an extremely low value before alpha shows any significant drop at all. In fact, transistors are often specified by their alpha cutoff frequency (f_α or $f_{h(fb)}$). This frequency is usually sightly higher than f_T, and is the point at which alpha drops to 0.707 of its midband value. Figure 9-7(a) shows a typical common-base high-frequency amplifier. The input step-down transformer is necessary to match the inherently low input impedance of the common-base circuit.

Frequency compensation: In addition to these general considerations, there are a number of specific techniques designed to compensate for the drop in gain of an amplifier at high frequencies. Several of these are illustrated in Fig. 9-7. The general idea behind each of these circuit modifications is to use an inductor or capacitor to increase the gain of the amplifier at precisely the frequency where it would normally begin to drop off.

In Fig. 9-7(b), R_{B2} presents a low enough impedance to ac to reduce the input impedance of the amplifier considerably at low frequencies, but at high frequencies L_1 begins to present enough reactance to raise the input impedance. With proper selection of the value of L_1, this effect can be made to compensate for the drop in r_b caused by the decrease in transistor beta at high frequencies.

C_2 and L_2 form a tuned circuit in the collector lead of the amplifier. This circuit is adjusted to be resonant (and have a high impedance) at a frequency just above the cutoff resulting from transistor output capacitance. The effect is to remove R_C from the collector at high frequencies and thus increase the amplifier gain to compensate for the normal drop in gain resulting from output capacitance.

C_3 in Fig. 9-7(b) is not the usual large-value emitter-bypass capacitor. Its value is quite small, so that at midband it has little effect and the amplifier gain is

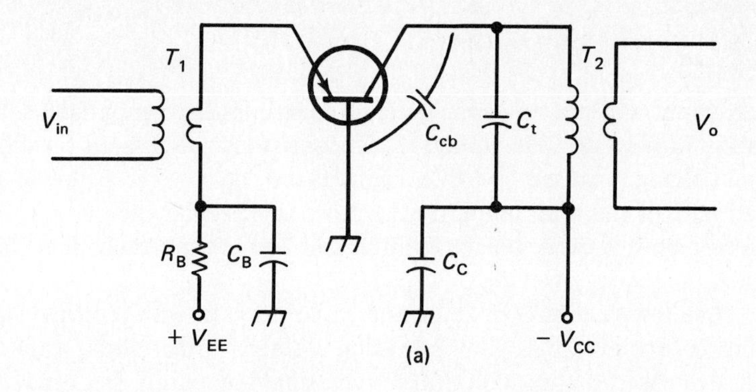

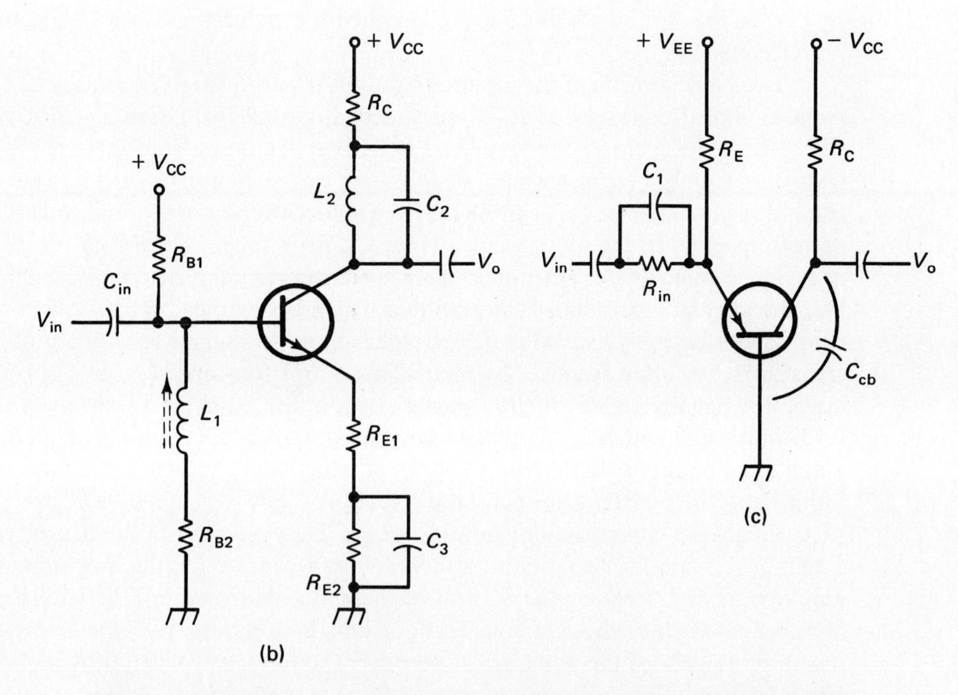

Figure 9-7 (a) The common-base amplifier generally holds its midband gain to a higher frequency than the common-emitter circuit. C_{cb} becomes part of the output tuned circuit C_t–T_2 and does not limit frequency response. (b) A high-bandwidth common-emitter amplifier. R_{E1} keeps Z_{in} high; L_1 isolates R_{B2} from the input at high frequencies; C_3 is small and just begins to shunt R_{E2} at high frequencies. L_2 and C_2 resonate just above the normal cutoff frequency raising $R_{c(line)}$. (c) A broad-band common-base amplifier. C_1 boosts the input signal to the emitter just as C_{cb} begins to load down the output.

approximately

$$A_v = \frac{R_C}{R_{E1} + R_{E2}} \tag{9-17}$$

At high frequencies the capacitor begins to have a low reactance, leaving R_{E_1}

alone in the ac emitter circuit, and the gain tends to increase to

$$A_v = \frac{R_C}{R_{E1}} \tag{9-18}$$

We say the gain "tends to increase," but the idea, of course, is to keep the gain constant. C_3 is chosen to give a tendency to increase gain at exactly the frequency when some other effect is tending to decrease it.

A final example of frequency compensation is shown in Fig. 9-7(c). In this common-base circuit, the approximate voltage gain at midband would be

$$A_v = \frac{R_C}{R_{in} + r_j} \tag{9-19}$$

At high frequencies, however, R_C is effectively shunted by the reactance of output capacitance C_{cb}, and gain tends to drop. Compensating capacitor C_1 is chosen to counteract this tendency by shunting R_{in}, thus providing more input signal current.

9.7 FREQUENCY-RESPONSE MEASUREMENTS

The total frequency-response curve for an *RC*-coupled amplifier may typically appear as shown in Fig. 9-8. The main points of interest are the upper and lower −3-dB points where the voltage gain drops to 0.707 of its midband value. Above and below these points the gain begins to drop off, slowly at first and then more rapidly as several frequency-dependent factors combine to limit the gain.

The most straightforward measurement of an amplifier's frequency response is to apply a sine-wave input from a signal generator and vary the frequency across a wide range while taking measurements of the amplifier output voltage with an oscilloscope. This process is somewhat time consuming unless a sweep-frequency generator

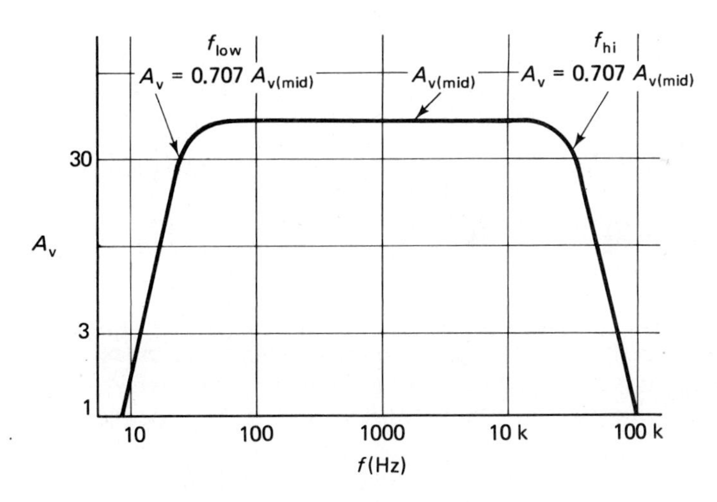

Figure 9-8 Typical hi-fi amplifier frequency-response curve.

is available. Fortunately, there is a way to obtain much of the same information that is contained in the frequency-response curve with a single measurement at one frequency. The technique is to observe the amplifier's response to a square-wave input signal.

Square-wave test: An amplifier that can respond to high frequencies must have the ability to follow the extremely fast changes in amplitude of the high-frequency waveform; that is, it must have a fast *response time*. A square-wave input signal requires an amplifier to change its output almost instantly, but, of course, no amplifier can actually respond instantly. The time that it takes for an amplifier's output voltage to rise in response to a very fast square-wave input can be measured directly from the oscilloscope face, and can be used to determine the approximate high-frequency cutoff point.

Risetime is defined as the time it takes for the output to rise from 10 to 90% of its full value, as shown in Fig. 9-9(a).

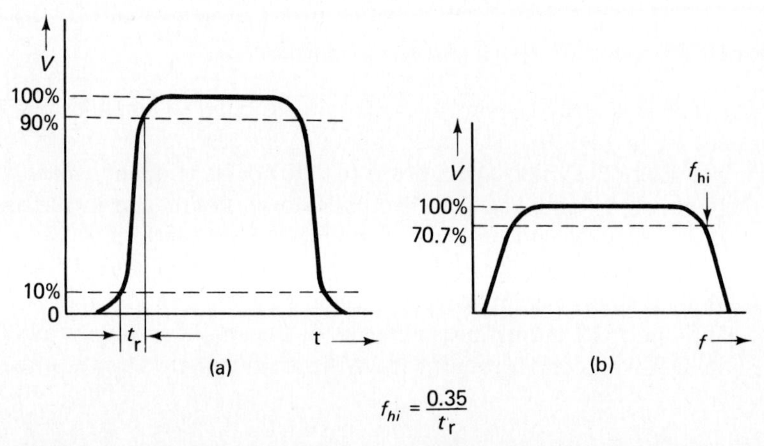

Figure 9-9 (a) Risetime on a square wave is defined as the time required to rise from 10 to 90% of full voltage. (b) f_{hi} is defined as the frequency at which voltage gain drops to 70.7% of full value.

If the amplifier responds to a square wave without overshoot, the relationship between risetime and the high-frequency limit (also called bandwidth) is fairly constant and can be computed as

$$f_{hi} = B_w = \frac{0.35}{t_r} \tag{9-20}$$

Sometimes an amplifier may be overfrequency compensated, resulting in excessively high gain at the high-frequency end of its response. Where feedback from output to input exists or where an *LC* network is used for frequency compensation, the amplifier may have excessive high-frequency gain and a tendency to oscillate or *ring* at one frequency. The frequency-response curve for these conditions, with two possible square-wave time-domain response traces, is shown in Fig. 9-10.

Overshoot and ringing are usually undesirable in an amplifier, and if the overshoot becomes greater than about 5%, equation (9-20) loses its validity.

In all square-wave tests of risetime, it is necessary that the square-wave generator and the oscilloscope have risetimes several times faster than the amplifier to be tested.

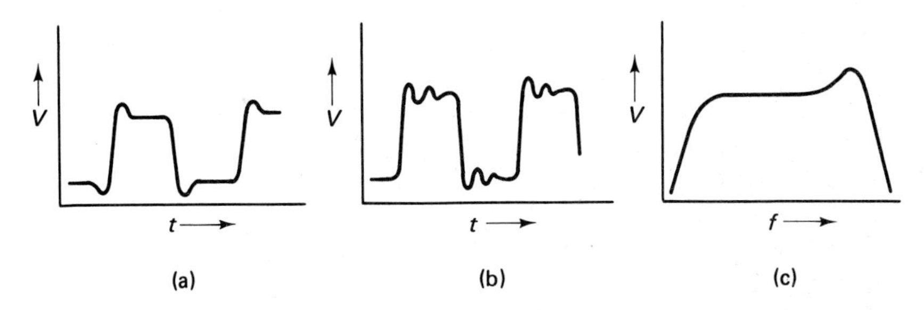

Figure 9-10 Overshoot (a) and ringing (b) on a square-wave test indicate an overpeaked high-frequency response (c).

Example 9-7

An amplifier is fed a 10-kHz square wave and the output risetime is measured as 3.5 μs (10 to 90%). What is the upper cutoff frequency of the amplifier?

Solution

$$f = \frac{0.35}{t_r} = \frac{0.35}{3.5 \times 10^{-6}} = 0.1 \times 10^6 = \mathbf{100\ kHz}$$

Example 9-8

What is the fastest risetime that can be accurately measured on an oscilloscope having a bandwidth of 2 MHz?

Solution

$$t_{r\,(\text{scope})} = \frac{0.35}{f_{\text{hi}}} = \frac{0.35}{2 \times 10^6} = 0.17 \times 10^{-6} = 0.17\ \mu\text{s}$$

This is the risetime of the scope itself. Accurate measurements can be obtained only for signal risetimes considerably longer than these. A factor of 5 usually provides enough accuracy:

$$t_{r\,(\text{sig})} \geq 5t_{r\,(\text{scope})} = (5)(0.17) = \mathbf{0.85\ \mu s}$$

CHAPTER SUMMARY

1. Decibels provide a comparison between two signals. Positive values indicate a gain, negative values indicate a loss.

2. Decibels are logarithmic, not linear. They tend to expand the low-ratio range and compress the high-ratio range. The relevant equations are

$$\alpha_{dB} = 10 \log \left(\frac{P_2}{P_1}\right) = 20 \log \left(\frac{V_2}{V_1}\right) \tag{9-1, 3}$$

$$\frac{P_2}{P_1} = \log^{-1} \left(\frac{\alpha_{dB}}{10}\right) = 10^{\alpha_{dB}/10} \tag{9-2}$$

$$\frac{V_2}{V_1} = \log^{-1} \left(\frac{\alpha_{dB}}{20}\right) = 10^{\alpha_{dB}/20} \tag{9-4}$$

3. Decibels can be used to represent voltage ratios only if V_1 and V_2 appear across equal impedances.
4. Adding decibel values is equivalent to multiplying gain factors.
5. Coupling and bypass capacitors limit amplifier response to frequencies higher than

$$f_{low} = \frac{1}{2\pi RC} \tag{9-7}$$

where

for input capacitor C_{in}, $R = Z_s + Z_{in}$
for output capacitor C_o, $R = Z_o + R_L$
for bypass capacitor C_E, $R = r_j + R_{E1}$

6. Stray capacitance limits amplifier response to frequencies lower than

$$f_{hi} = \frac{1}{2\pi RC}$$

where

$R = Z_s \| Z_{in}$ for Miller input capacitance $C_{be} + C_{cb}(A_v + 1)$ in the common-emitter configuration
$R = R_C \| R_L$ for output capacitance C_{cb}

7. High-frequency response may also be limited by a reduced beta. High-frequency beta is given by

$$h'_{fe} = \frac{f_T}{f} \tag{9-16}$$

provided that h'_{fe} cannot be higher than the low-frequency beta.
8. Amplifier circuits designed for high-frequency response use low-value circuit impedances, low gain factors per stage, and insert emitter followers between voltage-gain stages. Common-base circuitry and frequency-compensating inductors and capacitors are sometimes employed.
9. Bandwidth B_w of an untuned amplifier is the upper frequency at which gain drops to 0.707 of maximum. Risetime is the time it takes the output voltage to go from 10% to 90% of maximum in response to a fast square-wave input. The two are related by

$$B_w = \frac{0.35}{t_r} \tag{9-20}$$

QUESTIONS AND PROBLEMS

9-1. Convert the following power ratios to decibels: (a) 7/1; (b) 1/7; (c) 1/1; (d) 4/1; (e) 4000/1.

9-2. Convert the following decibel numbers to power ratios: (a) 3 dB; (b) −3 dB; (c) 10 dB; (d) 35 dB; (e) 60 dB.

9-3. Convert the following voltage gains to decibels: (a) 12.4; (b) 0.62; (c) 850; (d) 1.5. What assumption about Z_{in} and R_L is required?

9-4. Convert the following decibel numbers to voltage gains: (a) 0 dB; (b) −15 dB; (c) 35 dB; (d) 72 dB.

9-5. A TV master-antenna system has a 300-μV signal at the antenna input. A preamplifier gives 7-dB gain, the main amplifier gives 38-dB gain, total cable loss from antenna to end-of-line is 22 dB, and the tap-off boxes from the main line to each TV set drop the signal 15 dB. What is the voltage delivered to the last TV set at the end of the line?

9-6. In Fig. 9-2(a), $C_{in} = 0.16\ \mu$F. Sketch the Bode plot for this circuit, labeling the frequency and attenuation axes.

9-7. An amplifier has $Z_o = 600\ \Omega$ resistive. It feeds a 600-Ω load. What value of C_o will produce a low cutoff of 150 Hz?

9-8. In Fig. 9-3(a), what is the low cutoff frequency due to C_E?

9-9. In Fig. 9-3(a), what is the minimum transistor β that will keep f_{low} due to C_{in} above 200 Hz? $Z_s = 500\ \Omega$ resistive.

9-10. Find f_{low} in Fig. 9-3(b). The FET has $y_{fs} = 4$ mS.

9-11. For the circuit of Fig. 9-3(b) how many decibels down is the gain at 50 Hz?

9-12. The output of one common-emitter amplifier drives the input of another. C_o of the first stage is 8 pF. C_{in} of the second stage is 120 pF. The collector resistor of the first stage is 4.7 kΩ. Find f_{hi}.

9-13. A transistor has $C_{be} = 7$ pF and $C_{cb} = 5$ pF. What is the input capacitance of an amplifier using this transistor in a common-emitter circuit with a gain of −30?

9-14. What is the input capacitance of an amplifier using the transistor of Problem 9-13 in an emitter-follower circuit with a gain of 0.9?

9-15. A transistor has $\beta = 240$ at low frequencies and $f_T = 150$ MHz. At what frequency is β reduced to 0.707 of 240?

9-16. A transistor has f_T of 5 MHz. What is its β at 600 kHz?

9-17. Explain why common-base and common-collector circuits each have potentially a higher frequency response than common-emitter circuits.

9-18. In Fig. 9-7(c), $C_o = 8$ pF, $R_C = 10$ kΩ, and $R_{in} = 680\ \Omega$. What value of C_1 will give optimum frequency compensation?

9-19. An amplifier has a square-wave response time (10% to 90%) of 0.5 μs. What is its upper −3-dB cutoff frequency?

9-20. An oscilloscope has a bandwidth of 15 MHz. What is its risetime?

10

MULTISTAGE AMPLIFIERS

10.1 BROADBAND RC-COUPLED STAGES

A single transistor amplifier is generally limited to voltage gains in the vicinity of 100, yet it is commonplace for antenna signals of 10 μV to be amplified to a level of 10 V—a gain of 1000000. Of course, this is done by *cascading* amplifiers—feeding the output of one stage to the input of the next. Three gain-of-100 stages in cascade can, with certain precautions, yield a gain of 1000000.

RC coupling is popular in low-level audio-frequency amplifiers because it requires no expensive or bulky components and no adjustments. It can be designed to cover a frequency range from a few hertz to a few megahertz, or to deliberately attenuate frequencies above or below any selected cutoff frequency. Figure 10-1 shows an elementary two-stage *RC*-coupled amplifier. The first stage has its emitter resistance completely bypassed to maximize voltage gain. Currents are relatively low and resistances high because signal levels are low in this stage.

Stage two has an unbypassed emitter resistance to swamp out changes of r_j with signal swing, thus preserving constant A_v and input waveshape. This was not required in stage one because the signal swings were small. Note that currents are higher and resistances lower in stage two. C_4 is selected to limit the amplifier's high-frequency response.

Example 10-1

Find Z_{in}, A_v, f_{low}, f_{hi}, and $V_{o(p-p)max}$ for the circuit of Fig. 10-1.

Solution The bias for both stages is calculated first.

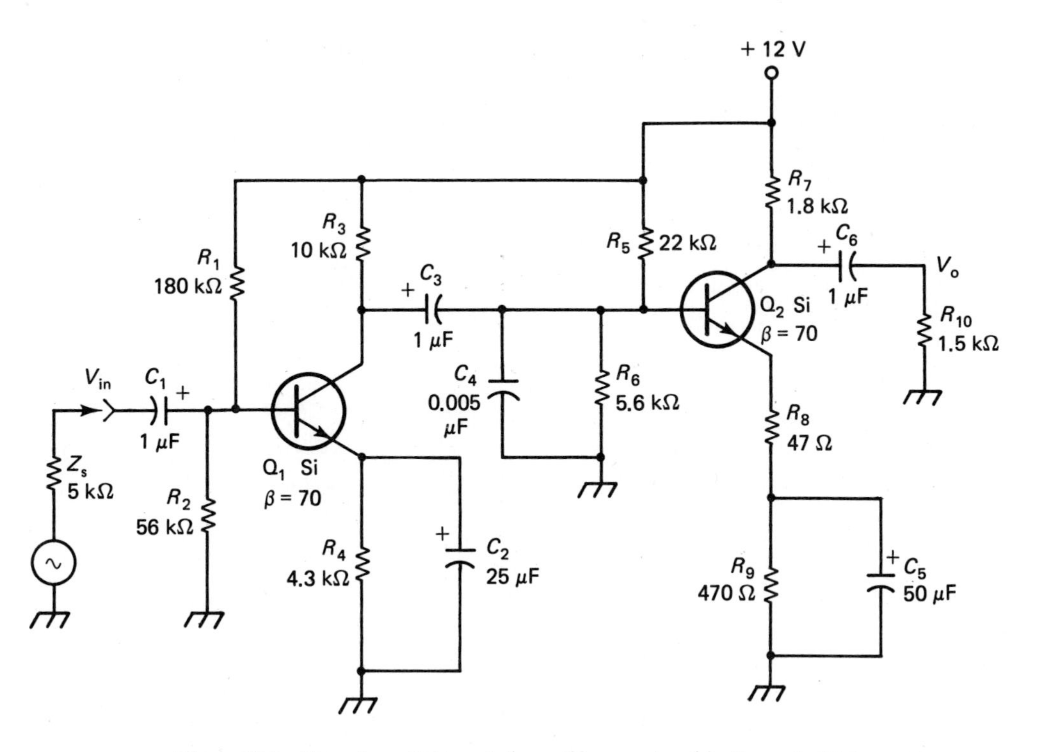

Figure 10-1 Two-stage *RC*-coupled amplifier, analyzed in Example 10-1.

$$V_{1B} = V_{CC}\frac{R_2}{R_1 + R_2} = (12)\frac{56}{180 + 56} = 2.8 \text{ V} \tag{6-15}$$

$$I_{1E} = \frac{V_E}{R_4} = \frac{V_B - V_{BE}}{R_4} = \frac{2.8 - 0.6}{4.3} = 0.5 \text{ mA} \tag{6-17}$$

$$r_{1j} = \frac{30 \text{ mV}}{I_{1E}} = \frac{30 \text{ mV}}{0.5 \text{ mA}} = 60\ \Omega \tag{6-2}$$

$$V_{2B} = (12)\frac{5.6}{22 + 5.6} = 2.4 \text{ V} \tag{6-15}$$

$$I_{2E} = \frac{2.4 - 0.6}{47 + 470} = 3.5 \text{ mA} \tag{6-17}$$

$$r_{2j} = \frac{30 \text{ mV}}{3.5 \text{ mA}} = 8.6\ \Omega \tag{6-2}$$

Now the input impedances and voltage gains are found.

$$Z_{1in} = R_1 \,||\, R_2 \,||\, \beta r_{1j} = 180 \text{ k}\Omega \,||\, 56 \text{ k}\Omega \,||\, (70)(60) = \mathbf{3.8\ k\Omega} \tag{6-23}$$

$$Z_{2in} = R_5 \,||\, R_6 \,||\, \beta r_{2j} = 22 \text{ k}\Omega \,||\, 5.6 \text{ k}\Omega \,||\, 70(8.6 + 47) = 2.1 \text{ k}\Omega$$

$$A_{1v} = \frac{R_3 \| Z_{2in}}{r_{1j}} = \frac{10\,000 \| 2100}{60} = 28.9 \tag{6-11}$$

$$A_{2v} = \frac{R_7 \| R_{10}}{r_{2j} + R_8} = \frac{1800 \| 1500}{8.6 + 47} = 14.7 \tag{6-25}$$

$$A_{v(total)} = A_{1v}A_{2v} = (28.9)(16) = \mathbf{425}$$

The capacitor which produces the highest f_{low} is found with the aid of a reactance chart (Appendix C).

C_1 vs. $Z_s + Z_{1in} \longrightarrow 1\ \mu$F vs. 5 kΩ + 3.8 kΩ $\longrightarrow$ 20 Hz

C_2 vs. $r_{1j} \longrightarrow 25\ \mu$F vs. 60 Ω $\longrightarrow$ 100 Hz

C_3 vs. $R_3 + Z_{2in} \longrightarrow 1\ \mu$F vs. 10 kΩ + 2.1 kΩ $\longrightarrow$ 15 Hz

C_5 vs. $r_{2j} + R_8 \longrightarrow 50\ \mu$F vs. 8.6 + 47 Ω $\longrightarrow$ 60 Hz

C_6 vs. $R_7 + R_{10} \longrightarrow 1\ \mu$F vs. 1.8 kΩ + 1.5 kΩ $\longrightarrow$ 50 Hz

Obviously, C_2 determines f_{low}. An exact calculation is

$$f_{low} = \frac{1}{2\pi C_2 r_{1j}} = \frac{1}{2\pi(25\ \mu\text{F})(60\ \Omega)} = \mathbf{106\ Hz} \tag{9-7}$$

C_4 produces a high cutoff when its reactance equals $Z_{1(out)}$ in parallel with Z_{2in}.

$$f_{hi} = \frac{1}{2\pi C_4(R_3 \| Z_{2(in)})} = \frac{1}{2\pi(5\ \text{nF})(10\ \text{k}\Omega \| 2.1\ \text{k}\Omega)} = \mathbf{18\ kHz} \tag{9-10}$$

Maximum peak V_o is limited by V_{CE} in the down direction and by $I_{2C}(R_7 \| R_{10})$ in the up direction.

$$V_{o(down)max} = V_{2CE} = V_{CC} - I_C R_C - V_E \tag{6-28}$$

$$= 12 - (3.5)(1.8) - (2.4 - 0.6) = 3.9\ \text{V pk}$$

$$V_{o(up)max} = I_{C2}(R_7 \| R_{10}) = 3.5(1.8 \| 1.5) = 2.9\ \text{V pk} \tag{6-29}$$

$$V_{o(p\text{-}p)max} = 2V_{o(up)max} = (2)(2.9) = \mathbf{5.8\ V\ p\text{-}p}$$

10.2 DIRECT-COUPLED STAGES

In cascaded amplifiers it is often possible to set the collector voltage of the low-level stage equal to the base voltage required by the high-level stage, thus eliminating a coupling capacitor and two base-bias resistors. Figure 10-2(a) shows an audio amplifier employing this technique.

A widely used improvement on this basic idea is shown in Fig. 10-2(b). R_3 provides the additional bias stability of collector self-bias (Section 7.3), since V_{2E} follows 0.6 V below V_{1C}. Miller-effect and input-impedance drop are avoided because R_6 is bypassed to ground by C_2.

The circuit of Fig. 10-2(c) uses complementary transistors to get the base of Q_2 closer to ground. This means less voltage loss across the stage-two emitter resistor R_5 for higher output, and more voltage across the stage-one emitter resistor R_3 for better bias stability. R_4 is the collector resistor for Q_1.

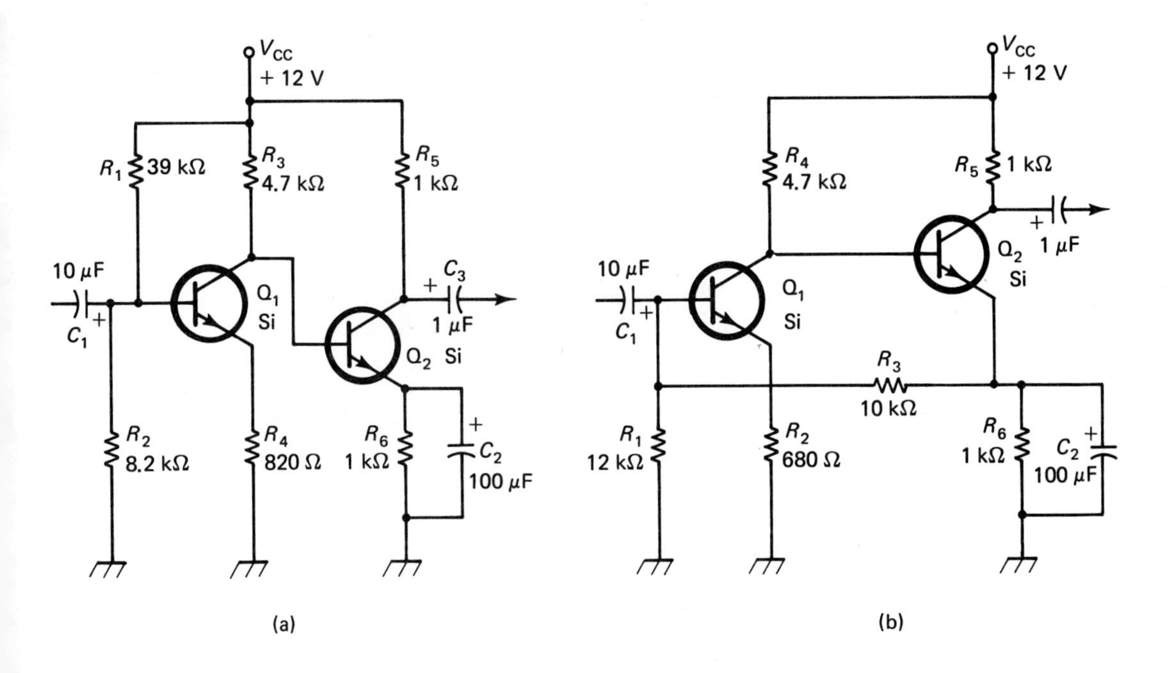

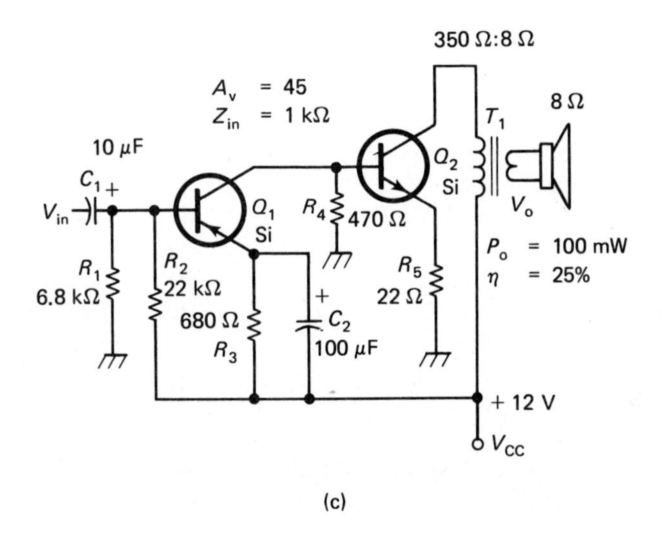

Figure 10-2 Three direct-coupled amplifiers. (a) Elimination of coupling capacitor and base-bias resistors. (b) Feedback resistor R_3 stabilizes bias. (c) *PNP–NPN* cascade permits lower base voltage at Q_2, minimizing waste across R_5.

10.3 THE DIFFERENTIAL AMPLIFIER

The differential amplifier is characterized by two input terminals, neither one of which is necessarily grounded, and two output terminals with the output waveform of one being equal to but inverted with respect to the other. The input voltage is the *difference* in potential between the two input terminals, and the output is the *difference* between the two output terminals. Thus if two signals are fed into the inputs of a differential amplifier, identical in every respect except that one signal has been slightly distorted, only the *difference* between the two signals (i.e., the distortion) will be amplified. The *common-mode* signal (i.e., signal components common to both input lines) will be ignored and not amplified.

The differential amplifier has numerous applications. Among the most common are

1. Compare two signals and detect any difference.
2. Amplify the signal between a pair of transmission wires while rejecting large common-mode noise signals that may have been picked up between both wires and ground identically.
3. Pick up and amplify the voltage across a circuit element, neither end of which is grounded (the differential-input oscilloscope uses such an amplifier).

The differential amplifier is a balanced dc amplifier. Single-ended (as opposed to differential) dc amplifiers are susceptible to output-voltage drift with temperature and supply-voltage changes. In the differential amplifier, any drift in one-half of the amplifier will be almost exactly balanced by an equal drift in the other half, resulting in a stable voltage between the two outputs. This stability has led to the widespread use of differential amplifiers in voltmeters, oscilloscopes, and other instruments where stability is crucial, even though the two *floating* inputs of the differential amplifier may not be required. It is worth mentioning that universal practice in designing operational amplifiers (Chapter 12) is to start with a differential amplifier.

Figure 10-3 shows a transistor differential amplifier. In analyzing the circuit it is helpful to trace electron flow from the negative supply to the positive supply. Q_3 is a current source which feeds the emitters of the two differential transistors, Q_1 and Q_2. The current delivered by the collector of Q_3 is fixed by bias resistors R_7 to R_9 and does not vary with the loads imposed by Q_1 and Q_2. With both inputs grounded the circuit is perfectly symmetrical and $I_1 = I_2$. The current source requires that $I_1 + I_2$ remain constant, so any increase in I_1 which is caused by applying a positive voltage at V_{IN1} will cause an exactly equal decrease in I_2. These current changes will cause a lowering of V_{O1} and an equal raising of V_{O2}.

If a common-mode signal (say $+1$ V) is applied to both inputs simultaneously, it will be impossible for both I_1 and I_2 to increase together. The result will be that the collector voltage of Q_3 will simply rise by 1 V and neither I_1 nor I_2 will change.

The differential gain of the amplifier can be analyzed by looking at the two halves of the amplifier separately. Let us assume that V_{IN2} is grounded and $+1$ V is applied

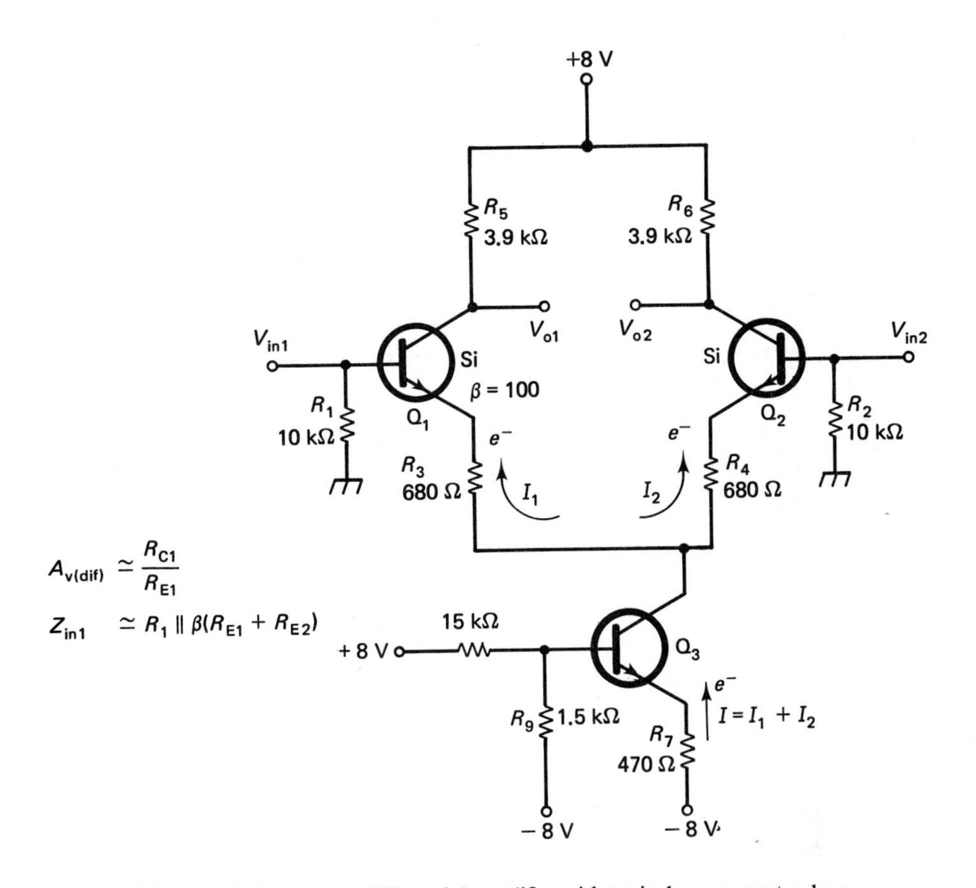

Figure 10-3 Transistor differential amplifier with typical component values.

to V_{IN1}. The differential input voltage is then 1 − 0, or simply 1 V. The emitter of Q_2 must always be 0.6 V below the base, but the base is grounded, so V_{E2} *is fixed at* −0.6 V. The application of +1 V at V_{IN1} thus causes an increase of 1 V across $R_{E1} + R_{E2}$. Ignoring the emitter-junction resistance of Q_1 and analyzing Q_1 as a simple common-emitter amplifier with an emitter resistance of $R_{E1} + R_{E2}$,

$$\frac{V_{O1}}{V_{IN1} - V_{IN2}} = \frac{R_{C1}}{R_{E1} + R_{E2}} \tag{10-1}$$

But the output voltage consists of $V_{O1} - V_{O2}$, and the change in V_{O2} has been shown to be equal to but opposite from the change in V_{O1}. The total differential gain of the amplifier is therefore

$$A_{v(dif)} = \frac{V_{O1} - V_{O2}}{V_{IN1} - V_{IN2}} = \frac{R_{C1} + R_{C2}}{R_{E1} + R_{E2}} \tag{10-2}$$

If the amplifier is balanced, $R_{C1} = R_{C2}$ and $R_{E1} = R_{E2}$, so the equation above reduces to

$$A_{v(dif)} = \frac{R_{C1}}{R_{E1}} \tag{10-3}$$

For a more precise calculation, it would be necessary to remember that R_{E1} and R_{E2} each have an emitter junction resistance r_j in series.

The input impedance of the amplifier from either input terminal to ground is the base resistor in parallel with the base-input resistance of the transistor:

$$Z_{in1} = R_{B1} \| \beta\left(r_{1j} + R_{E1} + R_{E2} + r_{2j} + \frac{R_{B2}}{\beta_2}\right) \tag{10-4}$$

The differential input impedance (i.e., the impedance seen by a source connected between the two inputs) is

$$Z_{in\,(dif)} = (R_{B1} + R_{B2}) \| \beta(r_{1j} + R_{E1} + R_{E2} + r_{2j}) \tag{10-5}$$

Example 10-2

Analyze the differential amplifier of Fig. 10-3 for bias point, A_v, and Z_{in}.

Solution The first requirement is to find the current delivered by current source Q_3. The key to this is to find V_{R9}. This is done by a voltage-division proportion:

$$\frac{(V_{R8} + V_{R9})}{R_8 + R_9} = \frac{8\text{ V} + 8\text{ V}}{15\text{ k}\Omega + 1.5\text{ k}\Omega} = \frac{V_{R9}}{1.5\text{ k}\Omega}; \qquad V_{R9} = 1.5\text{ V}$$

$$V_{R7} = V_{R9} - V_{BE} = 1.5 - 0.6 = 0.9\text{ V}$$

$$I = \frac{V_{R7}}{R_7} = \frac{0.9\text{ V}}{0.47\text{ k}\Omega} = 1.9\text{ mA}$$

The collectors of Q_1 and Q_2 divide this current equally if the amplifier is balanced and there is no signal input, so the collector voltage can be calculated:

$$I_{C1} = \tfrac{1}{2}I = 0.5 \times 1.9\text{ mA} = 0.95\text{ mA}$$

$$V_{R5} = I_{C1}R_5 = 0.95(3.9\text{ k}\Omega) = 3.7\text{ V}$$

$$V_{C1} = V_{CC} - V_{R5} = 8 - 3.7 = \mathbf{4.3\ V}$$

The emitter junction resistance is approximately 30 mV/0.95 mA, or 32 Ω. This is less than 5% of R_E, so we feel justified in neglecting it. Next, the differential voltage gain is determined:

$$A_{v\,(dif)} = \frac{R_5}{R_3} = \frac{3.9\text{ k}\Omega}{0.68\text{ k}\Omega} = \mathbf{5.7}$$

The impedance from each input to ground is

$$Z_{in} = R_1 \| \beta\left(R_3 + R_4 + \frac{R_2}{\beta}\right)$$

$$= 10 \| (100)(1.46) = \mathbf{9.4\ k\Omega}$$

The differential input impedance (between the two inputs) is

$$Z_{in\,(dif)} = (10 + 10) \| (100 \times 1.36) = \mathbf{17.4\ k\Omega}.$$

Variable gain: The gain of the differential amplifier can be varied quite conveniently without upsetting the bias point by simply connecting varying values of resistance between the emitters of Q_1 and Q_2. This is in contrast to single-ended dc amplifiers where any attempt to change gain invariably results in a dc bias point upset. If the gain-controlling resistor is designated R_G, the gain formula becomes

$$A_{v\,(dif)} = \frac{2R_C}{(2R_E) \| R_G + 2r_j} \tag{10-6}$$

Example 10-3

The amplifier of Fig. 10-4 is to have a differential gain of 50. The negative current source is to operate with -4 V at point A. The collectors of Q_1 and Q_2 are to be biased at $+6$ V. Find the current from the source, R_{E1} and R_{E2}, and R_G.

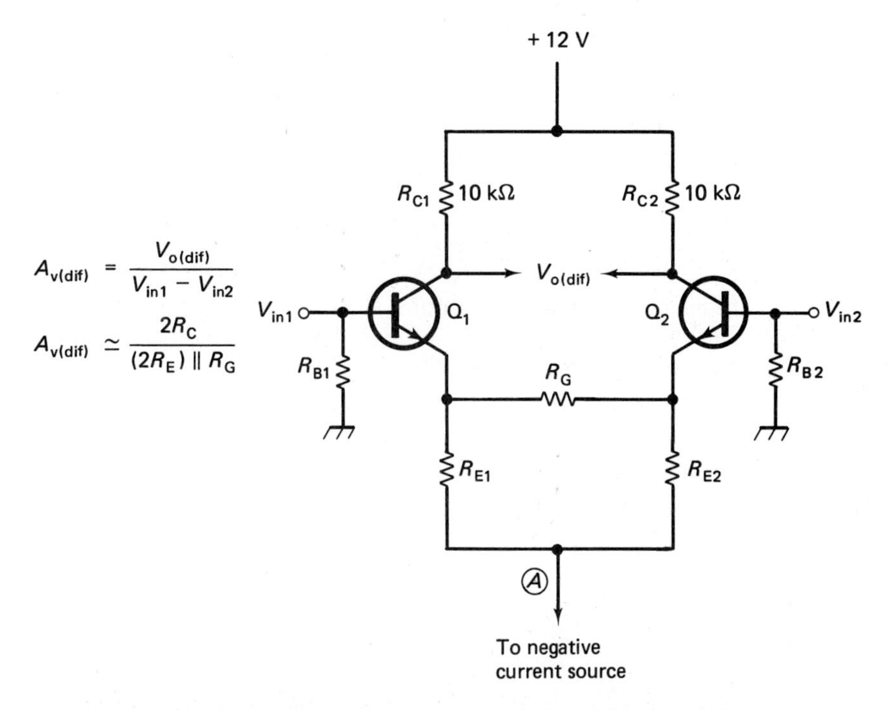

Figure 10-4 Differential amplifier with a gain-controlling resistor R_G.

Solution If the collector voltages are to be +6, there will be a 6-V drop across the collector resistors:

$$I_{R(C1)} = I_{R(C2)} = \frac{V}{R_C} = \frac{6\text{ V}}{10\text{ k}\Omega} = 0.6\text{ mA}$$

The source current is the sum of the collector currents:

$$I_S = I_{R(C1)} + I_{R(C2)} = 0.6 + 0.6 = \mathbf{1.2\text{ mA}}$$

The voltage at the top (emitter) end of R_{E1} is 0.6 V below the base voltage, which is zero at no signal:

$$V_{R(E1)} = V_{R(E2)} = V_{E1} - V_A = -0.6 - (-4) = 3.4\text{ V}$$

$$R_{E1} = R_{E2} = \frac{V}{I_{E1}} = \frac{3.4\text{ V}}{0.6\text{ mA}} = \mathbf{5.7\text{ k}\Omega}$$

The nearest standard value is 5.6 kΩ.

The emitter junction resistances are each about 30 mV/0.6 mA or 50 Ω. The value of emitter line resistance required to produce the desired gain of 50 can be found by transposing the gain formula.

$$A_{v(dif)} = \frac{2R_C}{(2R_E) \| R_G + 2r_j}; \qquad 2R_E \| R_G = \frac{2R_C}{A_{v(dif)}} - 2r_j$$

$$2(5.6\ \text{k}\Omega) \| R_G = \frac{2(10\ \text{k}\Omega)}{50} - (2 \times 50); \qquad 11.2\ \text{k}\Omega \| R_G = 300\ \Omega$$

Stated verbally, this last expression asks, What value of R_G must be paralleled with 11.2 kΩ to yield a total resistance of 2 kΩ? This can be found from the product-over-the-*difference* formula, which is related to the familiar product-over-the-sum formula, as shown below:

$$R_T = \frac{R_1 R_2}{R_1 + R_2}; \qquad R_2 = \frac{R_1 R_T}{R_1 - R_T}$$

Using the "difference" formula to find R_G,

$$R_G = \frac{(11.2\ \text{k}\Omega)(0.3\ \text{k}\Omega)}{11.2\ \text{k}\Omega - 0.3\ \text{k}\Omega} = \mathbf{308\ \Omega}$$

Common-mode rejection ratio: It was mentioned earilier that a differential amplifier responds only to the difference signal between the two inputs, and any *common-mode* signal appearing identically on the two inputs would be ignored and would not appear at the output. The degree to which this ideal is realized is expressed by the *common-mode rejection ratio* of the amplifier. Stated more exactly, the common-mode rejection ratio is defined as follows:

$$CMRR = \frac{A_{v(dif)}}{A_{v(com)}} \tag{10-7}$$

The technique for measuring this parameter is

1. Apply a signal to one input of the differential amplifier with the other input grounded.
2. Measure $V_{in(dif)}$ and $V_{o(dif)}$ (between the two outputs), and compute $A_{v(dif)} = V_{o(dif)}/V_{in(dif)}$.
3. Tie the two inputs together and apply an input signal to them (common mode). This signal will generally have to be quite a bit larger than the V_{in} applied in step 1 to produce an observable output.
4. Measure $V_{in(com)}$ and $V_{o(dif)}$, and compute $A_{v(com)}$.
5. Compute the ratio of differential to common-mode voltage gain.

Example 10-4

An amplifier connected for differential inputs has input terminal V_B grounded and $V_A = 50$ mV p-p, $V_o = 2$ V p-p. With V_A and V_B tied together and $V_A = V_B = 1$ V p-p, $V_o = 8$ mV p-p. Find the common-mode rejection ratio of the amplifier.

Solution

$$CMRR = \frac{A_{v(dif)}}{A_{v(com)}} = \frac{2000\ \text{mV p-p}/50\ \text{mV p-p}}{8\ \text{mV p-p}/1000\ \text{mV p-p}} = \mathbf{5000}$$

This number is often expressed as 5000 to 1, meaning that differential signals are amplified 5000 times as much as common-mode signals.

Prediction of *CMRR* from a knowledge of circuit-component values is generally not practical because it is the nonideal attributes of the components that make *CMRR* less than infinite.

Common-mode voltage limit. If both input voltages in Fig. 10-3 are raised together, a positive voltage will be reached at which Q_1 and Q_2 saturate. This will be near $V_{IN} = V_C$. Similarly, if both input voltages are brought lower, a negative voltage will be reached at which Q_3 saturates. This will be near $V_{IN} = (V_{E3} + I_E R_4)$. Beyond these limits the circuit loses its ability to amplify faithfully and to reject common-mode input signals.

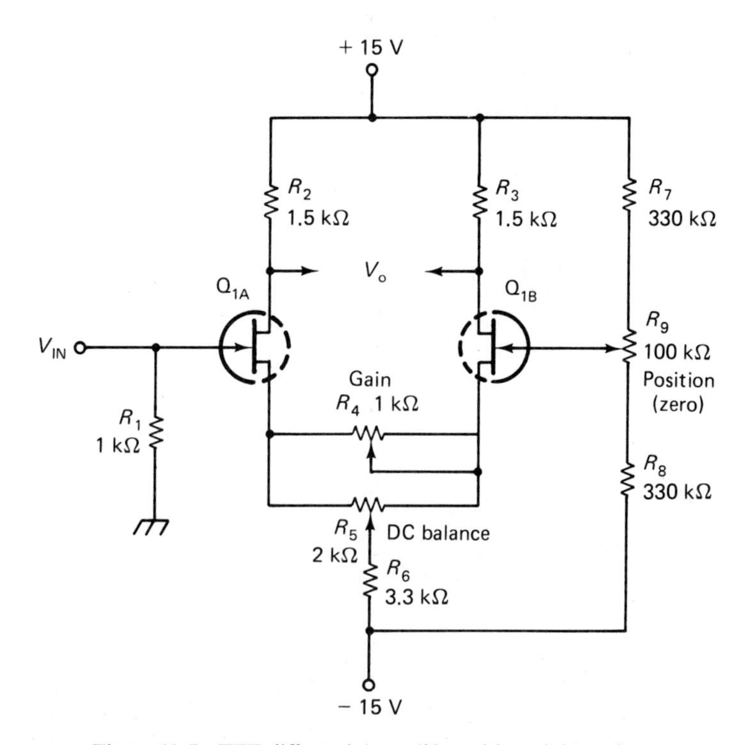

Figure 10-5 FET differential amplifier with variable gain.

To amplify dc, a differential amplifier is almost the universal choice. Figure 10-5 shows an FET differential amplifier that might be used in the input stage of an oscilloscope or voltmeter. There is only one signal input. The advantage of the differential circuit is freedom from dc drift in the output. The FETs are a matched pair mounted in the same six-lead package. Thus, any change in V_{GS} of one unit will be nearly identical in the other—a *common-mode* disturbance which the amplifier ignores.

Dc-balance pot R_5 is adjusted for zero voltage across R_4 when $V_{IN} = 0$. Under these conditions V_O is zero and remains zero as the gain is adjusted. With dc balance misadjusted, varying the amplifier's gain varies the dc voltage at V_O.

This circuit has the current source replaced by a single resistor, R_6. At zero input V_{R6} is almost 15 V. As long as the input voltage is no more than 5% of this we can expect reasonably constant current in R_6. Common-mode rejection suffers from this simplification, but high CMRR is not needed in this single-input amplifier.

10.4 NEGATIVE FEEDBACK

The characteristic curves of real transistors show that beta is not perfectly constant but varies with collector current and voltage. This is shown in exaggerated form in Fig. 10-6(a). Notice that the I_B lines become farther apart at high collector currents, indicating higher beta.

From the load line given, it is evident that large signal swings toward saturation will produce large output-voltage changes (distance A in Fig. 10-6), while equally large signal swings toward cutoff will produce smaller output voltage changes (distance B in the figure). The output wave will therefore be a distorted copy of the original, as shown in Fig. 10-6(b).

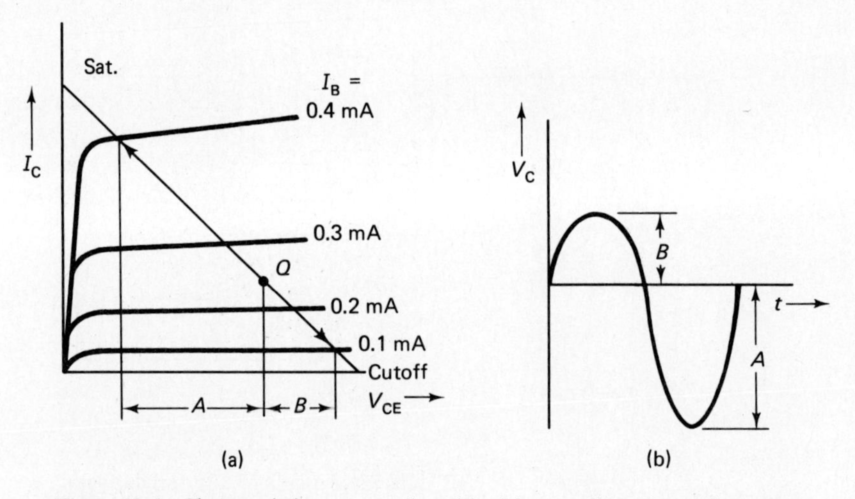

Figure 10-6 Characteristic curve nonlinearities (a) cause distortion of the output waveform (b) for large signal swings.

Other factors can contribute to the distortion of the input signal within an amplifier. The most common sources of distortion are

1. Characteristic-curve nonlinearities, as described above
2. Change of input impedance as the base-emitter-diode current is swung over a large range
3. Change of output impedance if the amplifier is driving a nonlinear load, such as the base of a following amplifier stage

4. Distortion caused by magnetization nonlinearities in coupling chokes and transformers
5. Sixty-hertz hum either coupled to the input by stray capacitance from the ac line or fed in through the power supply
6. Frequency distortion, limiting the high- or low-frequency response, as discussed in Chapter 9

The first of these problems can be overcome by *loafing* the final power amplifier stage, that is, designing it for, say, 100-W output, but using it for only 10-W output, so the signal never swings over the entire range of the curves. However, this is wasteful in terms of power efficiency and initial cost. A second solution to the first problem is to select transistors whose curves are as linear as possible.

A solution to the distortion problem which acts on *all* of the six causes listed, however, is *negative feedback*. Negative feedback is a scheme whereby a portion of the output signal is fed back to be compared with the input signal. If the comparison reveals any difference in wave shape, the *difference signal* is applied to the amplifier input. The phase of the fed-back signal must be inverted (negative) compared to the input signal. Any difference signal at the input will cause the output-signal distortion to be minimized.

This process is illustrated in Fig. 10-7. The figure shows an input triangle wave applied to an inverting amplifier with a large gain, $-A$. The amplifier, however, is assumed to have a nonlinearity which attempts to produce a flattening to 45% on the positive half of the ouput wave. A portion of this distorted wave is sent back through the feedback network to be combined with the input signal. The input signal

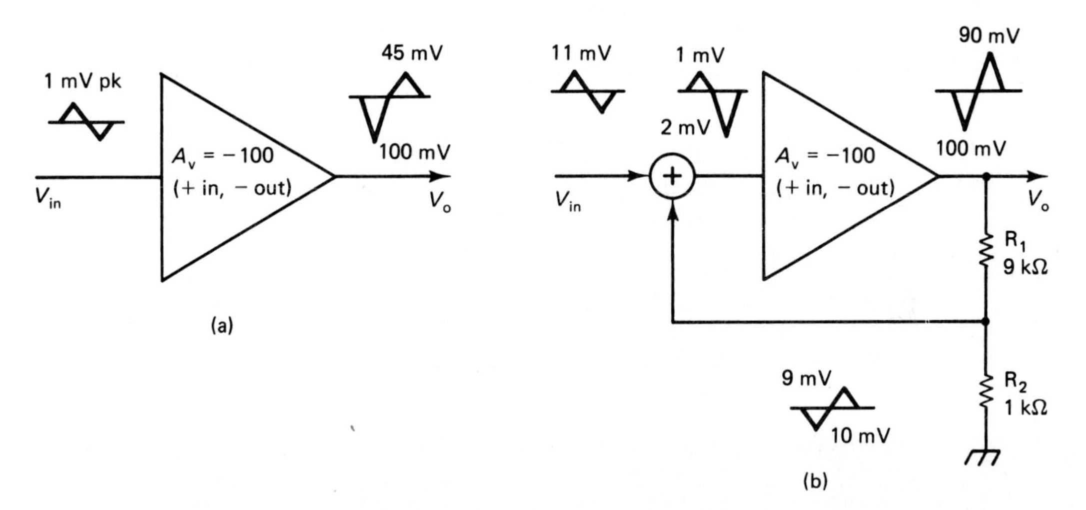

Figure 10-7 A distorting amplifier with $A_v = -100$ for negative outputs but $A_v = -45$ for positive outputs (a) produces nearly equal outputs with negative feedback. (b) A_v is reduced to about -9.

and the fed-back signal are out of phase and cancel each other to a large extent. However, the flattened positive half-cycle is not canceled by the input signal to as great an extent as is the larger negative half-cycle, so the negative half-cycle appears exaggerated at the amplifier input (2 mV compared to 1 mV).

This overemphasized negative input tends to compensate for the lower gain of the amplifier on the negative half-cycle, resulting in an improved output waveshape. Of course, the distortion can never be completely eliminated because the difference signal at the input of the amplifier would then show no emphasis on the flattened side and this emphasis is needed to keep the output from gross distortion.

Closed loop/open loop: The gain of the amplifier in Fig. 10-7(a) is called open-loop gain. We will give it the symbol A_{vo}. The gain of the circuit in Fig. 10-7(b) is called *closed-loop gain*, which we will identify as A_{vc}. The feedback factor, which is the voltage division ratio of R_1 and R_2, is called B (some texts use β). In any negative-feedback system, the closed-loop gain is given by

$$A_{vc} = \frac{A_{vo}}{1 + A_{vo}B} \tag{10-8}$$

Although A_v is negative, we treat it in this equation as an unsigned quantity. If A_{vo} is very much larger than A_{vc}, the overall gain does not depend on the amplifier's A_{vo} but only on the feedback factor B:

$$A_{vc} \approx \frac{1}{B} \tag{10-9}$$

It may seem wasteful to design an amplifier for an open-loop gain of 1000 and then deliberately knock it down to 100 with negative feedback, but it is actually a bargain. Low-level amplifier stages can be built for 50 cents. If sacrificing the gain of this low-cost stage can decrease distortion and provide gain stability for a system containing a $10 power amplifier, then the sacrifice is well worth it.

Improvement in distortion is, in fact, directly proportional to the amount of gain we "throw away." In Fig. 10-7, for example, a gain of 45/1 or 45 was reduced to 90/11 or 8.18 on the positive output half-cycle—a factor of 5.5. The distortion was reduced from a 100 − 45 or 55-mV deficit on the positive output to a 100 − 90 or 10-mV deficit—also a factor of 5.5. In general, the gain reduction ratio is

$$\frac{A_{vo}}{A_{vc}} = 1 + A_{vo}B \tag{10-10}$$

and distortion is also reduced by this factor. The specifications for an amplifier may include the amount of feedback used, which is the term $1 + A_{vo}B$, expressed in dB.

In addition to reducing distortion, negative feedback extends the high- and low-frequency response of an amplifier, and it can be used to raise Z_{in} and lower Z_o in certain connections.

10.5 TYPES OF NEGATIVE FEEDBACK

The signal derived from the amplifier's output *voltage* can be combined with the input signal in two distinct ways: it can be applied in *shunt* to subtract from the input-signal current, or it can be applied in *series* to subtract from the input-signal voltage. In both cases it is essential that the feedback phasing be negative, that is, subtracting from V_{in}. The feedback signal can also be derived from the amplifier's output *current*, instead of its output voltage. Thus there are four types of negative feedback: voltage-shunt, voltage series, current-shunt, and current-series.

Voltage-shunt feedback is shown in Fig. 10-8, which closely approximates the hypothetical situation of Fig. 10-7. Gross distortion is evident with V_{in} adjusted for $V_o = 3$ V p-p because the diodes turn on at ± 0.6 V, throwing on the heavy load R_L and flattening the signal peaks. The distortion is especially evident if a triangle-wave input signal is used. In addition to the nonlinearity of the load, nonlinearities of the transistors produce other distortion, although this is not so dramatically evident.

With the feedback loop closed (S_1 closed) V_o feeds a current back to the input

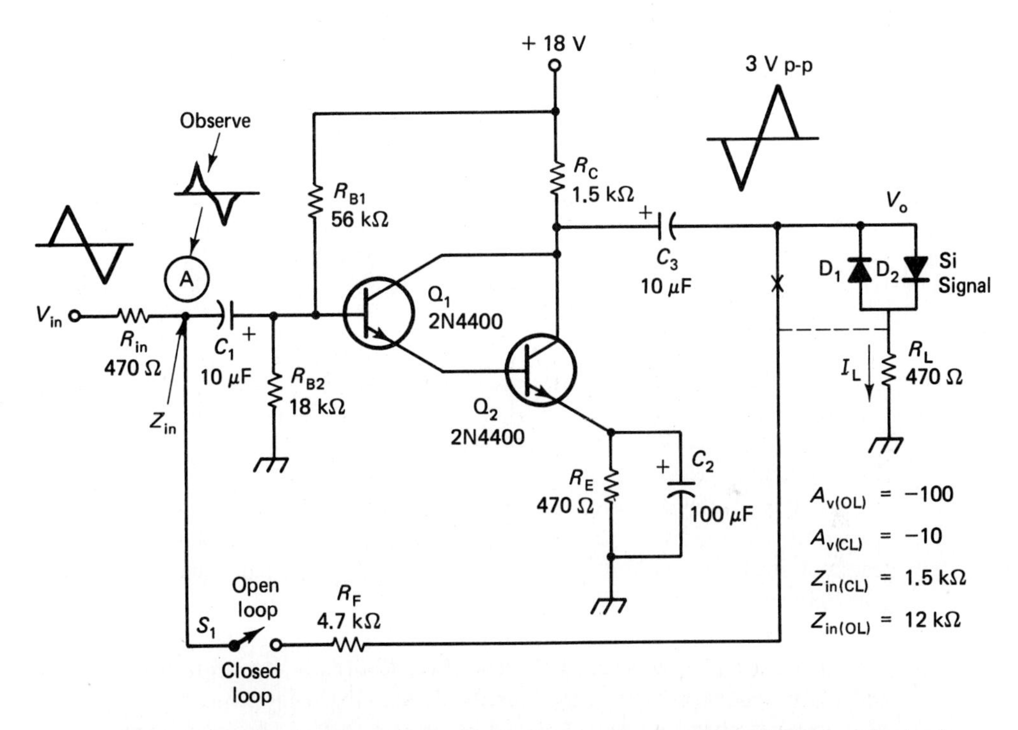

Figure 10-8 An amplifier demonstrating voltage-shunt negative feedback. The load consists of diodes and a resistor to deliberately produce distortion. Breaking the line at the X and connecting the dashed line converts to current-shunt feedback.

through R_F. This current is inverted from the input current which V_{in} feeds through R_{in}. The ratio R_{in}/R_F is the feedback factor B, in this example, 0.1. Note that R_{in} includes (and may consist entirely of) the resistance of the signal source.

It is interesting to build the circuit of Fig. 10-8 and observe the difference signal at A. The triangle waves fed to the base of Q_1 will have their peaks overemphasized to compensate for the flattening caused by the amplifier and nonlinear load.

Input impedance with voltage shunt feedback is lower than open-loop Z_{in}. This is because R_F shunts the normal input impedance $r_b \,||\, R_{B1} \,||\, R_{B2}$. The Miller effect makes R_F appear as $R_F/(1 + A_v)$, where A_v is the gain from point A to the output. In general, voltage-shunt feedback lowers Z_{in} at point A (R_{in} not included) by the factor $1 + A_{vo}B$. Closed-loop Z_{in} for the circuit of Fig. 10-8 is then

$$Z_{in\,(c)} = R_{in} + \frac{R_{B1} \,||\, R_{B2} \,||\, r_b}{1 + A_{vo}B} = R_{in} + \frac{Z_{in\,(A)}}{1 + A_{vo}B} \tag{10-11}$$

where

$$r_b = 2\beta^2 \frac{30 \text{ mV}}{I_{E2}} \tag{7-9}$$

Example 10-5

A circuit similar to the one in Fig. 10-8 has $A_{vo} = 1000$, $Z_{in\,(A)} = 10 \text{ k}\Omega$, $R_F = 8 \text{ k}\Omega$, and $R_{in} = 2 \text{ k}\Omega$. Find $Z_{in\,(c)}$ and A_{vc}.

Solution

$$B = \frac{R_{in}}{R_F} = \frac{2000}{8000} = 0.25$$

$$Z_{in\,(c)} = R_{in} + \frac{Z_{in\,(A)}}{1 + A_{vo}B} = 2000 + \frac{10\,000}{1 + (1000)(0.25)} \tag{10-11}$$

$$= \mathbf{2040\ \Omega}$$

$$A_{vc} = \frac{A_{vo}}{1 + A_{vo}B} = \frac{1000}{1 + (1000)(0.25)} = \mathbf{3.98} \tag{10-10}$$

These results illustrate that as A_{vo} becomes very large A_{vc} approaches $1/B$ and Z_{in} approaches R_{in}.

Voltage-series negative feedback is shown in Fig. 10-9. Again, distortion is produced deliberately with a nonlinear load, and is nearly eliminated by negative feedback in the *closed-loop* position of S_1. This time the gain sacrificed is a factor of approximately 50, rather than 10 as in Fig. 10-8. The remaining distortion is therefore noticeably less.

V_{in} is applied between Q_1 base and emitter in the open-loop position (since C_1 and C_2 are short circuits to the signal). In the closed-loop position a fraction (1/11) of V_o appears across R_B and is placed in *series* with V_{in}. The waveforms show that the fed-back voltage subtracts from V_{in}, leaving the difference from the base to emitter of Q_1. This difference signal with its overemphasized peaks can be viewed more conveniently at the Q_1 collector.

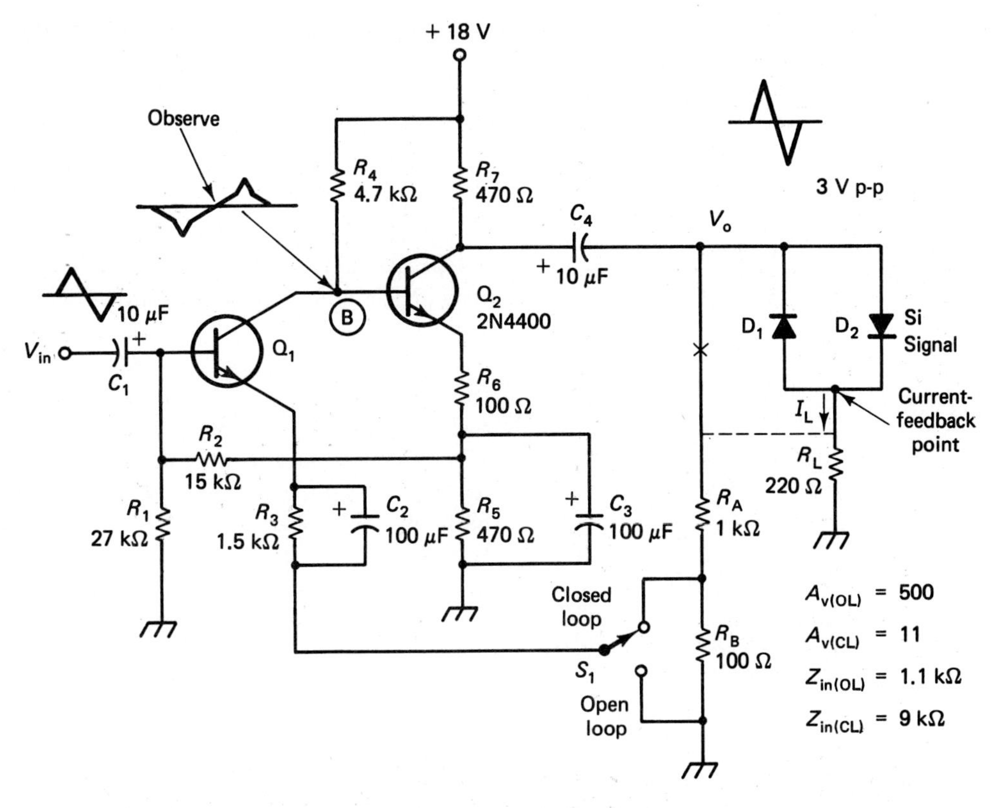

Figure 10-9 An amplifier demonstrating voltage-series negative feedback. The input-signal path is through C_1, $Q_{1(B\text{-}E)}$, C_2, and R_B to ground. $V_{R(B)} = \frac{1}{11}V_o$. The alternate connection (dashed line) produces current-series feedback.

The input impedance with voltage-series negative feedback is higher than open-loop Z_{in} because the fed-back voltage opposes V_{in}, reducing the input-signal current. In the circuit of Fig. 10-9 the fed-back voltage appears in series with the base-emitter signal path only, so it is only base-input resistance r_b that is increased (by the factor $1 + A_{vo}B$, of course). Bias resistors R_1 and R_2 still shunt the input, making the total input impedance

$$Z_{in(c)} = R_1 \,||\, R_2 \,||\, \beta_1 r_{j1}(1 + A_{vo}B) \tag{10-12}$$

Bootstrapping could be used to eliminate the shunting effect of R_1 and R_2. C_B would go from the top of R_3 to the junction of R_1 and R_2, with an added resistor in the line to the base.

Output impedance is the property of an amplifier which causes the output voltage to drop upon the addition of a load. If adding a 1-kΩ load causes V_o to drop to one half of its no-load value, Z_o is also 1 kΩ. Both voltage-feedback schemes tend to keep V_o constant in spite of changing loads. This is because a lower V_o produces less fed-back voltage, resulting in a larger difference voltage to the input of the amplifier.

Thus voltage negative feedback causes the effective amplifier output impedance to be lowered. The factor of decrease, as might be expected, is $1 + A_{vo}B$. For *voltage* feedback

$$Z_{o(c)} = \frac{Z_{o(o)}}{1 + A_{vo}B} \tag{10-13}$$

Current feedback. So far we have been assuming that the objective of an ideal amplifier is to produce an output voltage that is an exact copy of the input-voltage waveform. Most often this is so, but occasionally there may be need to make the output *current* waveform an exact copy of the input voltage, in spite of load changes or nonlinearities. If the load is a linear element with a constant value (such as a resistor) current is proportional to resistance and there is really no difference between voltage and current feedback.

If the load varies or is nonlinear we may wish to use current negative feedback. We derive a current-feedback signal by placing a resistor in series with the load. The voltage across this resistor is proportional to load current, and can be used to apply a shunt current to the input or it can be placed in series with the input voltage. The circuits of Figs. 10-8 and 10-9 can be converted to current negative feedback simply by moving the feedback line to the resistance R_L which appears in series with the nonlinear load. This will make the load-current waveform a near-perfect triangle, although V_o will be highly distorted.

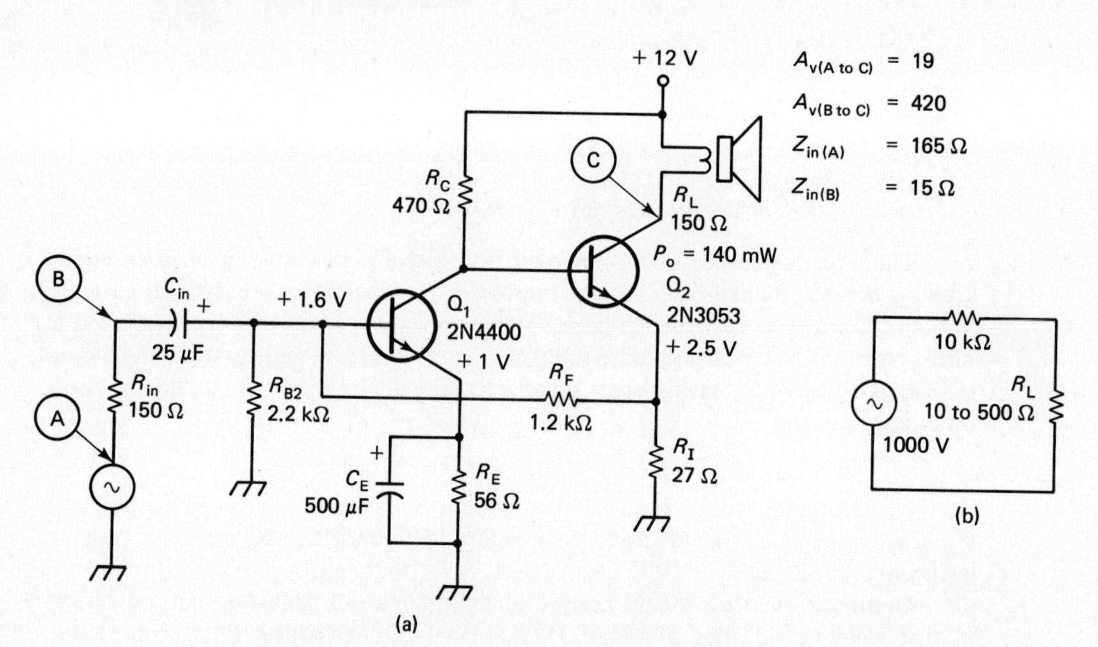

Figure 10-10 (a) Current feedback can be derived from an unbypassed emitter resistor in the output stage. (b) Current feedback makes the output impedance appear very high, so output current remains constant in spite of changing R_L.

Another, and perhaps more common method of implementing current negative feedback is shown in Fig. 10-10. The load is a loudspeaker which has inherent nonlinearities. The load current is equal to the current in R_I, so shunt feedback is taken from $V_{R(I)}$ through R_F. R_F also serves as base-bias resistor R_{B1} for Q_1.

Output impedance with current feedback. What if we had an amplifier with a no-load output 1000 V and a Z_o of 10000 Ω? I_L would be very nearly 0.1 A for any value of R_L from zero to several hundred ohms. This is illustrated in Fig. 10-10(b). Amplifiers with current-derived feedback behave like this (unless R_L becomes so high that $I_L R_L$ exceeds the maximum available V_o). We say that current feedback raises an amplifier's Z_o, making it look like a current source.

A summary of feedback types is presented in Table 10-1. In addition, notice the following points:

- For all types, A_{vc} is less than A_{vo} by the factor $1 + A_{vo}B$, and approaches $1/B$ as larger factors of gain are sacrificed. A_{vc} is the *system* gain (V_{in} is applied at the *left* end of R_{in} in Figs. 10-8 and 10-10.
- Voltage feedback is obtained by tapping across the load (shunt- or parallel-derived). It always lowers Z_o.
- Current feedback is obtained by a resistor in the load-current path (series-derived). It always raises Z_o.
- Feedback is applied in shunt by the junction of two resistances, R_F and R_{in} (current summing). It always lowers the Z_{in} at the junction.
- Feedback is applied in series by a voltage divider R_A-R_B. The voltage across R_B is placed in series with the input-to-ground signal path (voltage summing). It always raises Z_{in} of that signal path.
- Where input impedances are raised or lowered by $1 + A_{vo}B$ it is only the impedances within the feedback loop that are affected. In Fig. 10-8, Z_{in} at point A is lowered, but R_{in} appears unaltered in series with it. In Fig. 10-9, r_{b1} is raised by $1 + A_{vo}B$, but R_1 and R_2, unaltered, shunt this raised value.
- In the simple case where only one inductance or capacitance limits f_{hi} or f_{low}, negative feedback extends bandwidth by the factor $1 + A_{vo}B$. Where two reactive elements come into play together, the benefit factor is $\sqrt{1 + A_{vo}B}$. For three reactances it is $\sqrt[3]{1 + A_{vo}B}$.

TABLE 10-1 SUMMARY OF NEGATIVE FEEDBACK TYPES

Type	Z_{in}	Z_o
Voltage-derived, shunt-applied	Lowered	Lowered
Voltage-derived, series applied	Raised	Lowered
Current-derived, shunt-applied	Lowered	Raised
Current-derived, series-applied	Raised	Raised

10.6 TUNED RF VOLTAGE AMPLIFIERS

A common requirement of radio and other communications systems is for an amplifier to pass signals in a desired narrow range of frequencies, while rejecting all others. At first this may seem a simple task. The impedance of a parallel tuned circuit is high (theoretically, infinite) at resonance. The gain of a common-emitter amplifier is $Z_{c(\text{line})}/Z_{e(\text{line})}$. Simply placing a parallel tuned circuit in the collector line will give high A_v at resonance, as shown in Fig. 10-11.

Some relevant equations from basic circuit analysis are reproduced here for reference. The tuned (resonant) frequency is

$$f_r = \frac{1}{2\pi\sqrt{LC}} \tag{10-14}$$

The reactances of L and C (equal at resonance) are

$$X_{Lr} = 2\pi f_r L = X_{Cr} = \frac{1}{2\pi f_r C} \tag{10-15}$$

The Q (quality factor) of a parallel tuned circuit with parallel load resistance R_L, and the bandwidth B_w (span between upper and lower half-power points) are

$$Q = \frac{R_L}{X_{Lr}} = \frac{R_L}{X_{Cr}} \tag{10-16}$$

$$B_w = \frac{f_r}{Q} \tag{10-17}$$

At any frequency f the total impedance of L, C, and R_L in parallel is

$$Z = \frac{1}{\sqrt{\dfrac{1}{X_T^2} + \dfrac{1}{R_L^2}}} \tag{10-18}$$

where

$$X_T = \frac{1}{\dfrac{1}{X_L} - \dfrac{1}{X_C}} \tag{10-19}$$

The curve of Fig. 10-11(b) was obtained from these equations.

Realization of a curve with a Q and peak gain this high requires that the Q of the coil be much higher than the circuit Q, which is 20 in this example. Practical coils of reasonable size have Q typically in the range 50 to 200 at this frequency. Where Q_{coil} is less than 10 times the intended circuit Q, Q_{circuit} should be calculated from

$$Q_{\text{circuit}} = \frac{1}{\dfrac{X_{Lr}}{R_L} + \dfrac{1}{Q_{\text{coil}}}} \tag{10-20}$$

L/C ratio: Equation (10-16) would make it seem that higher circuit Q (hence, sharper selectivity) could be obtained by making X_L and X_C lower at resonance. This would be done by using a low L/C ratio—small inductor and large capacitor.

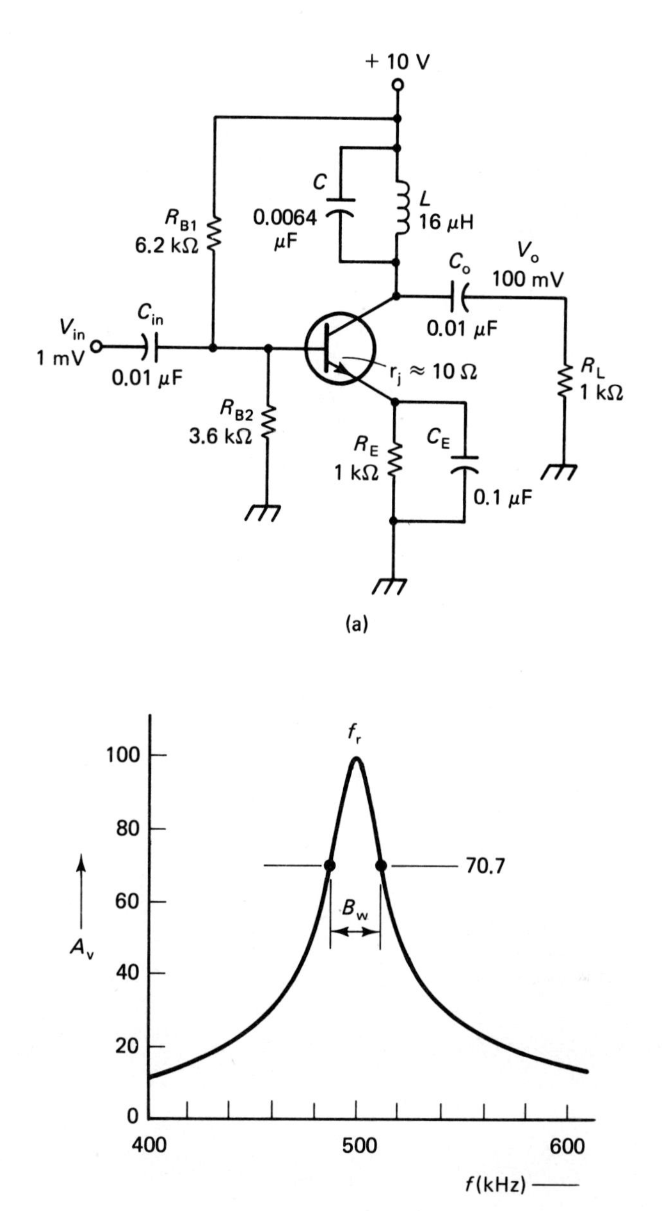

Figure 10-11 A tuned RF amplifier: A_v is collector-line impedance divided by r_j, and $Z_{c(\text{line})}$ is maximum where $X_L = -X_C$.

To an extent this is so, but for very low values of X_{Lr} the winding resistance of the coil itself places a low equivalent shunt resistance across the tuned circuit. This resistance limits the gain of the amplifier at f_r, and can have a greater effect than R_L if the L/C ratio is made too low. The approximate value of this shunt resistance is

$$R_P \approx Q_{coil} X_L \approx (Q_{coil})^2 r_w \tag{10-21}$$

where r_w is the winding resistance of the coil. In our example R_P is 2.5 kΩ if $Q_L = 50$; hardly negligible for the L/C ratio given, and disastrous if X_L is lowered to attempt a lower L/C ratio.

Transformer coupling. A better way to achieve sharp selectivity is to transform the load impedance to a higher value with a step-down transofrmer, as shown in Fig. 10-12(a). The base and emitter bias circuitry is the same as in Fig. 10-11. R_L is actually the base-input resistance of a following cascaded stage, as shown.

At first thought it would seem that a step-down transformer would reduce overall voltage gain. This is *not so*, however, because the transformer reflects the 1-kΩ load up by the factor $1/n^2$, so the voltage gain to the collector of Q_1 is increased by $1/n^2$. The transformer steps voltage to the load down by the factor n, so overall gain to the load is *increased* by $(1/n^2)(n)$ or a factor of $1/n$. This is 3.3 in Fig. 10-12(a).

A higher L/C ratio is necessary with transformer coupling to keep the effective shunt of the primary-winding resistance from swamping the reflected load resistance.

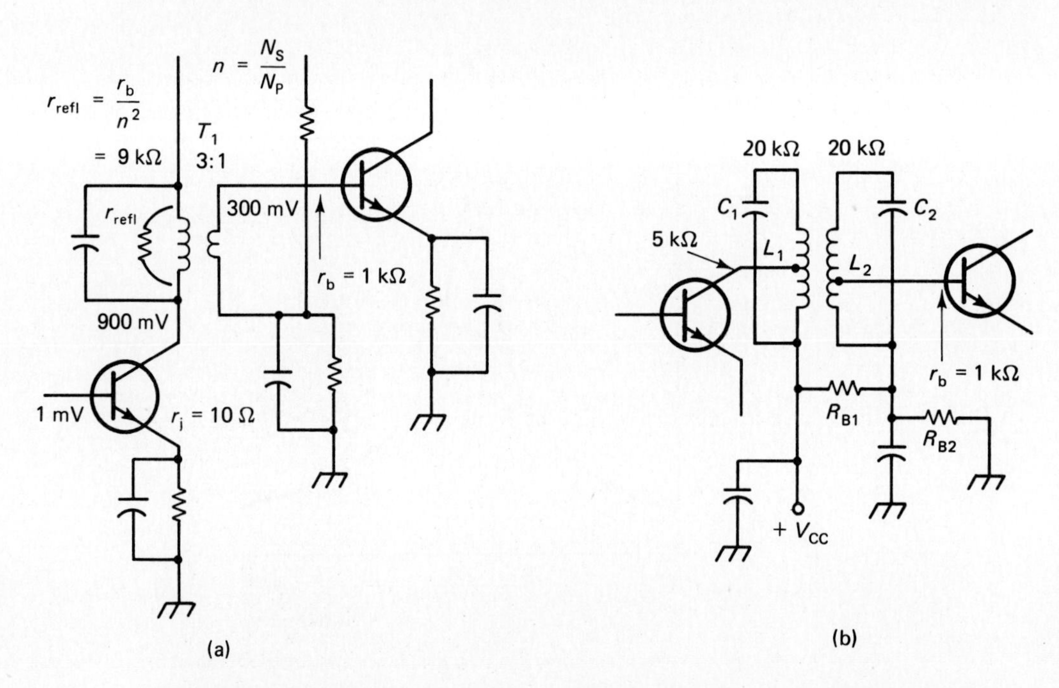

Figure 10-12 (a) A step-down transformer with untuned secondary is often used to raise the base-input resistance to a value that will not spoil the Q of the collector tuned circuit. (b) Tapped transformer coils permit tuning both sides as well as optimizing L/C ratio.

If we require the effective shunt to be 25 kΩ and the Q of the coil is 50 as before:

$$X_{Lr} = \frac{R_P}{Q_L} = \frac{25\ \text{k}\Omega}{50} = 500\ \Omega \tag{10-22}$$

This requires that $L = 160\ \mu\text{H}$ and $C = 640$ pF. Note that the L/C ratio is 100 times higher than in the example of Fig. 10-11, and the R_P value is 10 times higher.

A limit on the step-down ratio of T_1 is imposed by r_c the dynamic collector resistance of Q_1, which also appears in shunt with r_{refl}. A typical value of r_c for small-signal silicon transistors is 50 kΩ.

Tapping down on the transformer primary allows the L/C ratio to be optimized for the reflected resistance while transforming the transistor-output resistance up to a higher value. A tuned circuit can be added to the secondary for greater selectivity if the low base-input resistance is transformed up to a high value by using a tapped secondary winding. Both of these techniques are illustrated in Fig. 10-12(b), although they may be used independently.

Loose coupling is the norm with tuned RF transformers. This means that only a small part of the magnetic flux produced by the primary is intercepted by the secondary. This is because the transformers are not generally wound on closed-path magnetic cores, but on fiber or ceramic cylinders. They are made this way because it is more economical to tune them with a ferrite slug threaded inside the cylinder than it is to use a variable capacitor.

A complete treatment of loose-coupled RF transformers is quite an undertaking and will not be attempted here.* They do not behave like the idealized textbook transformers with $(N_S/N_P)^2 = Z_S/Z_P$. Several distinct effects are possible and the governing factors are things like closeness of coupling, skin-effect winding resistance, distributed winding capacitance, and transistor output resistance. It's not quite as well controlled an art as choosing a pair of 5% resistors for a voltage divider. Nevertheless, a few notes will be set forth to indicate some of the possibilities.

- A 1:1 transformer does not always have a V_{in}/V_o ratio of 1. With a 25% coefficient of coupling and primary tuning it can behave like a 4:1 step-down transformer.
- With a low source impedance, a very high load impedance, and double tuning (series tuning in the primary, parallel tuning in the secondary) a loose-coupled 1:1 transformer can step voltage up, commonly by a factor of 50.
- With a high Z_s and equally high Z_L a loose-coupled double-parallel tuned transformer can provide an effective circuit Q over 100 with small and inexpensive components.
- With somewhat tighter coupling and double parallel tuning, a double-peaked output response appears, as shown in Fig. 10-13(a). By adjusting the coupling, a flat-topped response curve can be achieved. This is ideal for passing a broader range of frequencies.

*For a complete treatment, see Chapter 5 of *Electronic Components, Instruments, and Troubleshooting*, by Daniel L. Metzger, Prentice-Hall, Englewood Cliffs, N.J., 1981.

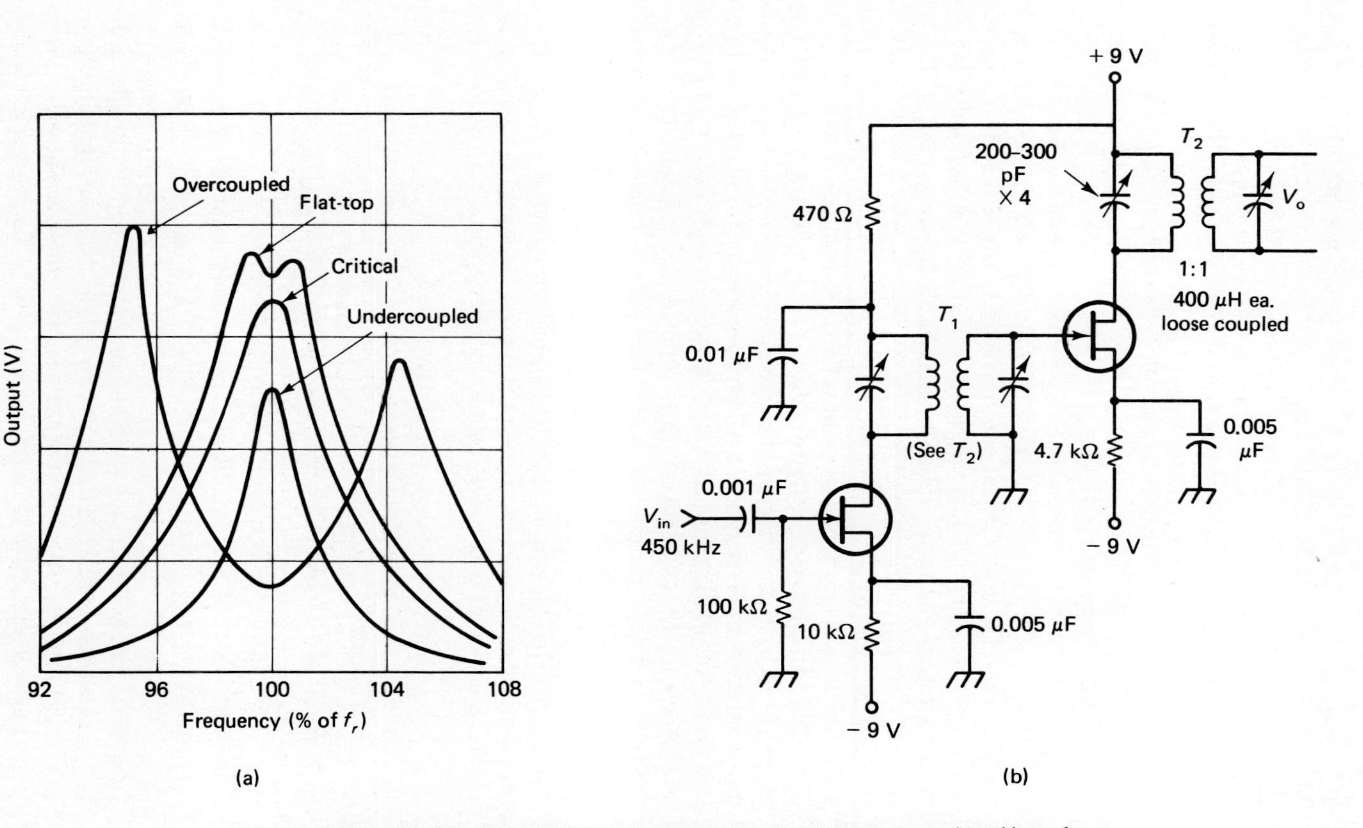

Figure 10-13 (a) Double-parallel-tuned transformers with high source and load impedances exhibit a variety of frequency-response curves depending upon the closeness of coupling. (b) FET amplifier using double parallel-tuned transformers.

Loose coupling is often used deliberately in cascading tuned FET amplifiers because the high load impedance presented to the driving stage permits high-Q and flat-topped response curves to be obtained. Figure 10-13(b) shows an FET cascade with loose-coupled double-tuned transformer. Cascaded tuned bipolar amplifiers are likely to use tighter coupling because the low base-input resistance loads the output and ruins the response curve if the coupling is too loose.

CHAPTER SUMMARY

1. The common-emitter amplifiers of Chapter 6 can be connected in cascade (output to input) to achieve very high voltage gains. If the collector of the first stage is biased at a low voltage, it can be coupled directly to the base of the second stage. An inverted-ground *PNP* stage driving an upright *NPN* stage is especially suited to direct coupling.
2. A differential amplifier has two inputs. It amplifies the *differential* signal between them but rejects any *common-mode* signal appearing between them both and ground. The dif amp uses no coupling capacitors and will respond to all low frequencies including dc.
3. The differential-amplifier circuit uses a balanced pair of transistors in the common-emitter configuration. A transistor current source supplies emitter-bias current. Voltage gain is calculated as

$$A_v = \frac{2R_C}{2r_j + (2R_E) \| R_G}$$

where R_C is one of two identical collector resistors, R_E is one of two emitter resistors, and R_G is a gain-control resistor between the two emitters.
4. Common-mode rejection ratio (*CMRR*) is a measure of a differential amplifier's ability to amplify differential signals and screen out common-mode signals. It is calculated as

$$CMRR = \frac{A_{v(\text{dif})}}{A_{v(\text{com})}} \qquad (10\text{-}7)$$

5. Negative feedback reduces amplifier distortion and extends bandwidth, while lowering voltage gain—all by the factor $1 + A_{vo}B$, where A_{vo} is the original gain and B is the fraction of output signal fed back to the input. The factor $1 + A_{vo}B$ also applies to:

- Lowering Z_{in} for shunt-applied, raising Z_{in} for series-applied feedback
- Lowering Z_o for voltage-derived, raising Z_o for current-derived feedback

6. Tuned amplifiers commonly use a parallel *LC* circuit in the collector. The *L*/*C* ratio and load impedance should be adjusted for optimum selectivity. Transformer coupling is often used to adjust load impedance. Loose-coupled tuned transformers provide a variety of selectivity and voltage-gain effects which differ markedly from the ideal close-coupled transformer.

QUESTIONS AND PROBLEMS

10-1. In Fig. 10-1, find the value of unbypassed resistance which could be placed in the Q_1 emitter line to reduce the stage-one gain to -10.

10-2. In Fig. 10-1, what value of C_4 would produce a high-frequency cutoff of 6 kHz?

10-3. In Fig. 10-1, the load R_{10} is changed to 600 Ω and a new bias current I_{2E} is to be established to maximize V_o. Find the new I_{2E} and $V_{o(max)p\text{-}p}$.

10-4. Find $V_{C(Q2)}$ in Fig. 10-2(a). Use 0.6 V for each V_{BE} and assume that both betas are high enough so I_B may be neglected.

10-5. Find $V_{C(Q2)}$ in Fig. 10-2(b). Equation (7-6) gives $V_{E(Q2)}$ for this circuit.

10-6. Does the Miller effect lower the effective value of R_3 seen by the input signal in Fig. 10-2(b)? Find Z_{in} for this amplifier.

10-7. Redraw the differential amplifier circuit of Fig. 10-3 with the following value changes: $R_3 = R_4 = 100\ \Omega$, $R_5 = R_6 = 680\ \Omega$, $R_7 = 330\ \Omega$, $R_9 = 2.2\ \text{k}\Omega$. Analyze this circuit for bias point V_C and differential gain A_v.

10-8. What value resistor placed between the emitters of Q_1 and Q_2 in Problem 10-7 would make $A_o = 15$?

10-9. Explain in your own words the terms *common-mode-rejection ratio* and *common-mode-voltage limit.*

10-10. A differential amplifier has 10 mV at input 1 with input 2 grounded. V_o is 3.7 V. In a second test, inputs 1 and 2 are tied together and both have an input signal of 500 mV to ground. V_o is 85 mV in this test. (a) What is A_v? (b) What is *CMRR*? (c) Express *CMRR* in decibels.

10-11. In the FET differential amplifier of Fig. 10-5, assume that $V_{GS} = -2.0$ V and $y_{fs} = 4$ mS for each FET. (a) Find the bias point V_D. (b) Find $A_{v(dif)}$ with R_4 set to 1 kΩ. (c) Find $A_{v(dif)}$ with $R_4 = 0$.

10-12. What problem of a differential amplifier is corrected by a dc-balance control?

10-13. What is sacrificed and what is achieved by using negative feedback?

10-14. Distinguish between the terms *closed loop* and *open loop.*

10-15. An amplifier has an open-loop gain of 140 and a feedback factor of $\frac{1}{20}$. What is its closed-loop A_v?

10-16. An amplifier has an open-loop gain of 27. What feedback factor will make $A_{vc} = 10$?

10-17. An amplifier has $B = 0.0100$. What is the minimum A_{vo} that will ensure A_{vc} not less than 99?

10-18. Analyze the circuit of Fig. 10-8 for (a) A_{vo}; (b) A_{vc}; (c) $Z_{in(c)}$; (d) $Z_{in(o)}$. Consider the diodes to be shorted when determining A_{vo}. Compare your answers with the measured values given with the figure.

10-19. Repeat Problem 10-18 for Fig. 10-9.

10-20. When is current-derived feedback preferred to voltage-derived feedback?

10-21. What is the advantage of series-applied feedback over shunt-applied feedback?

10-22. An amplifier has $A_v = 2000$, $Z_{in} = 1000\ \Omega$, and $Z_o = 1000\ \Omega$ open loop. Find A_v, Z_{in}, and Z_o closed loop if $B = \frac{1}{50}$ and feedback is voltage-derived, shunt-applied.

10-23. Repeat Problem 10-20 for current-derived, series-applied feedback.

10-24. In a cascaded amplifier (similar to Fig. 10-1) let us say that C_4 is chosen to produce $f_{hi} = 6$ kHz and that $A_v = 5000$. Now assume that voltage-derived, series-applied feedback is applied (as in Fig. 10-9) with $R_A = 1800$ and $R_B = 100\ \Omega$. What will be the new f_{hi} (closed-loop)?

10-25. A capacitance of 680 pF appears in parallel with an inductance of 3.3 μH. (a) What is the resonant frequency? (b) What is the total reactance at a frequency 1.1 times f_r? (c) If a resistance of 500 Ω is placed in parallel with the C and L values given, what is the total impedance at $1.1 f_r$?

10-26. A 2.5-mH inductor has a Q of 20 at 50 kHz. (a) What are X_s and R_s for the series-equivalent circuit? (b) What are X_p and R_p for the parallel-equivalent circuit?

10-27. What inductance is required to resonate a 365-pF capacitor at 550 kHz?

10-28. Calculate the gain of the tuned amplifier in Fig. 10-11(a) at 525 kHz. Use $r_j = 10\ \Omega$ and do not neglect R_L. Does your calculation confirm the 525-kHz point on the graph of Fig. 10-11(b)?

10-29. Why are step-down transformers usually used to drive common-emitter tuned amplifiers?

10-30. What is an advantage of double-parallel-tuned loose-coupled transformers?

11

POWER AMPLIFIERS

11.1 CHOKE COUPLING

Most low-power amplifier circuits use a resistor to supply dc collector current and isolate the collector from the supply for ac, because a resistor is small, light, inexpensive, and not frequency-dependent. The dc current through the collector resistor does cause wasted power, however; and in high-power amplifiers this waste becomes objectionable. One way to reduce this power loss is to use a choke in place of the collector resistor, as shown in Fig. 11-1(a). The collector is still isolated from the supply for ac, so the collector voltage is free to swing up and down as it follows the signal variations. The dc collector voltage, however, is fixed at V_{CC}.

The dc load line for an inductively coupled circuit is drawn essentially vertical from the supply-voltage point, V_{CC}, since the dc resistance of the choke is near zero. The Q point is established by R_{B1}, R_{B2}, and R_E, and the ac load line passes through the Q point with a slope determined by R_L alone. Notice from Fig. 11-1(b) that on low-base-current swings, it is possible for the ac collector voltage to have peaks which are as much as twice the dc supply voltage. This is because the inductor's self-induced voltage adds to the supply voltage when the inductor's current is decreased.

To realize maximum output from this circuit, the ac load line must "hit" the $I_C = 0$ and $V_C = 0$ ends of the curves at equal distances up and down from the Q point. This maximum output will be achieved when

$$I_Q = \frac{V_{CC}}{R_L} \qquad (11\text{-}1)$$

Lowering I_Q will reduce the maximum undistorted output, and raising I_Q will simply waste power.

While inductive coupling cuts power waste, it does present several other problems:

1. The choke begins to act like a short circuit at low frequencies, limiting the gain of the amplifier at low frequencies.
2. The stray capacitance between windings of the inductor shorts the collector to ground at high frequencies, limiting the high-frequency response of the amplifier.
3. The inductor has an impedance which approaches infinity at the frequency of parallel resonance with its stray capacitance. If R_L is removed, A_v then approaches infinity. If the transistor is suddenly turned off by a large negative input signal, the inductor's self-induced voltage can then become large enough to damage the transistor.
4. Chokes for use at audio frequencies are heavy, large, and expensive.

Analysis of inductively coupled amplifier circuits is quite similar to the analysis presented in Sections 6.9 through 6.12. An example will serve to point out the few differences.

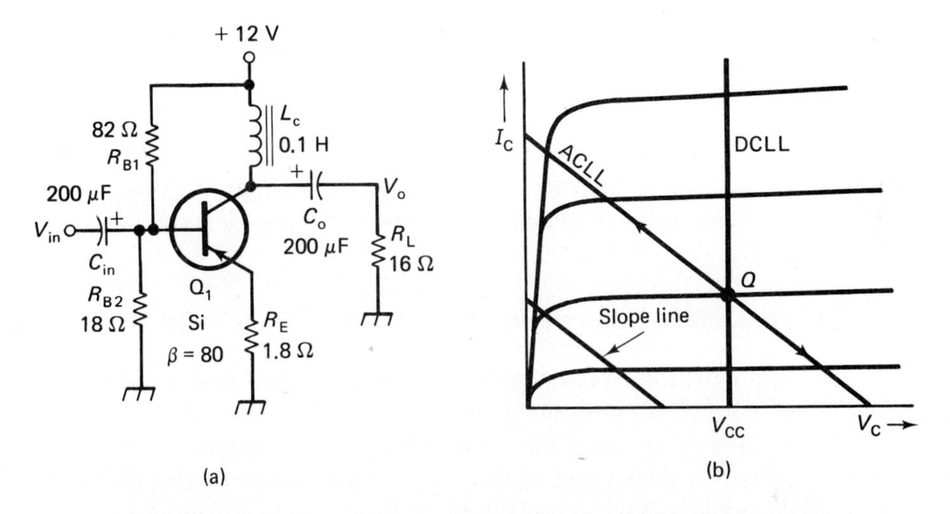

Figure 11-1 (a) Using a choke instead of a resistor in the collector line cuts power waste. (b) The dc load line for a choke-coupled amplifier is vertical from V_{CC}. The ac load line has a slope determined by R_L.

Example 11-1

Analyze the circuit of Fig. 11-1(a) to determine the input impedance, voltage gain, maximum power output, and power dissipated in the transistor. The choke L_C has a negligible dc resistance and a reactance that is many times R_L at the signal frequency.

Solution

$$V_B = V_{CC}\frac{R_{B2}}{R_{B1} + R_{B2}} = (12)\frac{18}{82 + 18} = 2.2 \text{ V}$$

$$V_E = V_B - V_{BE} = 2.2 - 0.6 = 1.6 \text{ V}$$

$$I_E = \frac{V_E}{R_E} = \frac{1.6}{1.8} = 0.89 \text{ A}$$

$$r_j \approx \frac{0.03}{I_E} = \frac{0.03}{0.89} = 0.03\ \Omega$$

$$A_v = -\frac{X_L \| R_L}{r_j + R_E} = \frac{-16}{1.8} = \mathbf{-8.9}$$

$$Z_{in} = R_{B1} \| R_{B2} \| \beta(r_j + R_E) = 82 \| 18 \| (80)(1.8) = \mathbf{13.4\ \Omega}$$

The maximum voltage swing toward ground occurs when the transistor saturates. Figure 11-1(b) shows that i_C will be about $2I_{C(Q)}$ at that point, so V_{RE} will be about 2(1.6) or 3.2 V. Allowing 0.5 V for $V_{CE(sat)}$ of the transistor, the minimum V_C is (3.2 + 0.5) = 3.7 V. This is a peak swing from the Q point of (12 − 3.7) = 8.3 V pk.

The maximum swing toward V_{CC} at cutoff is given by

$$V_{pk(up)} = I_Q(X_{LC} \| R_L) = (0.89)(16) = 14.2 \text{ V pk}$$

The saturation limit governs, and V_o = (2)(8.3) = 16.6 V p-p, which is 5.9 V rms. The maximum sine-wave power out is then

$$P_{R(L)} \frac{V^2}{R_L} = \frac{5.9^2}{16} = \mathbf{2.2\ W}$$

The power dissipated in the transistor at zero input signal is

$$P_Q = I_Q V_{CE} = I_E(V_{CC} - V_E) = 0.89(12 - 1.6) = \mathbf{9.3\ W}$$

The total power delivered by V_{CC} is constant since V_{CC} is fixed and $I_{C(avg)}$ remains the same even though I_C swings up and down. Therefore, at full signal $P_{Q'} = P_Q - P_{R(L)}$ = 9.3 − 2.2 = 7.1 W.

Some notes on the foregoing example might be instructive.

1. A_v is quite low. If it is desired to increase A_v, part of R_E could be bypassed, as as in Fig. 6-13.
2. Z_{in} is very low, even considering that this is a power amplifier. Since most of Z_{in} is due to R_{B1} and not to r_b, the bootstrap circuit of Fig. 7-2 could raise Z_{in} considerably.
3. The cutoff limit on V_o is much larger than the saturation limit. This means that I_Q is larger than necessary, and power is being wasted. The optimum bias current and transistor power would then be

$$I_Q = \frac{V_{pk(down)}}{X_{L(C)} \| R_L} = \frac{8.3}{16} = 0.52 \text{ A}$$

$$P_Q = I_Q V_{CE} = (0.52)(12 - 1.6) = 5.4 \text{ W}$$

Efficiency of a power amplifier is defined as power output to the load divided by power input to the collector circuit. It is designated by the Greek η (eta).

$$\eta = \frac{P_o}{P_{in}} = \frac{(V_{R(L)})^2}{R_L I_{C(av)} V_{CC}} \tag{11-2}$$

For the amplifier of Fig. 8-1 it is 2.2/9.3 = 24%, and if I_Q were readjusted for optimum bias it would be 2.2/5.4 = 41%.

It can be shown that for an amplifier using a resistor in the V_{CC}—collector line (as did all the circuits in Chapters 6 and 7)—the maximum possible efficiency is 25%. For an amplifier using choke feed, the theoretical maximum is 50%. These maxima are never reached in practice because they assume zero $V_{CE(sat)}$ loss, zero $V_{R(E)}$ loss, and zero resistance in the choke. Optimum bias is assumed too, but this, at least, is realizable in practice.

11.2 TRANSFORMER COUPLING

Many times the load impedance and load power requirement dictate a voltage quite different from the available supply voltage. For example, a 4-Ω speaker to be driven at 150 mW requires an rms voltage of

$$V = \sqrt{PR} = \sqrt{0.15 \times 4} = \sqrt{0.6} = 0.77 \text{ V} \tag{11.3}$$

which is a peak voltage of 1.41 × 0.77 or **1.1 V.** If the available supply is 12 V, a great deal of power would be wasted if such a small collector voltage swing were used. Maximum efficiency can be achieved in such cases by using a step-down transformer to step the maximum peak voltage available at the collector down to the maximum peak voltage required by the load. The maximum peak collector voltage $V_{C(pk)}$ is typically about 2 V less than the V_{CC} supply, the 2 V allowing for the emitter-resistor voltage and V_{CE} saturation voltage. The peak voltage required by the load is determined from the load resistance and load power as

$$V_{L(pk)} = 1.41\sqrt{P_{R(L)} R_L} \tag{11-4}$$

The transformer turns ratio is then found as

$$\frac{N_S}{N_P} = \frac{V_{L(pk)}}{V_{c(pk)}} \tag{11-5}$$

Of course, if the load resistance is comparatively high, a peak load voltage *higher* than V_{CC} will be required, and the ratio N_S/N_P will be greater than 1, indicating a step-*up* transformer.

Figure 11-2(a) shows a typical transformer-coupled amplifier. To analyze this circuit, it is only necessary to understand that the load resistance connected across the secondary is reflected across the primary winding as R_L divided by the *square of the transformer turns ratio.*

Figure 11-2(b) shows the effective circuit with R_L reflected across the primary. This circuit can be analyzed in a manner identical to the circuit of Fig. 11-1.

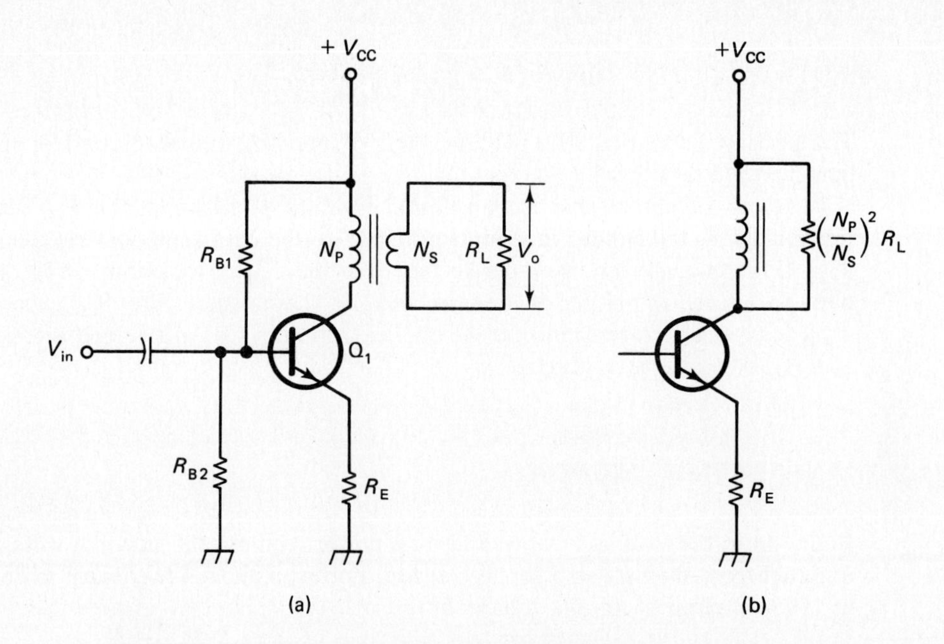

Figure 11-2 (a) A transformer in the collector line provides optimum power efficiency by matching the peak load voltage required to the V_{CC} supply. (b) The load impedance appears to be multiplied by the reciprocal of the transformer turns ratio squared and to be connected across the transformer primary.

Example 11-2

A transformer-coupled amplifier [Fig. 11-2(a)] is powered from a 12-V source. One volt is lost across R_E, and $V_{CE(sat)}$ is 1 V. The load resistance is 16 Ω. An output power of 30 W is required. Find the transformer turns ratio and dc Q-point current for maximum efficiency.

Solution

$$V_{c(pk)} = 12 - 2 = 10 \text{ V}$$

$$V_{L(pk)} = 1.41\sqrt{P_L R_L} = 1.41\sqrt{30 \times 16} = 31 \text{ V}$$

$$\frac{N_S}{N_P} = \frac{V_{L(pk)}}{V_{c(pk)}} = \frac{31}{10} = 3.1$$

A 1 : 3.1 step-*up* transformer is required.

The impedance reflected to the collector is

$$r_L = \left(\frac{N_P}{N_S}\right)^2 R_L = \left(\frac{10}{31}\right)^2 (16) = 1.6 \, \Omega$$

The peak collector current required to produce 30 W in this reflected load resistance is

$$I_{pk} = 1.41\sqrt{\frac{P}{R}} = 1.41\sqrt{\frac{30}{1.67}} = \mathbf{6.0 \text{ A}}$$

The dc Q-point current of the transistor must equal this peak signal current. The power dissipated in the transistor at no signal will be

$$P_Q = IV = (I_Q)(V_{CC} - V_{R(E)}) = (6.0)(11) = \mathbf{66 \text{ W}}$$

11.3 WINDING-RESISTANCE CONSIDERATIONS

The preceding two sections assumed that the winding resistance of the choke or transformer was low enough to be neglected. In general this is *not* true. In fact, for audio transformers in portable and miniature equipment the power lost in the transformer often exceeds that delivered to the load.

Circuit analysis including winding resistance follows the procedure outlined below.

1. R_L and r_S appear in series and are reflected across the primary X_L by a factor of $(N_P/N_S)^2$.
2. The reflected resistance $(R_L + r_S)(N_P/N_S)^2$ appears in series with r_P as the collector load impedance $R_{c(\text{line})}$.
3. The collector bias voltage V_C is less than V_{CC} by the drop across the dc primary resistance r_P.
4. In calculating the ac $V_{R(L)}$, first find V_c as $V_{in}\, R_{c(\text{line})}/R_{e(\text{line})}$. Then voltage-divide r_{refl} against r_P to get V_{pri}. Then transform V_{pri} to V_{sec} by N_S/N_P. Finally, voltage-divide R_L against r_S to get $V_{R(L)}$.

This procedure can be simplified somewhat by the application of the following equations, which are derived in Appendix E.

$$A_v = \frac{V_{R(L)}}{V_{in}} = \frac{R_L}{n(r_j + R_{E1})} \tag{11-6}$$

$$I_{Q(\text{optimum})} = \frac{V_{CC} - V_{CE(\text{sat})}}{\dfrac{r_S + R_L}{n^2} + R_{E2} + 2R_{E1} + 2r_P} \tag{11-7a}$$

or, if emitter-resistor drop $V_{R(E2)}$ is known but R_{E2} is not known,

$$I_{Q(\text{optimum})} = \frac{V_{CC} - V_{CE(\text{sat})} - V_{R(E2)}}{\dfrac{r_s + R_L}{n^2} + 2R_{E1} + 2r_P} \tag{11-7b}$$

$$V_{o(\max)\text{pk}} = \frac{I_Q R_L}{n} \tag{11-8}$$

where $n = N_S/N_P$, R_{E1} = unbypassed emitter resistance, and R_{E2} = bypassed emitter resistance. Surprisingly, a lower turns ratio n (greater step down) really does produce a *higher* A_v, assuming R_L and R_E remain fixed. This is because dividing the turns ratio by 2 reflects a *four* times higher load to the collector, increasing A_v (from input to collector) by a factor of 4.

Example 11-3

The amplifier of Fig. 11-3 is to deliver a 5-W sine wave to R_L with an A_v from V_{in} to R_L of 15. Four 5-W transformers are available with the following characteristics:

N_P/N_S	n	r_S (Ω)	r_P (Ω)
5:1	0.20	3	70
8:1	0.125	1.2	70
12:1	0.083	0.5	70
20:1	0.050	0.2	70

Select the transformer and determine the unspecified component values.

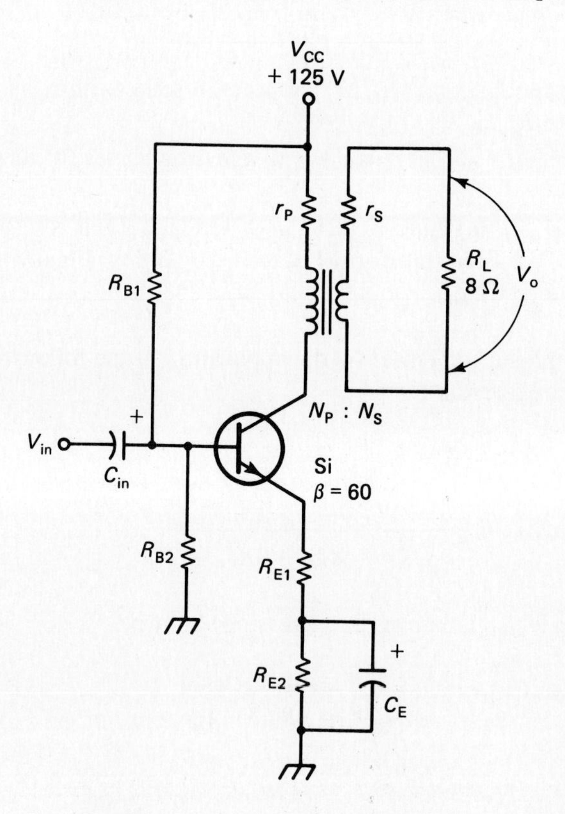

Figure 11-3 Transformer-coupled power amplifier for Example 11-3.

Solution We will need an estimate of I_E to calculate r_j, which is required by equation (11-6). If P_o is 5 W, $V_{CC}I_C$ must be at least 10 W, assuming maximum (50%) efficiency.

$$I_E \approx I_C \approx \frac{P_{CC}}{V_{CC}} = \frac{10}{120} = 83 \text{ mA}$$

$$r_j \approx \frac{30 \text{ mV}}{I_E} = \frac{30 \text{ mV}}{83 \text{ mA}} = 0.4\ \Omega \qquad (6\text{-}2)$$

For the 8:1 transformer,

$$R_{E1} = \frac{R_L}{nA_v} - r_j = \frac{8}{(0.125)(15)} - 0.4 = \mathbf{3.9\ \Omega} \qquad (11\text{-}6)$$

We will allow 5 V across R_{E2} for bias stability and 1 V for $V_{CE(sat)}$.

$$I_{Q\,(\text{optimum})} = \frac{V_{CC} - V_{CE(sat)} - V_E}{\left(\frac{r_S + R_L}{n^2}\right) + 2R_{E1} + 2r_P} \tag{11-7b}$$

$$= \frac{125 - 1 - 5}{\frac{1.2 + 8}{0.125^2} + 2(3.9) + 2(70)} = \mathbf{162\ mA}$$

$$P_{CC} = I_Q V_{CC} = (0.162)(125) = 20\ \text{W}$$

$$V_{o(\max)\,pk} = \frac{I_Q R_L}{n} = \frac{(0.162)(8)}{0.125} \tag{11-8}$$

$$= \mathbf{10.4\ V\ pk} = 7.33\ \text{V rms}$$

$$P_{o(\max)} = \frac{V^2}{R_L} = \frac{7.33^2}{8} = \mathbf{6.7\ W}$$

$$\eta = \frac{P_o}{P_{in}} = \frac{6.7}{20} = \mathbf{34\%}$$

Repeating these calculations for all four transformers reveals the following results:

Transformer	$V_{o(\max)\,pk}$ (V)	$P_{o(\max)}$ (W)	$I_{Q\,(opt)}$ (mA)	P_{CC} (W)	η (%)
5:1	11.4	8.1	284	36	23
8:1	10.4	6.7	162	20	34
12:1	8.3	4.3	86	10.7	40
20:1	5.5	1.9	35	4.3	44

The 12:1 transformer falls just short of meeting the 5-W output, so we choose the 8:1 transformer. I_Q is larger than expected, making r_j more likely 0.2 Ω, but the 3.9-Ω R_{E1} swamps out this difference.

$$R_{E2} = \frac{V_{R(E2)}}{I_Q} = \frac{5}{0.162} = \mathbf{31\ \Omega}$$

Let $R_{B2} \approx 10(R_{E1} + R_{E2}) = 10(3.9 + 31) = 349\ \Omega$. We choose $R_{B2} = 330\ \Omega$, the nearest standard value.

$$V_B = V_{R(B2)} + I_Q R_{E1} + V_{BE}$$

$$= 5 + (0.162)(3.9) + 0.6 = 6.2\ \text{V}$$

$$\frac{R_{B1}}{V_{CC} - V_B} = \frac{R_{B2}}{V_B}; \qquad \frac{R_{B1}}{125 - 6.2} = \frac{330}{6.2}$$

$$R_{B1} = \mathbf{6.3\ k\Omega}$$

$$Z_{in} = R_{B1} \,||\, R_{B2} \,||\, \beta(r_j + R_{E1})$$

$$= 6300 \,||\, 300 \,||\, 60(0.2 + 3.9) = \mathbf{138\ \Omega}$$

11.4 CLASS A VS. CLASS B

In all the amplifiers discussed so far, the objective has been to fix the Q point somewhere near the middle of the characteristic curves, so that the output voltage can swing equally up and down the load line without distortion. Amplifiers biased in this way are termed *class A*. Of course, class-A amplifiers waste considerable power because there is a dc collector current even when no ac signal is present.

For a class-A amplifier, the theoretical maximum efficiency is 50%, but practical class-A amplifiers *typically operate with collector efficiencies of 25%* or less, even when they are handling their maximum undistorted signal levels. In small-signal amplifiers this power waste amounts to a few tenths of a watt at most and is not a matter of great concern. In high-power amplifiers, however, such inefficiency is not easily tolerated, since it means bigger power transistors, bigger heat sinks, bigger power supplies, and consequently higher cost.

The *class-B* amplifier obtains greater efficiency by moving the bias point to the current-cutoff point, as shown in Fig. 11-4(a). Thus no current is drawn and no power is used until an input signal drives the operating point back up the load line. Even then, the current drawn depends on the strength of the input signal, so no more current than necessary is drawn.

The theoretical maximum efficiency of a class-B amplifier is 78.5%, but practical amplifiers run at *typically 50% efficiency*.This figure may be misleading, especially in voice-amplification systems, because the voice waveform is not at its peak amplitude 100% of the time, and, as stated above, a class-B amplifier draws current only in proportion to the amplitude of the input signal. An example will serve to illustrate the difference between class-A and class-B amplifier efficiency.

Example 11-4

An audio voice amplifier is to deliver 10 W of output power. Find the collector dissipation rating required of the output transistor in class A and class B.

Solution In the class-A amplifier, the efficiency is typically 25%, so the 10-W output represents 25% of the total collector input power, and the transistor itself must dissipate the other 75%:

$$0.25P_{\text{in}} = 10 \text{ W}; \qquad P_{\text{in}} = 40 \text{ W}$$

$$0.75P_{\text{in}} = P_C = \mathbf{30\ W} \qquad \text{for class A}$$

In class-B service, the efficiency is typically 50%, so the collector of the transistor must dissipate the remaining 50%:

$$0.50P_{\text{in}} = 10 \text{ W}; \qquad P_{\text{in}} = 20 \text{ W}$$

$$0.50P_{\text{in}} = P_C = \mathbf{10\ W} \qquad \text{for class B}$$

But for class-B *voice* operation the *average* power is less than half of this maximum peak power, so a transistor with a collector dissipation of **5 W** may be used. This is one-sixth the rating required for a class-A amplifier with the same power output. Notice also that the peak power required from the power supply is half as much for the class-B as compared to the class-A amplifier.

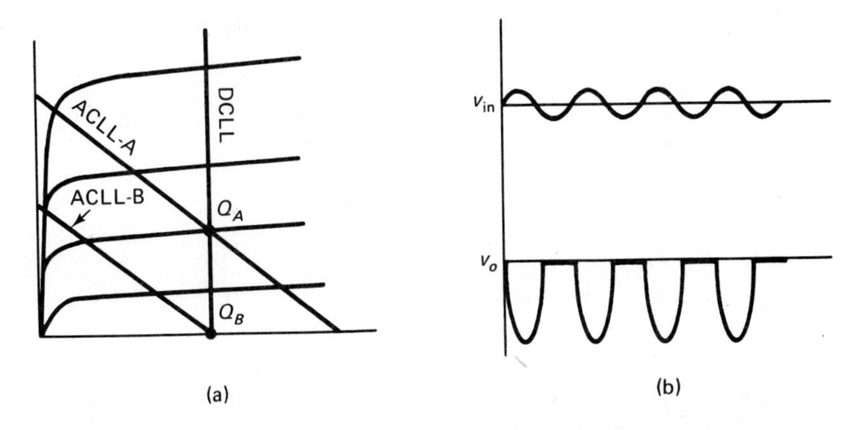

Figure 11-4 (a) A class-A amplifier is biased for optimum efficiency near the center of the load line. A class-B amplifier is biased at cutoff, so collector current flows only when an input signal is present. (b) A class-B amplifier reproduces only one-half of the input waveshape.

Of course, the great disadvantage of the class-B amplifier is that its output voltage can swing down in response to a positive input signal (assuming an *NPN* transistor) but it cannot swing up in response to a negative input signal, since it is already biased at the highest possible voltage on the load line. The result is that the output waveform contains only the negative halves of the total waveform, as illustrated in Fig. 11-4(b).

There are a few applications where such distortion of the input waveform is not objectionable. Some radio transmitters use what is called a class-B linear amplifier, which is linear (or nondistorting) for one-half of the input waveform; the other half of the output waveform is "filled in" by energy stored in an L-C resonant circuit. In audio amplifiers, however, the distortion of the elementary class-B amplifier cannot be tolerated.

11.5 PUSH-PULL AMPLIFIER

The advantages of high efficiency inherent in the class-B amplifier and the low distortion characteristic of the class-A amplifier can be obtained together by using two class-B amplifiers arranged so that one takes the positive half-cycles and the other handles the negative half-cycles. When these two outputs are combined, a complete undistorted reproduction of the input waveform will result. This scheme is widely used in audio and instrumentation amplifiers and is called the *push-pull* circuit.

The essential form of the push-pull amplifier is shown in Fig. 11-5(a). Q_1 is turned on by the positive half-cycle of V_{in}, and current is drawn from the V_{CC} supply upward through output transformer T_2, resulting in a negative half-cycle output across R_L. Meanwhile, the voltage at the base of Q_2 is negative, so Q_2 remains turned off.

When V_{in} swings to the negative half-cycle, the base of Q_2 goes positive with respect to ground, and Q_2 draws current from the supply downward through the transformer, resulting in a positive output across R_L. The resistances of the transformer windings and of the signal source limit the base input current to the transistor.

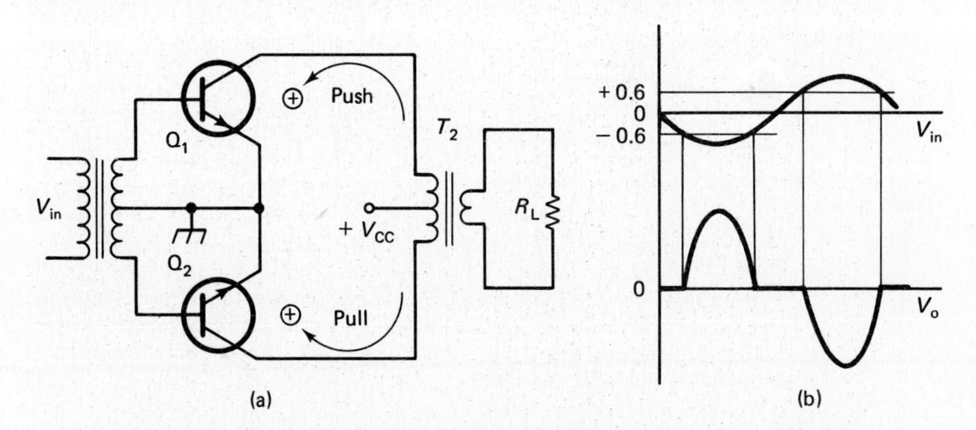

Figure 11-5 (a) Elementary (but impractical) push-pull amplifier. The half-wave outputs from two class-B amplifiers can be pieced together to reproduce the complete input waveshape. (b) The fact that the transistors do not begin to conduct until V_{in} reaches the V_{BE} threshold means that there will be a dead zone on the input signal, producing crossover distortion on the output waveform.

Crossover distortion: The output waveform of this basic circuit shows serious *crossover distortion* because the bases of the transistors do not turn on at 0 V but at 0.2 V for germanium or 0.6 V for silicon types. For silicon transistors, there is therefore a 1.2-V *dead zone* on the input signal, within which neither transistor is turned on and the output is zero. Figure 11-5(b) illustrates the resulting crossover distortion.

Another problem which may occur in the simple circuit of Fig. 11-5 is unequal amplitude of the two halves of the output signal because of unequal *betas* of the two transistors.

A fully stabilized push-pull amplifier is shown in Fig. 11-6. Voltage divider R_1-R_2 biases the bases of Q_1 and Q_2 just at the threshold of turn-on ($V_{BE} = 0.6$ V for silicon) to prevent crossover distortion. Emitter resistors R_3 and R_4 are low values (often less than an ohm in power amplifiers) which serve to stabilize the power gain of each half of the amplifier, thus eliminating unequal positive and negative half-cycles.

For added fidelity the transistors in Fig. 11-6 are often biased somewhere between the class-A and class-B points, allowing a little *idling current* with no signal present. Of course, the improvement in fidelity is paid for by a loss of power efficiency. Such circuits are often called class-AB amplifiers.

Analysis for the push-pull amplifier is quite similar to that for the single-ended transformer-coupled amplifier of Fig. 8-3. The same A_v formula applies.

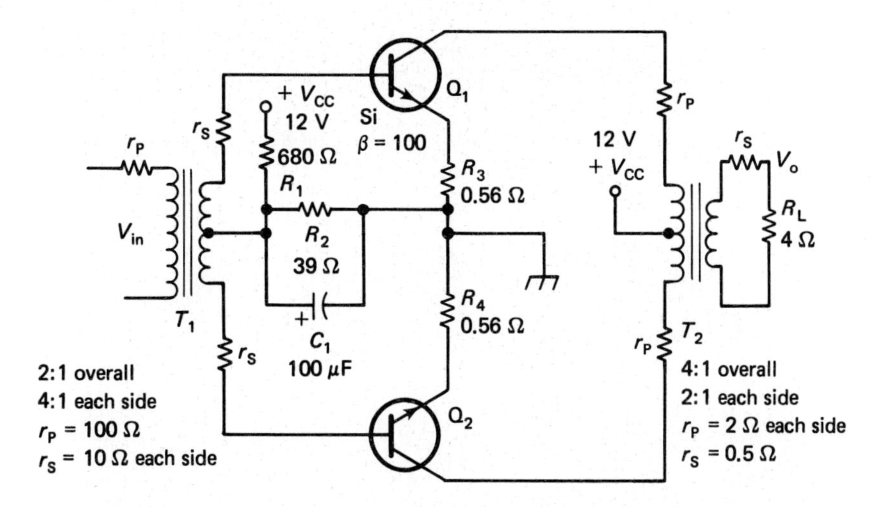

Figure 11-6 Practical push-pull amplifier biased slightly into conduction to prevent crossover distortion.

$$A_v = \frac{R_L}{n_2(r_j + R_E)} \quad \text{(base to load)} \tag{11-6}$$

where $n_2 = N_S/\frac{1}{2}N_P$, the ratio of secondary voltage to *one side* of the primary of the output transformer. The maximum output voltage for Fig. 11-6 is shown in Appendix E to be

$$V_{o(p\text{-}p)\max} = \frac{2R_L(V_{CC} - V_{CE(sat)})}{n_2(R_E + r_P) + \dfrac{r_S + R_L}{n_2}} \tag{11-9}$$

where r_P is the resistance of one side of the transformer primary winding.

The input-impedance calculation is similar to that presented in Section 7.7 for the common-base amplifier.

Example 11-5

Determine Z_{in}, A_v, and $P_{R(L)\max}$ for the push-pull amplifier of Fig. 11-6.

Solution Only one-half of the circuit operates at a time. We will assume that Q_2 and the bottom halves of the transformers draw no current during the half-cycle we will analyze. First we need an estimate of r_j. The 4-Ω load will be reflected up by $(2/1)^2$, presenting roughly 16 Ω to the collector. The 12-V supply will cause a peak current of $12/16 = 0.75$ A in the collector [see Fig. 11-4(a)]. The average current will be something over 0.3 A, so $r_j \approx 0.03/0.3 = 0.1$ Ω. This is an *estimate*, and r_j will vary with signal current, but R_E is large enough to swamp the total variation of $R_{e(\text{line})}$ to perhaps 20%.

$$r_b = \beta(r_j + R_E) = 100(0.1 + 0.56) = 66\ \Omega$$

$$r_{\text{refl}} = \frac{r_S + r_b}{n^2} = \frac{10 + 66}{0.25^2} = 1200\ \Omega$$

$$Z_{in} = r_P + r_{\text{refl}} = 100 + 1200 = \mathbf{1300\ \Omega}$$

$$A_v(\text{base to } R_L) = \frac{R_L}{n_2(r_j + R_E)} = \frac{4}{0.5(0.1 + 0.56)} = 12.1 \tag{11-6}$$

$$A_v(\text{input to } R_L) = \frac{r_{\text{refl}}}{r_P + r_{\text{refl}}}\left(\frac{N_S}{N_P}\right)\left(\frac{r_b}{r_S + r_b}\right)(A_{v(\text{b to } R_L)})$$

$$= \frac{1200}{100 + 1200}\left(\frac{1}{4}\right)\frac{66}{10 + 66}(12.1) = \mathbf{2.4}$$

$$V_{o(p\text{-}p)\max} = \frac{2R_L(V_{CC} - V_{CE(sat)})}{n_2(R_E + \frac{1}{2}r_P) + \dfrac{r_S + R_L}{n^2}} \tag{11-9}$$

$$= \frac{(2)(4)(12 - 1)}{0.5(0.56 + 2) + \dfrac{0.5 + 4}{0.5}} = \mathbf{8.56\ V\ p\text{-}p}$$

$$P_{R(L)} = \frac{(V_{rms})^2}{R_L} = \frac{(8.56/2.828)^2}{4} = \mathbf{2.3\ W}$$

Base bias for a class-B amplifier has a somewhat different objective than it did for the class-A amplifier. We are not trying to place the collector current at some optimum point between saturation and cutoff—we are trying to hold the base-emitter voltage just at the threshold of turn-on. The threshold can vary by ± 0.1 V from unit to unit or with temperature, making this objective difficult to meet.

If V_{BE} is too low, the result will be crossover distortion, as shown in Fig. 11-5(b). If V_{BE} is too high, the result is that the excess voltage will be dropped across R_E, causing wasteful *idling current* at zero signal. R_E must generally be kept to a very low value to keep A_v reasonably high, so even a slight excess voltage $V_{R(E)}$ will cause a fairly large idling current. A large-value bypassed resistor R_{E2}, dropping 2 V or more, would solve the problem (as it did in Fig. 6-13), but where V_{CC} is less than 25 V we are generally unwilling to lose 2 V on the emitter—we want it all on the collector side where it contributes to $V_{o(max)}$.

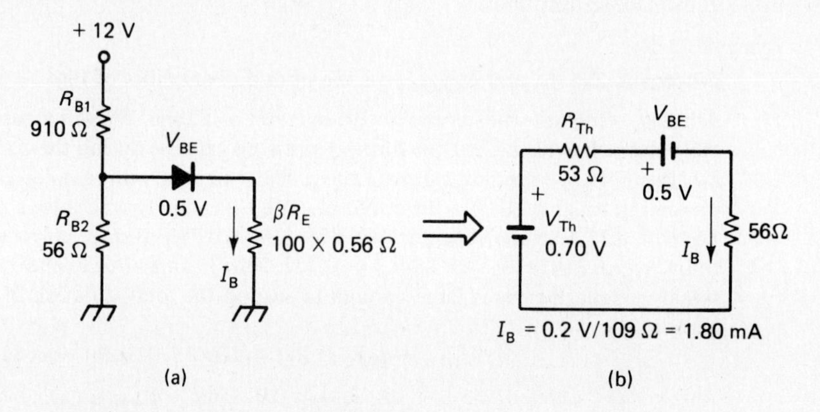

Figure 11-7 If R_{B2} is made $100R_E$ (rather than the usual $10R_E$), and β_{max} is limited to 100, idling current will not be excessive even if V_{BE} drops to 0.5 V.

Many solutions are in common use to solve the class-B bias dilemma. Among them:

- Select transistors for a narrow range of V_{BE}.
- Keep R_1 and R_2 fairly high ($R_{B2} \approx 100\ R_E$) and limit $\beta_{(max)}$ of the transistors (to 100 or less). This is the approach used in Fig. 11-6. Although the voltage divider provides 0.7 V if V_{BE} should go that high, I_E will not go above 180 mA even if $V_{BE} = 0.5$ V and $\beta = 100$. Figure 11-7 shows the analysis.
- Use a diode in series with R_{B2} to establish V_B. Select the diode for the same drop as the transistors, and mount it on the same heat sink to make temperature variations track. Figure 11-8(a) shows the circuit. Matched power-transistor and diode sets are sold for this purpose.
- Place a thermistor R_T, heat-sunk to the transistors, in the R_{B2} position. If the transistors heat up from too much current, R_T will decrease, lowering V_B and limiting the turn-on of the transistors. This circuit is shown in Fig. 11-8(b).
- Use a separate low-level dc amplifier to sense the bias voltage across R_E and alter the base voltage if it is not correct. This is the ultimate in power-amplifier biasing, but a presentation of the circuit will have to wait until the coverage of operational amplifiers in Chapter 12.

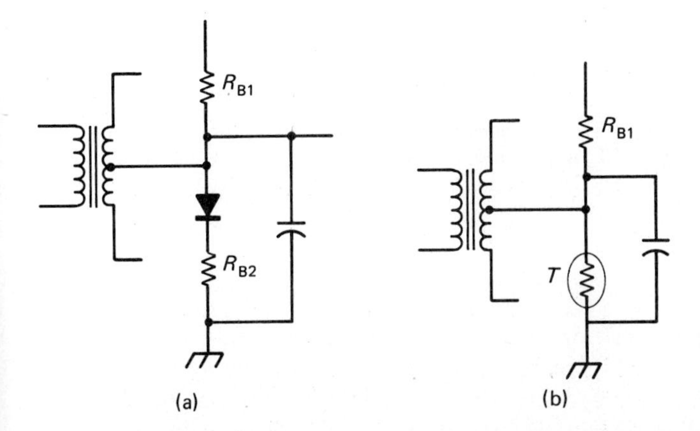

Figure 11-8 Diodes (a) and thermistors (b) are often used to stabilize the bias of class-B power amplifiers in the face of temperature changes.

A phase splitter can be used to eliminate the input transformer in some push-pull amplifiers, as shown in Fig. 11-9. This technique is especially effective with FETs, which require no input current, and with class-A bipolar push-pull stages in which the *average* (dc) current from the driving stage is zero. Driving a class-B bipolar stage with this circuit is difficult because the base-emitter diodes conduct in only one direction, requiring an average dc input current which cannot be supplied through the coupling capacitors C_3 and C_4. C_2 is for suppression of high-frequency spurious oscillations.

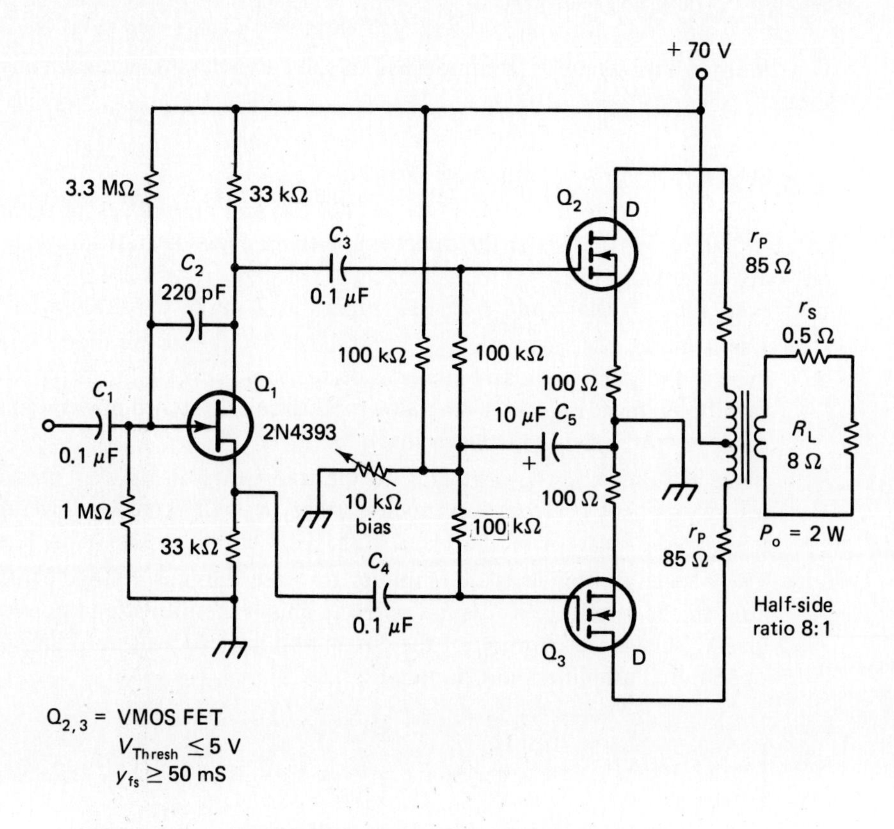

Figure 11-9 A phase-splitter Q_1 eliminates the input transformer of a push-pull amplifier employing VFETs in the output stage.

11.6 COMPLEMENTARY SYMMETRY

The possibility of obtaining transistors in complementary pairs (i.e., *PNP* and *NPN* types with similar characteristics), permits the design of a completely transformerless power amplifier whose efficiency and linearity are at least as good as in the conventional push-pull amplifier. Eliminating the transformer extends both high- and low-frequency response and reduces cost and weight.

An elementary *complementary-symmetry* amplifier is diagrammed in Fig. 11-10. In principle of operation it resembles a variable voltage divider, with Q_1 and Q_2 comprising the two resistances, and the load resistance connected at their common center point. A positive input signal tends to turn Q_1 on, connecting the output to the $+V_{CC}$ supply. A negative input tends to turn Q_2 on, connecting the output to the $-V_{CC}$ supply. Thus the circuit is a type of push-pull amplifier, since turning one transistor on turns the other transistor off.

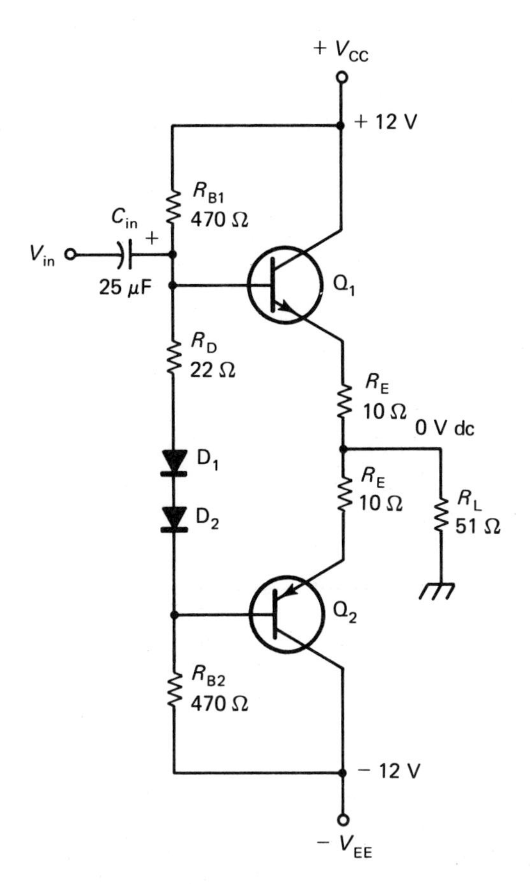

Figure 11-10 The complementary-symmetry amplifier is a push-pull pair of emitter followers.

The load resistance is connected in the emitter lead of the transistors, giving the circuit the essential characteristics of an emitter follower, that is, unity voltage gain, no phase inversion, and input impedance much higher than output impedance.

Resistors R_{B1} and R_{B2} are equal in value and are used to hold the bases of the transistors midway between $+V_{CC}$ and $-V_{CC}$ under zero-signal conditions. The two diodes, D_1 and D_2, are used to compensate for the base-emitter drops of the transistors, holding the two bases just at the threshold of conduction. If these diodes were omitted, there would be a 1.2-V difference between the point where Q_1 turns off and the point where Q_2 turns on, resulting in crossover distortion of the same type found in the unbiased push-pull amplifier of Fig. 11-5.

An additional dropping resistor R_D is used to fix the voltage appearing across R_{E1} and R_{E2} and thus to fix the idling current of the amplifier. For example, if R_D is calculated to drop 0.2 V and R_{E1} and R_{E2} are 1-Ω values, the idling current would be

$$I = \frac{V}{R} = \frac{0.1\ \text{V}}{1\ \Omega} = 0.1\ \text{A} = \mathbf{100\ mA}$$

Both the diodes and the resistor are necessary to ensure the temperature stability of the dc bias point of the amplifier. The diodes are generally mounted directly on the same heat sink as the power transistors, so that as the operating temperature rises and V_{BE} drops, the voltage across the diodes will drop by an equal amount to compensate for the change. Special compensating diodes with controlled voltage-temperature characteristics are available, and only one of these may be required to compensate for both transistors.

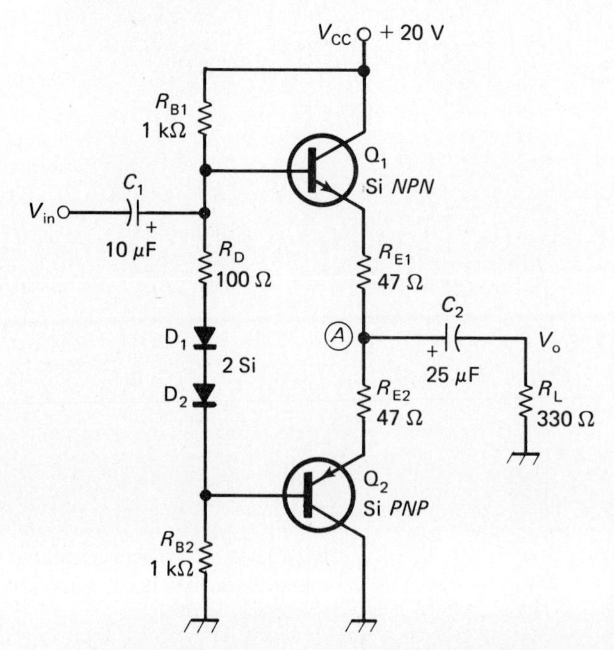

Figure 11-11 Single-supply complementary-symmetry amplifier. The voltage gain is slightly less than 1, and the input impedance is much higher than R_L.

A single-supply complementary-symmetry amplifier is shown in Fig. 11-11. The only addition is the coupling capacitor C_2, which attains a dc charge of $\frac{1}{2}V_{CC}$. Point A is held at $\frac{1}{2}V_{CC}$ by the base-bias resistors.

The relevant equations for both complementary-symmetry amplifiers are

$$Z_{in} = R_{B1} \,||\, R_{B2} \,||\, \beta(r_j + R_{E1} + R_L) \tag{11-10}$$

$$A_v = \frac{R_L}{r_j + R_{E1} + R_L} \tag{11-11}$$

$$V_{o(p\text{-}p)\max} = \frac{(V_{CC} - V_{EE} - V_{BE})\beta R_L}{R_B + \beta(R_E + R_L)} \tag{11-12}$$

The term $V_{CC} - V_{EE}$ indicates the total difference between the two supplies (i.e., 24 V for dual $\pm$12-V supplies).

The driver amplifier for a complementary-symmetry stage is invariably direct coupled, the collector-emitter of the driver taking the place of R_{B2}. R_{B1} (R_5 in Fig. 11-12) serves as the collector resistor for driver stage Q_1. One of the compensating diodes has been omitted and R_{E1} and R_{E2} have been combined to reduce component count.

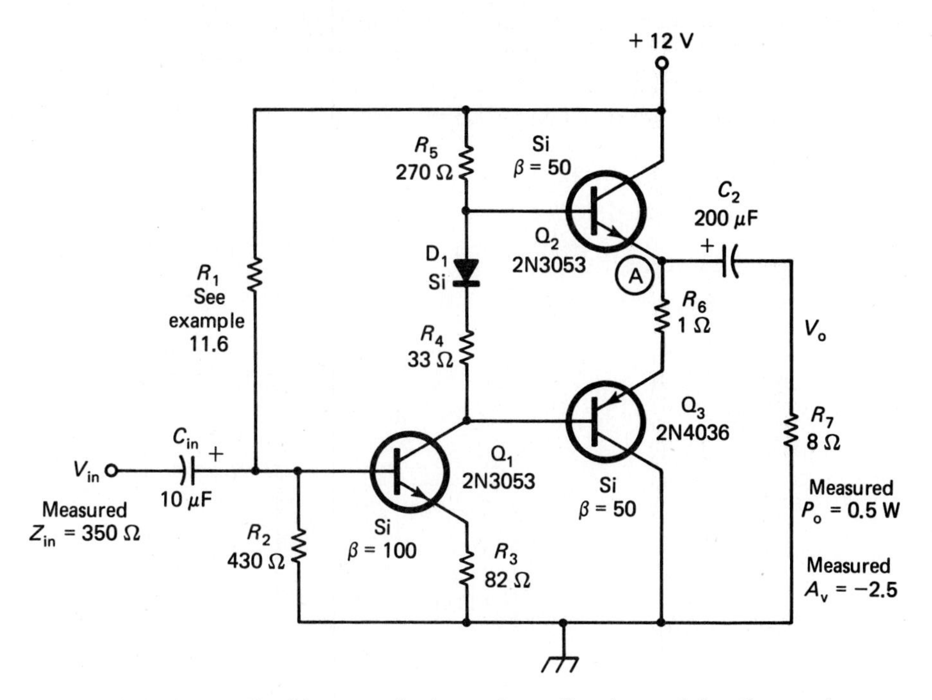

Figure 11-12 The driver stage is almost always directly coupled to the complementary-symmetry input.

Example 11-6

Find the value of R_1 to bias the circuit of Fig. 11-12 for 6 V at point A. Then find Z_{in}, A_v, and $V_{o(p\text{-}p)max}$.

Solution

$$V_{B2} = V_A + V_{BE} = 6.0 + 0.6 = 6.6 \text{ V}$$

$$I_{C1} = I_{R5} = \frac{V_{CC} - V_{B2}}{R_5} = \frac{12 - 6.6}{270} = 20 \text{ mA}$$

$$V_{R3} = I_{C1}R_3 = (0.02)(82) = 1.64 \text{ V}$$

$$V_{B1} = V_{R3} + V_{BE} = 1.64 + 0.6 = 2.24 \text{ V}$$

$$\frac{V_{B1}}{R_2} = \frac{V_{CC} - V_{B1}}{R_1}; \qquad \frac{2.24}{430} = \frac{12 - 2.24}{R_1}$$

$$R_1 = \mathbf{1.9 \text{ k}\Omega}$$

$$Z_{in} = R_1 \,||\, R_2 \,||\, \beta_1 (r_{j1} + R_3) = 1900 \,||\, 430 \,||\, 100(1.5 + 82) \tag{11-10}$$

$$= \mathbf{337 \ \Omega}$$

$$r_{b2} = \beta_2(R_7 + r_{j2}) \approx (50)(8) = \mathbf{400 \ \Omega}$$

$$A_v = \frac{R_5 \,||\, r_{b2}}{r_{j1} + R_3} = \frac{270 \,||\, 400}{1.5 + 82} = \mathbf{1.93} \tag{11-11}$$

$$V_{o(p\text{-}p)max} = \frac{(V_{CC} - V_{BE})\beta R_7}{R_5 + \beta(R_6 + R_7)} \tag{11-12}$$

$$= \frac{(12 - 0.6)(50)(8)}{270 + (50)(1 + 8)} = \mathbf{6.3\ V\ p\text{-}p}$$

Two problems of the circuit of Fig. 11-12 yield to easy solutions, as explained below and illustrated in Fig. 11-13.

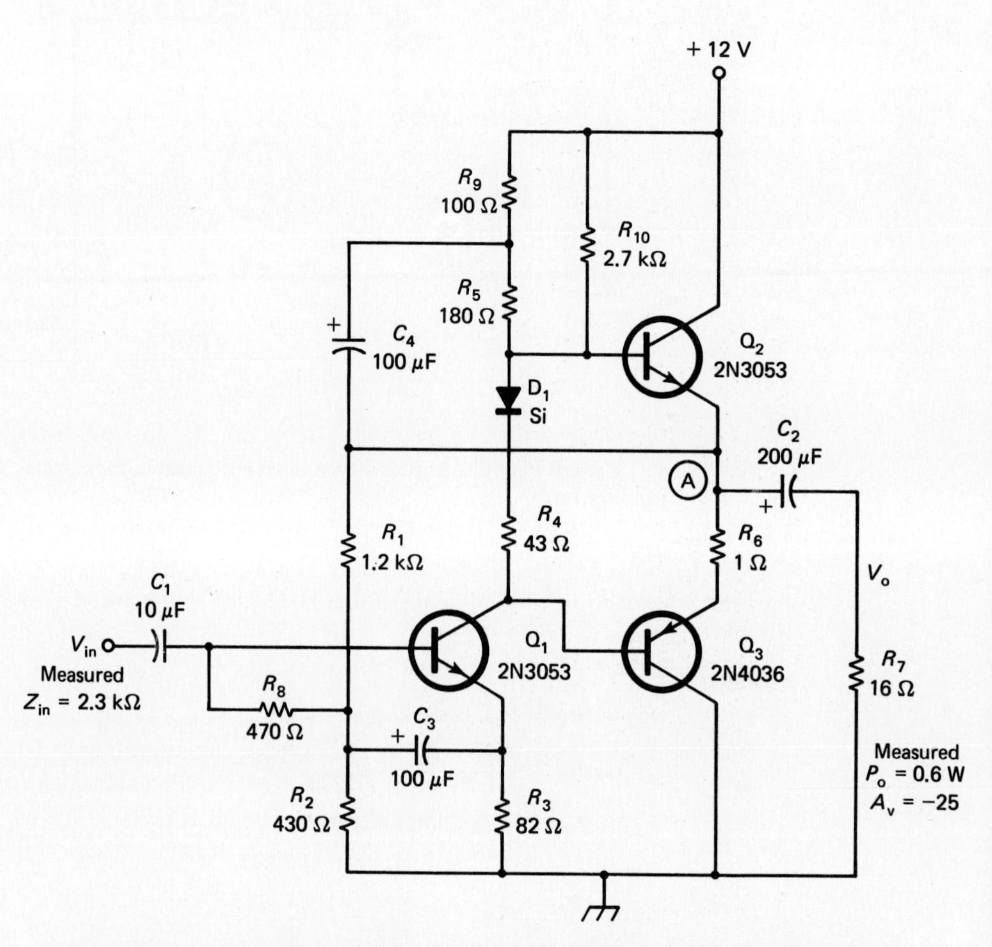

Figure 11-13 Bootstrapping capacitor C_4 raises the effective value of base-bias resistor R_5 many times. R_{10} restores linearity to the collector load for Q_1.

1. R_5 must be quite low or an excessive voltage will be required across it to develop the base current required to turn Q_2 on. However, a low value of R_5 loads down the input impedance of the complementary-symmetry stage, which is R_L for the driver. This decreases A_v. The solution is to bootstrap the input of Q_2. (See

Section 7.2 for a discussion of bootstrapping.) C_4 couples a signal nearly equal to V_{2B} to the top of R_5, multiplying the effective value of R_5 by a factor $(r_{2j} + R_7)/r_{2j}$. Since r_{2j} is small and varies wildly with signal swings, the effective value of R_5 varies wildly, too. R_5 is the collector load for Q_1, so this means that A_v becomes very large and unstable. To tame the gain of Q_1 we add a fixed load R_{10}, much larger than the original R_5 but not so large as to allow the variations of the botstrapped R_5 to cause instability.

2. The bias voltage at point A in Fig. 11-12 is quite critical, yet it is determined by R_1 and R_2 which are two stages removed. Changes in β of Q_1 change the loading of $R_1 - R_2$. Changes in V_{BE} of Q_1 and changes in β of Q_2 also cause bias-point changes. The solution is dc feedback from the point A to the base of Q_1, achieved by the new position of R_1 in Fig. 11-13. (See Section 7.3 for a discussion of this technique.) The Miller effect lowers the effective resistance of R_1 in this case, so bootstrap components R_8 and C_3 are added to remove R_1 and R_2 from the input of Q_1 for ac.

11.7 ADVANCED COMPLEMENTARY-SYMMETRY DESIGNS

Power is proportional to the square of voltage, so in a power amplifier it is important to get as much of the available V_{CC} to the load as possible. The basic complementary-symmetry circuits, for which Fig. 11-10 is the prototype, all lose considerable voltage across R_{B1} in supplying base current to the output transistors. If the transistor had a very high beta, much less I_B would be required and a smaller voltage would be lost across R_{B1}. The Darlington compound (see Section 7.6) provides just such a super-beta transistor, at the sacrifice of only one additional V_{BE} drop from the V_{CC} supply to the load.

Quasi-complementary-symmetry: PNP power transistors are a bit more expensive then *NPN* types. They tend to have higher $V_{CE(sat)}$ at high currents and are difficult to match to complementary *NPN* types at high power levels. The *quasi-complementary-symmetry* circuit of Fig. 11-14(b) uses a low-power *PNP* transistor (Q_4) to drive a high-power *NPN* transistor (Q_5). The pair looks like a single *PNP* transistor with a current gain $\beta_4\beta_5$. This permits two identical *NPN* power transistors (Q_3 and Q_5) to be used in the output stage. Figure 11-14(a) shows the current paths in the *PNP–NPN* pair.

Example 11-7

Find the maximum sine-wave output to the load in the circuit of Fig. 11-14.

Solution

$$V_{o(p\text{-}p)\max} = \frac{(V_{CC} - V_{BE})\,\beta R_L}{R_B + \beta(R_E + R_L)} \tag{11-12}$$

$$= \frac{(12 - 1.2)(2500)(8)}{4700 + (2500)(1 + 8)} = = \mathbf{7.9\ V\ p\text{-}p}$$

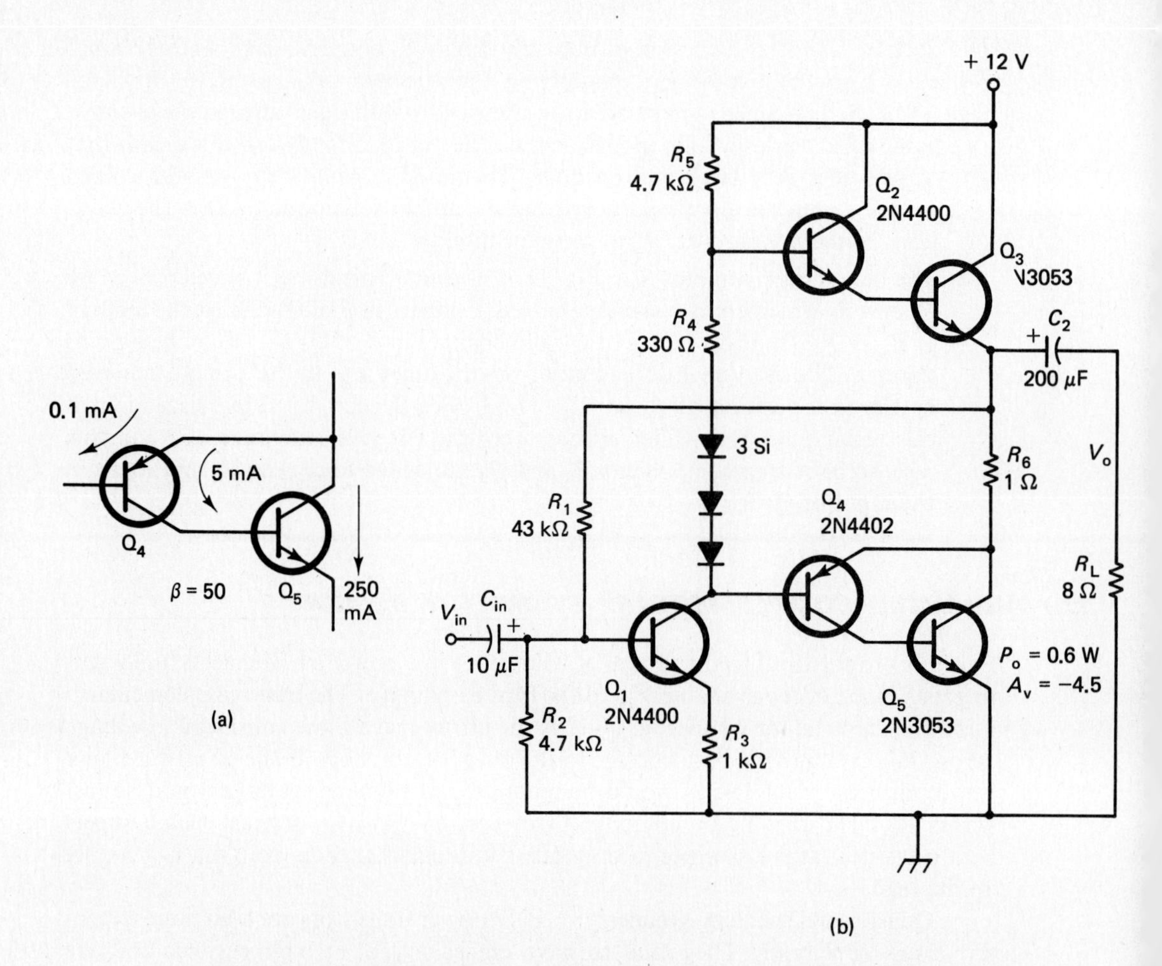

Figure 11-14 Quasi-complementary symmetry reduces the driving current requirement and permits larger signal swings: (a) current path; (b) complete amplifier circuit.

Bridge power amplifier: Even if we are able to salvage every millivolt in an automotive electrical system, 12 V p-p still develops only 4.5 W in a standard 4-Ω loudspeaker. This, unfortunately, falls short of the requirements of those who are intent upon doing serious damage to their eardrums. Must we then resort to transformers in the quest for higher power? No, the ingenious circuit of Fig. 11-15 offers a fourfold power increase over complementary-symmetry designs without transformers. The power level of the example circuit is modest to facilitate lab construction, but the principles are the same in the high-power commercial units.

Q_4 and Q_5 form one complementary-symmetry amplifier while Q_6 and Q_7 form another, which is driven *in inverted phase* by phase splitter Q_1. The load, connected between these two outputs, experiences twice the voltage, hence four times the power, of the previous designs. Q_4 to Q_7 are arranged in the shape of a Wheatstone bridge circuit, hence the name "bridge amplifier."

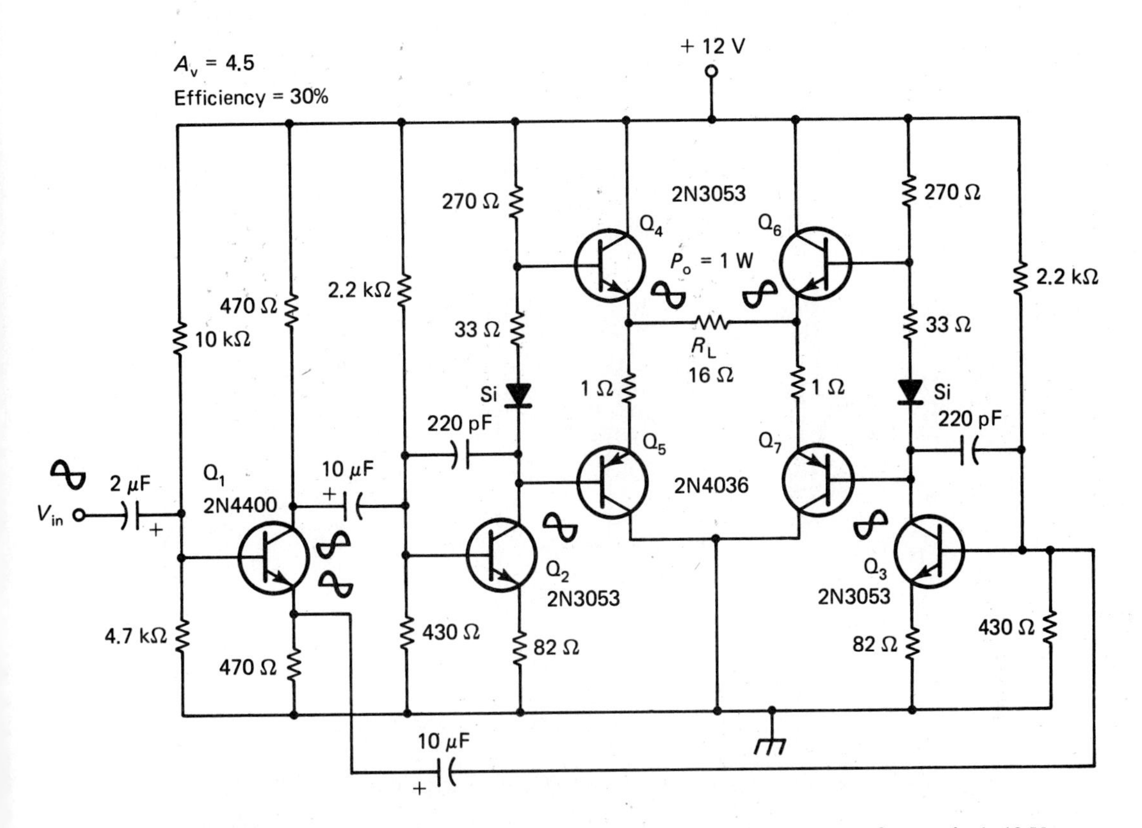

Figure 11-15 The bridge amplifier permits (ideally) a 24-V p-p output from a single 12-V supply, without using transformers.

11.8 CLASS-C AMPLIFIERS

Class-C amplifiers are not amplifiers at all in the sense that we have been using the word. That is, they are not capable of producing a high-power copy of a low-power input-signal waveshape. Rather, they are used to develop large amounts of high-frequency ac power, without regard for the waveshape of the input or driving signal. The most common application for the class-C amplifier is in radio transmitters, which are nothing more than devices for generating high-power high-frequency energy.

We cover class-C amplifiers after class-B amplifiers for reasons of tradition, but they might more properly be called power switchers, and the techniques used to analyze their behavior are more similar to the methods of Chapter 5 than to those of this chapter. Three distinct concepts will be required to understand class-C amplifiers—input switching, output switching, and impedance matching.

Input switching must be accomplished as quickly as possible. Since power equals IV, the transistor dissipates no power when it is switched off ($I = 0$) and very little power when it is full on ($V = V_{CE(sat)}$). However, during the transition from full-on to full-off, both V and I are present, and power in the transistor is significant. The goal is to produce a great amount of power in the load, but very little in the transistor, so it is important that the input signal carry the base from nonconduction to full conduction quickly. Since we are not concerned about waveshape, we apply the largest possible input signal through the lowest possible source impedance, short of any danger of destroying the base-emitter junction.

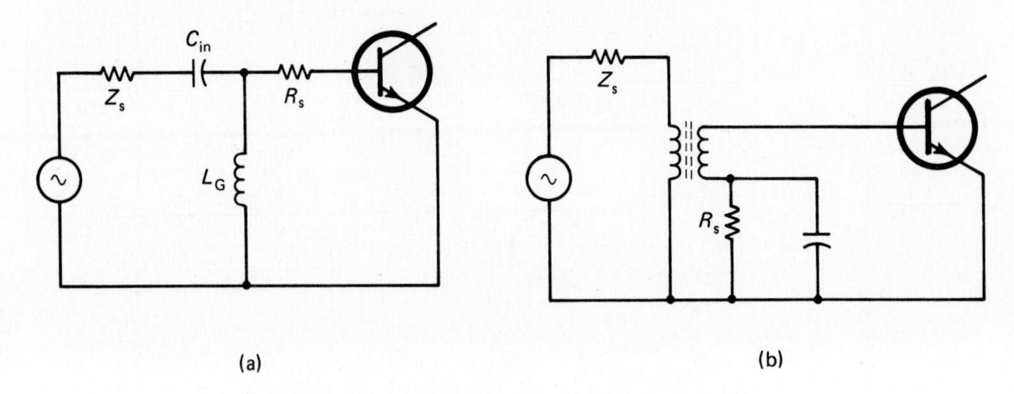

Figure 11-16 Typical class-C base-drive circuits.

The input signal is usually a capacitively coupled ac sine wave. It is necessary to provide a dc path around the base to discharge the capacitor on the half-cycles when the base does not conduct. Figure 11-16 shows some possible circuits. The series resistors R_s are to protect the base from destruction by excessive drive levels. To the extent that switching time and $V_{CE(sat)}$ can be reduced to zero, a class-C amplifier can approach the theoretical maximum efficiency which is 100%. Practical amplifiers may be only 75% efficient due to losses in circuit components.

Output switching is really at the heart of class-C amplifier analysis. Consider the circuit of Fig. 11-17(a). The switch represents the collector-emitter of the transistor. When it is closed there is 10 V across R and current starts to rise in L. When it opens the current from L starts flowing through R, producing a reverse voltage across it. If the *on* time is equal to the *off* time, and if L is chosen so that $\tau = L/R$ is much longer than the off time, the voltage across R will be a 20-V p-p square wave. This is so because the dc voltage across L must be zero, which is to say that the average value of the negative half-cycle must equal the average value of the positive half-cycle. Also, if τ is long compared to T (period) the top of the negative-half-cycle curve will be flat, showing essentially no discharge.

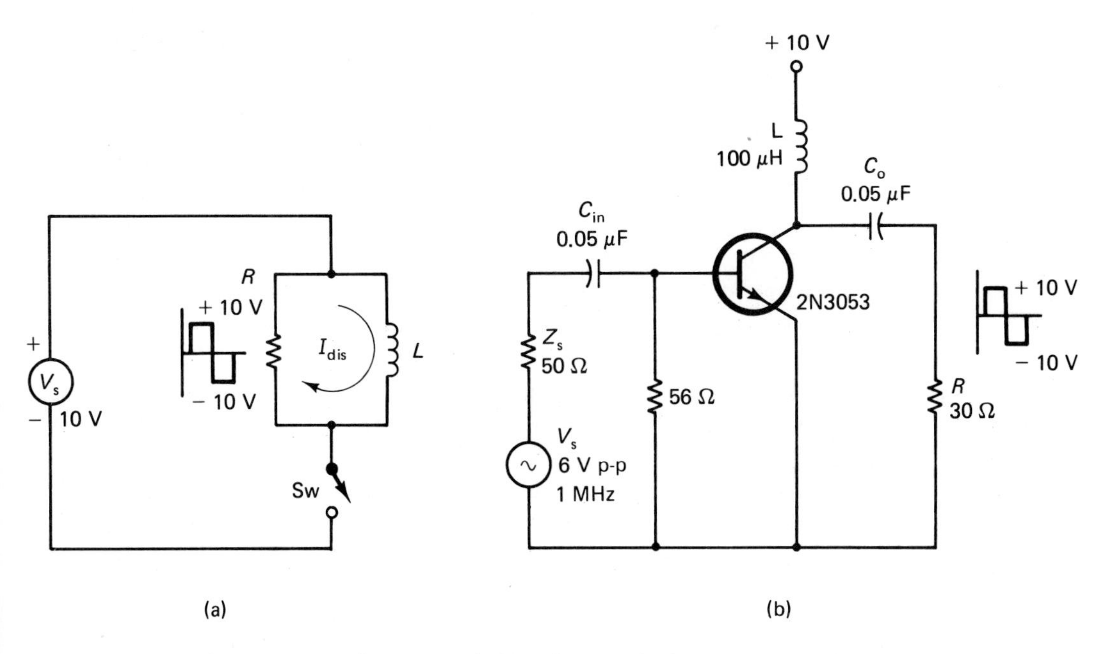

Figure 11-17 A simple *RL* switching circuit (a) is the prototype for the class-C "amplifier" (b).

Figure 11-17(b) shows an implementation of the idealized switching circuit with a transistor as the switch. *R* has had one end moved from V_{CC} to ground, but both points are ac "grounds" and there is no dc across *R*, so this makes no difference. C_o is required to keep dc off resistor *R*. The average supply current from V_{CC} for this circuit is

$$I_{CC} = \frac{V_{CC}}{R} \tag{11-13}$$

During the transistor's *on* time instantaneous supply current i_{CC} is twice this, and during the *off* time i_{CC} is zero. The power delivered to *R* is

$$P_R = \frac{(V_{CC})^2}{R} \tag{11-14}$$

These equations are ideal, neglecting $V_{CE(sat)}$ and the transistor's turn-on and turn-off times.

Impedance matching: The circuit of Fig. 11-17(b) can be built and operated, but it has two serious disadvantages.

1. The output is a square wave, and square waves contain harmonics. In transmitter applications the radiation of all these harmonics is sure to get one in trouble with the government.

2. If the supply voltage and load resistance are both fixed, there is no choice of output power level. By equation (11-14), if $V_{CC} = 12$ V and R is a 50-Ω antenna, P must be 2.9 W.

In the days of vacuum tubes these problems were solved, respectively, by tuned circuits and RF transformers. With transistors it is more common to use L-section impedance-matching networks. Here are some of the reasons.

- Transistors operate at lower impedances (low V, high I) than tubes. A parallel tuned circuit across R in Fig. 11-17(b) would require a variable tuning capacitor of 0.05 μF. We don't have such things.
- Air-core RF transformers have leakage inductance which must be cancelled out by critically tuned capacitors. Ferrite-core transformers can be used at low power levels, but saturate at high levels.
- L sections are broad-band so they can be fixed-tuned, requiring no variable components. They require no transformers or ferrite-core coils, yet they eliminate harmonics as well as tuned circuits.

L sections may be analyzed and designed quickly once a simple equivalent-circuit conversion technique is mastered.

Series-parallel *RX* equivalents. For any parallel capacitor and resistor there is a series capacitor and resistor that is exactly equivalent *at one particular frequency*. The parallel combination of Fig. 11-18(a) is primarily capacitive, with 10% resistive current. It is equivalent to the series combination which is also primarily capacitive,

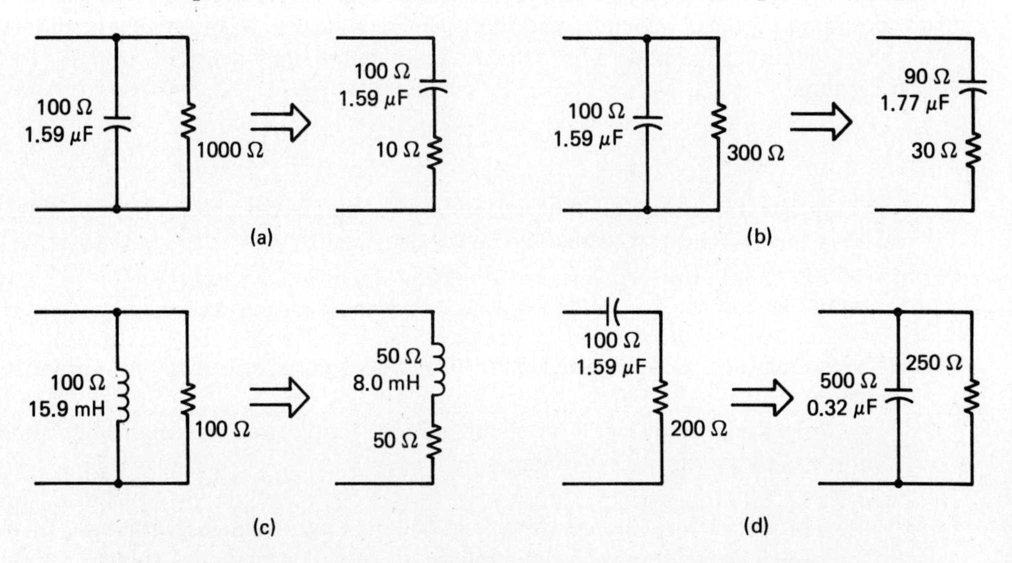

Figure 11-18 Every series *RX* circuit has an equivalent parallel *RX* circuit (and vice versa) at a given frequency. This fact is at the heart of class-C output-network analysis. Four examples are shown here for $f = 1000$ Hz.

with 10% resistive voltage. In Fig. 11-18(b) the resistive element is more noticeable and has a significant effect on the capacitive element. In (c) the two elements are equal (one is an inductor this time, just for variety), and in (d) a series-to-parallel transformation is made, this time with the resistive effect predominant. The component values given are for a frequency of 1 kHz, so the equivalencies can be checked out on a standard impedance bridge.

Three equations are sufficient to transform any series RX circuit to a parallel equivalent, or vice versa.

$$Q = \frac{X_s}{R_s} = \frac{R_p}{X_p} \tag{11-15}$$

$$R_p = R_s(1 + Q^2) \tag{11-16}$$

$$X_p = X_s\left(1 + \frac{1}{Q^2}\right) \tag{11-17}$$

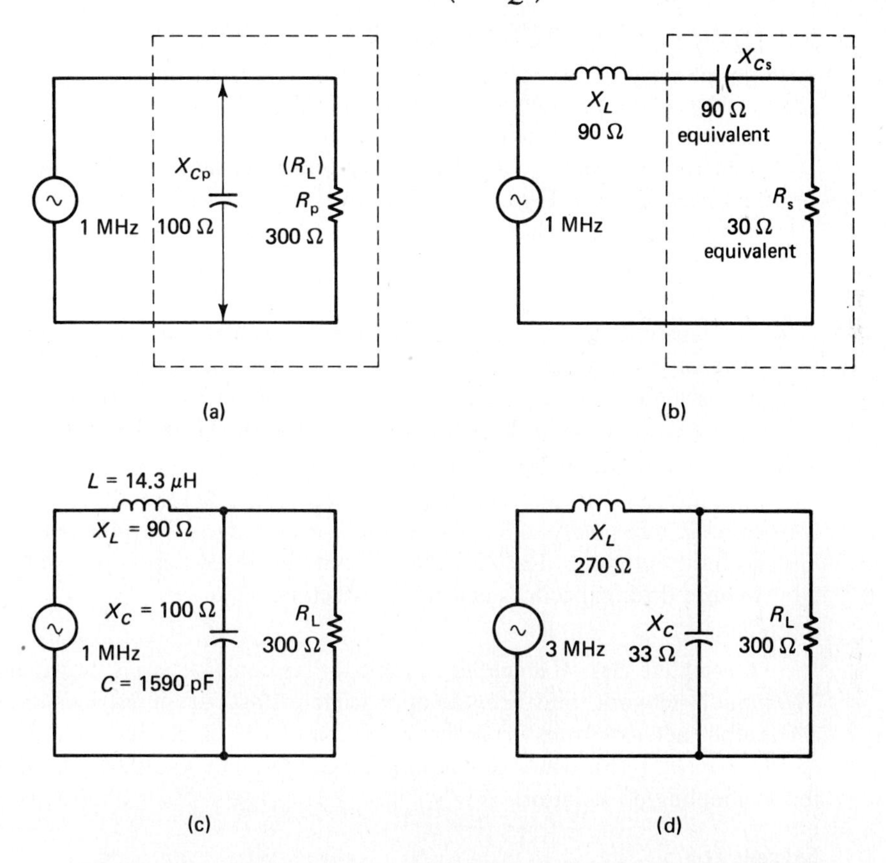

Figure 11-19 A 300-Ω resistor, when paralleled with a 100-Ω X_C (a) looks like a 30-Ω resistor in series with a 90-Ω X_C. (b) A 90-Ω series X_L cancels the 90-Ω X_C (c) leaving the source looking at a 30-Ω resistance. (d) X_L and X_C voltage-divide the third harmonic down to about $\frac{1}{24}$ of the fundamental voltage.

Impedance transformation: Let us say that we want to transform a 300-Ω resistive load into a 30-Ω resistive load at 1 MHz. Figure 11-19 shows that if we parallel 100 Ω of capacitance with the load, the combination will be equivalent to a series combination with $R_s = 30\ \Omega$. It is now only necessary to cancel out the 90-Ω equivalent C_s with a real series inductor of 90 Ω, and the source sees 30 Ω resistive—at the chosen frequency. The component values given in Fig. 11-19(c) are for 1 MHz.

How did we know that choosing $X_{Cp} = 100\ \Omega$ would give the desired 10:1 transformation? Transposing equation (11-16):

$$\frac{R_p}{R_s} = \frac{300}{30} = 1 + Q^2; \qquad Q = \sqrt{9} = 3$$

$$X_p = \frac{R_p}{Q} = \frac{300}{3} = \mathbf{100\ \Omega} \tag{11-15}$$

Here are a few more notes on impedance transformation with L networks.

- In Fig. 11-19 we could have used a parallel 100-Ω *inductor* and a series 90-Ω capacitor, but we usually prefer to keep the capacitor across the load and the inductor in series with it. Thus higher-frequency harmonics are prevented from reaching the load. Figure 11-19(d) shows that the third harmonic in the square-wave output of a class-C amplifier is voltage-divided by about $\frac{1}{8}$. Third-harmonic amplitude is $\frac{1}{3}$ the fundamental for a square wave, so the third-harmonic voltage is about $\frac{1}{24}$ the fundamental at the output.
- To transform a high resistance to a lower value, parallel the high resistance with an X_C and then add an X_L in series. To transform a low resistance to a higher value, add X_L in series and then add X_C in parallel with the series combination. This low-to-high transformation is shown in the input of Fig. 11-20.
- A broader bandwidth can be achieved by cascading two or three L sections. A commercial transmitter might transform its 300-Ω load first to 140 Ω, then to 64 Ω, and finally to 30 Ω to permit a wide range of frequencies to be covered without retuning. The cascaded circuit for three-step transformation would require three capacitors and three inductors.

A complete class-C amplifier suitable for lab construction is shown in Fig. 11-20. The input L network transforms an approximate 20-Ω base-input resistance to 300 Ω. The output network transforms the 300-Ω load to 30 Ω, producing a circuit equivalent to Fig. 11-17(b), while eliminating harmonics. The two 0.05-μF capacitors are for ac coupling/dc isolation.

Example 11-8

Find L and C in Fig. 11-20 to match the 20-Ω base input to the 300-Ω source. Also find P_o assuming that $V_{CE(sat)} = 1$ V.

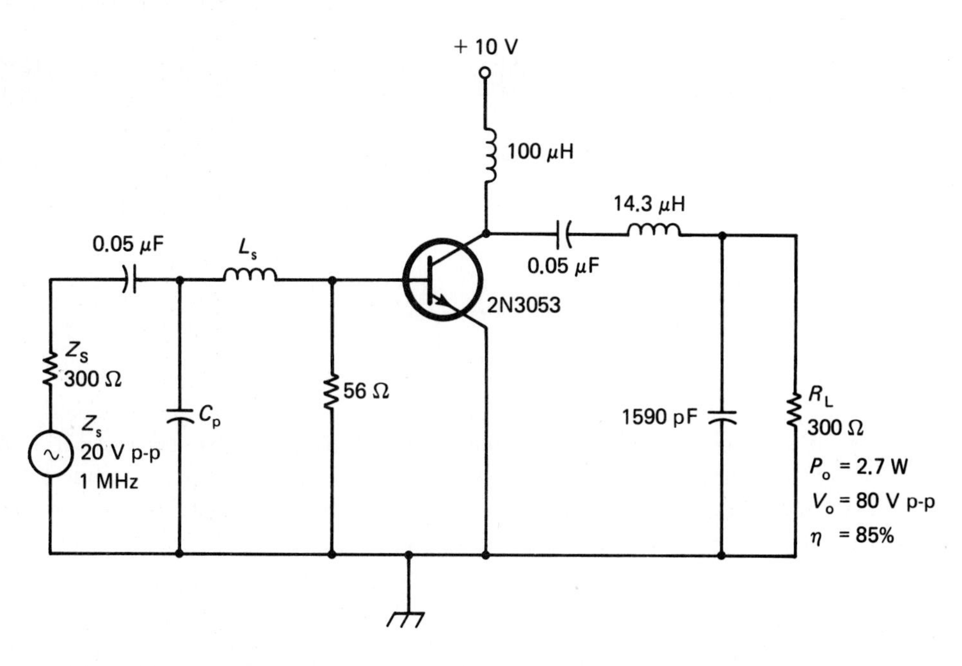

Figure 11-20 Complete class-C amplifier for 1 MHz. L_s and C_p are found in Example 11-8.

Solution Figure 11-21 shows the input transformation.

$$Q = \sqrt{\frac{R_p}{R_s} - 1} = \sqrt{\frac{300}{20} - 1} = 3.74 \tag{11-16}$$

$$X_{Ls} = QR_s = (3.74)(20) = 75\ \Omega \tag{11-15}$$

$$L = \frac{X_{Ls}}{2\pi f} = \frac{75}{2\pi(1 \times 10^6)} = \mathbf{11.9\ \mu H}$$

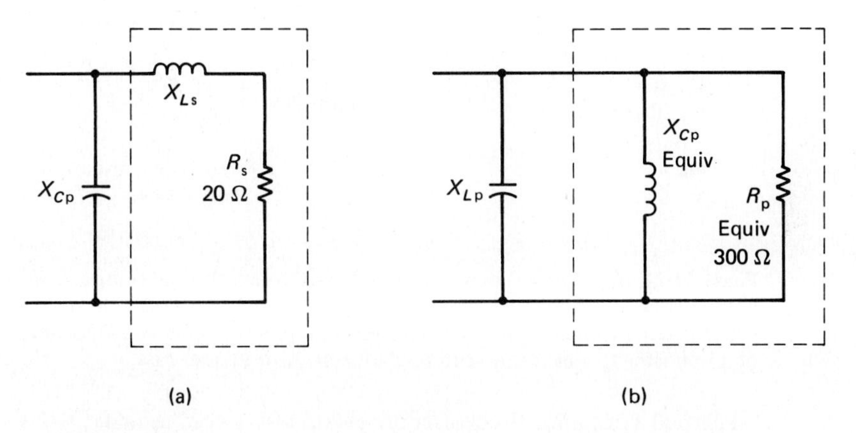

Figure 11-21 Design of an input-matching network for the circuit of Fig. 11-20, Example 11-8.

$$X_{Cp} = \frac{R_p}{Q} = \frac{300}{3.74} = 80\ \Omega \tag{11-15}$$

$$C_p = \frac{1}{2\pi f X_{Cp}} = \frac{1}{2\pi(1 \times 10^6)(80)} = \mathbf{2000\ pF}$$

$$P_o = \frac{(V_{CC} - V_{CE(sat)})^2}{R} = \frac{(10 - 1)^2}{30} = \mathbf{2.7\ W}$$

11.9 HEAT SINKING

When energy is dissipated in a transistor, its temperature rises until the rate of thermal energy flow from the transistor to the environment equals the rate of electrical energy input to the transistor. "Rate of energy" is, of course, power, expressed in watts. We might visualize power P flowing through thermal resistance θ producing a temperature drop ΔT. This is analogous to current I flowing through resistance R producing a voltage drop V. The thermal "Ohm's law" is then

$$P = \frac{\Delta T}{\theta} \tag{11-18}$$

Figure 11-22 illustrates this concept.

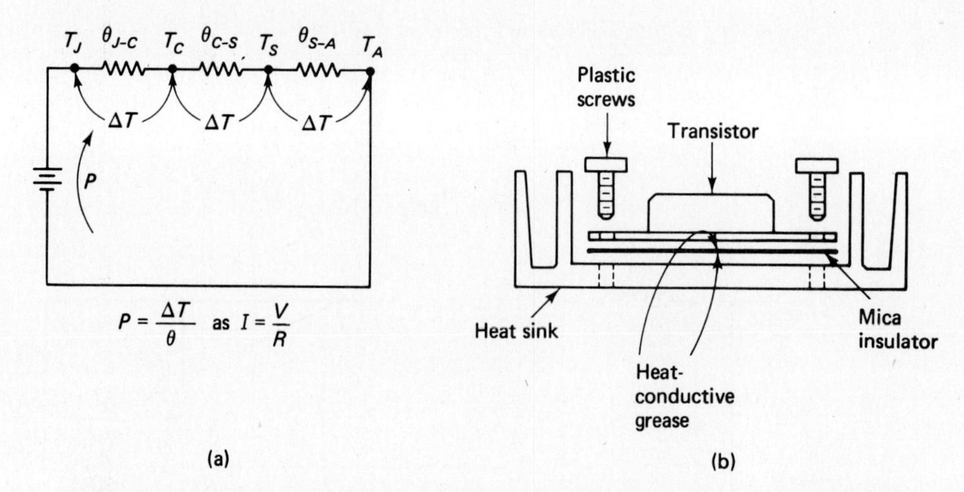

Figure 11-22 (a) Thermal power (watts) flows through thermal resistance (θ) producing a temperature differential ΔT. The subscripts refer to junction, case, sink, and ambient. The use of an electric-circuit schematic to represent the process is common. (b) Typical mounting of a transistor to a heat sink with insulation of the transistor case.

Thermal resistance θ consists of several series components: transistor junction to case (θ_{JC}), case to heat sink (θ_{CS}), and heat sink to ambient air (θ_{SA}). The first of these three is part of the specifications for any power transistor. The second (θ_{CS}) is zero unless a heat-sink insulator is used, in which case it is about 0.35°C/W for a

1-in.[2] insulator area. The third (θ_{SA}) is specified by the heat-sink manufacturer, or read from the graph of Fig. 11-23.

Where transistors are used without heat sinks, the following table may be used to estimate the case-to-ambient resistance.

Case style	θ_{CA} (°C/W)
TO-92 (plastic)	350–200
TO-18 (mini TO-5)	300
TO-5 (standard)	150
TO-60 (stud mount)	70
TO-66 (mini TO-3)	60
TO-220 (power tab)	50
TO-3 (standard power)	30
TO-36 ($1\frac{1}{4}$ in. round)	25

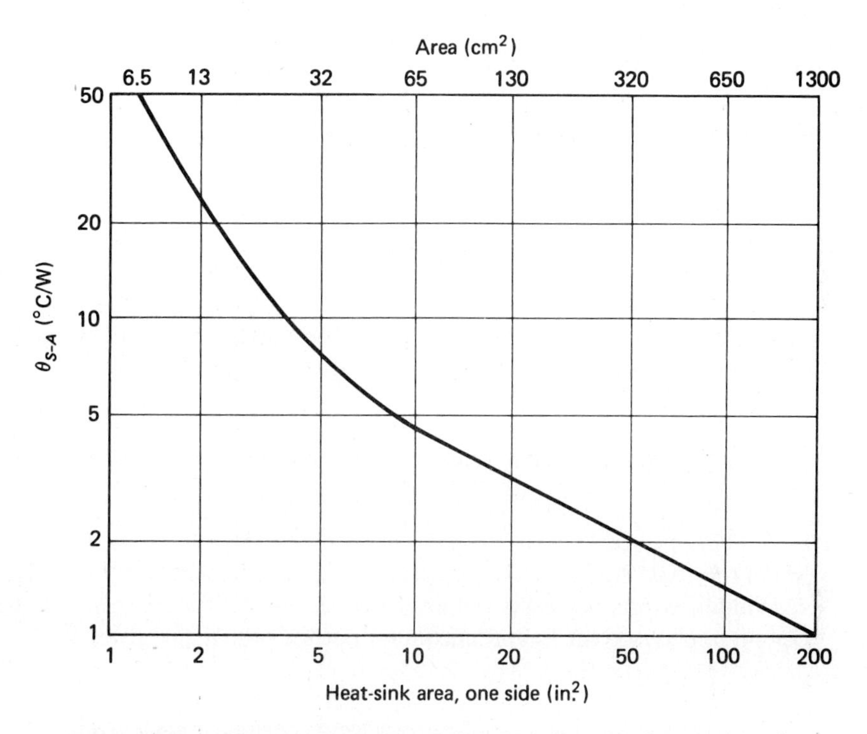

Figure 11-23 Thermal resistance versus total surface area for a square sheet painted black and mounted vertically in free air.

Example 11-9

A TO-220-case transistor has a specified θ_{JC} of 10°C/W, and is operated without a heat sink. Maximum operating temperature is specified as 150°C. Ambient temperature may reach 80°C. What is the maximum power dissipation of the transistor?

Solution

$$P = \frac{\Delta T}{\theta} = \frac{150 - 80}{10 + 50} = \mathbf{1.2\ W}$$

Example 11-10

A TO-3-case transistor has a specified θ_{JC} of 1.2°C/W and $T_{J(max)}$ of 180°C. It is to be used with a heat-sink insulator having θ_{CS} of 0.3°C/W. Power dissipation is to be 30 W. Ambient temperature may reach 75°C. Find the required heat sink area.

Solution

$$\theta_T = \frac{\Delta T}{P} = \frac{180 - 75}{30} = 2.5\text{°C/W}$$

$$\theta_{SA} = \theta_T - \theta_{JC} - \theta_{CS}$$

$$= 3.5 - 1.2 - 0.3 = 2.0\text{°C/W}$$

Figure 11-23 indicates that the plate area must be **50 in.²** to achieve this value.

CHAPTER SUMMARY

1. Efficiency of an amplifier is defined as ac power out divided by dc power input from the supply.

$$\eta = \frac{P_o}{P_{in}} = \frac{(V_{R(L)})^2}{R_L I_{C(av)} V_{CC}} \tag{11-2}$$

Where power levels above 100 mW are involved, an inductor is often used in place of collector resistor R_C to improve efficiency.

2. A transformer is often used to match the voltage required by the load to that available from the supply. Required peak load voltage is

$$V_{L(pk)} = 1.41\sqrt{P_L R_L} \tag{11-4}$$

3. Voltage gain for a transformer-coupled amplifier is

$$A_v = \frac{V_{R(L)}}{V_{in}} = \frac{R_L}{n(r_j + R_{E1})} \tag{11-6}$$

where $r_j = 0.03/I_E$, R_{E1} is unbypassed emitter resistance, and $n = N_S/N_P$.

4. Winding-resistance losses reduce the efficiency of a transformer-coupled amplifier. Optimum bias current I_Q and maximum output voltage are given by

$$I_{Q(opt)} = \frac{V_{CC} - V_{CE(sat)} - V_{R(E2)}}{\dfrac{r_S + R_L}{n^2} + 2R_{E1} + 2r_P} \tag{11-7b}$$

$$V_{o(max)p\text{-}p} = \frac{I_Q R_L}{n} \tag{11-8}$$

where r_P and r_S are primary and secondary winding resistances and R_{E2} is the unbypassed emitter resistance.

5. Class-A amplifiers are biased with a collector current well above zero. Input signals are not permitted to be large enough to cause collector-current saturation or cutoff. Output waveshape is essentially undistorted. Maximum theoretical efficiency is 25% with a collector resistor and 50% with a choke or transformer.
6. Class-B amplifiers are biased with collector current at or near zero. Positive input-signal swings (for *NPN* transistors) cause collector current to rise during that half of the cycle. Output waveshape reproduces the positive input half-cycle but clips off the negative half-cycle. Maximum theoretical efficiency is 78.5%.
7. The push-pull amplifier uses a pair of class-B transistors and a center-tapped transformer to provide reproduction of both half-cycles with class-B efficiency. Each transistor operates for only one half of the input cycle. Bias current may be increased toward the class-A point (class-AB operation) to minimize *crossover distortion.*
8. A phase splitter is a combination emitter follower and gain-of-1 common-emitter amplifier. It provides two outputs, equal in amplitude but one inverted from the other. It can be used to drive a push-pull stage if the driving-current requirement is minimal.
9. A complementary-symmetry amplifier is a pair of emitter followers, one *NPN* and one *PNP*, operating in push-pull class-B or class-AB. Voltage gain is slightly less than 1. The advantage is that no transformers are required. The disadvantage is that supply voltage available cannot be matched to load voltage required.
10. Complementary-symmetry amplifiers are usually direct-coupled to their driver at the input. They often use bootstrapping to increase Z_{in} and a form of collector self-bias to improve bias stability. The four-transistor quasi-complementary-symmetry circuit uses a Darlington-like connection to permit the two output power transistors to be identical *NPN* types. The bridge amplifier uses two complementary-symmetry stages fo feed either end of an ungrounded load out of phase; this provides twice the voltage and four times the power for a given supply voltage.
11. Class-C "amplifiers" are actually square-wave power switchers. The output square wave is usually filtered to a sine wave by *LC* circuits. The load resistance required for a given power level is

$$R_L = \frac{(V_{CC})^2}{P} \tag{11-14}$$

12. The available load resistance R_p can be transformed by an *L* network to the required resistance R_s by (a) adding a capacitive reactance X_{Cp} in parallel with R_p:

$$Q = \sqrt{\frac{R_p}{R_s} - 1} \tag{11-16}$$

$$X_{Cp} = \frac{R_p}{Q} \tag{11-15}$$

and (b) adding an inductive reactance X_{Ls} in series with the parallel combination:

$$X_{Ls} = QR_s \tag{11-15}$$

13. The maximum power dissipation of a power semiconductor may be calculated from

$$P = \frac{\Delta T}{\theta} \tag{11-18}$$

where ΔT is temperature differential (°C) and θ is thermal resistance (°C/W). The equation may be applied for θ_{JC} (junction to case), θ_{CS} (case to sink), or θ_{SA} (sink to ambient). Solved for θ_{SA} the equation may be used to find the size of heat sink required (see Fig. 11-23).

QUESTIONS AND PROBLEMS

11-1. In the choke-coupled circuit of Fig. 11-1, what is the maximum V_{CE} that the transistor may have to withstand, assuming that $V_{R(E)}$, $V_{CE(sat)}$ and the resistance of the choke are all near zero?

11-2. Why is the dc load line in Fig. 11-1(b) nearly vertical? What circuit component determines the slope of the ac load line?

11-3. What is the transformer turns ratio N_S/N_P required if a 200-V p-p primary sine wave is to produce 2 W in an 8-Ω secondary load? Assume no losses.

11-4. What is the ideal collector-bias current $I_{C(Q)}$ for Problem 11-3? Refer to Fig. 11-2(a), but assume no loss on R_E.

11-5. An amplifier has a collector supply of 36 V dc delivering 480 mA. It drives a 16-Ω load with a 9.4-V p-p sine wave. (a) What is the amplifier efficiency? (b) What is the power dissipated in the collector at full signal?

11-6. Redraw the circuit of Fig. 11-3 and label the following component values: $R_{B1} = 620\ \Omega$, $R_{B2} = 150\ \Omega$, $R_{E1} = 3.3\ \Omega$, $R_{E2} = 15\ \Omega$, $r_P = 12\ \Omega$, $r_S = 0.6\ \Omega$, $N_P/N_S = 5$, $R_L = 4\ \Omega$, and $V_{CC} = 12$ V. Find (a) A_v; (b) Z_{in}; (c) $V_{o(p\text{-}p)max}$; (d) efficiency at maximum sine-wave signal output.

11-7. What is the advantage of a class-B over a class-A amplifier? Why are two transistors required in a class-B audio amplifier, where one is sufficient for class-A?

11-8. Redraw the push-pull amplifier of Fig. 11-6 with the following component changes: $V_{CC} = 70$ V, $R_1 = 5.6$ kΩ, $R_2 = 82\ \Omega$, $R_E = 12\ \Omega$, $R_L = 16\ \Omega$; $T_1 = 1{:}2$ each side, $r_P = 45\ \Omega$, $r_S = 120\ \Omega$; $T_2 = 3{:}1$ each side, $r_P = 10\ \Omega$, $r_S = 1\ \Omega$. Find (a) $I_{Q(idle)}$; (b) A_v; (c) $P_{R(L)max}$; (d) Z_{in}.

11-9. Find A_v for the phase splitter in Fig. 11-9. Assume that $y_{fs} = 3$ mS.

11-10. Find $V_{o(p\text{-}p)max}$ for the phase splitter in Fig. 11-9. Assume that $V_{GS} = -2$ V and $V_{DS(sat)} = 3$ V. *Hint:* Find V_D and V_S at zero signal, with Q_1 saturated, and with Q_1 at cutoff.

11-11. For the complementary-symmetry amplifier of Fig. 11-10, find (a) idling current I_Q; (b) A_v; (c) Z_{in}. *Hint:* R_{B1} and R_{B2} always conduct. The transistor bases conduct alternately.

11-12. State one advantage and one disadvantage of the complementary-symmetry compared to the push-pull class-B audio amplifier.

11-13. For the bootstrapped complementary-symmetry amplifier of Fig. 11-3, find (a) Z_{in}; (b) the input resistance of the second stage; (c) the overall voltage gain. Assume that all $\beta = 100$. Refer to Section 7.2 for (a) and (b). Compare answers (a) and (c) with measured values given with Fig. 11-3.

11-14. Find A_v for the quasi-complementary-symmetry amplifier of Fig. 11-14. Assume that all $\beta = 100$.

11-15. The bridge power amplifier of Fig. 11-15 is obviously more complex than a complementary-symmetry circuit. What is the advantage gained by this complexity? Would there be any sense in designing a push-pull bridge amplifier? Explain your answer.

11-16. The purpose of L_G and R_s in Fig. 11-16 and the 56-Ω resistor in Fig. 11-17 is the same. What is this purpose?

11-17. What is the power delivered to a 50-Ω resistive load by an ideal class-C amplifier operating on a 12-V dc supply and switching 50% on, 50% off? Refer to Fig. 11-17(b).

11-18. What is the resistance needed to cause the amplifier in Problem 11-17 to deliver a power of 40 W?

11-19. What is the series equivalent of 100-Ω inductive reactance and 200-Ω resistance in parallel?

11-20. What is the parallel equivalent of 1-kΩ resistance and 5-kΩ inductive reactance in series?

11-21. What is the series equivalent (resistance and reactance) of a 75-Ω resistance and a 295-pF capacitance in parallel at 27 MHz?

11-22. Follow the steps of Fig. 11-19(a)–(c) to determine the values of L and C that will transform $R_L = 50\ \Omega$ to a 6-Ω resistance at $f = 52$ MHz.

11-23. In Fig. 11-20, what is the time constant of the 100-μH choke and the transformed resistance at the collector? How many times longer than half a period is this?

11-24. A power transistor is expected to dissipate 1.3 W. Three types are available: a TO-3 with $\theta_{JC} = 8°C/W$, a TO-220 with $\theta_{JC} = 12°C/W$, and a TO-5 with $\theta_{JC} = 25°C/W$. All have $T_{J(max)}$ specified at 175°C. Which, if any, of the available transistors may be used without a heat sink?

11-25. A TO-60 transistor with $\theta_{JC} = 10°C/W$ is mounted directly on a 4 cm × 5 cm heat sink, so θ_{CS} is negligible. $T_{J(max)}$ is 170°C and $T_{ambient}$ is 85°C. Find the maximum safe power dissipation.

12

LINEAR INTEGRATED CIRCUITS

12.1 INTEGRATED-CIRCUIT TECHNIQUES

The silicon planar process of transistor fabrication (described in Section 4.11) consists of alternately diffusing *P*- and *N*-type impurities into a single silicon wafer, using photomasking techniques to control the areas of impurity diffusion. Thousands of transistor "chips" are commonly fabricated together on a wafer the size of a silver dollar and then sawed apart for separate packaging. From this process, the logical next step was to interconnect several of these transistors with a deposited metallization layer, so that they would not have to be sawed apart, packaged separately, and then wired together externally. This was the beginning of the integrated circuit.

Resistors can be fabricated on the same wafer with the transistors by simply diffusing a line of impurity dopant into the silicon so that it becomes semiconductive. Resistors ranging from a few ohms to several tens of kilohms can be integrated in this way.

Diodes are easily fabricated by the planar process, since they are essentially one-junction devices, where the transistor is a two-junction device. To make an integrated diode it is only necessary to leave out one of the vapor-diffusion steps, or connect the collector and base of a transistor together by metallization and use the base-emitter junction as a diode.

Field-effect transistors are easier to integrate than bipolars because they have one *P-N* junction instead of two. They also take up less "real estate" (chip area), and are consequently popular for large-scale integrated circuits (LSI), such as computer memories and processors.

Capacitors can be fabricated only with great difficulty by the planar process, so they are used much less frequently in ICs than in discrete-component circuitry. When integrated capacitors are used, their values are normally limited to a few tens of picofarads because they take up a much larger area of the chip than transistors, diodes, and resistors.

Inductors are practically impossible to integrate, at least in any reasonable range of values, and integrated circuitry is generally designed to do without them.

Because of the above considerations, the schematic of the *internals* of an IC is likely to appear very different from a schematic for a discrete-component circuit to serve the same function. In discrete circuitry resistors go for about 2 cents apiece, coupling and bypass capacitors for about 10 cents, while diodes and transistors may typically cost 20 and 50 cents, respectively. Therefore, if four resistors can be used to eliminate one transistor, it is good practice, in discrete-circuit design, to do so.

In IC design, transistors and diodes are easiest to fabricate, resistors are somewhat harder, and capacitors are extremely difficult in small values and impossible in large values. An IC chip is therefore likely to contain more transistors than resistors, and an IC which contains any capacitors at all is more the exception than the rule.

Fortunately, a detailed understanding of the internal actions of integrated circuits is not crucial to the casual user. A knowledge of the external behavior of the devices is sufficient as a starting point for beginning to use integrated circuits.

Before proceeding further, however, it might be wise to clear up some confusion that has arisen as to just what an "integrated circuit" is. There are really several different types deserving the name, but the one which we have been discussing, and the one which is automatically implied by the verbal use of the term IC, is the *monolithic* integrated circuit. The word means *single stone*, and indeed monolithic ICs are fabricated from a single block of silicon material. Two other less widely used types of ICs are *thick-film* and *thin-film* types. Thick-film ICs use silk-screen printing to actually paint resistors, conductive paths, capacitor plates, and dielectrics on a glass or ceramic base. The transistors are then mounted, often as chips with no package, directly on the circuit substrate. Thin-film ICs are formed on a glass or ceramic base by depositing extremely thin layers of conductive, insulative, and semiconductive material in a high-vacuum chamber.

12.2 OPERATIONAL AMPLIFIERS

Linear integrated circuits can produce a *continuous* range of output voltages in response to an analog input signal, as contrasted with the two possible output levels of digital integrated circuits. Linear ICs may be characterized as amplifiers and signal processors, whereas digital ICs are essentially switches.

By far the most popular type of linear IC is the *operational amplifier*. This name was given because, as first applied in vacuum-tube analog computers, they were used to perform the basic mathematical *operations* required by the computer. The essential characteristics of all "op amps" are

1. A very high voltage gain, typically in excess of 10 000.
2. Differential inputs: At the *inverting* input, a *positive* signal tends to drive the output *negative*. At the *noninverting* input a *positive* signal tends to drive the output *positive*.
3. Direct-current coupling throughout: No coupling capacitors are used, so the op amp will respond equally well to ac or dc.
4. Low *offset* voltage and current: This means, ideally, that if the input voltage is zero, the input current will be zero and the output voltage will be zero.

The op-amp symbol is shown in Fig. 12-1. Internally this symbol may represent 20 transistors, 10 resistors, and a capacitor integrated on a single silicon ship, but the internals need not be of concern to us at this point.

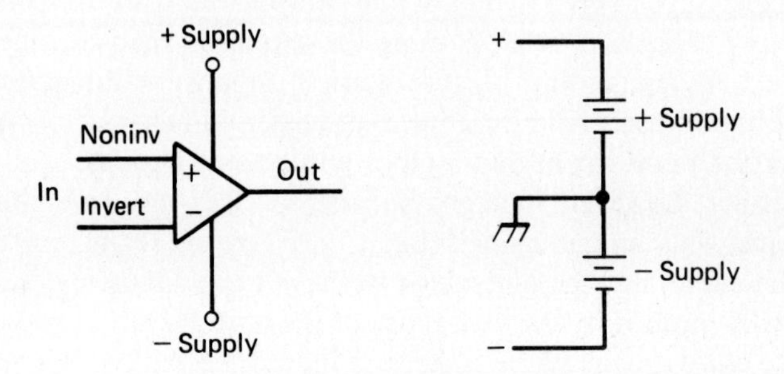

Figure 12-1 The operational amplifier requires positive and negative supplies referenced to ground at their common point.

Notice that two power supplies are required for the op amp—one positive with respect to ground, and the other negative with respect to ground. Typically these supplies may be ± 6 or ± 12 V. The op amp itself may or may not have a ground-terminal connection. If it does not, it is still referenced to ground through the two supplies. In most schematic diagrams the supply connections to the op amp are not shown, since it is assumed that anyone who works with op amps knows that they must be connected.

12.3 THE DIFFERENTIAL COMPARATOR

The value of input voltage required to drive the output of an op amp to saturation is very small and can be calculated from the voltage gain of the op amp, as shown in Fig. 12-2.

The values given in Fig. 12-2 are typical for the popular *741*-type operational amplifier. The 10-kΩ input resistor is strictly for protection of the IC in the event that V_{IN} goes higher than the supply voltage. Practically all of V_{IN} appears across the op-amp input terminals for lower input voltages. Assuming the output saturation voltage

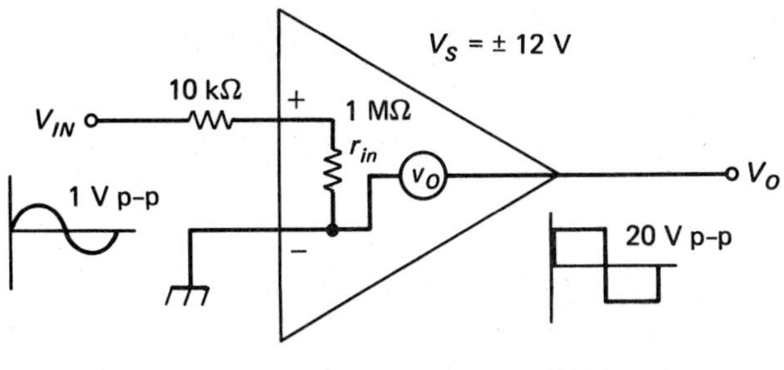

$A_v = 200{,}000$ $\quad V_{IN\,(sat)} = \frac{V_O}{A_v} = \frac{10\text{ V}}{200{,}000} = 50\ \mu\text{V}$

$V_{O(MAX)} = \pm 10$ V

Figure 12-2 Equivalent circuit for an operational amplifier.

to be about 10 V (for a $\pm$12-V supply), the input required to saturate is

$$V_{\text{IN}} = \frac{V_{\text{O(sat)}}}{A_{\text{V}}} = \frac{10\text{ V}}{200 \times 10^3} = 0.05 \times 10^{-3} = 0.05\text{ mV}$$

Because of this extremely low input-voltage requirement, it is almost impossible to balance the output of a differential comparator between positive and negative saturation. A simple use for this circuit is a square-wave shaper. If a sine wave (say 1 V p-p) is applied at V_{IN} (Fig. 12-2), V_o will be a square wave of the same frequency.

In Fig. 12-3(a), the output voltage will go to its full positive value for even a small positive voltage at V_{IN}. A small negative V_{IN} will produce the maximum negative output voltage. The maximum output voltages (also termed the *saturation* voltages) are typically 1 or 2 V less than the supply voltages, depending on the type of IC and the load resistance it must feed.

The circuit of Fig. 12-3(b) is identical to that just discussed, except that the input is applied to the *inverting* input of the op amp. A small positive input voltage will thus saturate the output in the *negative* direction. Used as a sine-to-square wave shaper, the circuit of Fig. 12-3(b) produces output square waves which are inverted (180° out of phase) from the input sine waves.

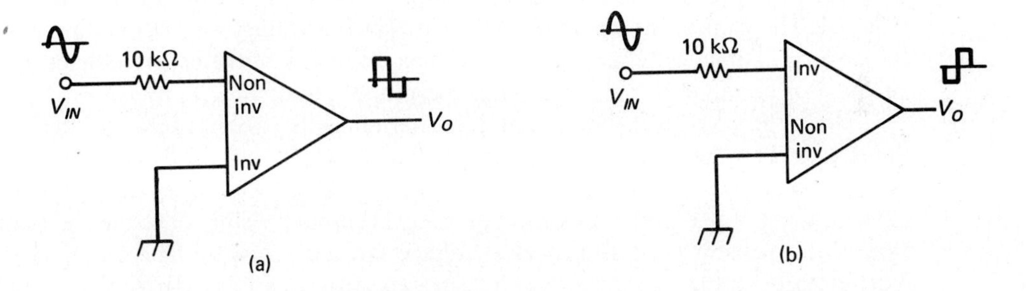

Figure 12-3 Differential-comparator circuits. (a) Positive V_{IN} produces positive V_{O}. (b) Positive V_{IN} produces negative V_{O}.

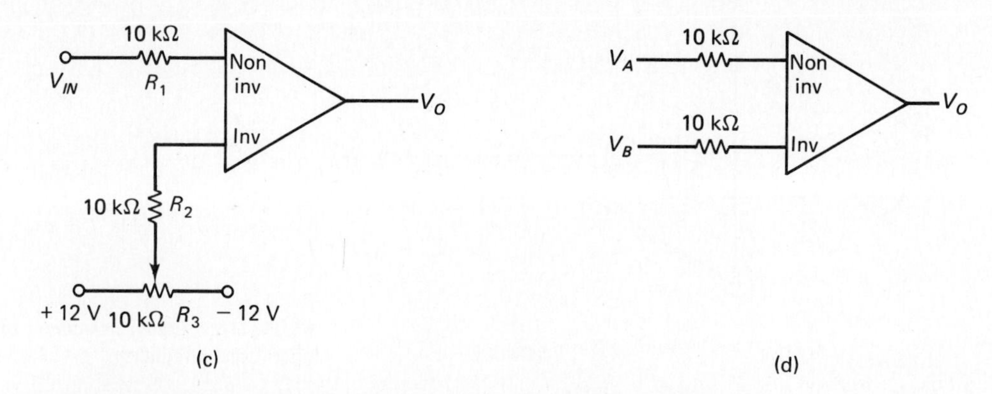

Figure 12-3 Continued. (c) V_{IN} compared to a reference voltage from R_3. (d) V_A compared to V_B. If V_A is more positive than V_B, V_O goes positive.

In Fig. 12-3(a) and (b), V_{IN} is compared to ground level (0 V) and the output switches accordingly. In Fig. 12-3(c), V_{IN} is compared to a variable voltage level. The output voltage now goes positive if V_{IN} is more positive than some level set by R_3 and negative if V_{IN} is more negative than this set point voltage.

In Fig. 12-3(d) two signals, V_A and V_B, are being compared. The output will be positive if V_A is the more positive and negative if V_B is the more positive of the two. This circuit is called a differential comparator, because it responds to the *difference* between V_A and V_B.

12.4 THE INVERTING OP-AMP CIRCUIT

The most widely used op-amp circuit is undoubtedly the inverting amplifier, or some variation of it. The voltage gain and input impedance of this circuit are almost completely fixed by two resistors, as shown in Fig. 12-4(a).

Virtual ground: To understand this circuit it is necessary to keep clearly in mind that, with the noninverting input grounded, the voltage at the inverting input need never be greater than a fraction of a millivolt, even if the full output voltage is required. The gain of the op amp is so great that the voltage at the inverting pin of the op amp is practically zero (ground) compared to any of the other voltages in the circuit. It is common to refer to the inverting-input pin as *virtual ground.*

Since the voltage at the op-amp input terminals is virtually zero and the input impedance is on the order of a megohm, the input current is virtually zero also. Now let us assume that +1 V is applied at V_{IN} in Fig. 12-4(a). This is represented in Fig. 12-4(b). The 1 V will push 1 mA through the 1-kΩ input resistor, since the right-hand side of this resistor is at almost zero voltage (virtual ground). This 1-mA current must go through the 10-kΩ resistor, R_F, since the current into the op amp is virtually zero. Therefore, the voltage across the 10-kΩ resistor must be 10 V, with the polarity negative at the right as in Fig. 12-4(b). This voltage is identical to V_O since the left-hand side of R_F is virtually grounded.

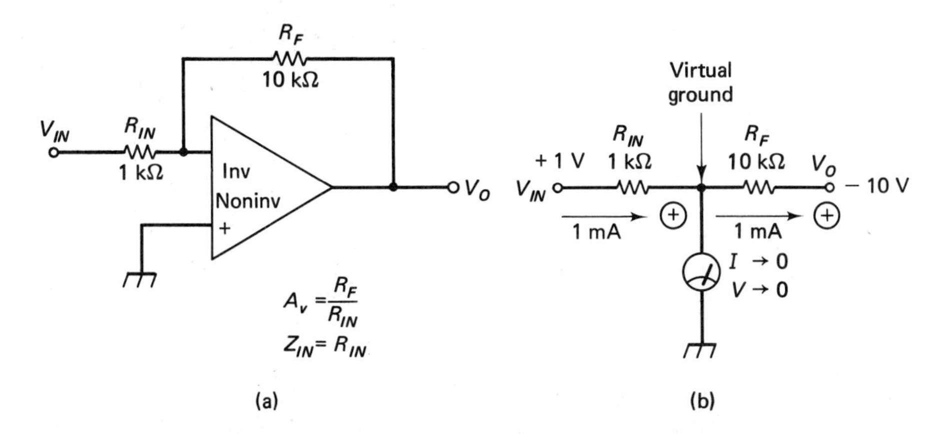

Figure 12-4 (a) Basic inverting amplifier circuit with gain formula. (b) Since $I_{\text{IN}} = 0$ and $V_{\text{IN}} = 0$, $I_{R(\text{IN})} = I_{\text{R(F)}}$ and the circuit gain must be $-R_{\text{F}}/R_{\text{IN}}$.

You may ask: How does the amplifier output voltage come to be -10 V if the op amp input voltage and current are zero? The answer is that the input voltage and current are not zero; they are *virtually* (very nearly) zero. The $+1$ V at V_{IN} causes a very small positive input current to the inverting terminal of the op amp, which causes the output voltage to go negative. The amount of current required by the op-amp input is negligible compared to the 1 mA through R_{IN}. The output voltage must rise to exactly -10 V, because only this value will make the input current virtually zero. If, for example, the output voltage tried to assume a value of -9 V, the current through R_{IN} would still be 1 mA, but the current through R_{F} would be only 0.9 mA. This would leave 0.1 mA going into the op-amp input, but the input current is supposed to be virtually zero (less than a microampere). The attempt to push $+0.1$ mA into the inverting input would immediately cause the output to go more negative until $I_{R(\text{IN})} = I_{RF}$ and $I_{(\text{OP AMP})}) = 0$.

In a similar fashion, the output voltage could not go to -11 V, because this would require a current of -0.1 mA into the op amp. Such a negative input current would immediately drive the output more positive (back to -10 V).

Since the right-hand end of R_{IN} is at virtual ground, the input impedance seen by V_{IN} is simply the value of R_{IN}.

The gain of the amplifier circuit is found as

$$A_v = \frac{V_{\text{O}}}{V_{\text{IN}}} = \frac{-R_{\text{F}}}{R_{\text{IN}}} \tag{12-1}$$

Values of A_v from -1 (or lower) to over -1000 can be obtained with commonly available IC op amps. For most low-cost op amps in simple circuits the value of R_{F} cannot be raised much beyond 1 MΩ without incurring problems of stability, however.

Example 12-1

In the circuit of Fig. 12-4(a), $R_{\text{F}} = 220$ kΩ, $R_{\text{IN}} = 4.7$ kΩ, and the op amp has a voltage gain of 45 000. Find A_v and Z_{in}.

Solution The voltage gain of the circuit is independent of the gain of the op amp:

$$A_v = -\frac{R_F}{R_{IN}} = \frac{-220\text{ k}\Omega}{4.7\text{ k}\Omega} \mathbf{-46.8}$$

$$Z_{in} = R_{IN} = \mathbf{4.7\text{ k}\Omega}$$

12.5 SUMMING CIRCUITS

An operational amplifier can be used to add two (or more) voltages or waveforms, as shown in Fig. 12-5(a). The voltage gain from V_1 to V_O is -1, as is the gain from V_2 to V_O. The voltage applied at V_1 does not affect the voltage applied to V_2 because the right-hand ends of both resistors are at virtual ground. The output voltage is actually the negative of $V_1 + V_2$ because the circuit inverts both signals, but it can be made positive by following the summing op amp with an inverting amp having a gain of -1, if necessary.

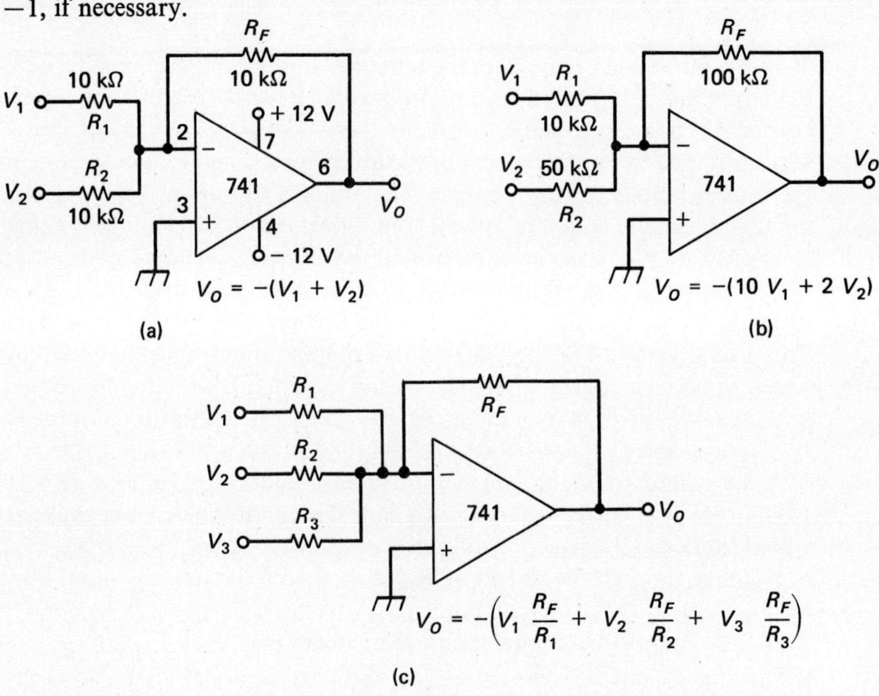

Figure 12-5 Op-amp summers and scaling summers: (a) two signals added and inverted; (b) two signals multiplied by two constants, added, and inverted; (c) three signals scaled, added, and inverted.

Figure 12-5(b) shows how a summing amplifier can also multiply the input signals by a constant factor. The gain from V_1 to V_O is -10, but the gain from V_2 to V_O is -2.

Figure 12-5(c) offers a general equation for multyplying three signals by constant factors and summing the products. The signals may be positive or negative dc levels, or ac waveforms

Example 12-2

In Fig. 12-5(b), what voltage must be applied at V_2 to make the output voltage zero when V_1 equals +0.75 V dc?

Solution

$$V_O = -(10V_1 + 2V_2)$$
$$0 = -(10)(0.75) - 2V_2$$
$$2V_2 = -7.5$$
$$V_2 = \mathbf{-3.75\ V\ dc}$$

Example 12-3

In Fig. 12-5(c), $R_1 = 400\ \text{k}\Omega$, $R_2 = 50\ \text{k}\Omega$, $R_3 = 10\ \text{k}\Omega$, $R_F = 100\ \text{k}\Omega$, $V_1 = +6.0$ V, $V_2 = -1.2$ V, and $V_3 = +0.20$ V. Find V_O.

Solution

$$V_O = -\left(V_1\frac{R_F}{R_1} + V_2\frac{R_F}{R_2} + V_3\frac{R_F}{R_3}\right)$$
$$= -\left(6.0\frac{100}{400} - 1.2\frac{100}{50} + 0.2\frac{100}{10}\right)$$
$$= -(1.5 - 2.4 + 2)$$
$$= \mathbf{-1.1\ V}$$

12.6 THE NONINVERTING OP-AMP CIRCUIT

The noninverting op-amp circuit can provide high voltage gains with high input impedance and no inversion of the signal waveform. The basic circuit configuration is shown in Fig. 12-6. In analyzing this circuit, it is necessary to recall that the op amp responds not to the voltage between either of the input terminals and ground but to the voltage between the inverting and noninverting inputs, without respect for the voltage to ground.

Resistors R_F and R_{IN} form a voltage divider which places a certain fraction of the output voltage at the inverting input of the op amp. The voltage division is seen more clearly in the redrawn circuit of Fig. 12-6(b). For the resistor values shown, the voltage fed back to the input is one-tenth of the output voltage. The voltage at the noninverting input (V_{IN}) must be virtually equal to the voltage at the inverting input, so the output voltage V_O must be 10 times the input voltage V_{IN} [for the values of R_{IN} and R_F in Fig. 12-6(b)]. If the output voltage tried to assume any other value, a voltage difference would appear between the inverting and noninverting inputs of the op amp. The polarity of this voltage difference would be such as to return the output voltage to 10 times the input voltage. The gain of the circuit is thus dependent only on the values of R_F and R_{IN}.

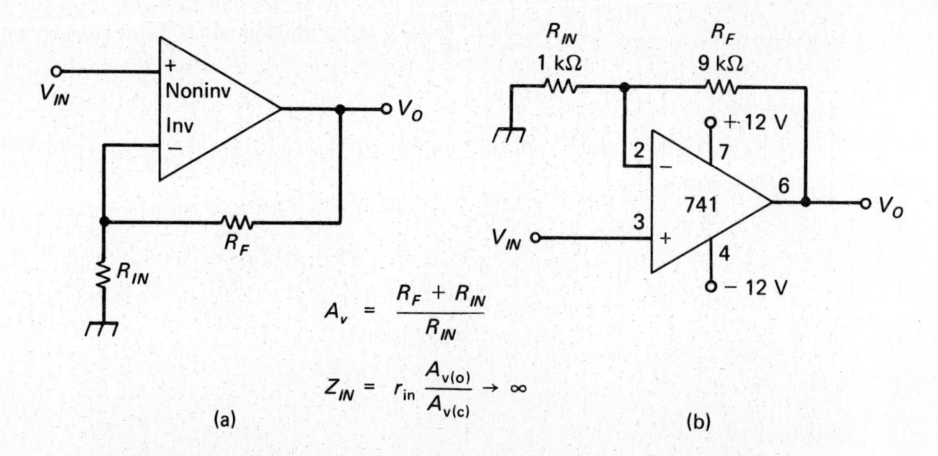

Figure 12-6 Two ways of drawing the noninverting operational-amplifier circuit. $A_{v(o)}$ is op-amp open-loop gain, $A_{v(c)}$ is circuit closed-loop gain, and r_{in} is op-amp input resistance.

$$A_v = \frac{R_F + R_{IN}}{R_{IN}} \tag{12-2}$$

The input impedance of this amplifier is very high—many megohms—but at high frequencies stray capacitance tends to lower the input impedance somewhat.

A special case of the noninverting amplifier is the voltage follower shown in Fig. 12-7. It is simply a noninverting amplifier with $R_{IN} = \infty$ and $R_F = 0$. The voltage gain is therefore unity, and the input impedance is extremely high.

This circuit is similar in function to the emitter follower (Section 7.1), except that it generally has a higher input impedance and a lower output impedance. In addition it is responsive to dc with no offset voltage between the input and outputs. It is useful for driving low-impedance loads with high-impedance signal sources.

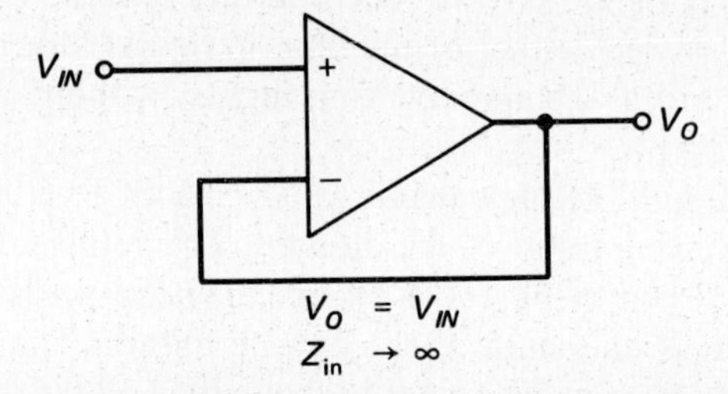

Figure 12-7 The voltage follower has very high Z_{in} and very low Z_o.

12.7 NONIDEAL EFFECTS IN OP AMPS

Op-amp frequency compensation: An op amp has (internally) typically four stages of amplification. Unavoidably, there will be stray capacitance associated with the transistors and resistors making up these stages, and at high frequencies these capacitors will cause appreciable phase shifts. A shift of 45° in each stage would produce a total phase shift (within the amplifier) of 180°, which is tantamount to signal inversion.

The negative-feedback resistor, which promotes stability at low frequencies, thus completes a *positive* feedback loop at the high frequency where stray capacitance causes a 180° phase shift within the amplifier.

As pointed out in Section 13.1, positive feedback and a total gain greater than 1 are the only requirements for an oscillator. Operational amplifiers in the inverting or noninverting amplifier configuration do, in fact, oscillate at a high frequency if *frequency compensation* is not employed.

The idea behind frequency compensation is to reduce the overall gain of the amplifier to unity or below at the frequency where the total phase shift amounts to 180°. This is usually done by placing a capacitor from collector to base of a high-gain common-emitter stage within the amplifier. Some op amps (such as the type *741*) have the compensating capacitor integrated on the chip. In other types, R and C must be added externally. The advantage of external frequency compensation is that it can be tailored for the particular gain being used to provide the highest possible frequency response for the amplifier. Low-gain amplifiers use a high degree of feedback and require higher-value capacitors to prevent self-oscillation. High-gain amplifiers which feed back only a small portion of their output signal do not need to have their high-frequency gain trimmed back so severely, and smaller compensating capacitors can therefore be used. If the op amp is undercompensated for the particular gain factor used, it will self-oscillate. If it is overcompensated, it will show a reduction in bandwidth.

Internally compensated op amps must be fixed-compensated for the worst possible case, which is the gain-of-1 circuit. For gains greater than 1, an internally compensated op amp is overcompensated and will show a reduced bandwidth. The upper frequency limit for the 741 op amp at various gains is given below to illustrate the point:

A_v	f_{max}
1	1 MHz
10	100 kHz
100	10 kHz
1000	1 kHz

Slew rate is an op amp's maximum rate of change in output voltage, expressed in V/μs. For large output signals it may limit the amplifier to frequencies lower than expressed by the *bandwidth* specification. For sine waves an approximate value of the upper-frequency limit due to slew rate is

$$f_{max} = \frac{S}{\pi V_{o(p\text{-}p)}} \tag{12-3}$$

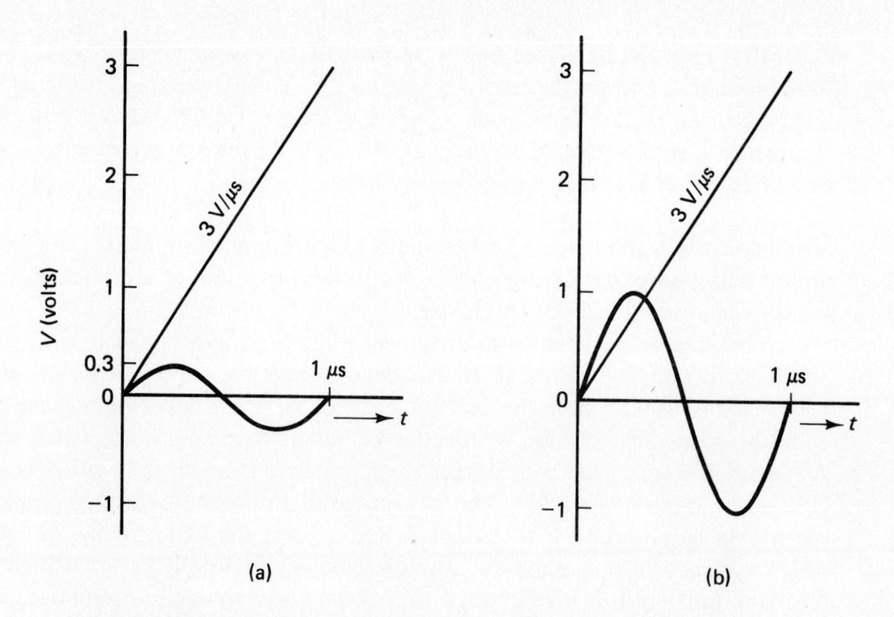

Figure 12-8 A slew rate of 1 V/μs is not exceeded by a 0.3-V 1-MHz wave (a), but is exceeded by a 1.0-V pk, 1-MHz wave (b).

where S is the slew rate in V/s. Figure 12-8 shows the relationship between slew rate and bandwidth for small and large signals.

Input bias current: In op-amp circuits where a very high value of feedback resistance is used, it may be noticed that the output dc level is *offset* from zero, even through the input signal is zero. This is due to the *input bias current* required by the bases of the op-amp input transistors. The amount of bias current for a given op amp can be determined from the dc output offset voltage and feedback resistance, as shown in Fig. 12-9(a). If the bias-current specification is known, the highest permissible value of R_F for a given output offset-voltage specification can be determined.

Example 12-4

A type 741 operational amplifier has a maximum input bias current requirement of 0.08 μA. The maximum tolerable output offset voltage is 0.1 V. What is the highest value of R_F which can be used? If the op amp is to have a voltage gain of −10, what will be its input impedance?

Solution

$$R_F = \frac{V_{OFFSET}}{I_{BIAS}} = \frac{0.1}{0.08 \times 10^{-6}}$$

$$= \mathbf{1.25\ M\Omega} \tag{12-4}$$

$$R_{IN} = \frac{R_F}{A_v} = \frac{1.25\ M\Omega}{10} = \mathbf{125\ k\Omega}$$

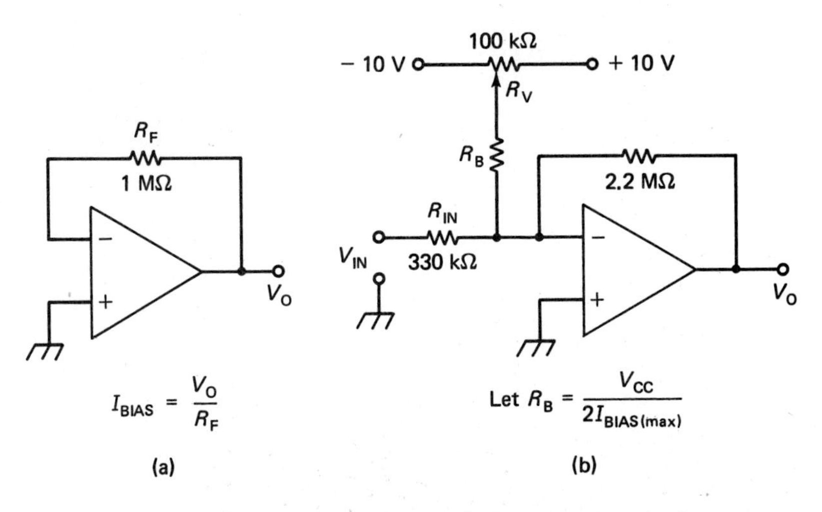

Figure 12-9 (a) Test circuit for determining the input-bias current of an op amp. The op amp should be over-frequency-compensated. (b) Adjustable resistor R_V is adjusted to minimize output voltage offset resulting from bias current.

Many op amps have offset-null pins to which a trim pot may be connected for elimination of offset. An externally supplied adjustable bias circuit, as shown in Fig. 12-19(b), can also be used to bring the offset voltage to nearly zero.

Bias-current compensation: The input bias current is assumed to be 1 μA for each input in the gain-of-10 circuit of Fig. 12-10(a), although modern op amps are available with bias currents tens and even hundreds of times less than this. Notice that, with the input grounded, +1 V is required at the output to supply 1 μA through R_F to the − input. The current through R_{IN} is zero since the − input is virtually ground. The 1 μA to the + input comes from ground but causes no voltage drop, since there is no resistance in this line.

The +1-V output-voltage offset could be reduced by decreasing R_F, but this would require a corresponding decrease of R_{IN}, resulting in lower input impedance. An op amp with lower bias current could be used, and indeed low-cost/low-speed op amps are available with bias currents on the order of 10 nA, and at a higher price 1 nA is obtainable. High-speed op amps generally have higher bias-current requirements, however.

Figure 12-10(b) shows how a bias-compensating resistor R_B can be added at the + input to eliminate output offset if the input bias currents are equal. The − input bias current is obtained through R_{IN} and R_F, which are in parallel if $V_O = 0$ and V_{IN} is grounded. R_B is chosen equal to $R_{IN} \| R_F$, so the voltage developed at the − input (across $R_{IN} \| R_F$) equals the voltage developed at the + input (across R_B). The differential-input voltage is then virtually zero. The value of R_{IN} in this determination includes the resistance of the signal source if it is not negligible.

Input offset current: The + and − input transistors of an IC op amp are fabricated on the same chip, so their bias-current requirements tend to be equal. To

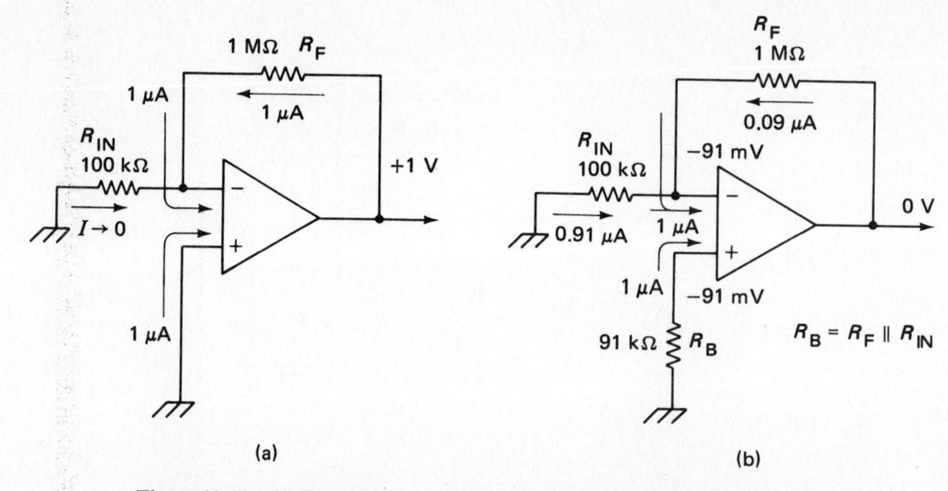

Figure 12-10 (a) Input bias-current requirements force the output to assume a nonzero value even when the input is zero. (b) Bias-compensating resistor R_B minimizes output offset.

the extent that they are not equal, the compensating resistor R_B will be ineffective in eliminating output offset. The difference between the input bias currents is called input offset current, and is typically $\frac{1}{10}$ to $\frac{1}{3}$ of bias current for most op amps.

Input offset voltage is the voltage difference that must be applied between the two inputs of an op amp to force its output to go to zero. Ideally, both transistors of the input differential amp would turn on at the same voltage, but slight differences do exist. Input offset voltage is specified at 0.5 to 10 mV for commonly available op amps. In Fig. 12-11(a), input offset voltage is represented as a voltage source in

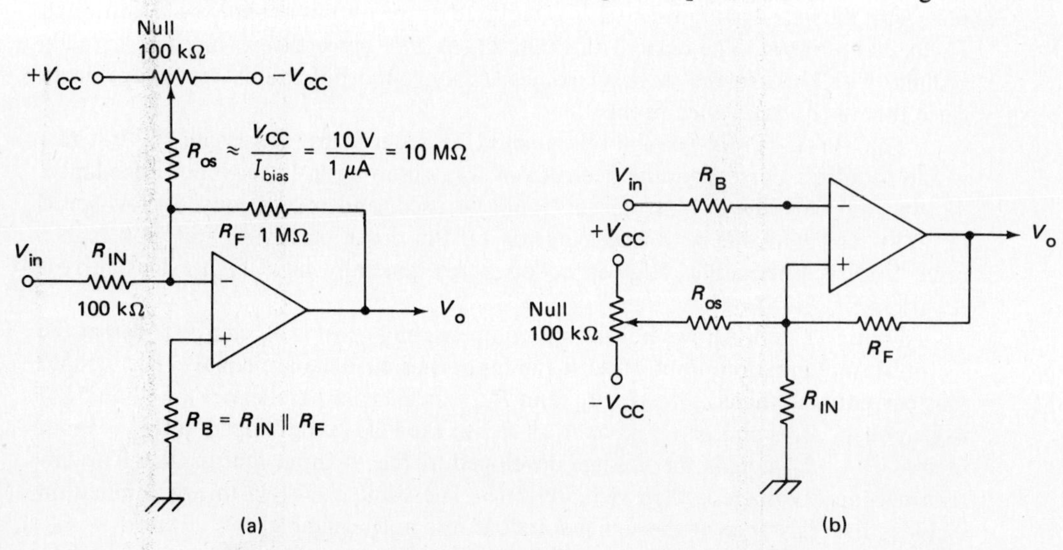

Figure 12-11 (a) Input offset voltage also causes output offset; (b) input offset voltage null circuit.

series with the + input in an inverting gain-of-10 circuit. Notice that the output voltage must assume a value $V_{OS}(A_V + 1)$ in order to bring the ideal + and − inputs to zero differential voltage. (Bias and offset currents are being neglected for the moment.) Decreasing R_F and and R_{IN} will not reduce the output offset due to V_{OS} as it will in the case of I_{BIAS} and I_{OS}. Offset voltage is usually the predominant problem in high-gain circuits, whereas offset current tends to predominate in high-impedance circuits. Figure 12-11(b) shows a circuit designed specifically to null the input offset voltage in an inverting op amp.

12.8 THE OP-AMP INTEGRATOR

Integration is a mathematical operation of calculus which may be descriptively understood as taking the summation of an infinitely large number of infinitely small parts. As applied to electrical waveforms on a voltage vs. time graph, integration can be viewed as a process of *accumulation.*

Figure 12-12(a) shows a positive dc waveform x and the result of integrating that waveform. The integral of x with respect to time ($\int x\,dt$) is seen to be the accumulation of x as time goes on. The value of x is positive and constant, so the rate of accumulation (slope) of $\int x\,dt$ is positive and constant.

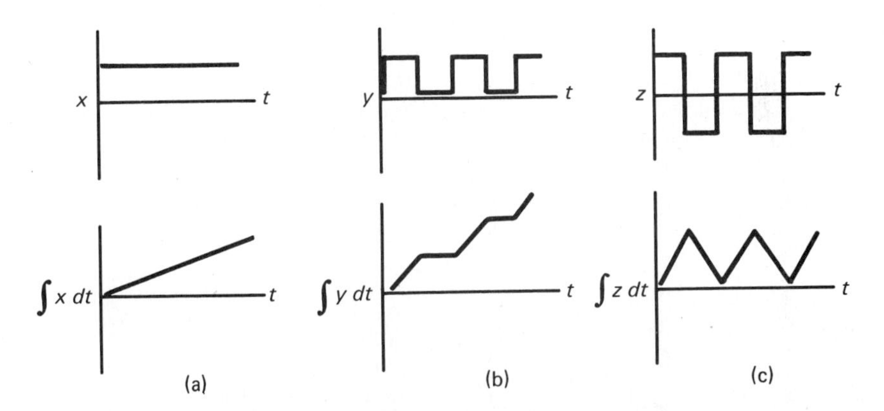

Figure 12-12 Three waveforms and their integrals.

Figure 12-12(b) shows the result of integrating a string of rectangular positive pulses. The value of $\int y\,dt$ increases (accumulates) as long as y has a positive value. However, when the value of y goes to zero, the value of $\int y\,dt$ simply holds at the value it has accumulated to that point.

Figure 12-12(c) shows the integral of a square wave having both positive and negative values. The positive values of z cause a positive accumulation (increase), and the negative values of z cause a decrease in the value of $\int z\,dt$. Thus the integral of a square wave is a triangle wave.

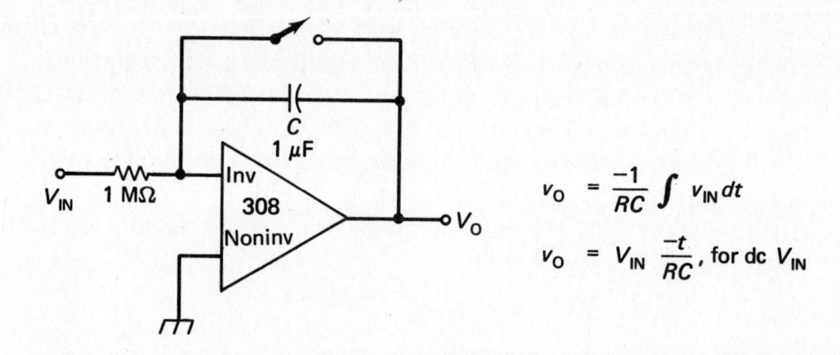

Figure 12-13 Op-amp integrator with the basic transfer function equation. For dc inputs this equation reduces to equation (12-4).

The operational amplifier circuit which performs the mathematical operation of integration on the input signal is shown in Fig. 12-13. The general equation for the output voltage is given for reference, but for a constant dc input the equation reduces to

$$v_O = V_{IN}\frac{-t}{RC} \tag{12-5}$$

where t is in seconds, R is in ohms, and C is in farads.

The output wave shape for a dc input is a ramp, with the above equation defining the value of the ramp voltage at any time t after switch S_1 is opened.

A number of additional waveforms of signals and their integrals are given in Fig. 12-14. You should be able to verify these waveforms from an understanding of

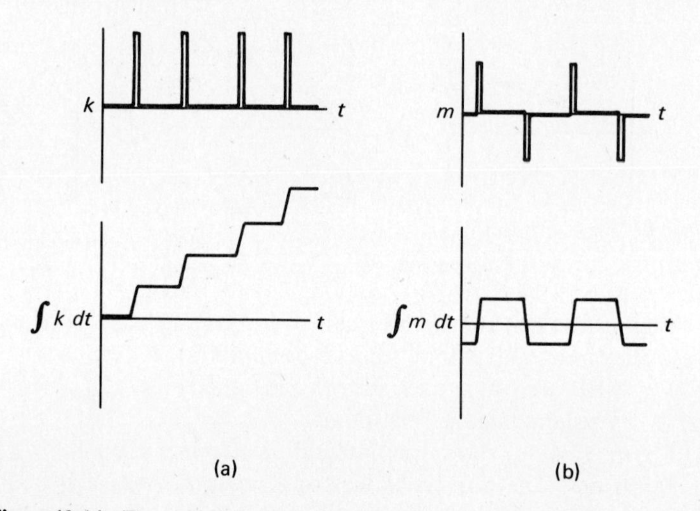

Figure 12-14 Four waveforms and their integrals. Integration can be thought of as a process of accumulation.

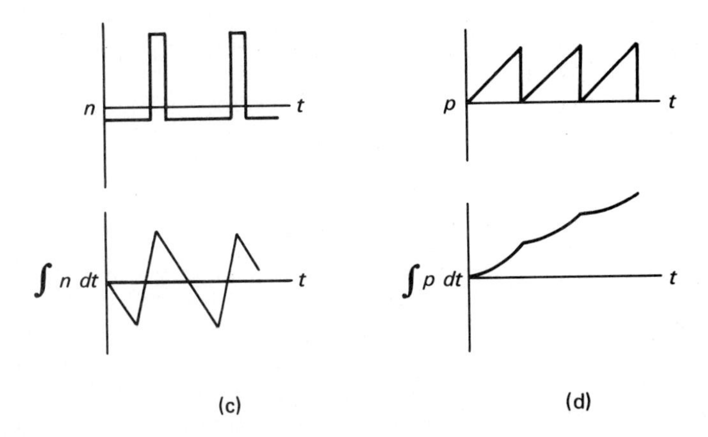

Figure 12-14 Continued

integration as a process of accumulation. Notice that the general tendency of the integrator is to smooth off the waveforms, making them less peaked.

A simple RC filter network [Fig. 12-15(a)] provides a rough approximation to the integration process if the time constant RC is long compared to the period of the input waveform. This circuit is therefore often referred to as an integrator. For mathematically precise integration, however, the circuit of Fig. 12-13 should be used.

The differentiator circuits: Most mathematical operations have associated inverse operations: addition and subtraction, multiplication and division, logarithms and antilogs. The inverse of integration is *differentiation*. The operation of differentiation produces a quantity called the *derivative*. The derivative of a quantity expresses the *rate of change* of that quantity with respect to some other variable. Applied to voltage

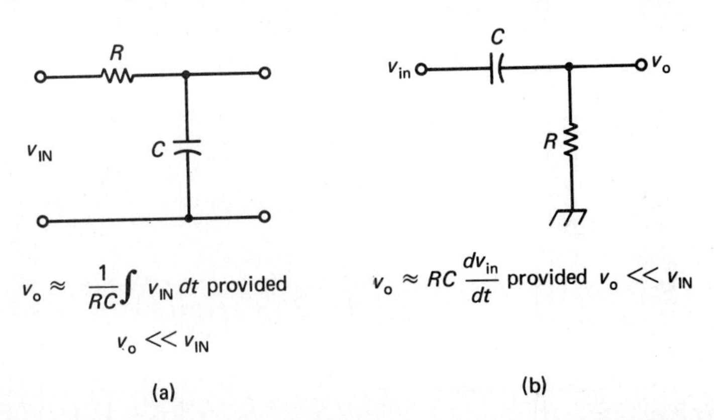

Figure 12-15 (a) Simple circuit giving an approximate integral output provided that τ ($= RC$) is much *longer* than the period of the input waveform; (b) circuit giving a derivative output, provided that τ is much *shorter* than the time period during which v_{IN} is changing.

vs. time waveforms, the derivative of a wave at any point in time is simply the *slope* (rate of change) of the wave at that point.

The simple *RC* circuit of Fig. 12-15(b) provides a fair approximation to a derivative output if the time constant *RC* is short compared to the period of the input signal waveform. This simple *RC* circuit is widely used, whereas op-amp differentiators are used relatively infrequently because of noise problems.

Figure 12-16(a) shows a positive dc waveform w and its derivative with respect to time, dw/dt. The rate of change (slope) of a dc signal is zero; hence the value of the derivative is zero. In general, the derivative of a constant is zero.

Figure 12-16(b) shows the derivative of a positive ramp signal. The slope of q is a positive constant, so the value of the derivative dq/dt is a positive constant. Now refer back to Fig. 12-12(a), which shows the *integral* of a positive constant to be a positive-going ramp. This comparison demonstrates that integration and differentiation are inverse operations.

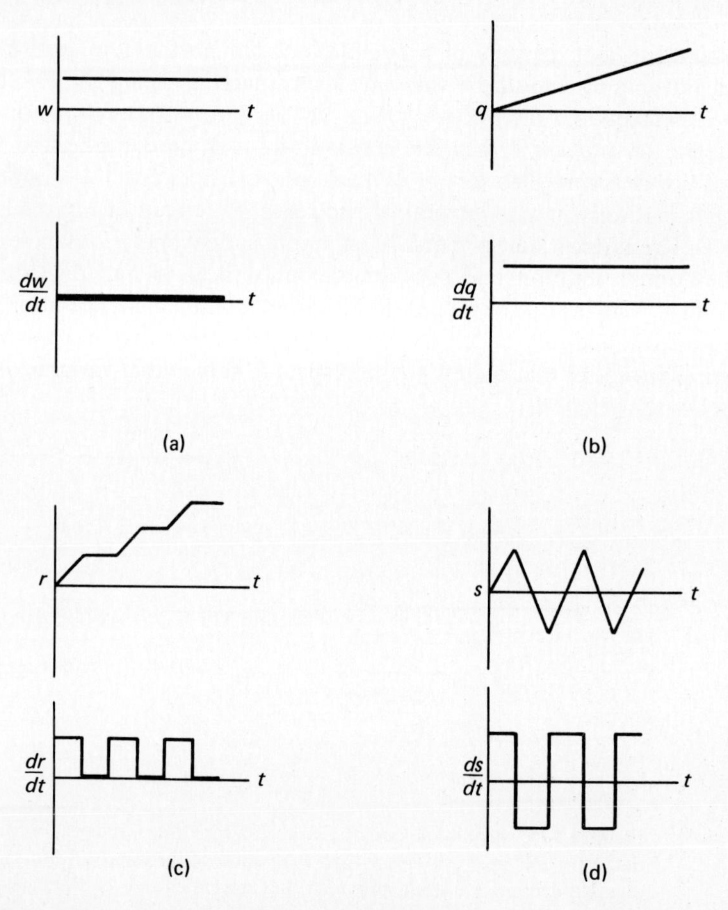

Figure 12-16 Four waveforms and their derivatives. The derivative can be thought of as the rate of change of a function.

In Fig. 12-16(c) and (d) there are further demonstrations of how the derivative waveform can be found as the slope of the original waveform. These two can also be compared with Fig. 12-12 (b) and (c), respectively, to show that differentiation and integration are inverse operations.

12.9 THE HYSTERESIS SWITCH

With positive feedback (from output to noninverting input) the op amp becomes an electronic switch, changing from positive saturation to negative saturation in response to the input signal. The hysteresis switch is an IC version of the Schmitt trigger (Section 5.3) with controllable triggering point and dead zone.

The basic hysteresis switch is shown in Fig. 12-17(a). If the output is positive, current through the feedback resistor to the noninverting input will hold the output saturated positive until a negative input at V_{IN} overcomes the current from R_F and drives the output to negative saturation. Notice that there are two critical switching voltages, one positive and the other negative. Low switching voltages can be obtained with high values of R_F and low values of R_{IN}. Frequency compensation is not required in this, or any similar switching circuit, because the output is always saturated, making self-oscillation impossible.

Figure 12-17(b) shows an op-amp circuit which functions as a Schmitt trigger. Variable resistor R_T sets the trigger point, R_{IN} sets the input impedance, and R_F controls the dead zone.

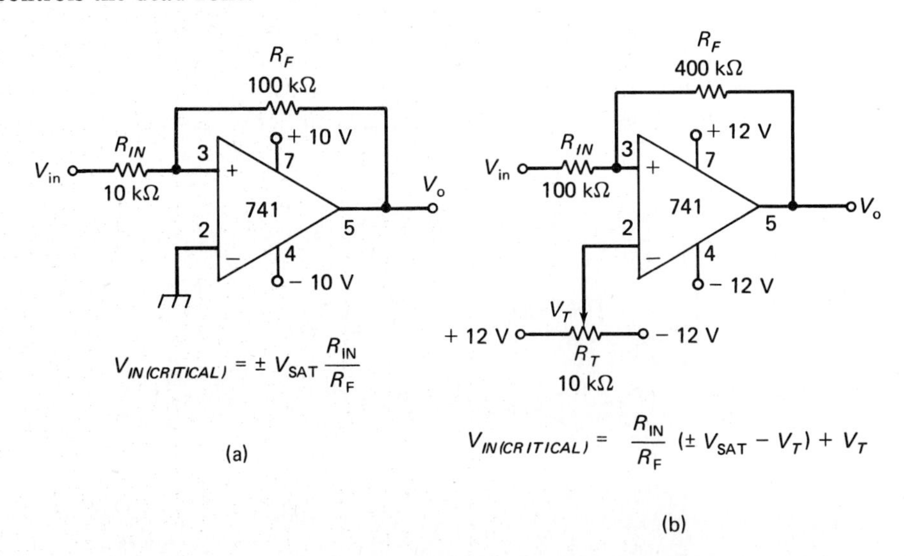

Figure 12-17 (a) Op-amp hysteresis switch. V_O switches from negative saturation to positive saturation when the critical positive V_{IN} is reached. (b) Op-amp Schmitt trigger. V_T sets the trigger voltage, and the ratio of R_F/R_{IN} sets the dead zone.

Example 12-5

In Fig. 12-17(b), $R_{IN} = 100\ k\Omega$, $R_F = 400\ k\Omega$, $R_T = 10\ k\Omega$, and V_T is set to -2 V dc. The supply voltages are ± 12 V, and the saturation voltages are ± 10 V. Find the upper and lower trigger points.

Solution

$$V_{IN(CRITICAL)} = \frac{R_{IN}}{R_F}(\pm V_{SAT} - V_T) + V_T \qquad (12\text{-}6)$$

$$= \frac{100\ k\Omega}{400\ k\Omega}(\pm 10 + 2) - 2$$

$$V_{UPPER} = 0.25(12) - 2 = \mathbf{+1\ V}$$

$$V_{LOWER} = 0.25(-8) - 2 = \mathbf{-4\ V}$$

The output of the hysteresis switch will go positive when the input voltage goes more positive than +1 V. It will go negative when the input goes more negative than −4 V. In the dead zone between +1 V and −4 V, the output will remain in the saturated state to which it was previously set.

12.10 OP-AMP APPLICATIONS

The function generator circuit: The hysteresis switch and the integrator (Section 12.8) can be combined to form a very useful variable-frequency signal generator circuit, as shown in Fig. 12-18.

The basic circuit produces square and triangle output waveforms, but the triangle wave can be shaped into a remarkably pure sine wave by a diode and resistor network, as will be shown in Chapter 17.

To understand the operation of the circuit, let us assume that the supplies are

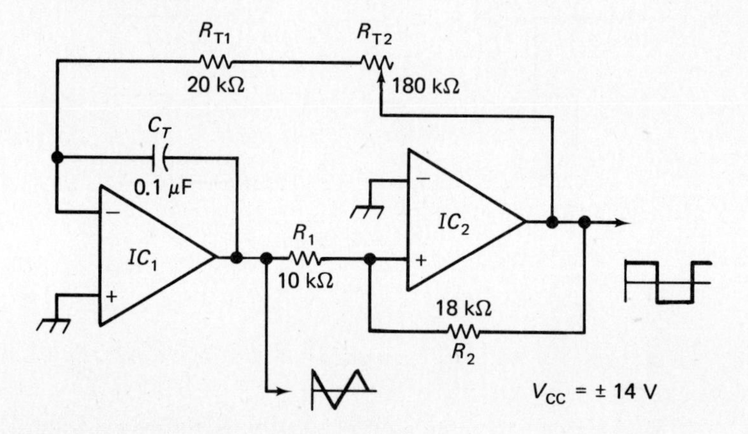

Figure 12-18 An elementary function generator circuit. IC_1 is an integrator and IC_2 is a hysteresis switch.

± 14 V, so the saturated output voltage of IC_2 is typically ± 12 V. Assuming that the IC_2 output is $+12$ V, R_T will feed positive current into the inverting input of IC_1, causing a negative-going ramp at the IC_1 output. When this negative ramp reaches the switching voltage of the hysteresis switch, the IC_2 output will switch to -12 V and a positive ramp will be generated by integrator IC_1. This ramp will eventually become positive enough to switch the hysteresis switch to the $+12$ V state again, and the cycle will repeat. The frequency of the generated signal can be adjusted by varying the timing resistance R_T. The frequency range covered can be changed by switching to a different value of C_T.

Commercial function generators which produce triangle, square, ramp, pulse, sine, and other complex waveforms at frequencies from the cycle-per-hour range to the several-megahertz range are generally elaborations and refinements of the basic circuit of Fig. 12-18.

Example 12-6

In Fig. 12-18, $C_T = 0.1\ \mu F$, $R_T = 100\ k\Omega$, $R_1 = 10\ k\Omega$, $R_2 = 18\ k\Omega$, and the IC output-saturation voltage is ± 12 V. Find the generated frequency and the peak-to-peak triangle wave voltage from IC_1.

Solution The peak triangle voltage is the critical voltage of the hysteresis switch. From Section 12.7,

$$V_{CRITICAL} = \pm V_{SAT}\frac{R_{IN}}{R_F} = \pm 12\ V\frac{10\ k\Omega}{18\ k\Omega} = \pm 6.7\ V$$

$$V_{p\text{-}p\ triangle} = (2)(6.7) = \mathbf{13.4\ V\ p\text{-}p}$$

The time required for the ramp to run up a 13.4-V span from -6.7 to $+6.7$ V is calculated as explained in Section 12.8:

$$V_O = V_{IN}\frac{-t}{RC}$$

$$t = \frac{-V_O}{V_{IN}}RC = \frac{-13.4}{-12}(100 \times 10^3)(0.1 \times 10^{-6})$$

$$= 1.12 \times 10^{-2}\ s$$

This is the time for one-half of a cycle. The period for a full cycle is $2t$, or 2.24×10^{-2} s. The signal frequency can be calculated from the period:

$$f = \frac{1}{T} = \frac{1}{2.24 \times 10^{-2}} = 0.45 \times 10^2$$

$$= \mathbf{45\ Hz}$$

Bias regulation of a power amplifier may be achieved using the dc voltage gain of an op amp. The circuit is given in Fig. 12-19. Only 0.2 V is wasted across the emitter resistor, but this is compared with the desired 0.2 V at the wiper of R_2. Any difference is amplified by the op amp, whose gain is $-R_6/(R_4 + R_5) = 110$ in this example. Q_1 buffers the op-amp output, which should not be required to deliver more

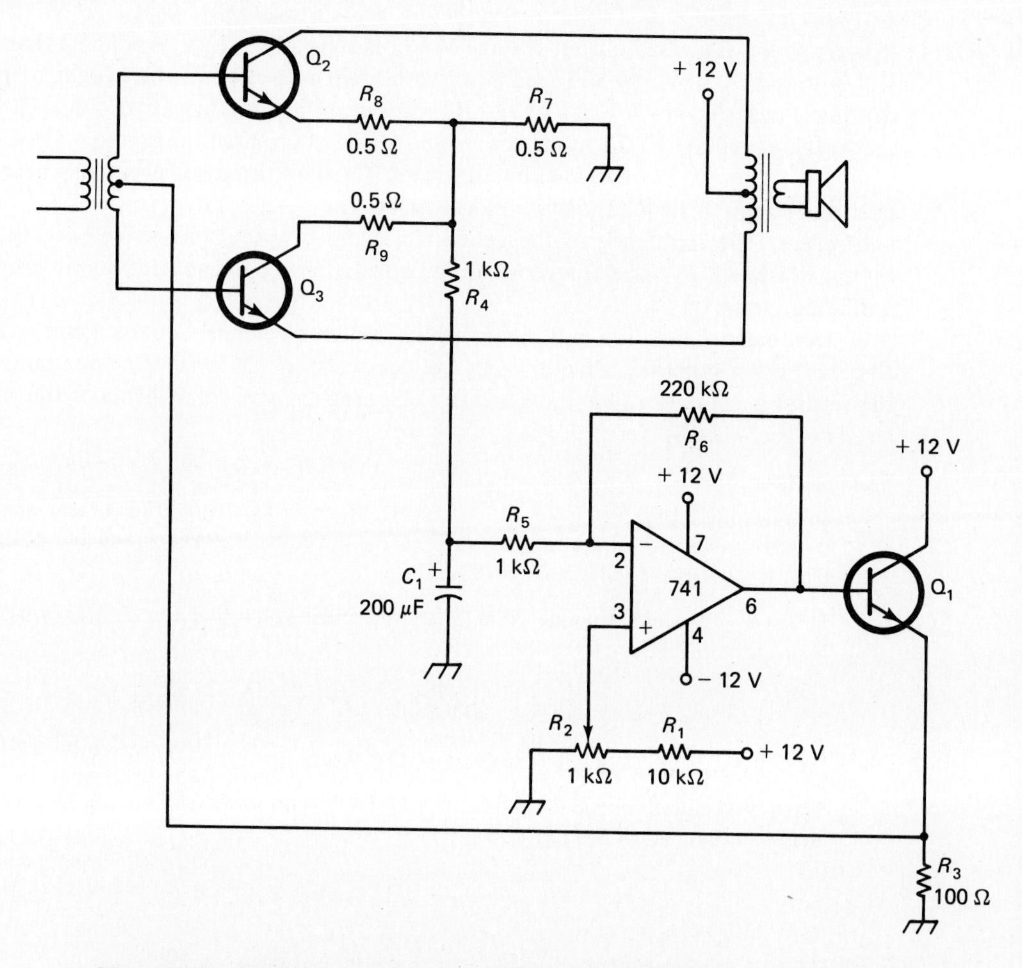

Figure 12-19 An op amp senses emitter current and controls bias point while wasting much less voltage across the emitter resistor than conventional bias circuits.

than about 3 mA, to meet the base-current requirements of Q_2 and Q_3. If V_{R7} tends to increase, the op-amp output decreases, I_E of Q_1 decreases and the base bias decreases, thus restoring V_{R7}. C_1 filters audio variations from V_{R7} so that the op amp feeds back only dc voltages.

The perfect diode: The circuit of Fig. 12-20 gives a half-wave negative- or positive-rectified output with virtually no forward diode drop V_D. Current summing as illustrated in Fig. 12-4 requires that $v_O/v_{IN} = R_F/R_{IN}$, but the diodes permit only one polarity of v_O across each R_F. Amplification of v_O is possible by simply raising the ratio R_F/R_{IN}. Fast signal diodes should be used because power-supply types take too long to turn off. Where a few millivolts of output offset are not detrimental, the offset pot and 10-MΩ resistor may be omitted. This circuit is the basis of many electronic ac voltmeters.

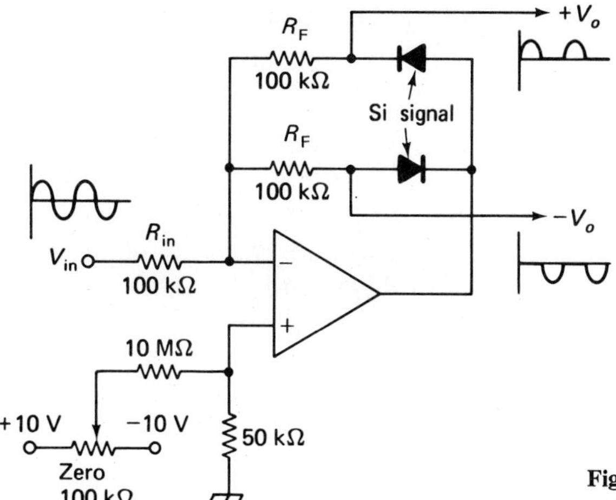

Figure 12-20 Precision rectifier with positive and negative half-cycle outputs.

CHAPTER SUMMARY

1. Monolithic integrated circuits are fabricated by the silicon planar process. This involves photoreduction techniques, vapor diffusion of impurity elements into silicon wafers, and deposition of metal interconnections on the wafer. Diodes, FETs, and bipolar transistors are fabricated easily in this process. Resistors are somewhat harder to form, capacitors of even a few tens of picofarads are difficult, and inductors are all but impossible.

2. The operational amplifier is the most important linear IC. Op amps have differential inputs, very high voltage gain, and dc coupling throughout. Most often they require separate positive and negative power supplies.

3. In the differential-comparator circuit the op-amp output switches fully positive if the noninverting (+) input is more positive than the inverting input (—). If the (—) input is more positive the output switches to negative saturation.

4. The inverting op-amp circuit has R_{IN} feeding the (—) input and R_F feeding back from the output to (—) input. The (+) input is grounded. The (—) input is *virtual* ground (less than a 1-mV signal). Voltage gain and input impedance are

$$A_v = -\frac{R_F}{R_{IN}} \tag{12-1}$$

$$Z_{in} = R_{IN}$$

5. A summing circuit can be made using the inverting op-amp circuit with multiple R_{IN} resistors.

$$V_O = -\left(V_1\frac{R_F}{R_1} + V_2\frac{R_F}{R_2} + \cdots\right)$$

6. The noninverting op-amp circuit has an input impedance of hundreds of megohms. V_{IN} is fed to the (+) input. Feedback to the (−) input is via a voltage divider R_F–R_{IN}. Voltage gain is

$$A_v = \frac{R_F + R_{IN}}{R_{IN}} \tag{12-2}$$

7. Op amps must be frequency compensated with external or integrated capacitors to prevent self-oscillation in amplifier circuits. Bias currents and offset voltage of the input-transistor bases cause dc output voltage shifts, especially in high-gain circuits and in circuits where R_F is high.

8. Integration is a process of accumulation. An op-amp integrator uses a capacitor from the output to the (−) input. For a positive dc input voltage the output is a negative ramp.

$$v_O = V_{IN}\frac{-t}{R_{IN}C} \tag{12-4}$$

9. An op-amp Schmitt trigger or hysteresis switch is formed by connecting R_F from the output to the (+) input. The output voltage "snaps" from the positive to the negative saturation levels. The input triggering voltages are

$$V_{IN(crit)} = \pm V_{SAT}\frac{R_{IN}}{R_F} \tag{12-5}$$

10. A function generator is made by connecting the output of an integrator to the input of a hysteresis switch, and feeding the output of the hysteresis switch back to the input of the integrator. Triangle waves are taken from the integrator and square waves from the hysteresis switch.

QUESTIONS AND PROBLEMS

12-1. Arrange in order from the easiest to the hardest to fabricate in a silicon monolithic IC: capacitor; inductor; resistor; transistor.

12-2. List four characteristics of operational amplifiers.

12-3. An op amp operates from ±5-V supplies. What is the potential difference between the positive and negative supply pins?

12-4. In Fig. 12-3(c), the wiper of R_3 is set to 0.5 V and V_{in} is a 2-V p-p sine wave. Sketch V_{in} and V_o one above the other showing the timing relationship.

12-5. In Fig. 12-3(d), V_A is a 1-V p-p sine wave and V_B is a 1-V p-p sine wave which lags V_A by 45°. Sketch V_A, V_B, and V_o.

12-6. Define the following terms: monolithic; differential; virtual.

12-7. An amplifier with $A_v = -28.3$ and $Z_{in} = 5\ \text{k}\Omega$ is required. Use the circuit of Fig. 12-4(a) and find the required values of R_{IN} and R_F.

12-8. In the circuit of Fig. 12-5(a), $V_1 = 170$ mV and $V_2 = 80$ mV. Find V_O.

12-9. The circuit of Fig. 12-5(c) is to be used to implement the expression $V_O = -2.83V_1 - 1.11V_2 - 0.707V_3$. The lowest input resistor value is to be 100 kΩ. Find R_1, R_2, R_3, and R_F.

12-10. Draw a schematic diagram of a circuit which will implement the expression $V_O = -2V_1 + 5V_2$. *Hint:* You will need two op amps.

12-11. In the circuit of Fig. 12-6(a), $R_F = 27$ kΩ, $R_{IN} = 2.2$ kΩ, A_{vo} of the op amp is 20 000, and the input resistance of the op amp is 500 kΩ. Find circuit gain A_{vc} and Z_{in}.

12-12. What is the result of using too large a compensating capacitor in an op-amp circuit? What is the result of using too small a capacitor?

12-13. Which requires a larger compensating capacitance—a gain-of-10 or a gain-of-100 circuit? What is the disadvantage of fixed internal compensation of an op amp?

12-14. An op amp has a slew-rate specification of 4 V/μs. It is required to deliver 2 V rms. What is the frequency limitation imposed by slew rate?

12-15. In the circuit of Fig. 12-9(b), what value of R_B will permit R_v to compensate for bias currents up to 1 μA?

12-16. In the circuit of Fig. 12-9(a), $V_O = 0.18$ V. What is I_{BIAS}?

12-17. Redraw the circuit of Fig. 12-9(b) adding an offset-current compensating resistor and marking its value.

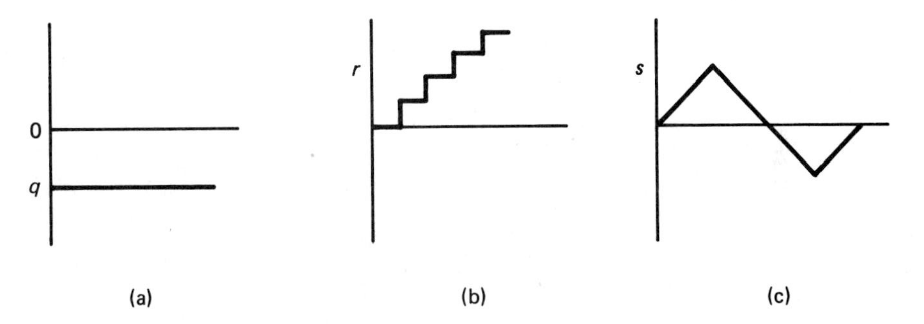

Figure 12-21 Waveforms for Problems 12-18 and 12-21.

12-18. Sketch the waveshape of the integrals for the waveforms in Fig. 12-21.

12-19. For the op-amp integrator of Fig. 12-13, $V_{IN} = 2.7$ V, $R = 470$ kΩ, and $C = 0.018$ μF. How long after S_1 is opened does it take v_O to saturate at -10 V?

12-20. In Fig. 12-13, what value of C is needed to produce $v_O = 1$ V/s for $V_{IN} = 1$ V dc, if $R = 100$ kΩ?

12-21. Sketch the waveshapes of the derivatives for the waveforms in Fig. 12-21.

12-22. In the hysteresis switch of Fig. 12-17(a), $R_{IN} = 10$ kΩ, $R_F = 47$ kΩ, and $V_{SAT} = \pm 10$ V. What v_{IN} makes v_O switch to $+10$ V? What v_{IN} makes v_O switch to -10 V?

12-23. In Fig. 12-17(b) the input-trigger points are to be $+2.0$ and $+2.5$ V. $R_{IN} = 3.3$ kΩ and $V_{SAT} = \pm 10$ V. Find R_F and V_T.

12-24. In Example 12-6 (Fig. 12-18) R_2 is changed to 27 kΩ and R_T to 500 kΩ. Find the new $V_{o(p\text{-}p)}$ and signal frequency.

12-25. In the bias-stabilized power amplifier of Fig. 12-19, what power will be dissipated in Q_2 at zero signal if R_2 is set to maximum voltage at its wiper? Assume zero transformer-winding resistance, but do not neglect drops across R_7 and R_8.

12-26. Draw a schematic diagram of a precision rectifier using a gain-of-1 inverting op amp and a scaled summer to yield a $\times 10$ full-wave-rectified output on a single line ($v_O = 10\,|v_{IN}|$).

13

OSCILLATORS

13.1 OSCILLATOR REQUIREMENTS

An oscillator is a device which generates an ac output signal without requiring any externally applied input signal. One class of oscillator, the *relaxation oscillator*, depends on the charge and discharge time constant of an *RC* (or *RL*) network in a switching circuit to produce its signal.

Another class of oscillator produces ac signals by using a basic amplifier circuit (as discussed in Chapter 6 through 8) with the input signal derived from the amplifier's own output. Two conditions must be met if such a *feedback oscillator* is to continue to generate a signal:

1. The feedback signal to the input must be strong enough to reproduce the original signal at the output. For example, if the amplifier has a voltage gain of 10, *at least* one-tenth of the output voltage must be fed back to the input, or oscillation will not continue.
2. The feedback signal must arrive at the input of the amplifier in a reinforcing phase relationship. For example, a +1-V pulse at the input of a standard common-emitter amplifier might produce a −10-V pulse at the output (*minus* because of the inverting character of this amplifier). If part of this −10-V pulse were fed directly to the input, this negative input would tend to cancel, rather than reinforce, the original positive input. However, if the −10-V output pulse were applied through a transformer to the input, and the secondary leads of the transformer were switched so as to invert the signal applied to the input, the positive-feedback input would reinforce the original positive input signal, and

oscillation would be possible. The first situation (canceling inputs) is called negative feedback, while the second (reinforcing) condition is termed positive feedback.

The Armstrong feedback oscillator, shown in Fig. 13-1(a), illustrates the basic oscillator requirements. A more complete explanation of the phenomenon of oscillation can be undertaken if reference is made also to the characteristic curves of Fig. 13-1(b).

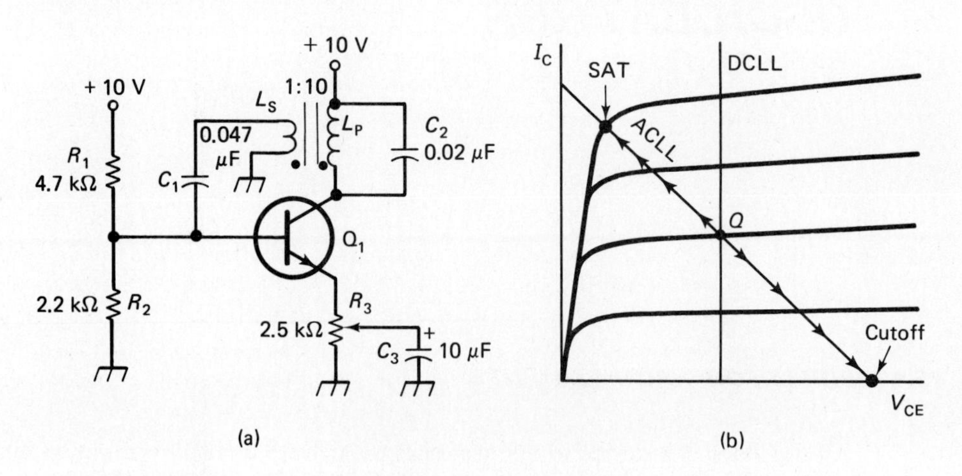

Figure 13-1 (a) Armstrong feedback oscillator; (b) positive feedback drives the transistor from saturation to cutoff along the load line.

Initially R_1 and R_2 bias the transistor at a Q point near the middle of the characterstic curves, as shown. The output (collector of Q_1) should ideally be 0 V ac, but there is bound to be a small noise signal present (a millivolt of ripple from the power supply, or a few microvolts of random thermally generated noise). For the purpose of analysis, let us assume that the noise is a -1-mV pulse at the collector of Q_1. The transformer will invert this pulse and step its voltage down by a factor of 10, so a $+0.1$-mV pulse will be applied through C_1 to the base of Q_1.

Now let us further assume that the gain of the transistor from base to collector is -100 (minus because it inverts the input signal). The $+0.1$-mV input will thus cause a -10-mV output. This output will be inverted and stepped down by the transformer and sent back to the base as a $+1$-mV pulse. The amplifier will now produce a -100-mV output in response to the $+1$-mV input. This upward spiral will continue, producing output pulses of -0.1-V, -1.0-V, and finally -10 V. At this point, the transistor has been driven back along the dynamic load line all the way to saturation [see Fig. 13-1(b)] and no further reduction in output voltage is possible.

With no voltage change across the transformer primary, the secondary output drops to zero, and the bias resistors R_1 and R_2 return the transistor collector voltage to the Q point. This positive-going voltage (from saturation to Q point) is fed back via the transformer to the input as a *negative-going* voltage, which continues to drive the transistor past the Q point toward cutoff. This spiral of *output causes input* and *input causes more output* continues until cutoff is reached and the transistor's collector voltage cannot increase further. The transformer then ceases to supply a reinforcing input signal, and the transistor heads back the load line toward the Q point. This negative-going voltage at the collector starts another feedback spiral which ends with the transistor reaching the saturation point again.

The frequency of the oscillations produced by the circuit of Fig. 13-1(a) is primarily determined by the resonant frequency of L_P and C_2, because it is only near this frequency that the feedback exactly reinforces the input signal. This is explained in detail in Section 13.4.

The connection of the transformer windings is critical. If one pair of windings is reversed, the feedback will be negative rather than positive, and oscillation will not result, or will occur at some spurious frequency, and then only when excessive feedback is used.

The waveform of the output signal can be changed by varying the amount of R_3 that is not bypassed to ground, and thus varying the gain of the amplifier. At full gain (wiper at top end of range) the feedback is so heavy that very distorted waves, approaching square waves, are produced. At reduced gain the wave shape becomes more sinusoidal, until finally the feedback is insufficient to maintain oscillation.

For most types of feedback oscillators the pure-sine-wave condition is difficult to maintain, so a feedback somewhat greater than the minimum required is used to ensure oscillation under all conditions. This results in distortion of the sine wave produced, which may be objectionable in some cases. Where high-purity sine waves are required, special current-sensitive resistors are often included in the oscillator circuit to balance the feedback at just the minimum required to sustain oscillation without producing distortion.

The Armstrong oscillator is rather critical as to the amount of feedback used, and it may go into a *blocking* mode if the feedback is too heavy. In this mode, several cycles of ac will be produced, but then the oscillator will be driven to a bias point where it cannot oscillate for several cycles, and a broken chain of *on-again-off-again* oscillations will result (see Fig. 13-3).

13.2 THE HARTLEY OSCILLATOR

The Hartley oscillator is similar to the Armstrong feedback-type described in Section 13.1, except that a tapped coil (autotransformer) is used instead of the full transformer. The resonating capacitor is placed in parallel with the entire coil winding, as shown in Fig. 13-2. The oscillator frequency can be determined by the formula

$$f = \frac{1}{2\pi\sqrt{LC}} \tag{13-1}$$

where L is the inductance of the total coil L_1 in henrys and C is the value of C_1 in farads.

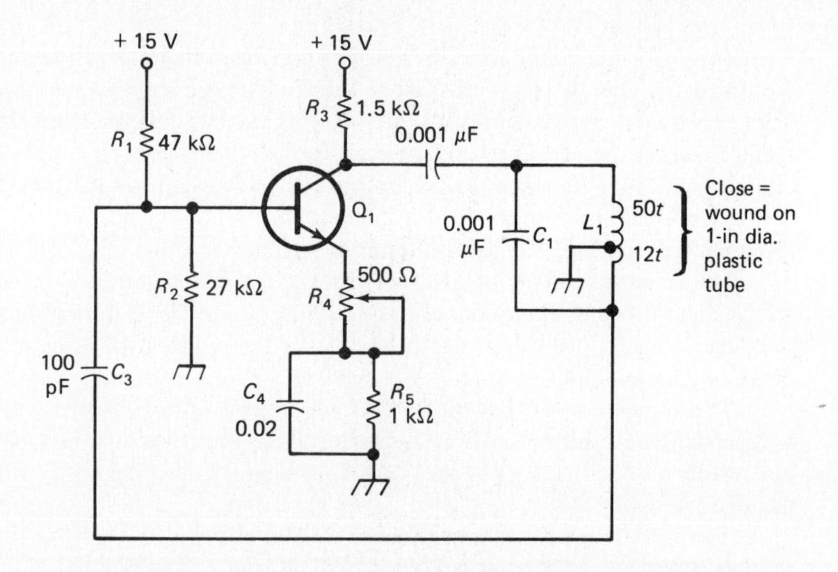

Figure 13-2 The Hartley oscillator is recognized by its tapped coil.

Feedback factor: As with the Armstrong oscillator, the degree of feedback affects the output waveform. The approximate amount of feedback is determined by the placement of the tap on L_1. The *feedback factor* is the ratio of the number of turns below the center tap of L_1(base side) to the number of turns above the center tap (collector side). In Fig. 13-2 this factor is $\frac{12}{50}$ or 0.24.

Typical feedback factors range from 0.5 to 0.1 for transistor Hartley oscillators. Variable resistance R_4 is left unbypassed so the amplifier gain can be reduced to control the oscillator waveshape. Low values of R_4 will cause a distorted output wave; higher values will produce a more sinusoidal wave, until eventually a point is reached where the amplifier's gain is too low to permit oscillation.

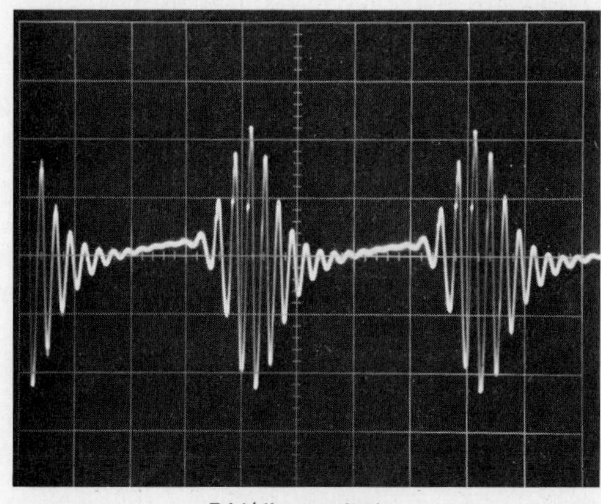

5 V/div. vertical
2 μs/div. horizontal

Figure 13-3 The 2-MHz Hartley oscillator of Fig. 13-2 is made to "block" at a 125-kHz rate by changing C_3 from 100 to 1000 pF.

Blocking mode: The Hartley oscillator can also be put into a blocking mode, as explained at the end of Section 13.1. The oscillator of Fig. 13-2 can be made to block by simply changing C_3 from 100 to 1 000 pF. Figure 13-3 shows the blocking waveform produced by the Hartley oscillator.

The Hartley oscillator circuit can be greatly simplified if the dc collector supply is fed through the coil L_1 rather than through a separate resistor. Substituting an inductor for the collector resistor makes the bias point much less critical, so a single 100-kΩ resistor can be used to provide the base bias current. The simplified Hartley oscillator is shown in Fig. 13-4.

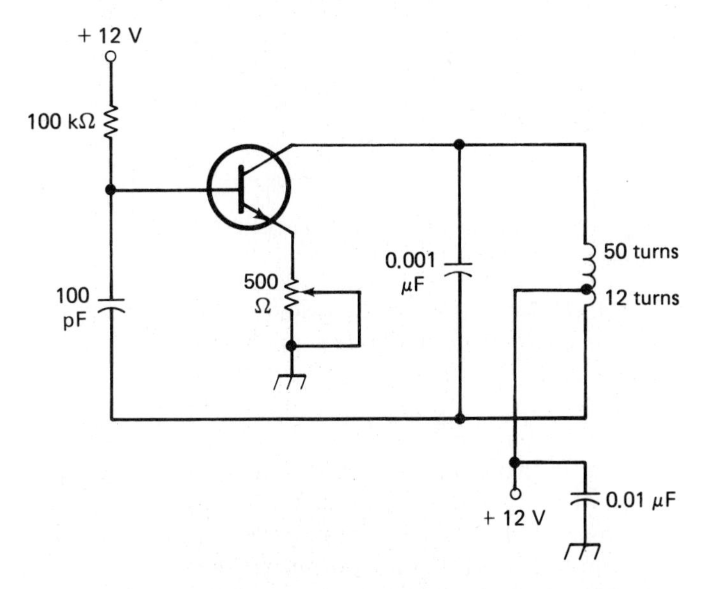

Figure 13-4 The Hartley oscillator of Fig. 13-2 can be greatly simplified without degrading its performance.

13.3 THE COLPITTS OSCILLATOR

The Colpitts oscillator is similar to the Hartley circuit just described, except that the inductive voltage divider is replaced by a capacitive voltage divider, thus eliminating the need for a tapped coil. An elementary Colpitts oscillator is shown in Fig. 13-5. The frequency of oscillation is determined from the formula

$$f = \frac{1}{2\pi\sqrt{LC_T}} \tag{13-1}$$

where C_T is the series combination of C_1 and C_2.

$$C_T = \frac{1}{\dfrac{1}{C_1} + \dfrac{1}{C_2}}$$

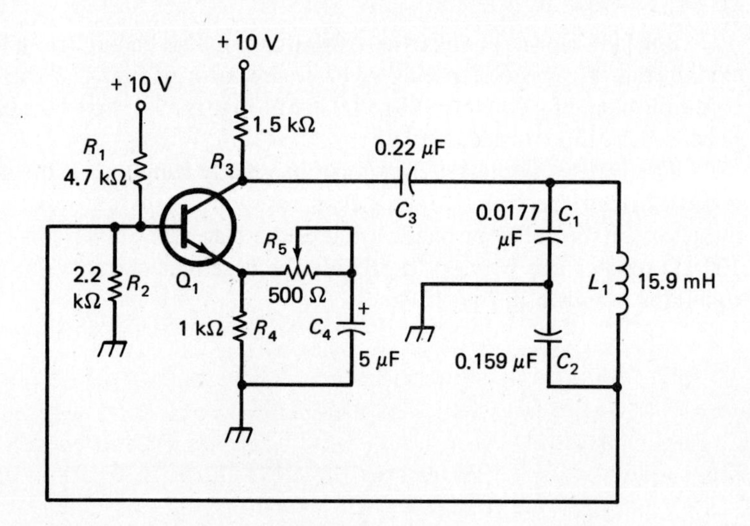

Figure 13-5 A Colpitts oscillator uses a two-capacitor network (C_1–C_2) to produce the inverted feedback voltage.

The feedback factor is set by the ratio of C_1 to C_2, which is typically in the range 0.1 to 0.5. The collector output voltage appears across C_1 (high reactance), and the voltage fed back to the base appears across C_2 (low reactance). Feedback can be increased by lowering the value of C_2 or by raising the value of C_1.

Although both C_1 and C_2 affect the oscillator frequency, C_1 has a more pronounced effect. If slight variations in frequency are desired, it is feasible to change C_1 by 10 or 20% without seriously altering the oscillator waveform.

The value of R_5 can be increased until the gain of the amplifier is lowered enough to produce a pure sine wave, if a pure output waveform is desired. If the shape of the output wave is not important, R_5 may be omitted and C_4 connected to the emitter.

Example 13-1

For the circuit of Fig. 13-5, find the frequency of oscillation and estimate the value of R_5 which will produce the purest sine wave output.

Solution The series value of C_1 and C_2 is determined first:

$$C_T = \frac{1}{\dfrac{1}{C_1} + \dfrac{1}{C_2}} = \frac{1}{\dfrac{1}{0.0177} + \dfrac{1}{0.159}}$$

$$= 0.0159\ \mu\text{F}$$

Now the resonant frequency is found:

$$f = \frac{1}{2\pi\sqrt{LC}} = \frac{1}{2\pi\sqrt{(0.0159)(0.0159 \times 10^{-6})}} = \mathbf{10.0\ kHz}$$

The feedback factor is

$$\frac{C_1}{C_2} = \frac{0.0177}{0.159} = 0.11$$

The gain of the amplifier must be enough to just overcome this voltage reduction if oscillation is to be maintained:

$$A_{v\,(\mathrm{min})} = \frac{1}{0.11} = 9$$

With 1.5 kΩ of collector resistance, the emitter resistance required to produce a gain of 12 can be estimated:

$$A_v \approx \frac{R_C}{R_E}$$

$$R_5 = R_E \approx \frac{R_C}{A_v} = \frac{1500\ \Omega}{9} = \mathbf{167\ \Omega}$$

This estimate does not take into account the losses in the coil and the loading effect caused by the reflection of the amplifier input impedance back to the collector of the transistor. However, we can be sure that the required value of R_5 is not greater than 167 Ω. Much lower values of R_5 will produce heavier feedback, resulting in a distorted output waveform.

13.4 OSCILLATOR STABILITY

The elementary Armstrong, Hartley, and Colpitts oscillators using bipolar transistors in the common-emitter configuration may be expected to experience frequency shifts of 1% or more in response to temperature changes from 50 to 100°F or dc supply voltage changes of 10%. This is primarily due to changes in the beta, and hence in the input impedance, of the transistor amplifier.

Figure 13-6(a) shows an idealized equivalent circuit for the Colpitts oscillator of Fig. 13-5. The amplifier input impedance and the Q of L_1 are both infinite. The vector diagram shows that there is a 180° phase lag between the collector-output and the base-input voltages. C_1 cancels the remaining 900-Ω inductive reactance so V_s sees an infinite impedance, there is no current in amplifier output resistance r_o, and $V_c = V_s$. (The factor μ is the voltage gain of the amplifier with the load of the feedback circuitry removed.)

Input impedance affects phase shift. A more realistic equivalent circuit is given in Fig. 13-5(b), where the total amplifier input impedance r_{in} is assumed to be 1 000 Ω. This corresponds to a transistor β of about 20, which is extreme—but we are merely making an example. A parallel-to-series conversion of $X_2 \| r_{in}$ to $X_s + R_s$ (review Fig. 11-18) shows that the actual phase shift from V_c to V_b is less than 180°. This is further illustrated in the second vector diagram, which cannot reasonably be rendered in true scale. The actual phase shifts, leading to a 6.3° phase deficit, can easily be confirmed by right-angle trigonometry.

Output impedance and detuning correct phase shift: This 6.3° lag must be made up by a phase shift from V_s to V_c, across amplifier output resistance r_o. This means that C_1 must have a lower reactance and L a higher reactance, so r_o sees a capacitive load. Thus the actual frequency of oscillation must be slightly higher than the resonance frequency given by equation (13-1). Figure 13-6(c) shows the equivalent circuit, the

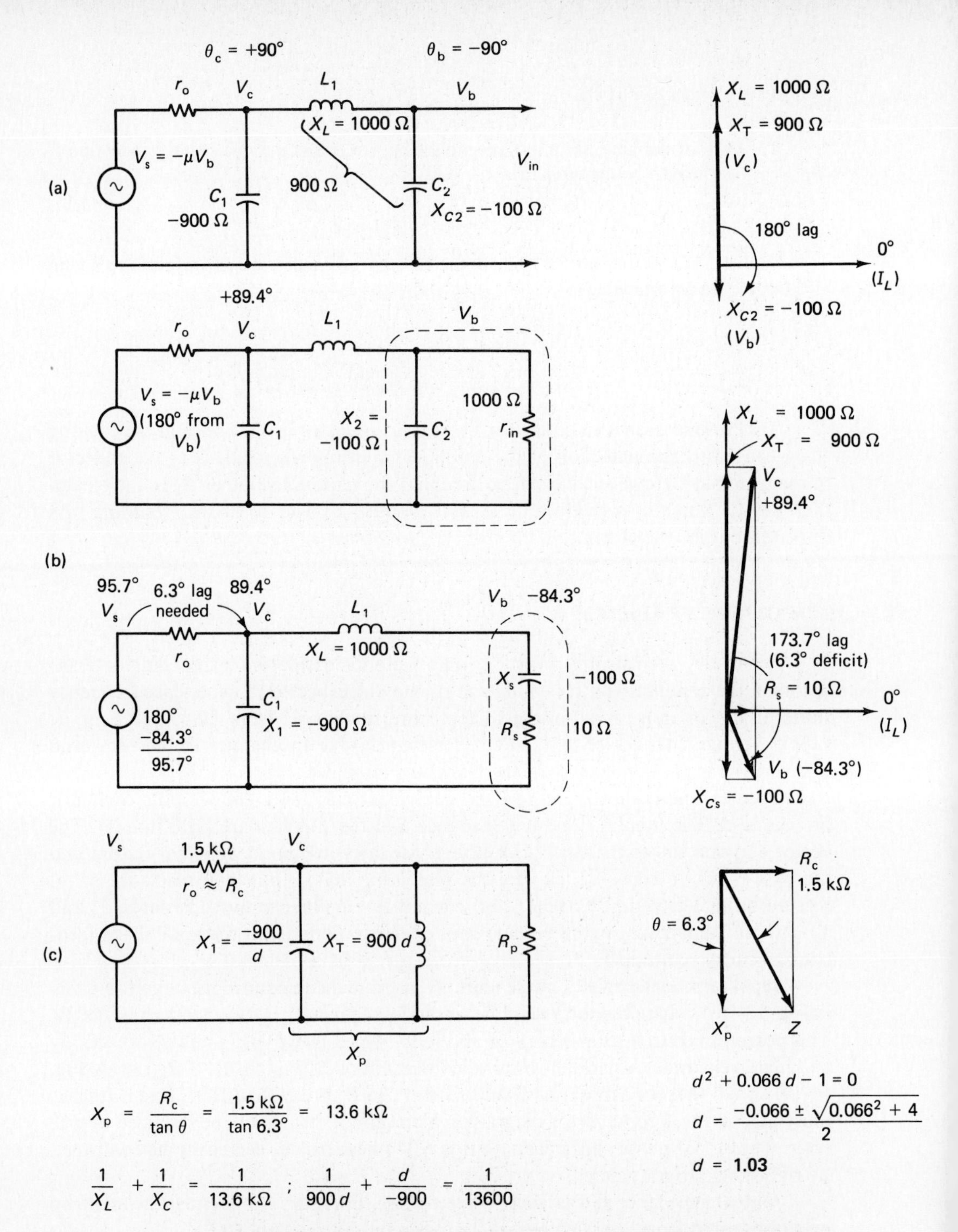

Figure 13-6 Stability analysis for the oscillator of Fig. 13-5. Exactly 180° phase shift is required (a), and whatever is lost by input loading and coil resistance (b) must be made up by detuning from resonance so the amplifier output sees a capacitive reactance (c).

vector diagram, and the calculations for the factor of detuning d. The final solution for d is accomplished by the quadratic formula. Note that for the circuit of Fig. 13-5 we would expect the oscillator frequency to be 3% higher than predicted by the resonance formula. We might reasonably estimate that a high-beta transistor would increase r_{in} by 33% and that this would result in a 33% smaller frequency shift; 2% above f_r rather than 3%. Frequency shifts on the order of 1% are, in fact, observable when changing transistors in this circuit.

Since X_s is small compared to X_L the fact that X_s also decreases by 3% has been neglected. Coil resistance r_w has also been neglected, but it can be seen that r_w and R_s appear in series in Fig. 13-6(b), so the effect of r_w is to reduce the V_c–V_b phase shift still further below 180°.

Improving oscillator stability: Figure 13-5 is a fine one for demonstrating oscillator *instability*. Here are some pointers for improving oscillator *stability*.

- Keep amplifier input impedance as high as possible. FETs, especially as source followers, are desirable in this regard.
- Use the highest-Q coil possible. This helps keep the phase-shift deficit low. In Fig. 13-6(b), an r_w of 100 Ω ($\frac{1}{10}$ of X_L) causes about the same amount of trouble as an r_{in} of 1000 Ω ($10X_2$).
- Use the lowest feedback factor required to sustain oscillation. In Fig. 13-6(b) if C_2 were twice as large, X_2 would be −50 Ω, R_s would be 2.5 Ω, and the phase-shift deficit would be about one-fourth of its original value. Of course, A_v would need to be twice its original value to sustain oscillation.
- Use the lowest practical L/C ratio (small L, large C). This will keep C_2 large, yielding the benefit described in the previous note. It will also keep X_1 and X_L small [see Fig. 13-6(c)] so that a sufficiently low X_T will result from a smaller frequency shift above resonance.
- Keep amplifier output impedance as high as possible. Replacing the collector resistor in Fig. 13-5 with a high-Q choke could raise r_o by a factor of 10 or more, reducing the detuning necessary to make up the phase-shift deficit by the same factor. Circuits which feed the supply current through the tuning coil (as in Fig. 13-4) have this advantage inherently. In such cases r_o consists almost entirely of r_c, the dynamic collector resistance. Since r_c in the common-base configuration is about β times r_c in the common-emitter configuration, the common-base oscillator has a frequency-stability advantage.
- Do not load either the base or the collector by taking the oscillator output from them. Take the output from the emitter, and use an emitter follower to keep the load on the oscillator light.

Figure 13-7 shows a common-base Colpitts oscillator. Notice that signal ground is at the bottom end of C_2 and the feedback is taken from the C_1–C_2 junction. The common-base amplifier is noninverting, so the feedback circuit must be noninverting to provide positive feedback.

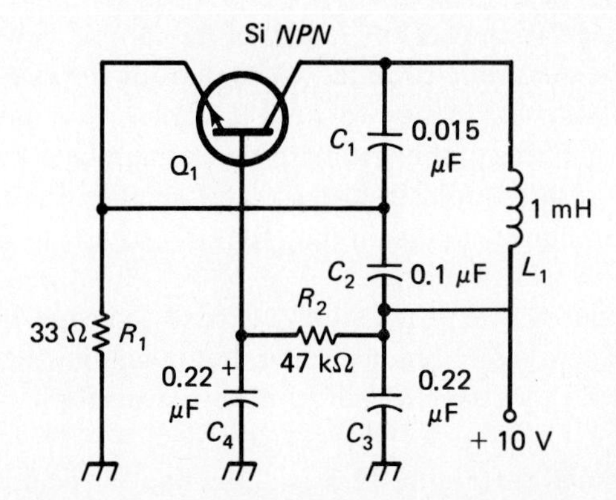

Figure 13-7 The common-base circuit generally makes a more frequency stable oscillator than the common-emitter circuit.

A basic FET common-source oscillator is shown in Fig. 13-8(a). The common-drain circuit of Fig. 13-8(b) has the advantage of being easier to bias. The feedback network provides a noninverted and slightly larger signal for the drain, since the source-follower amplifier has a noninverting gain slightly less than unity.

The high input impedance of the source follower and the high output impedance of the common-base amplifier team up in the circuit of Fig. 13-8(c) to provide an oscillator with a frequency shift of less than 0.1% for any FET and transistor ($\beta_{min} = 30$), supply variations of $\pm 10\%$, and temperatures from 32 to 100°F.

Component precautions: Oscillator stability is also affected by the mechanical rigidity of the coils, capacitors, and connecting wiring. Large inductors should be solidly encapsulated to minimize vibration. Small coils should be wound tightly on a form which will not expand with moisture or temperature. Special attention should be paid to the frequency-determining capacitors. Mica- or polystyrene-dielectric types should be used at high frequencies, and mylar-dielectric types should be used where large capacitance values are required. Temperature shifts from 50 to 150°F may be expected to cause capacitance changes on the order of 0.2 to 1% in these types, producing frequency changes of 0.1 to 0.5%. The popular ceramic disk types do not generally have as stable a capacitance value as the aforementioned types. Electrolytic types should be avoided at all costs if stability is a factor, as their capacitance may vary widely with temperature and voltage.

13.5 CRYSTAL OSCILLATORS

Where an exceptionally high degree of frequency stability is required, a crystal oscillator is often used. The "crystal" is a finely ground wafer of translucent quartz stone held between two metal plates and housed in a package about the size of a postage stamp. Crystals which are so ground and packaged behave like resonant circuits with Q in the neighborhood of 10 000—a hundred times better than LC-tuned circuits. The stability of this oscillator is therefore about 100 times better, even without considering the greater mechanical rigidity of the crystal as opposed to the

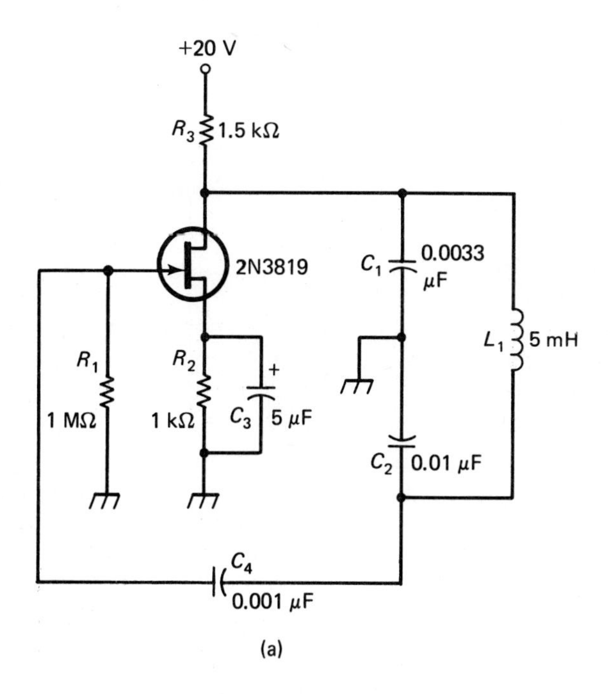

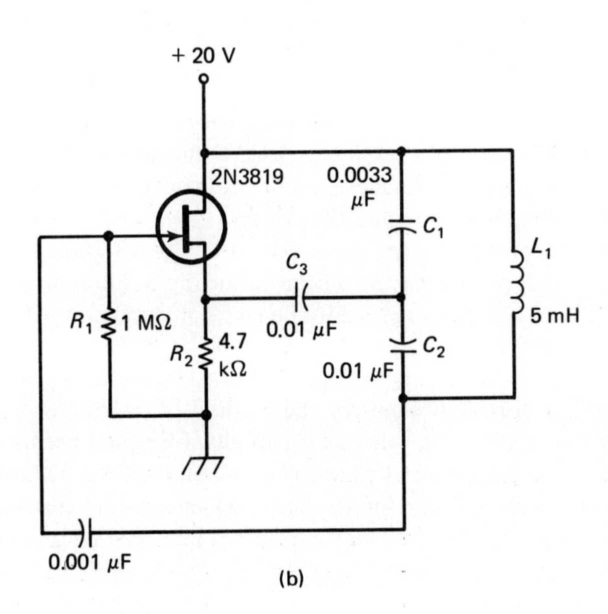

Figure 13-8 (a) The high input impedance of the FET permits the design of a highly frequency stable oscillator. (b) The source-follower circuit is more bias stable.

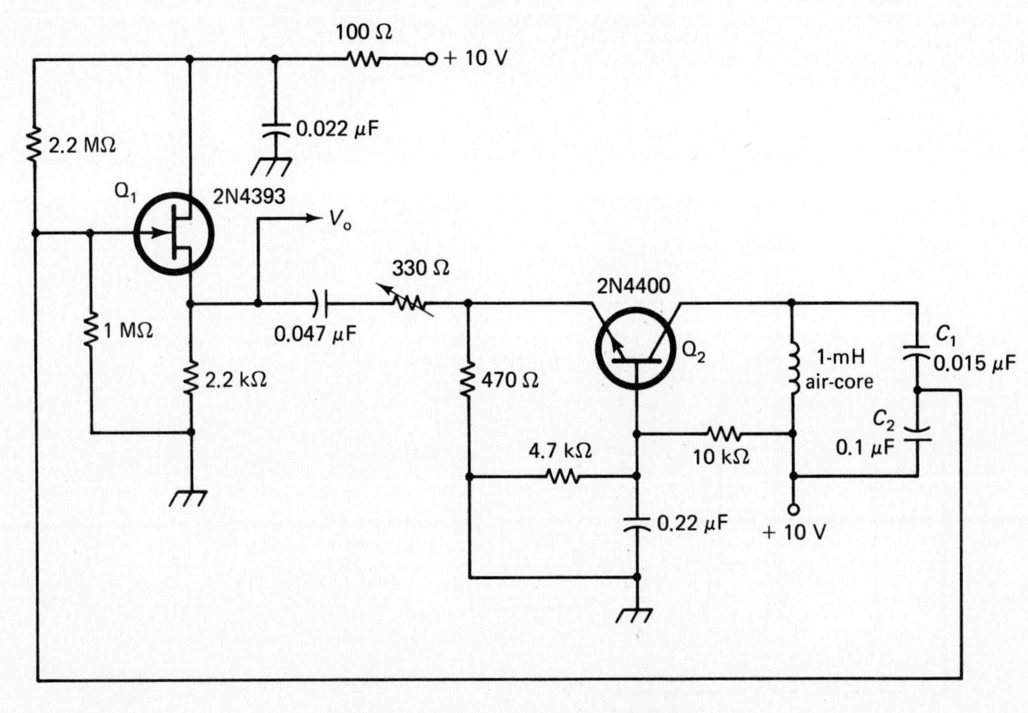

Figure 13-8 Continued. (c) When combined with the high output impedance of the common-base amplifier, even better frequency stability is achieved. Stability of the L and C components is the primary determiner of frequency stability in this oscillator.

coil-and-variable-capacitor assembly. Thinner crystals produce higher resonant frequencies. Crystals are commonly available for frequencies from about 50 kHz to about 50 MHz. Of course, changing the frequency of a crystal-controlled oscillator, even by a fraction of a percent, requires changing the crystal.

The simplest crystal oscillator is the Pierce circuit shown in Fig. 13-9(a) and (b). The crystal Y_1 provides a feedback path from the output to the input, along with the necessary phase shift. The Pierce circuit is especially convenient because no tuned circuit, other than the crystal itself, is required.

Most of the basic oscillator circuits can be crystal-controlled by replacing part of the original tuned circuit with a crystal. Sometimes the method of operation is not obvious, as in the circuit of Fig. 13-9(c). This circuit is actually a Colpitts oscillator, the capacitive voltage divider being furnished by the stray capacitance of the transistor junctions. If the crystal is kept in a temperature-controlled oven, crystal oscillators can be used to produce frequencies which vary less than 1 part per million over a period of several weeks.

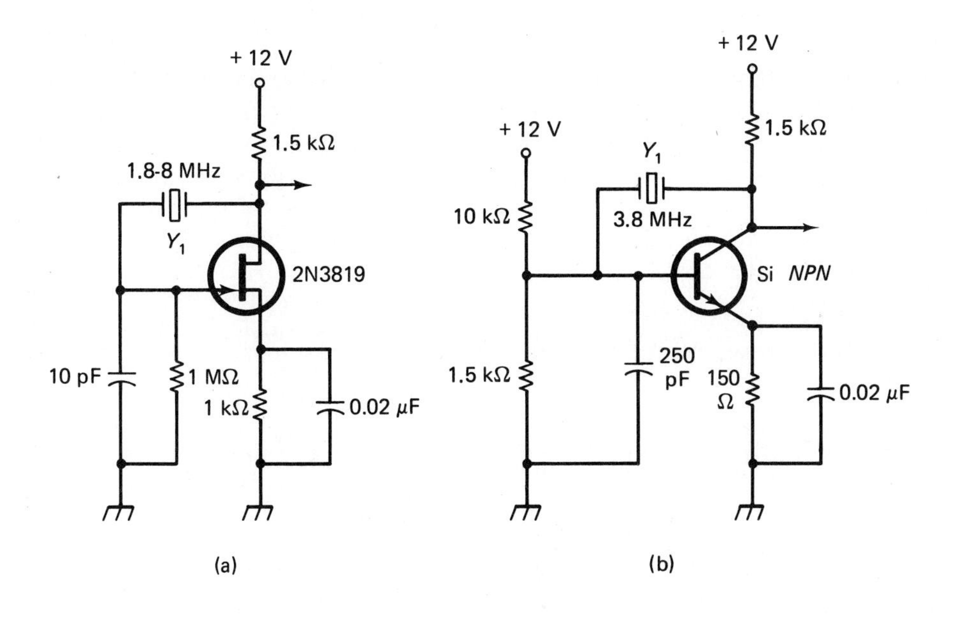

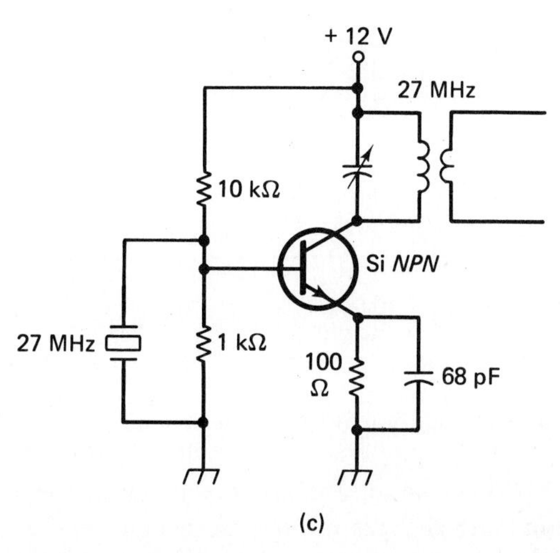

Figure 13-9 Crystal oscillator circuits: (a) Pierce FET; (b) Pierce transistor; (c) Colpitts oscillator for the 27-MHz citizens' radio band.

13.6 PHASE-SHIFT OSCILLATORS

The phase-shift oscillator demonstrates convincingly that the frequency of oscillation is that frequency for which the total phase shift is 180°. A cascading of three *RC* phase-shift networks, each providing a phase shift of 60°, can produce the 180° shift necessary for a common-emitter oscillator. Such an oscillator does not require inductors and can therefore be built for low frequencies with considerably less weight and expense than the *LC*-type oscillator. In addition, its frequency can be varied continuously by a variable resistor. Variable inductors or variable capacitors for use at low audio frequencies are not practical, so variable low-frequency oscillators almost always use *RC* rather than *LC* circuitry. The stability of this oscillator is poor since the feedback network does not provide an abrupt phase-shift change with frequency shift as does an *LC*-tuned circuit near resonance.

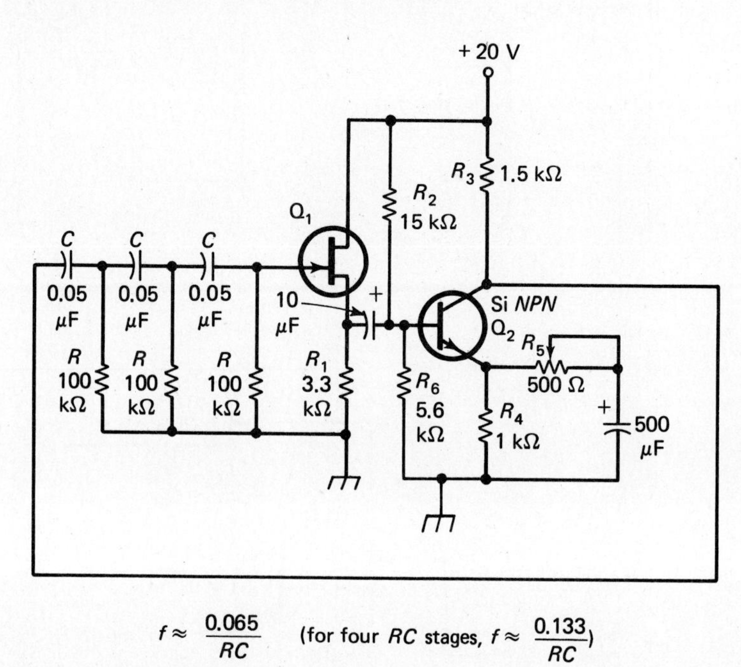

$f \approx \dfrac{0.065}{RC}$ (for four *RC* stages, $f \approx \dfrac{0.133}{RC}$)

Figure 13-10 A reliable phase-shift oscillator. No inductors are required, making this circuit easy to miniaturize.

Figure 13-10 shows an *RC* phase shift oscillator which uses an FET source follower to eliminate changes in input impedance and a bipolar transistor for high gain. A single bipolar transistor can be used, but the frequency stability of the oscillator suffers. Furthermore, not all bipolar transistors provide enough gain to be used alone in this circuit (a voltage gain of about 30 to 50 is required). Few FET amplifiers can provide this much gain in one stage either, so the most reliable circuit uses two stages of amplification, as shown.

The frequency of the oscillator of Fig. 13-10 is given approximately by

$$f \approx \frac{0.065}{RC} \tag{13-2}$$

Four *RC* sections may be used instead of the three sections shown in the figure. In this case, the frequency of oscillation is given by

$$f \approx \frac{0.133}{RC} \tag{13-3}$$

Frequency adjustments on the order of 10% shift can be made by inserting a variable resistor (value = R) in series with any one of the existing resistors R in Fig. 13-10.

Waveshape adjustment is provided by variable resistor R_5. Low values of R_5 produce high gain and distorted output waveforms.

13.7 FREE-RUNNING MULTIVIBRATOR

An example of the relaxation-oscillator class is the *astable multivibrator*, so called because it is never at rest or stable. It is used to generate a continuous string of output pulses. The typical television circuit contains two of these circuits, which are used to drive the beam horizontally and vertically across the face of the screen. If the time for which the astable remains in each state is made quite long (several seconds), it can be used to control an on-off cycling process, such as flashing a light or alternately running and stopping a conveyer.

The astable circuit uses two RC timing networks to turn its two transistors on alternately. The end of one charging ramp is used to trigger the start of the next so that the pulse train is self-sustaining. There are two outputs (X and $\bar{X}$), one always being the complement of the other.

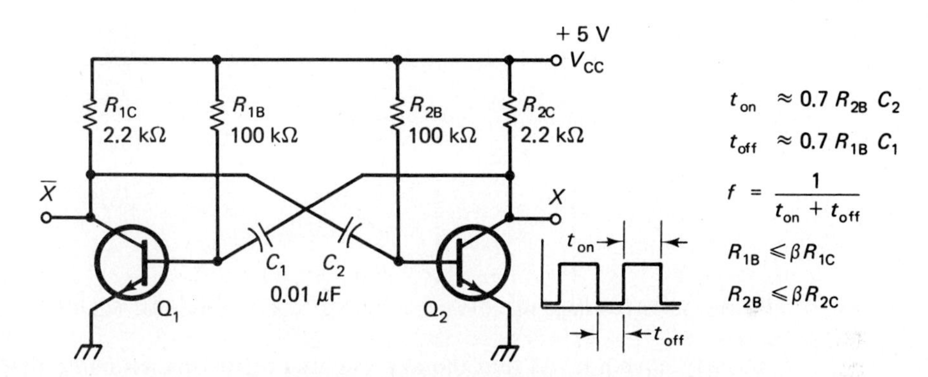

Figure 13-11 Astable or free-running multivibrator.

In examining the operation of the circuit (Fig. 13-11), let us assume that Q_1 is initially turned on. This causes V_{1C} (the voltage at the collector of Q_1) to drop from +5 to 0 V (assuming that $V_{CC} = +5$ V). C_2 couples this drop to the base of Q_2, and since V_{2B} was initially +0.6 V, it drops to −4.4 V. Q_2 is now turned off, and its collector voltage rises. Note that V_{2C} cannot rise instantly, however, because C_1 is coupled from that point to the base of Q_1, and V_{1B} cannot rise above 0.6 V. C_1 must, therefore, be charged through R_{2C}, and the rise of V_{2C} follows this time-constant curve.

C_2 will be discharged through R_{2B} from -4.4 toward $+5$ V. This rise will be clamped at $+0.6$ V by the base of Q_2, and Q_2 will be turned on after a time of $t_{on} = 0.7R_{2B}C_2$.

When Q_2 is turned on, its collector voltage drops from $+5$ to 0 V, and this drop is coupled to the base of Q_1 through C_1. V_{1B} now starts its rise from -4.4 toward $+5$ V as C_1 is discharged through R_{1B}. The rise to 0.6 V takes a time $t_{off} = 0.7R_{1B}C_1$, at which time Q_1 turns back on, and the process goes through the same cycle. Figure 13-12 gives the waveforms at various points in the multivibrator circuit.

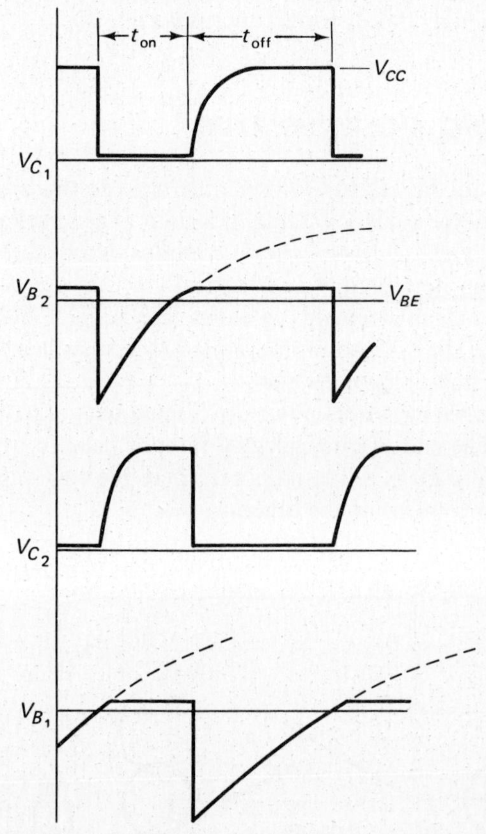

Figure 13-12 Voltage waveforms for the astable multivibrator of Fig. 13-11.

You may have noticed that the analysis was started by assuming that "somehow" Q_1 turned on. When the power is first applied, one of the transistors is almost certain to turn on a little faster and a little harder than the other, and the cycle starts with whichever one saturates first. Some astable circuits are designed with a larger base resistor on one side or with some other unbalancing device to ensure that a particular transistor will turn on first. It is possible to produce a stalemate by turning both transistors on through external base bias resistors. Both transistors will then remain saturated, and no output will appear. This situation does not often come about in normal operation, however, especially if R_{1B} and R_{2B} are kept near their maximum value to keep the transistors from saturating too heavily.

CHAPTER SUMMARY

1. An oscillator is a device that generates an output signal without requiring an input signal. One common type of oscillator uses an amplifier with its output fed back to its input through a frequency-determining feedback network. The fed-back signal must be in reinforcing phase and of adequate amplitude to sustain oscillation. Excessive feedback causes distortion of the sine-wave output signal.
2. Three popular oscillator circuits use *LC*-tuned circuits to govern the frequency of oscillation. The Armstrong circuit uses a transformer with tuned primary, the Hartley uses a tapped coil, and the Colpitts uses two series capacitors across an untapped coil. In each circuit oscillation frequency is

$$f = \frac{1}{2\sqrt{LC}} \qquad (13\text{-}1)$$

3. Amplifier input and output resistances, load resistance, and coil-winding resistance cause phase shifts which offset the actual frequency of oscillation slightly from that given by equation (13-1). Changes in these resistances with supply voltage and temperature cause oscillator frequency to shift. High-Q coils, emitter followers to isolate the load, and FETs for high input impedance promote frequency stability. Coils and capacitors must be selected for highest thermal and mechanical stability to keep oscillator frequency stable.
4. A specially cut and ground quartz crystal behaves like a very high-Q tuned circuit. Crystal oscillators are fixed-frequency but have a stability over 100 times better than *LC*-tuned oscillators.
5. Three *RC* networks, each providing a 60° phase shift, can be used as a feedback network which requires no inductors. This *RC* phase-shift circuit is popular for audio-frequency oscillators where the inductors required by other types would be heavy and expensive. Frequency of oscillation is

$$f \approx \frac{0.065}{RC} \qquad (13\text{-}2)$$

6. The astable multivibrator uses two *RC*-charging networks and two switching transistors to produce square-wave oscillations. The frequency is given by

$$f \approx \frac{1}{0.7(R_{1B}C_1 + R_{2B}C_2)}$$

QUESTIONS AND PROBLEMS

13-1. What two conditions are necessary for feedback oscillation?

13-2. What is the usual result of excessive feedback in an oscillator? What is the result of insufficient feedback?

13-3. The Hartley oscillator of Fig. 13-2 is built using a 1-mH inductor. The frequency of oscillation is to be 50 kHz. What value of C_1 is required?

13-4. In the Colpitts circuit of Fig. 13-7, what is the frequency of oscillation? What value of C_1 will change the frequency to 35 kHz?

13-5. In the Colpitts circuit of Fig. 13-5, higher transistor β causes higher amplifier input resistance, and oscillation frequency (increases or decreases) slightly. Changing from a germanium to a silicon transistor would probably raise r_o, causing oscillator frequency to (increase or decrease) slightly.

13-6. What are the advantages of using high-Q coils and FETs (rather than bipolar transistors) in oscillator circuits?

13-7. What type of oscillator is associated with each of the following? (a) Tapped coil; (b) two capacitors series-connected across a coil; (c) transformer feedback; (d) sine-wave output but no inductors.

13-8. State the major advantage and the major disadvantage of crystal over *LC*-tuned oscillators.

13-9. Calculate the frequency of the phase-shift oscillator of Fig. 13-10.

13-10. Calculate the approximate frequency of the multivibrator of Fig. 13-11.

14

MODULATION AND MIXING

14.1 AMPLITUDE MODULATION

Modern communication systems rely heavily on the ability of a single transmission medium to carry a large number of signals. A single TV-antenna line delivers a dozen channels. A single pair of telephone wires carries dozens of conversations. A single satellite transmitter relays hundreds of audio and video channels. This is accomplished by assigning a separate *carrier frequency* to each channel, transmitting all the carriers via one medium, and selecting the desired carrier with tuned circuits at the receiving end. The separate signals (let's say they are audio waveforms) ride on their respective carriers by a process called modulation.

Modulation is the process of varying the amplitude (strength), frequency, or phase of a higher-frequency signal (carrier) in proportion to the amplitude of a lower-frequency signal (program). Figure 14-1 shows a 1-kHz audio-frequency signal (program), a 20-kHz carrier signal, and a 20-kHz signal whose amplitude has been modulated (controlled) by the 1-kHz modulating signal.

Twenty kilohertz is above the audio range, so if the modulated signal were delivered to a loudspeaker, nothing would be heard. With a little imagination you should be able to visualize other carriers at 30 kHz, 40 kHz, and so on up to several hundred kilohertz, each amplitude modulated with its own audio-frequency program material. Of course, the program material will not actually consist of a single 1-kHz tone, unless someone is sustaining a long whistle. The actual audio signals will be complex waveforms containing frequencies from about 200 Hz to 5 kHz for voice signals. All of these modulated carriers may be transmitted via a single wire pair.

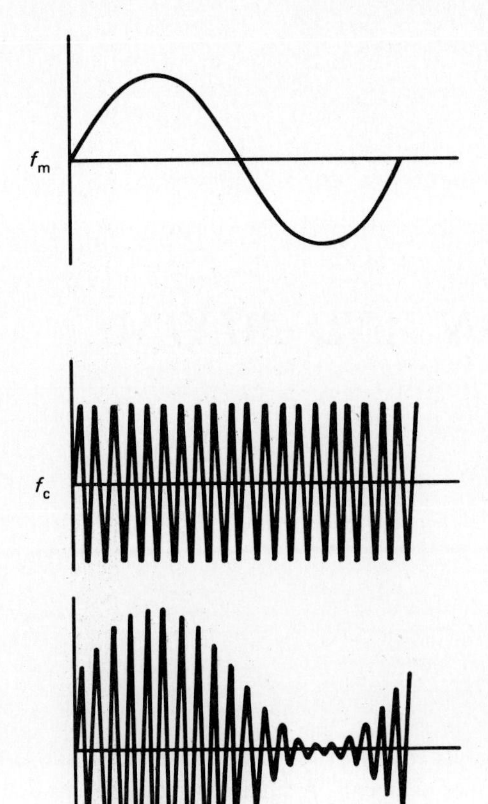

Figure 14-1 Amplitude modulation: modulating signal (top trace), carrier (center), modulated signal (bottom).

Yet nothing will be heard at the receiving end since all the signals are above the audio range.

A diode modulator: To produce amplitude modulation (AM) we must have a means of controlling the strength of one signal (the carrier) with another signal (the modulating signal). A very simple way to do this is with the *diode modulator* shown in Fig. 14-2.

The dc source biases the diode at about 2 mA. Its resistance to ac (slope) is about 15 Ω at this point. The 1-kHz signal moves the diode roughly 1 mA up and down from this point, changing its ac resistance from about 10 Ω to about 30 Ω. These points are marked *A* and *B*, respectively, on Fig. 14-2(b). The 20-kHz carrier causes additional diode-current swings of about 1 mA p-p as shown by the arrowheads about *A* and *B*. This current results in diode-voltage swings that vary from 10 mV to over 30 mV p-p as the 1-kHz signal swings the diode from point *A* to point *B*. Thus the 20-kHz signal is modulated by the 1-kHz signal. The tuned circuit *L* and *C* picks the 20-kHz signal from the diode but rejects the 1-kHz signal and dc.

Collector modulation: The diode modulator is nonlinear. Swings toward point

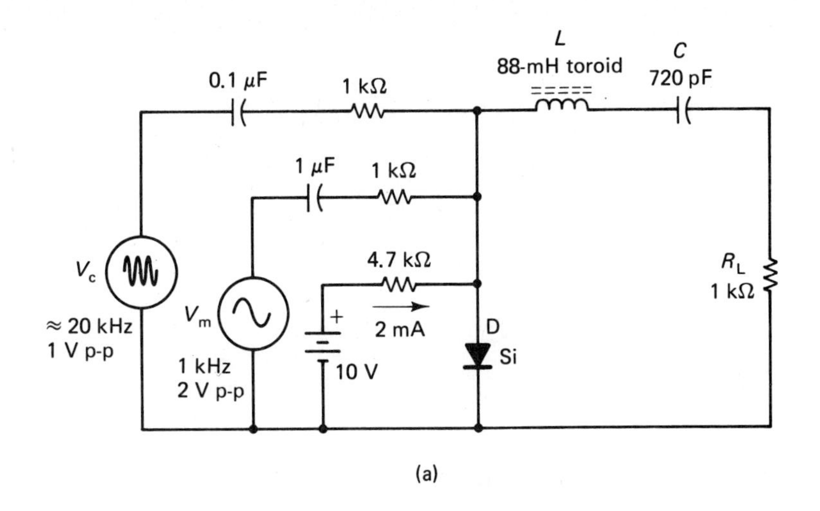

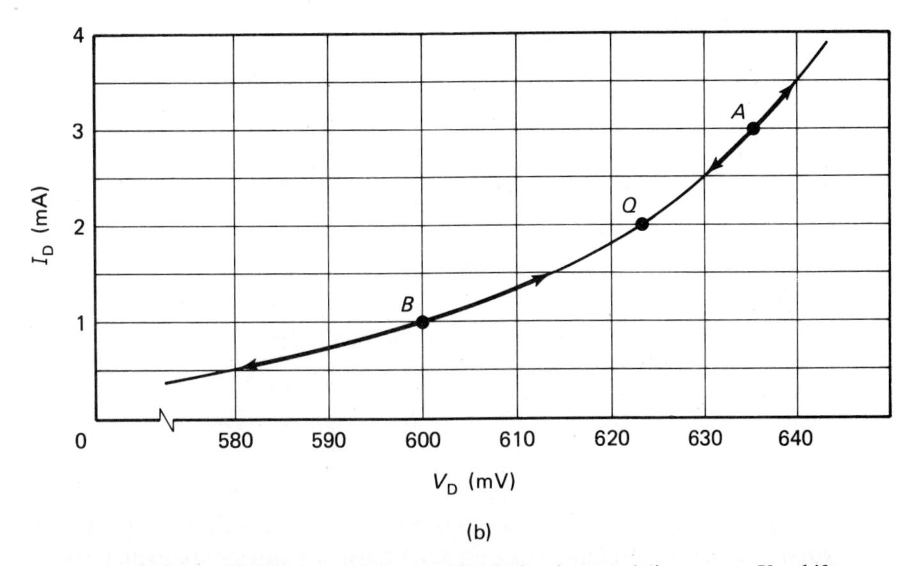

Figure 14-2 (a) Diode modulator; (b) diode characteristic curves. V_m shifts diode operating point from A to B, making amplitude of V_c swing smaller and larger, respectively.

B in Fig. 14-2 raise the 20-kHz output greatly while swings toward point A lower this output only slightly. A more widely used modulator is shown in Fig. 14-3. Q_1 is a class-C RF amplifier (see Section 11.8). Q_2 and Q_3 are push-pull audio amplifiers (see Section 11.5). The carrier-frequency input to Q_1 switches it on and off completely on each cycle, but the Q_1 output voltage is limited by the voltage of the collector supply. This supply is the dc voltage V_{CC} *in series with the audio voltage* from the secondary of the transformer. Thus the output of Q_1 varies according to the audio signal.

It can be shown that the maximum audio power required to completely modu-

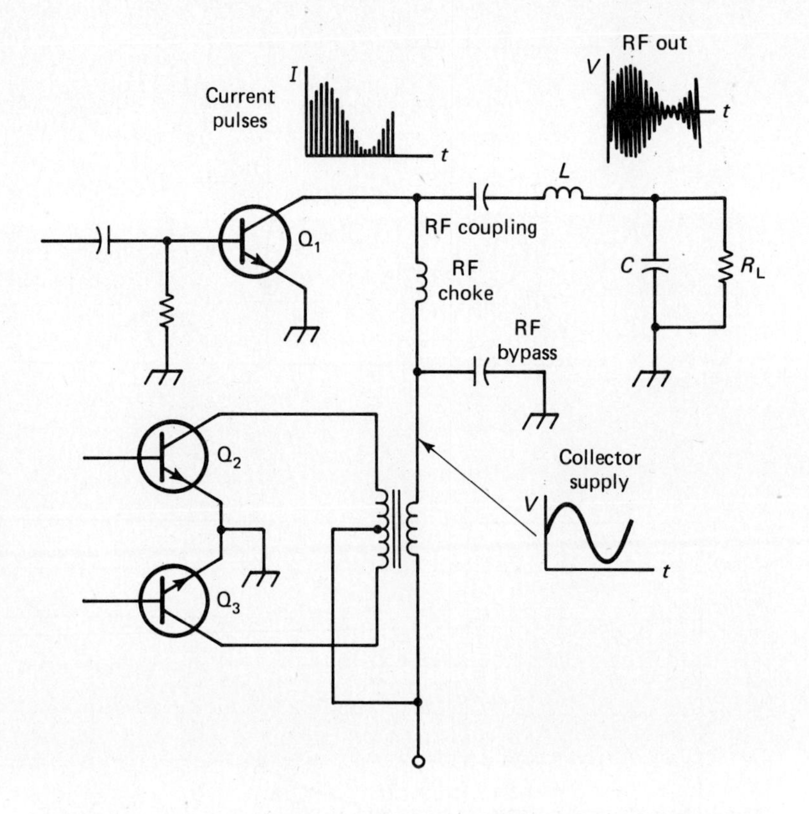

Figure 14-3 Typical amplitude-modulation circuit for low- and medium-power radio transmitters.

late the class-C stage (bring the collector-voltage to zero at the negative peaks) is one-half the normal dc power input to the class-C stage. Thus 5-W of audio is all that is required to modulate a 10-W-input class-C amplifier.

The circuit of Fig. 14-3 is the basis of most small AM radio transmitters, such as the 5-W citizens'-band types. In that case, the carrier frequency would be about 27 MHz and R_L would be the radiation resistance of the antenna.

14.2 DEMODULATION

Demodulation is the process of recovering the program signal from a modulated signal. It is often called *detection*, a term from the early days of radio. A simple AM detector is shown in Fig. 14-4(a). The circuit is identical to the half-wave rectifier with capacitor filter presented in Chapter 3. Here the filter capacitor must be chosen to filter out the carrier-frequency pulsations, but not the program-material variations. The filtered waveform has the shape of the modulation *envelope*, which is the shape of the original modulating signal.

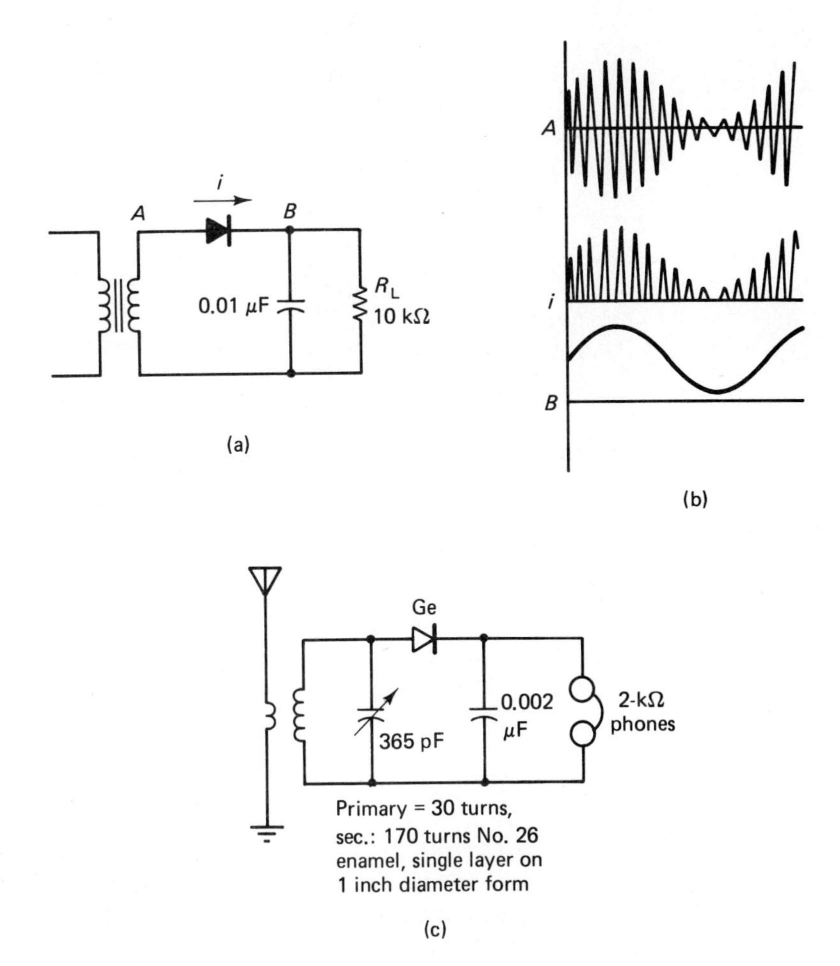

Figure 14-4 (a) AM-detector circuit; (b) modulated, rectified, and filtered waveforms; (c) simple diode AM-radio receiver.

AM radio: If the detector is fed from an antenna and tuned circuit, as shown in Fig. 14-4(c) the result is a simple AM radio—the "crystal set" which dates back to the early years of this century. Don't expect too much if you build this radio. The antenna will have to be long, a good earth ground (such as a cold-water-pipe connection) will be needed, and only local stations will be heard.

14.3 SIDEBANDS

Figure 14-5 shows a most interesting series of waves. The upper trace is a 20-kHz sine wave. The center trace is a 19-kHz sine wave of the same amplitude. The lower trace shows the algebraic point-by point sum of these waves, obtained with a summing oscilloscope (channel A + channel B). Notice that the two top waves fall in and then

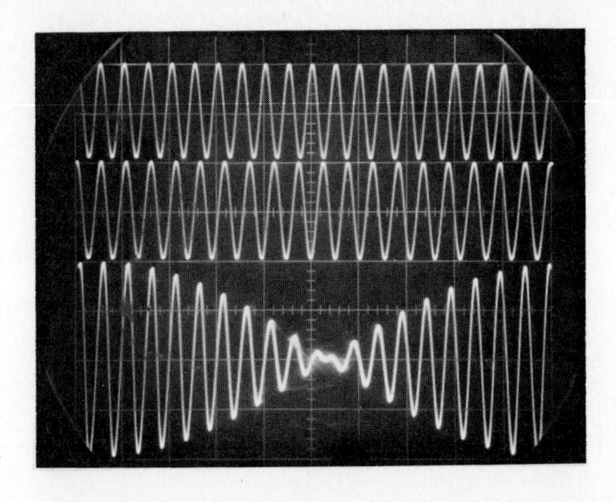

Figure 14-5 A 20-kHz signal (top), a 19-kHz signal of the same amplitude (center), and the sum of these two (bottom). A modulated wave actually contains more than one frequency.

out of phase with each other at a 1-kHz rate, alternately reinforcing and then canceling each other. Their sum is a waveform that looks just like the amplitude-modulated wave.

Let us use the subscript c to refer to carrier frequency and m to refer to modulating frequency. Since two waves of frequencies (f_c) and $(f_c - f_m)$ when added produce a modulated wave, may we conclude that a modulated wave actually contains both frequencies, f_c and $(f_c - f_m)$? Almost true. In fact, a modulated wave contains *two* other frequencies besides f_m; these are $(f_c + f_m)$ and $(f_c - f_m)$. Thus, modulating a 20-kHz carrier with a 1-kHz audio wave produces new frequencies at 19 kHz and 21 kHz. These two new frequencies are called *sidebands*. If the original wave is fully modulated the amplitudes (voltages) of $(f_c + f_m)$ and $(f_c - f_m)$ are each one-half of the amplitude of (f_m).

Sideband hunting: The existence of the sidebands can be confirmed easily if a spectrum analyzer or wave analyzer is available, but these instruments are expensive and not every lab is so well equipped. Sidebands can be isolated and measured with common test equipment using the circuit of Fig. 14-6. The circuit is the same diode modulator of Fig. 14-2, but the modulating frequency has been changed to 5 kHz to place the sidebands at 15 kHz and 25 kHz—well separated from the carrier. The tuned circuit has been changed to parallel-resonant for higher selectivity and it has been made variable to cover the range from about 14 kHz to 26 kHz. It should be possible to tune in each sideband and the carrier separately, view them on the oscilloscope, check their frequencies on a counter, and note how their amplitude and frequency varies with that of the modulating signal.

Bandwidth: When a carrier is amplitude modulated by a voice or music signal the sidebands will consist not of two discrete frequencies, but of two bands containing numerous and constantly changing frequencies. These bands will extend $\pm f_{mm}$ around the carrier frequency, where f_{mm} is the maximum modulating frequency. For voice work f_{mm} may be only 3.5 kHz; for good fidelity it should be 5 kHz, and for

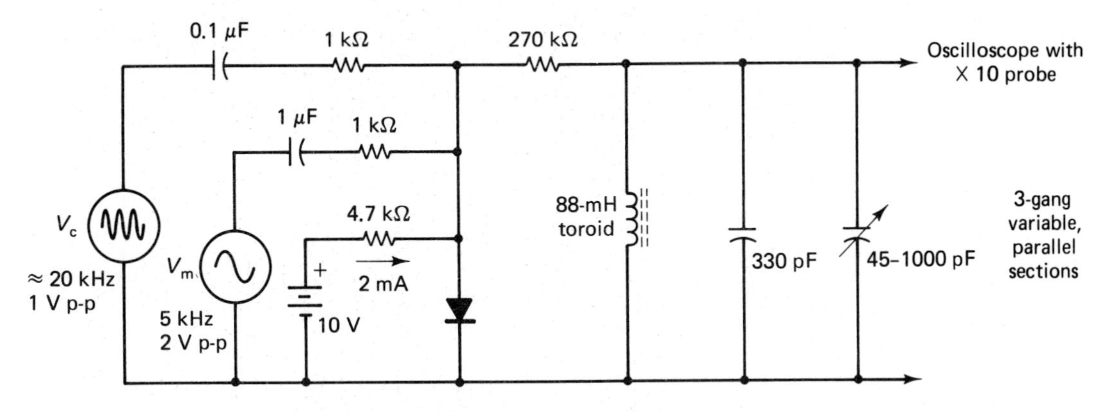

Figure 14-6 Diode modulator with high-Q tuned circuit capable of picking off carrier and sidebands separately.

high fidelity it should be 10 kHz. The bandwidth of the transmitted AM signal is then 7, 10, or 20 kHz, respectively, depending on the fidelity required.

Carrier signals should not be spaced closer than the bandwidth because sidebands from one signal would interfere with those of the adjacent signal. Bandwidth thus limits the number of signals that can be packed into a given spectrum of frequencies. The AM-broadcast band can accommodate 107 stations in the 1.07 MHz from 535 to 1605 kHz because each station has a bandwidth of 10 kHz.* However, television signals require a bandwidth of 6 MHz per channel, so the 12 VHF channels require 72 MHz of spectrum space (54 to 216 MHz, with a big space in between channels 6 and 7 for other services).

14.4 FREQUENCY CONVERSION

The fact that new frequencies are produced when one signal modulates or controls the amplitude of another can be used to change the frequency of a signal and its associated sidebands. This process is called *heterodyning*, *mixing*, or frequency conversion. It has been basic to most radio and TV receivers for half a century, and is now widely used in radio transmitters and a variety of other instruments.

Although the theory behind frequency conversion is the same as that behind modulation, the objectives and terminology are different. The carrier frequency is now called the input frequency f_i. This is the signal whose frequency is to be changed, and it may include sidebands or it may actually consist of a whole band of separate signals. The modulating frequency is now called the local oscillator f_o, and it is a pure, stable sine wave. As f_m changed the gain of the device passing f_c, so f_o changes the gain of the device passing f_i. Therefore, f_o is generally much stronger than f_i.

*Stations in the same region are kept at least 20 kHz apart.

In modulation, frequency f_c is usually 10 times or more higher than f_m. In frequency conversion f_i is usually on the same order as f_o, and may be either higher or lower than f_o. Figure 14-7 shows typical spectrum graphs for the signals involved in modulation and in heterodyning.

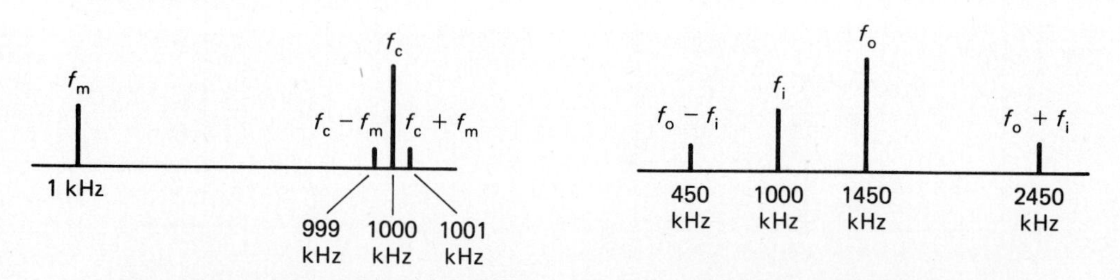

Figure 14-7 Typical frequency spectra for amplitude modulation (a) and heterodyning (b). The processes are essentially the same, but the frequency relationships involved are usually quite different.

14.5 MIXER CIRCUITS AND INTERMODULATION

Amplifier circuits are generally preferred over diodes as mixers because the output signal (usually $f_o - f_i$) can be many times larger than the input signal f_i. To fill this function an amplifier must be nonlinear—its gain must change considerably over its range of operation.

A bipolar-transistor mixer is shown in Fig. 14-8(a). The voltage gain is $R_{c(\text{line})}/r_j$, and r_j varies inversely with emitter current. For maximum effectiveness as a mixer it is desirable that:

1. The emitter be completely bypassed, so changes in r_j cause maximum change in A_v
2. The local-oscillator signal be large enough to swing the emitter current over a large range, but not quite large enough to cause saturation or cutoff
3. The input signal be fed from a low-impedance source so maximum nonlinearity is obtained (see Fig. 6-12 for a review of this concept)

Undesired mixing may take place in an amplifier that handles several input frequencies, at least one of which is large. This is called intermodulation (IM) distortion, and it can be a problem in radio-receiver amplifiers and audio amplifiers. To minimize IM distortion, the opposite of the three steps listed above should be followed: leave some of the emitter resistance unbypassed, keep signal levels low, and feed signals from high-impedance sources. Negative feedback also helps in audio amplifiers.

Square-law mixers: In general, when a large sine wave f_1 is passed through a nonlinear device, the output will consist of f_1 and all its harmonics: Af_1, Bf_2, Cf_3,

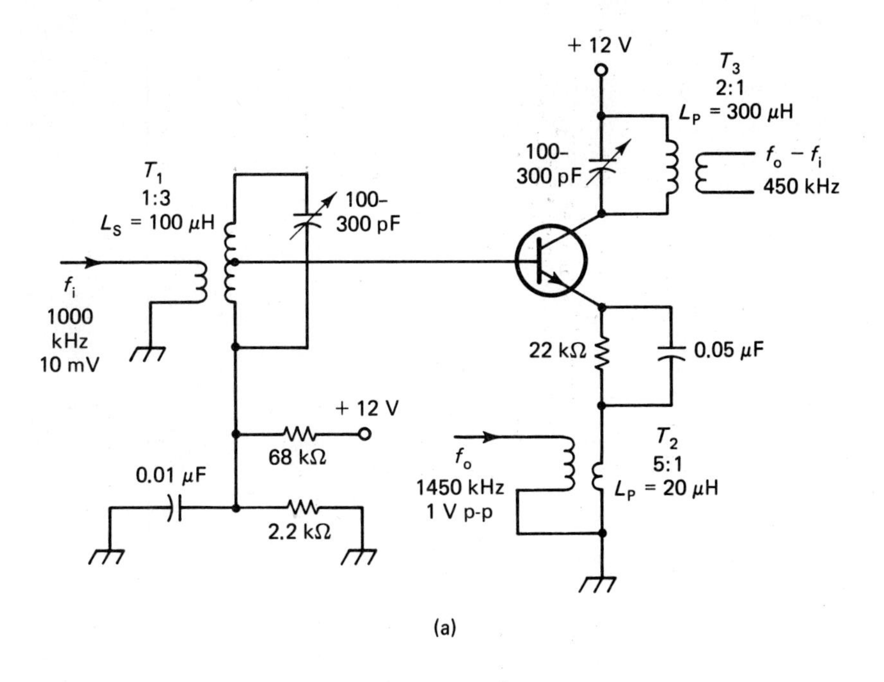

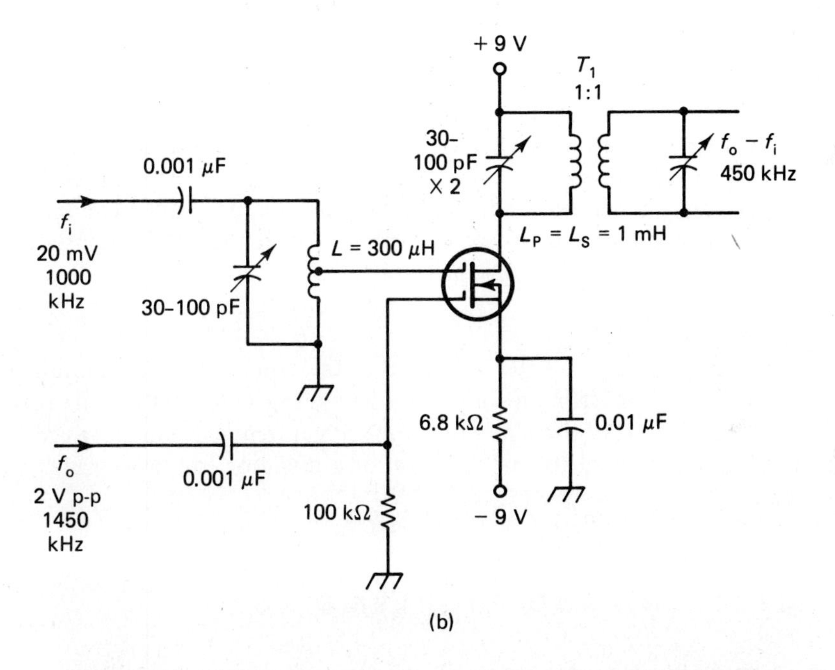

Figure 14-8 Typical mixer circuits: (a) bipolar transistor; (b) FET. Because of their square-law gain characteristic, FETs produce fewer spurious intermodulation products.

Df_4, etc., where A, B, C, and D are coefficients representing generally decreasing amplitudes. There is one amplifier characteristic which meets the criterion of nonlinearity yet produces only the second harmonic f_2 of the input signal f_1. This is the *square-law* characteristic, and it is possessed by the FET amplifier as expressed in equation (8-2).

Stated verbally, FET gain y_{fs} varies as a function of the square of input voltage, so for large signals, V_o contains the fundamental signal and its second harmonic only. This is a consequence of the trigonometric identity

$$2 \cos^2 \alpha = 1 + \cos 2\alpha \tag{14-1}$$

This limitation of the number of harmonics developed makes the FET amplifier superior in applications where unwanted or spurious signals must be minimized. An FET mixer is shown in Fig. 14-8(b).

CHAPTER SUMMARY

1. Modulation is the process of varying the amplitude, frequency, or phase of a higher-frequency carrier wave with a lower-frequency program waveform. It is used to allow a single medium to carry multiple programs.
2. Amplitude modulation is accomplished with a diode by using the program signal to vary the dynamic resistance of the diode. A class-C amplifier can be amplitude modulated by combining the program voltage with the collector-supply voltage.
3. The program waveform can be recovered by a process called demodulation or detection. A simple detector consists of a half-wave rectifier followed by a capacitor sized to filter out the carrier frequency but not the program frequency.
4. The distortion of the carrier sine waves inherent in the amplitude-modulation process causes new *sideband* frequencies to appear at $(f_c - f_m)$ and $(f_c + f_m)$, where f_c is the carrier frequency and f_m is the modulating frequency. The *bandwidth* of an AM signal is thus $2f_{m(max)}$.
5. Frequency mixing or heterodyning uses the same process as modulation, but for the purpose of converting a signal to a new frequency. This new frequency may be $(f_o - f_i)$ or $(f_o + f_i)$, where f_i is the input-signal frequency and f_o is the frequency of a local oscillator used to vary the gain of a diode or transistor which passes the f_i signal.

QUESTIONS AND PROBLEMS

14-1. Define the following terms: amplitude; carrier; modulation; detection.

14-2. Would the modulation scheme of Fig. 14-3 work if Q_1 were biased in class A? Explain why or why not.

14-3. What is the time constant of the *RC* filter in Fig. 14-4(a)? How many times longer (or shorter) is this than the period between 20-kHz input pulses? How many times longer (or shorter) than the period of the 1-kHz modulation envelope?

14-4. What is the bandwidth of a 1210-kHz radio station which is amplitude modulated by audio frequencies from 150 Hz to 4.5 kHz?

14-5. An amateur-radio station using AM is authorized to emit signals in the band from 7.2 to 7.3 MHz. If the modulating frequencies are limited to a maximum of 5 kHz, what are the maximum and minimum allowable carrier frequencies for this station?

14-6. An incoming signal of 60 MHz is mixed with a local oscillator of 105 MHz. What new output frequencies are produced?

14-7. An input signal of 94.5 MHz is to be converted to 10.7 MHz. If the local-oscillator frequency is higher than f_i, find f_o.

14-8. Why is the FET preferred as a mixer over diodes and bipolar transistors?

14-9. Can mixing be accomplished in resistive networks or air-core transformers? How, or why not?

14-10. How large should the amplitude of f_o be in an ideal mixer? Explain the consequences if this amplitude is (a) too large; (b) too small.

15

REGULATED POWER SUPPLIES

15.1 DIODE REGULATORS

Zener-diode characteristics: The reverse current through a silicon-junction diode is usually less than 1 μA, even at relatively high reverse voltages. However, at some reverse voltage the diode's back resistance will break down and current will *avalanche* at the slightest increase in reverse voltage, as shown in Fig. 15-1(b).

This *zener voltage*, V_Z, is the limiting reverse voltage across the diode, and the device will actually destroy itself by drawing excessively large currents before it allows a reverse voltage greater than V_Z to exist across its terminals. All silicon-junction diodes have such a zener or avalanche breakdown point, but by special manufacturing techniques, diodes can be produced with controlled zener points as low as 3 V and as high as 100 V and above.

Actually, there are two different internal mechanisms which may produce a reverse-breakdown effect. True *zener conduction* occurs in diodes with breakdowns of 7 V and below, and has a negative temperature coefficient (higher temperature, lower V_Z). *Avalanche breakdown* is a separate effect, occurring above 7 V, and exhibiting a positive temperature coefficient. Both types of diode are commonly called "zeners." Regulator diodes with V_Z near 7 V may have a temperature coefficient near zero, making this voltage a popular choice.

Physically, zener diodes appear quite the same as other diodes, but they are used with current flowing in the reverse direction. It must be remembered that zener diodes will conduct readily in the forward direction with about 0.6-V drop, like any other silicon diode. Types are currently available with power dissipation ratings from $\frac{1}{4}$ to 50 W and above.

As evidenced by the steep vertical slope of the graph [Fig. 15-1(b)], the dynamic resistance, r_z, of a zener diode in the reverse-conduction region is quite low, typically about 10 Ω for a 10-V, $\frac{1}{2}$-W zener, 30 Ω for a 30-V zener, etc. This low resistance to ac makes the zener useful in eliminating ripple as well as for stabilizing power-supply dc output voltage.

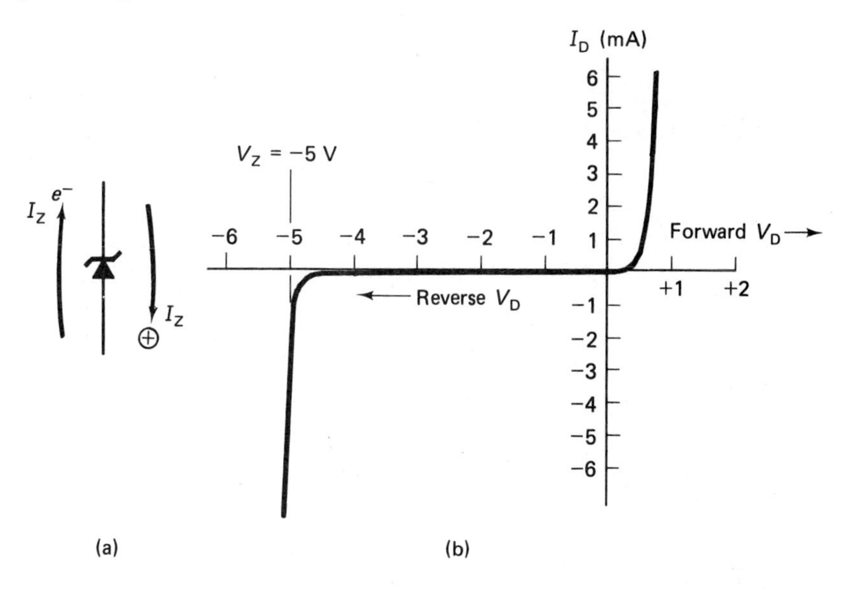

Figure 15-1 Zener-diode symbol (a), and characteristics for a 5-V zener (b). A zener will conduct at 0.6 V forward, but in the zener mode current is in the reverse direction through the diode.

Zener-regulated supplies: A zener diode placed across a load resistance will limit the dc load voltage to V_Z. If the power-supply rectifier and filter provide a voltage somewhat higher than V_Z, a dropping resistor in series with the line from the filter will take up the excess voltage. Any fluctuations in the voltage supplied will then be absorbed by the dropping resistor, since the zener voltage is constant.

If the value of the dropping resistor is chosen so that there is a substantial dc current through the zener, any changes in load current will be compensated for by changes in zener current. The voltage across, and hence the current through, the dropping resistor must remain constant if the supply output voltage and V_Z are constant. Therefore, whatever current is *not* drawn by the load *will* be drawn by the zener, so that the IR drop across R_d will remain constant. This is illustrated in Fig. 15-2. The supply output is fixed at 20 V, and the zener voltage is limited at 14 V. The voltage across the dropping resistor must then be 20 − 14, or 6 V. The current through this resistor is therefore constant and must be sufficient to supply the load current and the current to keep the zener "turned on."

An example will show how the value of the dropping resistor is determined in a practical zener regulator.

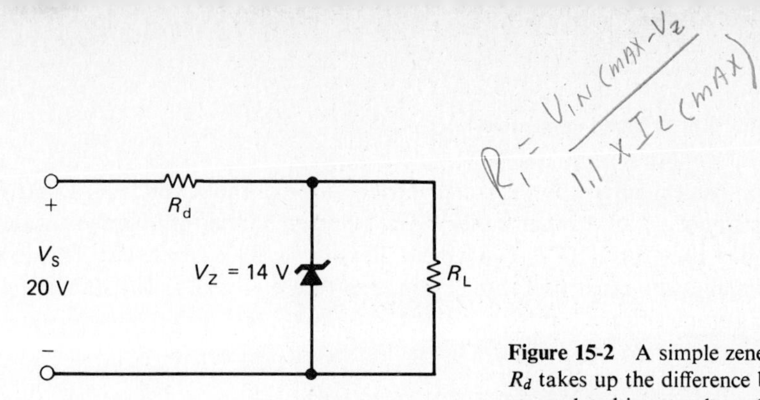

Figure 15-2 A simple zener regulator. R_d takes up the difference between the unregulated input voltage (20V) and the regulated output voltage (14 V).

Example 15-1

Find a suitable value of R_d in the circuit of Fig. 15-2. $I_{R(L)}$ varies between 20 and 50 mA.

Solution To assure operation well down on the zener curve, zener current should be 10 mA even at maximum load current:

$$I_{R(d)} = I_{R(L)} + I_Z = 50 + 10 = 60 \text{ mA} \tag{15-1}$$

$$V_{R(d)} = V_S - V_Z = 20 - 14 = 6 \text{ V} \tag{15-2}$$

$$R_d = \frac{V}{I} = \frac{6}{0.06} = \mathbf{100\ \Omega}$$

$$P_{R(d)} = IV = (0.06)(6) = \mathbf{0.36\ W}$$

A $\frac{1}{2}$-**W** resistor would be adequate.

The power dissipation of the zener must also be determined. This will be largest when $I_{R(L)}$ is smallest and I_Z is largest, since the diode will draw any current not drawn by the load:

$$I_{Z(\text{max})} = I_{R(d)} - I_{R(L)\text{min}} = 60 - 20 = 40 \text{ mA} \tag{15-3}$$

$$P_Z = I_Z V_Z = (0.04)(14) = \mathbf{0.56\ W}$$

A **1-W** zener would be required.

Zener effects on ripple: The effects of a zener regulator on ripple can be determined by simply analyzing a series-parallel circuit, recognizing that, insofar as the ac ripple voltage is concerned, the zener acts like a resistor of value r_z. Thus, the zener resistance r_z and the series dropping resistor R_d form a voltage divider, with the input ac ripple appearing across $R_d + r_z$ and the output ripple appearing across r_z alone. In a well-designed zener regulator both R_L and R_d will be much larger than r_z, so that a large reduction in ripple is realized.

Example 15-2

The +20-V input to the regulator of Fig. 15-2 contains 100-mV p-p ripple when $I_{R(L)}$ = 50 mA. Find the percentage ripple at the load if r_z = 15 Ω.

Solution The equivalent circuit seen by the ac ripple is given in Fig. 15-3.

$$R_L = \frac{V_{R(L)}}{I_{R(L)}} = \frac{14 \text{ V}}{0.05 \text{ A}} = 280\ \Omega$$

$$R_T = R_d + (R_L \| r_z) = 100 + (280 \| 15) = 114\ \Omega \tag{15-4}$$

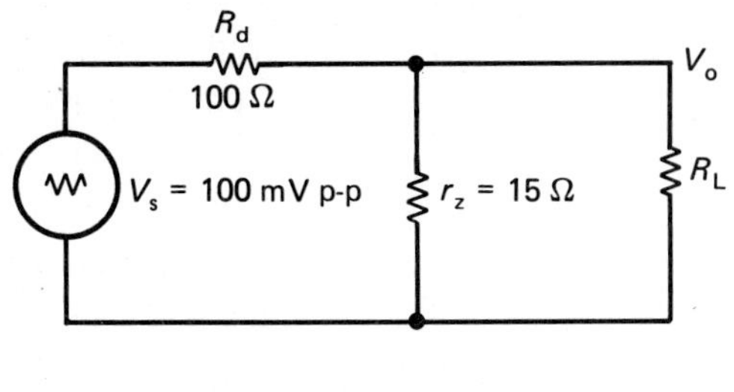

Figure 15-3 The equivalent of the regulator of Fig. 15-2 as seen by the ac ripple voltage.

$$\frac{V_s}{R_T} = \frac{V_o}{R_L \parallel r_z}; \quad \frac{100 \text{ mV}}{114\,\Omega} = \frac{V_o}{14\,\Omega} \tag{15-5}$$

$$V_o = \mathbf{12\ mV\ p\text{-}p}$$

$$\text{ripple factor} = \frac{V_{o(p\text{-}p)}}{2.82 V_o} = \frac{0.012 \text{ V}}{2.82(14\text{V})} = \mathbf{0.03\ \%}$$

Zener effects on regulation: Load regulation is generally improved by using a zener regulator. The equivalent circuit of Fig. 15-4 represents the regulator of Fig. 15-2 when the load is removed. The source resistance that the load "sees" looking back into the supply is very nearly equal to r_z because R_d is very much larger than r_z. Where regulator-diode power dissipation is limited to less than $\frac{1}{10}$ of its rated maximum, temperature rise will be minimal and a general approximation can be made that

$$\text{load-regulation factor} \approx \frac{r_z}{R_L} \tag{15-6}$$

Where temperature rise is not negligible, ΔV_Z and hence load regulation may be on the order of 2 to 5%, provided that care is taken to ensure that the zener is still in reverse conduction at full load and that its power limits are not exceeded at no load.

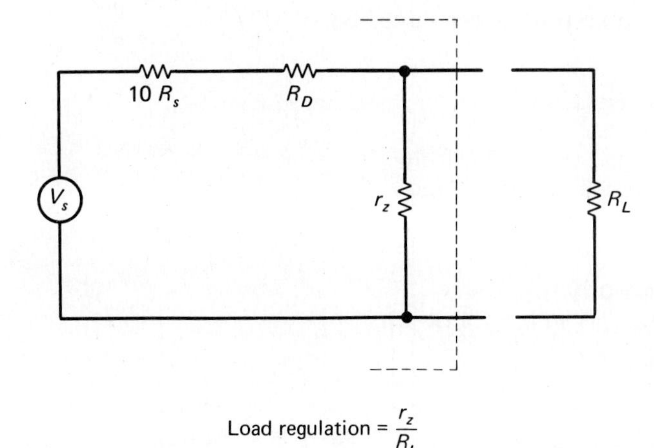

Figure 15-4 Equivalent circuit of a power supply and zener regulator as seen by the load. The low dynamic resistance of the zener r_z improves load regulation.

15.2 LINE REGULATION

The output voltage of a zener-regulated power supply can remain constant in the face of large variations in the ac line voltage. In an era of brownouts where line voltages from 105 to 125 V may be expected, and where equipment must be able to operate with a high degree of stability from a portable emergency power generator with a wildly fluctuating output voltage, zener regulation of critical power supplies is standard practice.

The *line regulation factor* specifies how much the output voltage of a supply varies as a result of a specified change in ac line input voltage. Lower line regulation factors indicate an output which is more stable in the face of line voltage variations.

Line regulation is determined experimentally from the following relationship:

$$\text{line regulation factor} = \frac{\%\ \text{load voltage change}}{\%\ \text{line voltage change}} \tag{15-7}$$

$$= \frac{(V_{OH} - V_{OL})V_{IL}}{(V_{IH} - V_{IL})V_{OL}}$$

where O and I stand for output and input, and H and L stand for high and low voltage, respectively. The usual test is to vary the line voltage by 10% (say from 110 to 121 V) and note the percentage change in output voltage. For unregulated power supplies, line regulation is near 100%; that is, there is no regulation at all. For a zener regulator the line regulation factor can be obtained from the approximation formula

$$\text{line regulation factor} = \frac{V_S r_z}{V_Z R_d} \tag{15-8}$$

provided that regulator-diode power is limited to less than one-tenth of its maximum rating to keep temperature effects from dominating. This formula assumes that the zener is conducting at all times, even with the lowest anticipated line voltage. Example 15-3 illustrates calculations for both line and load regulations and shows the necessity of ensuring ample current to keep the zener turned on.

Example 15-3

Find the line and load regulation factors for the circuit of Fig. 15-5.

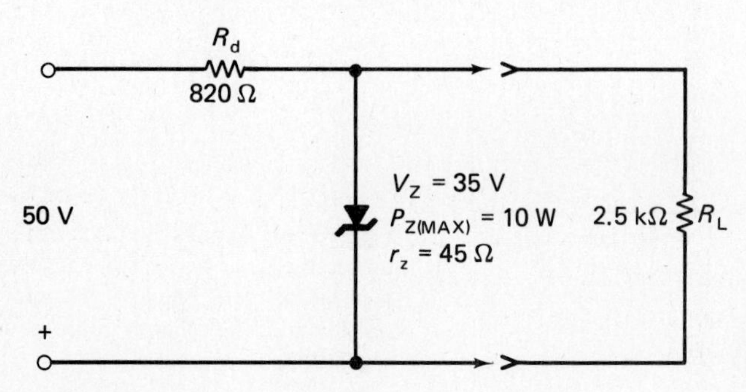

Figure 15-5 Example 15-3.

Solution First, it is necessary to determine that the zener is still conducting at full-load current:

$$I_{R(\mathrm{L})} = \frac{V_{R(\mathrm{L})}}{R_\mathrm{L}} \frac{35\ \mathrm{V}}{2.5\ \mathrm{k\Omega}} = 14\ \mathrm{mA}$$

$$I_{R(\mathrm{d})} = \frac{V_{R(\mathrm{d})}}{R_\mathrm{d}} = \frac{50\ \mathrm{V} - 35\ \mathrm{V}}{0.82\ \mathrm{k\Omega}} = 18.3\ \mathrm{mA}$$

The full-load I_Z is then 18.3 − 14, or **4.3 mA**. The no-load I_Z is the full 18.3 mA. The zener is conducting at all times.

$$P_{Z(\max)} = I_{Z(\mathrm{NL})} V_Z = (0.0183)(35) = 0.64\ \mathrm{W}$$

If the diode is properly heat sunk temperature rise should be minimal.

$$\text{load regulation factor} = \frac{r_z}{R_\mathrm{L}} = \frac{45\ \Omega}{2500\ \Omega} = 1.8\% \qquad (15\text{-}6)$$

This figure means that the output voltage will increase by 1.8% if the load is removed.

$$\text{line regulation factor} = \frac{V_S}{V_Z}\frac{r_z}{R_\mathrm{d}} = \left(\frac{50\ \mathrm{V}}{35\ \mathrm{V}}\right)\left(\frac{45\ \Omega}{820\ \Omega}\right) = 7.9\% \qquad (15\text{-}8)$$

This figure means that a line voltage change of 10% will cause a load voltage change of 0.079 × 10% or 0.79%, provided that the zener is still conducting.

It is interesting to note that in the example, a line voltage drop of 10% would leave V_S at 45 V and $V_{R(\mathrm{d})}$ at only 10 V. We would then expect a current $I_{R(\mathrm{d})}$ of 10/0.82 or **12.2 mA**. This is less than the current required by the load **(14 mA)** and indicates that the zener current is zero and that regulation has been lost. If such line voltage drops were anticipated, a redesign using a higher V_S or a lower R_d would be necessary.

It should be kept in mind that r_z is not a constant but varies with current I_Z. Ripple and regulation calculations using r_z must therefore be taken as indications of the orders of magnitude to be expected rather than as accurate predictions.

15.3 SERIES-TRANSISTOR REGULATOR

Zener regulation offers a considerable improvement in power supply performance, but several things are left to be desired. Zeners are inherently constant-voltage devices, whereas a *continuously variable*, but well-regulated, voltage is often required. Zeners have a limited *power-handling capability*, and will show voltage drifts on the order of a few percent if this power is even approached. Finally, since r_z is not zero, there is a limit to the *degree of regulation and ripple reduction* they can provide. All these limitations can be overcome by using a zener at low currents as the *reference element*, and a power transistor as the high-current *control element*. The simple series-transistor regulator shown in Fig. 15-6 can be analyzed as an emitter-follower amplifier which increases the load resistance seen by the zener regulator by a factor of β. Since the regulator is required to supply only the base current of Q_1, load regulation and ripple are reduced by a factor of β.

Line regulation is not a function of load resistance and so will not be directly affected by addition of the transistor. The decreased current demand of the base makes possible larger values of R_d and lower zener power dissipation, however, so some improvement in line regulation may be realized. The output voltage will be less than V_Z by the V_{BE} drop of the transistor.

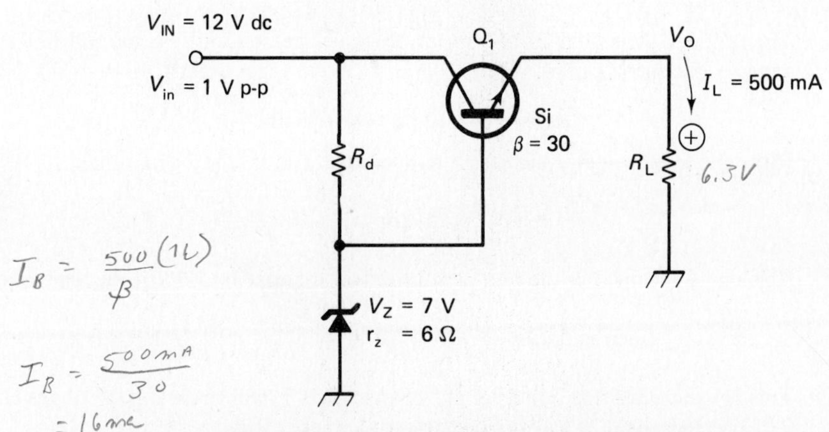

Figure 15-6 Series transistor regulator (Example 15-4).

The economics of the transistor regulator of Fig. 15-6 makes an interesting study. Let us assume that a maximum ripple of 0.1 V p-p is specified for a 60-mA, 12-V power supply. Using a half-wave rectifier and single-capacitor filter (as described in Sections 3.2 through 3.5), the filter capacitor required would be 10 000 μF. However, if the transistor regulator is used, the ripple at the filter capacitor could be 1 V p-p, requiring only a 1000-μF capacitor. The transistor and zener diode could then easily cut the 1-V ripple down to well below the 0.1-V specification at the emitter of the transistor and provide better line regulation and load regulation in the bargain. Following is a cost comparison of the two systems:

Single capacitor	Transistor regulator
10 000 μF/25 V........**$2.50**	1000 μF/25 V $0.70
	12-V zener 0.38
	1-W transistor........ 0.48
	R_d resistor 0.03
	$1.59

In addition, the transistor regulator system is likely to be smaller than the single large capacitor, and the problem of high surge currents when the supply is first turned on is eliminated.

The analysis for the transistor regulator circuit is similar to that for the simple zener regulator of the previous section, except that the "load" is the base of the transistor, and the base current is given by

$$I_B = \frac{I_L}{\beta}$$

Example 15-4

Find R_d and the ripple and regulation factors for the circuit of Fig. 15-6.

Solution The output voltage will be the base voltage minus the silicon base-emitter drop:

$$V_O = V_Z - V_{BE} = 7 - 0.6 = \mathbf{6.4\ V}$$

The base current is approximately the emitter current divided by beta:

$$I_B = \frac{I_E}{\beta} = \frac{500\text{ mA}}{30} = 16.7\text{ mA}$$

A considerable portion of the dropping resistor's current should be apportioned to the zener to assure zener conduction at all times. Let $I_{R(d)} = 25$ mA so that I_Z will be $25 - 16.7$, or 8.3 mA.

$$R_d = \frac{V_{R(d)}}{I_{R(d)}} = \frac{12\text{ V} - 7\text{ V}}{0.025\text{ A}} = \mathbf{200\ \Omega}$$

The power requirements of each device can be readily found:

$$P_{R(d)} = I_{R(d)} V_{R(d)} = (0.025\text{ A})(5\text{ V}) = \mathbf{0.125\ W}$$

$$P_Z = I_{Z(max)} V_Z = (0.025\text{ A})(7\text{ V}) = \mathbf{0.175\ W}$$

(If the load is removed, $I_B = 0$ and all of $I_{R(d)}$ must flow through the zener; hence $I_{Z(max)} = I_{R(d)}$.)

$$P_Q = I_E V_{CE} = (0.5)(12 - 6.4) = \mathbf{2.8\ W}$$

Half-watt resistors and diodes could be used, but the transistor should be capable of dissipating 5 W, leaving a reasonable safety margin.

In calculating ripple and regulation, the base of Q appears as a resistor of value

$$R_B = \frac{V_B}{I_B} = \frac{7\text{ V}}{0.0167\text{ A}} = 420\ \Omega$$

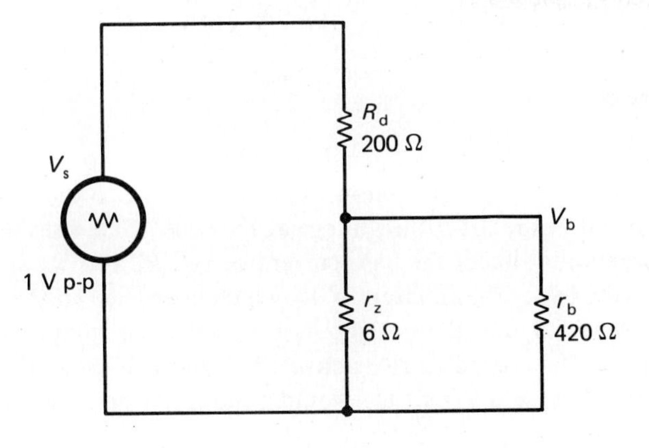

Figure 15-7 Equivalent of Fig. 15-6 as seen by the ac ripple.

Analyzing the equivalent circuit of Fig. 15-7, the ripple at the base (V_b) is found to be **29 mV p-p.** The output ripple will be nearly the same since the base-emitter junction resistance is nearly zero at the high emitter current used.

Similarly, using Fig. 15-7, load regulation is found:

$$\text{load regulation factor} = \frac{r_z}{R_B} = \frac{6\ \Omega}{420\ \Omega} = \mathbf{1.4\%}$$

$$\text{line regulation factor} = \frac{V_S}{V_Z}\frac{r_z}{R_d} = \frac{12\ \text{V}}{7\ \text{V}}\frac{6\ \Omega}{200\ \Omega} = \mathbf{5.2\%}$$

These factors assume negligible self-heating of the regulator diode. A 2-W device would meet this assumption. If a $\frac{1}{4}$- or $\frac{1}{2}$-W diode is used, much poorer regulation will result from thermal drift of V_Z.

Variable output voltage can be obtained from the series transistor regulator if a circuit such as the one in Fig. 15-8 is used. Load regulation will suffer somewhat because the resistance seen looking back from the base of the transistor is not the low dynamic zener resistance r_z, as it is in the fixed voltage circuit, but is raised considerably by the adjustment potentiometer R_{B1}–R_{B2}. Changes in base current will cause changing voltage drops from the zener to the base of the transistor. It is therefore advisable to use a low value for R_{B1}–R_{B2} and to ensure that a high beta transistor is used for Q_1. In calculating R_d, it is necessary to observe that $I_{R(d)} = I_Z + I_{R(B)} + I_B$.

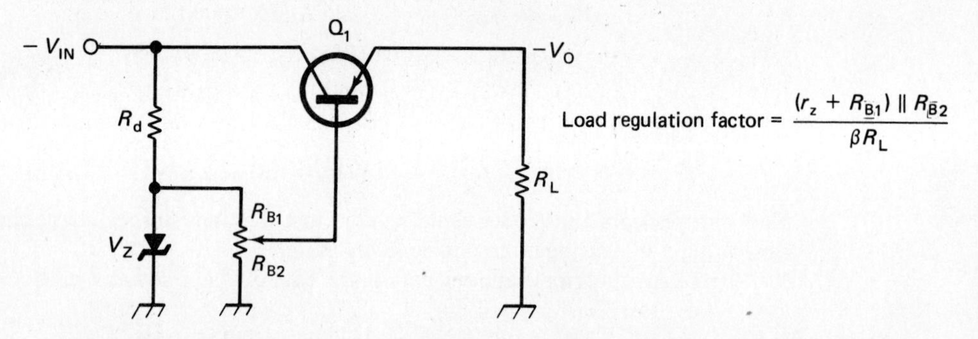

Figure 15-8 Potentiometer R_{B1}–R_{B2} permits the series regulator to provide a variable output voltage, but the load regulation suffers.

15.4 AMPLIFIED REGULATORS

Figure 15-9 shows an adjustable power-supply regulator which uses two transistors to obtain improved regulation and ripple reduction. R_B feeds current to the base of Q_2 which is in the main current path, but if the voltage at the base of Q_1 rises above $V_Z + V_{BE}$, Q_1 conducts more and reduces the base current of Q_2. R_1 and R_2 are the halves of a potentiometer used to set the fraction of V_O which is applied to the base of Q_1. For example, if V_Z were 6 V and the pot R_1–R_2 were set to its midpoint, V_O would be approximately 12 V. If V_O tried to rise above 12 V, the voltage at the top of the pot would tend to rise above 6 V and Q_1 would conduct more. This would tend to turn Q_2 off, restoring the output to 12 V. C_1 is used to provide a direct path for ac ripple from the output to the base of Q_1. The reactance of C_1 must be much lower than R_1 at the ripple frequency.

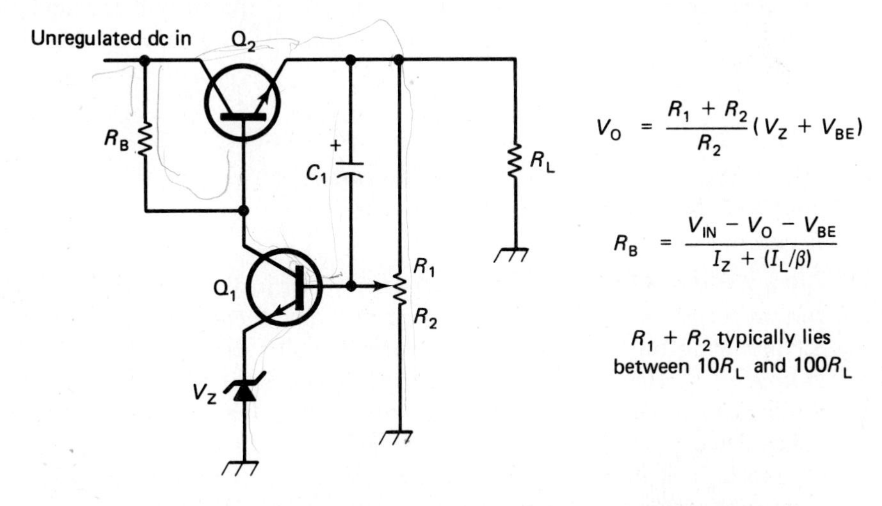

Figure 15-9 Amplified series regulator with variable output voltage.

Balanced-supply regulator: The ripple and regulation of the circuit of Fig. 15-9 can be made quite good, but it has a disadvantage in that the output voltage cannot be adjusted to a value lower than V_Z. The amplified regulator circuit of Fig. 15-10 eliminates this problem, at the expense of another diode and filter capacitor.

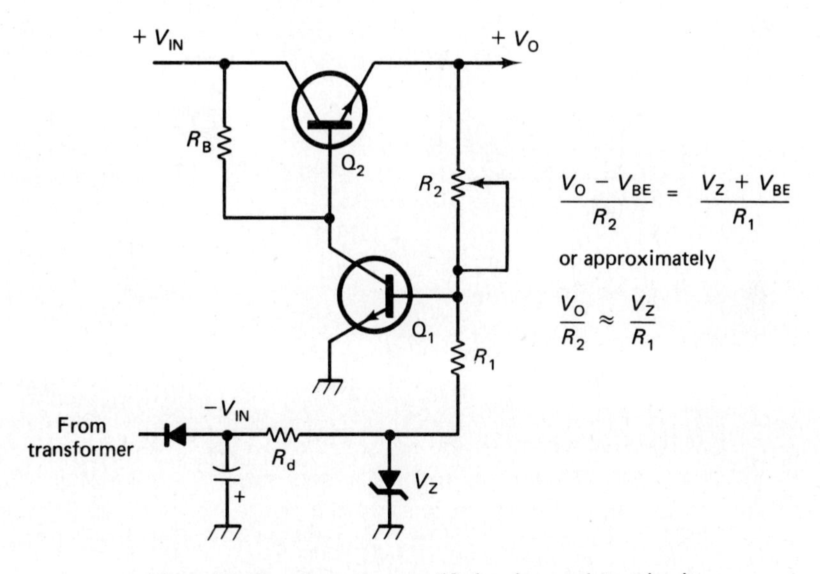

Figure 15-10 Alternative amplified-series-regulator circuit.

These extra components are used to provide a separate *negative* supply input which is regulated by R_d and the zener diode. The negative current through R_1 must be exactly equaled by the positive current through R_2, so Q_1 will be maintained at the threshold of conduction. If V_O tends to increase, the current through R_2 tends to turn Q_1 on, thus turning Q_2 off and restoring the original output voltage. Higher values of R_2 demand higher output voltages, and if R_2 is made zero, V_O will be reduced to the V_{BE} required to turn Q_1 on.

15.5 CURRENT REGULATION

Often it is desired to protect a power supply or the device driven by it against large current surges resulting from short circuits, component failure, etc. A simple means for providing this protection is the current limiter circuit, shown in Fig. 15-11(a). Normally, R_1 supplies base current to Q_1, keeping Q_1 on. However, if the current through R_2 is sufficient to cause a drop greater than 0.6 V, the voltage between points B and C of of the circuit will attempt to exceed 1.2 V, and the two silicon diodes D_1 and D_2 will turn on. Q_1 will thus be robbed of its base current and will turn off, limiting the output current.

V_{IN} must be several volts larger than V_O for this circuit to be successful, and since Q_1 must carry the entire output current, it must usually be an expensive power transistor. The current limiter can be combined with a series regulator (such as the one in Fig. 15-10) to eliminate both of these difficulties, however. This is illustrated in Fig. 15-11(b).

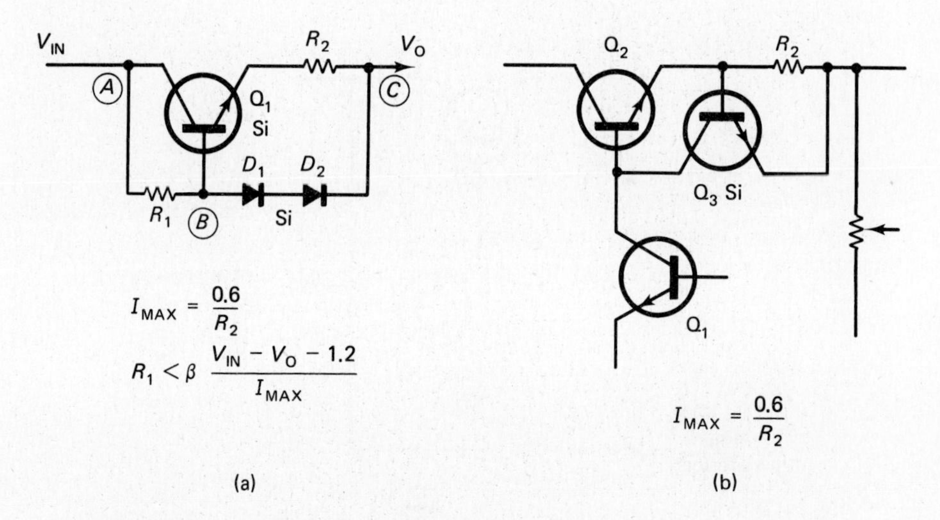

Figure 15-11 (a) Current-limiter circuit; (b) a current limiter (Q_3 and R_3) combined with the series regulator circuit of Fig. 15-10.

15.6 IC REGULATORS

An **operational amplifier** is used in the circuit of Fig. 15-12(a) to compare V_O with a fraction of the reference voltage V_Z. If V_O is larger than V_{IN}, the inverting input of the op amp predominates and V_B drops, restoring V_O to V_{IN}. V_O is limited to V_Z with this circuit, and we would like to keep V_Z much less than V_{RAW}, both to keep R_d large and to permit use of a 7-V zener with its near-zero temperature coefficient.

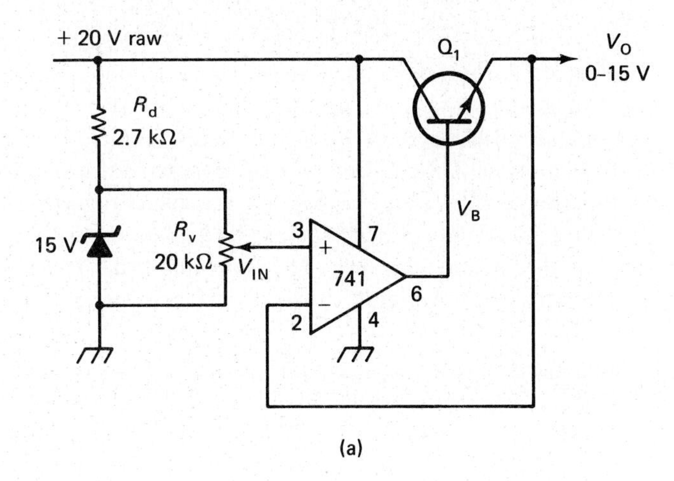

(a)

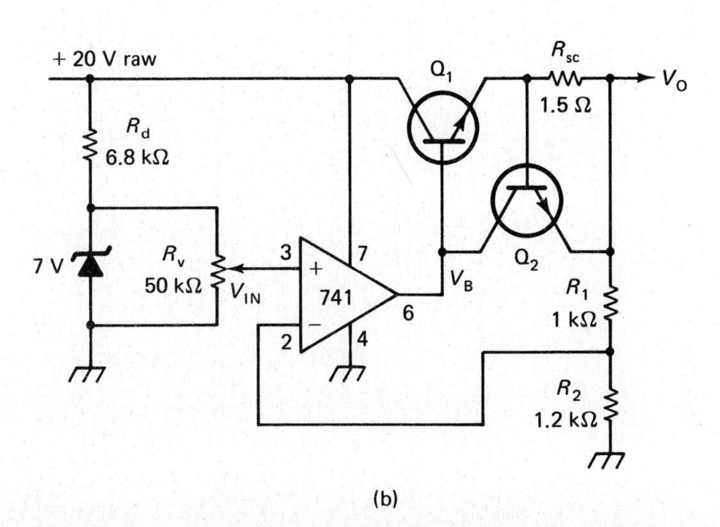

(b)

Figure 15-12 (a) An op amp can be used to compare V_O to a portion of V_Z, turning Q_1 *on* to hold V_O at the desired level. (b) If current-limiting transistor Q_1 is added and V_O is divided down before comparison to V_{REF}, the system duplicates the components of packaged IC regulators.

Figure 15-12(b) shows how a voltage divider at the output permits V_O to attain a maximum value several times V_Z. The current-limiting circuit of Fig. 11-11(b) is included. R_{sc} is chosen to limit the short-circuit current to prevent damage to Q_1. For Fig. 15-12(b)

$$I_{SC} = \frac{V_{BE}}{R_{sc}} \approx \frac{0.6 \text{ V}}{R_{sc}} \tag{15-9}$$

$$V_{O(max)} = V_Z \frac{R_1 + R_2}{R_2} \tag{15-10}$$

Packaged IC regulators, such as the popular type *723*, contain the elements of Fig. 15-12(b) in a single silicon chip. R_d is replaced with a current source so I_Z remains constant in spite of changes in V_{RAW}. R_v, R_1, R_2, and R_{sc} must be added externally. Figure 15-13 shows a *723*-regulator circuit. An external transistor Q_3 has been added in a Darlington configuration to increase the current capability of the regulator to over 1 A. Pins 2 and 10 are tied together if Q_3 is not used. C_1 is for frequency compensation of the internal op amp. The following table gives some abbreviated specifications for the *723* regulator.

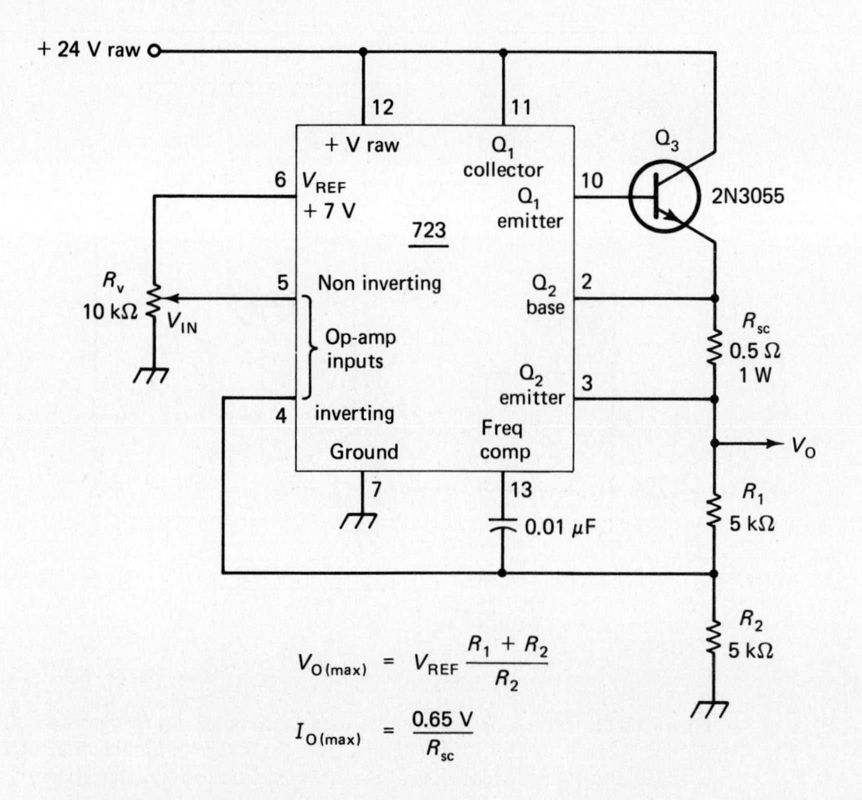

Figure 15-13 Popular *723* IC regulator with an external series-pass transistor for added current capability.

Rating (subscripts indicate pin numbers)	Value
Output current, I_{10}	100 mA max
Voltage difference, $V_{12} - V_{10}$	3 V min, 38 V max
Power dissipation, $I_{10}(V_{12} - V_{10})$	1 W max
Input voltage, V_{12}	9.5 V min, 40 V max
Load regulation, I_L = 50 mA to 1 mA	0.2% max
Line regulation, V_O = 5 V, V_{IN} 12 to 15 V	0.4% max

Three-terminal regulators are IC devices having *input* (V_{RAW}), *output* (V_O) and *common* terminals, and supplying a fixed output voltage. Commonly available types deliver 5, 6, 8, 12, 15, 18, and 24 V, either positive or negative with respect to the common pin. Maximum currents range from 100 mA in tiny plastic-package devices to 1 A in metal heat-sunk devices. They have become extremely popular because

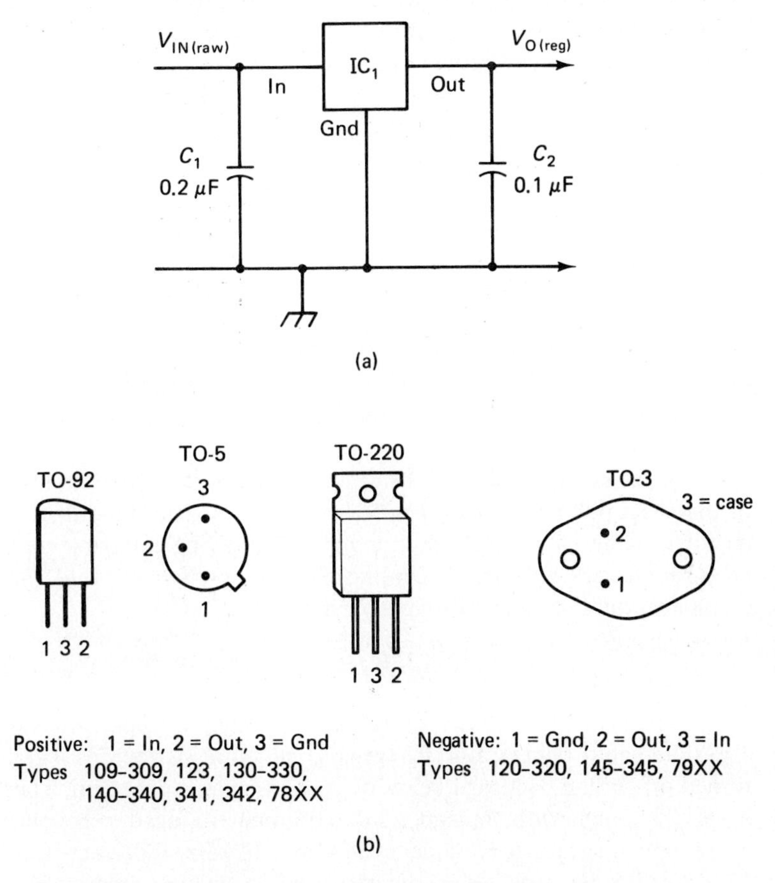

Figure 15-14 (a) Three-terminal fixed-voltage IC regulator in circuit; (b) popular case styles, type numbers, and pinouts.

they are easy to use and nearly indestructible. Most are internally protected from short circuits—they simply shut off upon overheating.

On the minus side, three-terminal regulators are a bit more expensive than a zener/transistor combination (Fig. 15-6), and their load and line regulation specifications are about 10 times poorer than those of general-purpose regulators such as the *723*.

Figure 15-14 shows case styles, type numbers, and pinouts of popular three-terminal regulators. C_1 is often required to keep the regulator from self-oscillating. C_2 is optional for filtering high-frequency noise from the output.

15.7 HIGH-FREQUENCY POWER SUPPLIES

All of the regulators discussed so far have used a resistive device (either a dropping resistor or a series-pass transistor) in the main current path. The power wasted by this device is $I_{OUT}(V_{IN} - V_{OUT})$, which is typically 50 to 200% of the load power. For load powers above a fraction of a watt this waste is likely to be objectionable from the standpoint of battery drain or system heating. There has been a strong trend in recent years to avoid this problem by using high-frequency switching power supplies.

A mechanical switch dissipates no power because $P = IV$ and either the current through it or the voltage across it is always zero. To the extent that we can turn a power transistor *on* and *off* instantly (with zero $V_{CE(sat)}$ when *on* and zero I_C when *off*) the transistor will dissipate essentially no power also. Diodes, inductors, and capacitors ideally dissipate no power, but they can be used to filter a train of pulses from a switching transistor to essentially pure dc.

Switching power supplies typically operate in the range 5 to 50 kHz using low-loss toroid or pot-core inductors, fast-switching power diodes (often of the hot-carrier type), and low-dissipation tantalum filter capacitors. They achieve efficiencies of typically 70 to 90%.

Switching regulators can be designed to step the input voltage down, step it up, or invert its polarity. Figure 15-15(a) shows the basic step-down circuit. When Q is *on* current flows through L, charging C and applying voltage across R. When Q turns *off*, L discharges through D, continuing to supply current to C and R. The ratio of *on* time to total *on* plus *off* time determines V_O.

$$V_O = V_{IN} \frac{t_{on}}{t_{on} + t_{off}} \tag{15-11}$$

Figure 15-15(b) shows a practical but simple step-down switching regulator. The differential comparator IC turns Q *on* until V_O exceeds V_{REF}, whereupon Q is turned off. When V_O drops below V_{REF}, Q is again turned on. The IC is set up as a hysteresis switch with R_{in} and R_F determining the dead zone, which is $V_{o(p\text{-}p)ripple}$. The supply operates at a frequency of about 12 kHz. Efficiency is only 65% at $V_{IN} =$ 12 V, but this one was designed with simplicity as the primary objective. Line regulation is 1.2%, load regulation 0.3%, and ripple is 0.8%. Power-supply diodes, RF

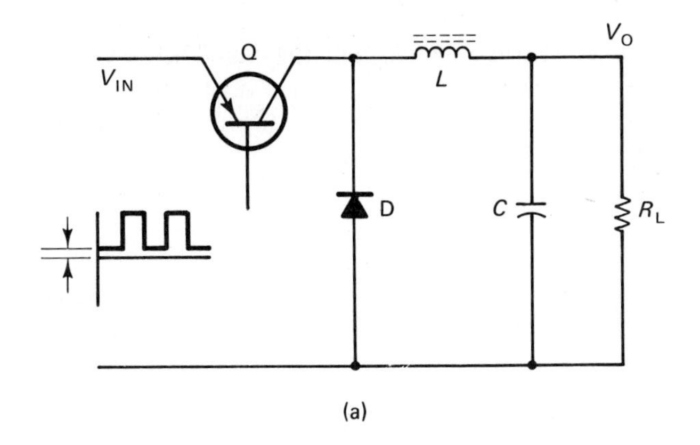

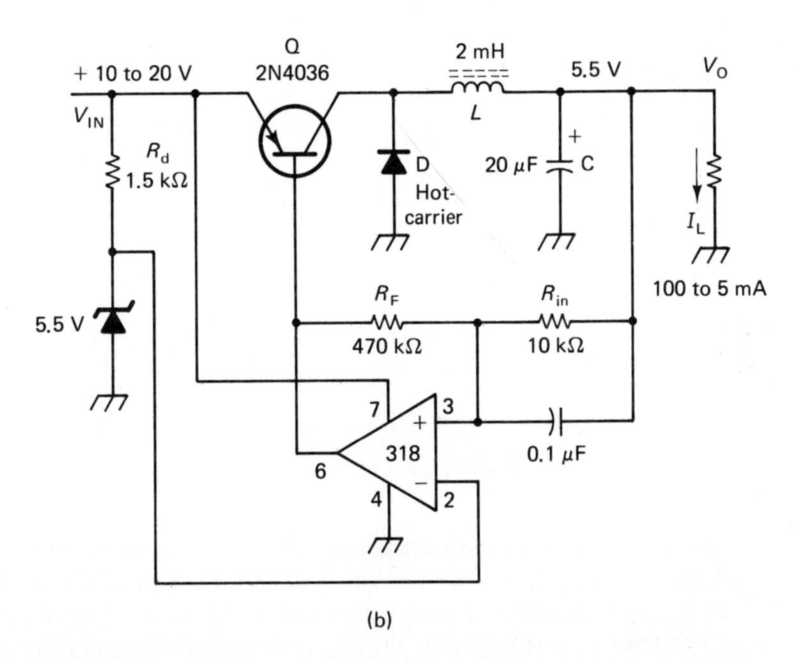

Figure 15-15 (a) Prototype switching regulator. Ideally, Q, D, L, and C waste no power. (b) Practical switching regulator. The IC is a hysteresis switch. When V_O drops, pin 6 goes low, turning Q *on*. The switching frequency is about 12 kHz.

chokes, and electrolytic capacitors will not do for D, L, and C—all are too lossy at the frequencies involved.

High-frequency transformers are much smaller and lighter then those designed for 60-Hz ac operation, and it is becoming common practice to replace the latter with the former in ac-line-powered equipment. Figure 15-16 shows a simple circuit illustrating the technique.

First the ac line is rectified and filtered. The resulting 170 V dc powers an oscillator which switches the primary of the high-frequency transformer at a rate of about 20 kHz. The transformer's secondary voltage (ac pulses at 10 V pk) is rectified and easily filtered with relatively small capacitors. A similar technique is commonly used to convert 12-V mobile power to higher voltages.

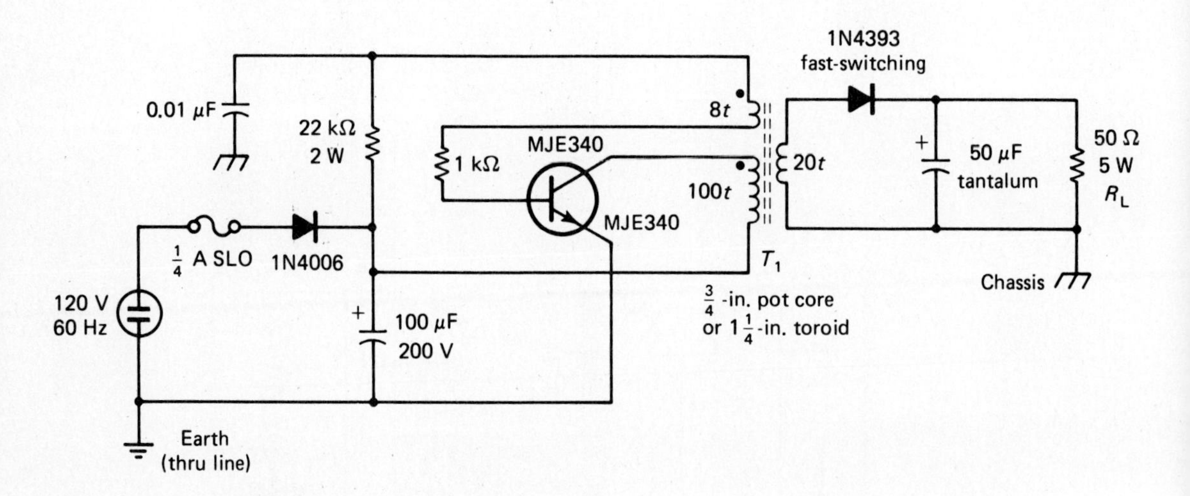

Figure 15-16 High-frequency switching of a dc line permits use of a much smaller transformer than is possible at 60 Hz.

15.8 ZENER REGULATOR MISCELLANY

An extremely stable reference voltage can be obtained by using two zener regulators in cascade, as shown in Fig. 15-17. The load regulation is not improved by this circuit, so it is not advisable for use where large output currents are required. However, where low currents are required, such as in amplified transistor-regulated supplies, the cascaded regulator gives line regulation and ripple reduction factors which are roughly equal to the square of the factors for a single-stage regulator. (For example, if line regulation is 0.1 or 10% for a single-stage regulator, it will be approximately 0.1^2, which is 0.01 or 1% for a two-stage regulator.) The input voltage must be several times higher than the output voltage for best success with this technique.

The values of voltage, current, and resistance given in Fig. 15-17 are typical for a 7-V reference supply with negligible output current drain.

A second low-current reference voltage source using only a single zener is shown in Fig. 15-18. In this circuit, R_2 is chosen to match the dynamic resistance r_z of the

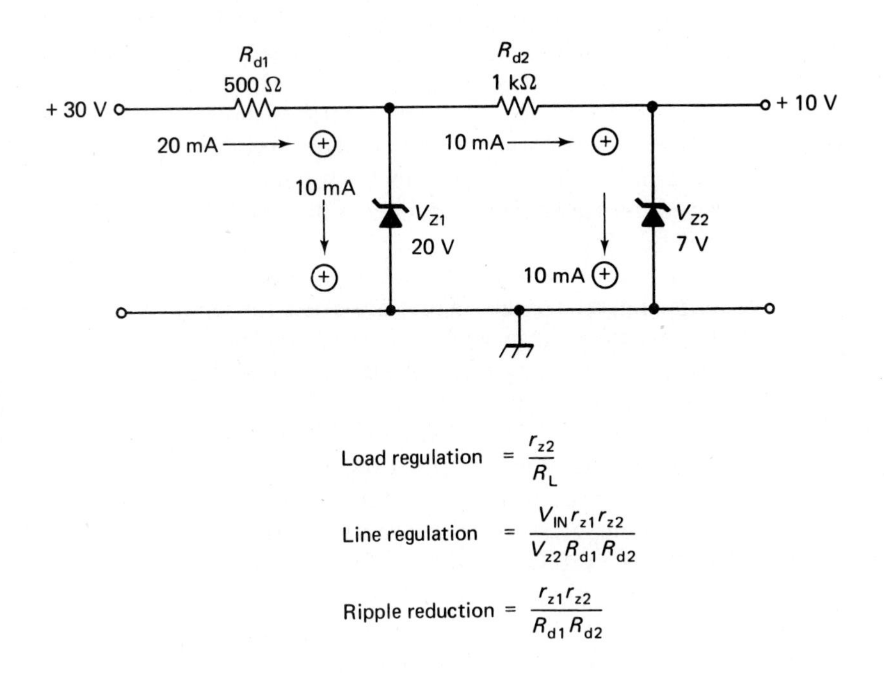

Figure 15-17 Two-stage zener regulator for extremely critical line regulation and ripple reduction.

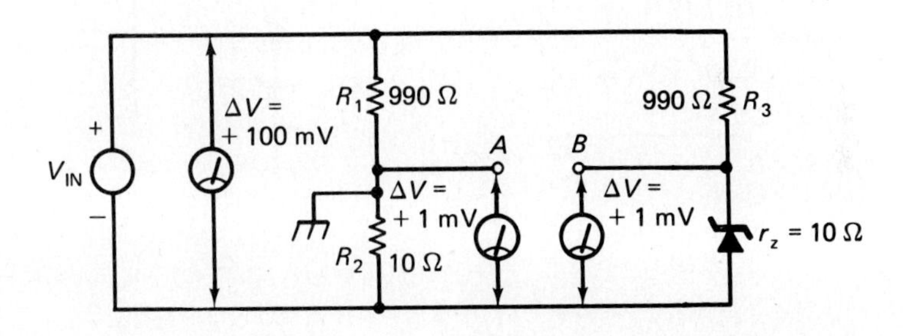

Figure 15-18 Highly stable voltage reference source. Voltage divider R_1–R_2 has the same ratio as R_3–r_Z, so V_{AB} is unaffected by small changes in V_{IN}.

zener, so that any voltage drop across r_z will be exactly compensated for by an equal drop across R_2. The voltage between points A and B in Fig. 15-18 will therefore remain constant in spite of ripple or level changes in V_{IN}. For example, if V_{IN} should increase by 100 mV because of an ac line change, the voltage division of R_3 and r_z would cause a 1-mV increase across the zener. However, the voltage division of R_1 and R_2 would cause an identical 1-mV increase across R_2, leaving V_{AB} unchanged. The disadvantage of this system is that neither side of the V_{IN} supply can be grounded to the same point as the reference-supply ground.

Zener diodes can be connected in series to provide a higher effective V_Z, as shown in Fig. 15-19(a). They *cannot* be placed in parallel to obtain higher power ratings, however, as one diode will invariably hog most of the current.

Voltage clipper: Zener diodes are often connected in series but opposing in polarity to limit the voltage across some critical element. In Fig. 15-19(b), two 10-V zeners are shown protecting the sensitive input of an FET from accidental overload. Any voltage in excess of 10.6 V in either polarity will be dropped across R_1.

Transistor zeners: Zener diodes are constructed by using a much higher concentration of impurity dopant than is used in a conventional diode. In the silicon-planar process of transistor manufacture, the emitter-base diode has a very high concentration of impurity and thus becomes a sort of zener itself. This is illustrated in Fig. 15-19(c). The breakdown voltage of the silicon emitter-base junction is typically 5–10 V. The dynamic resistance and power dissipating capabilities of this "zener" are not particularly good, but there have been circuits designed to take advantage of this zener characteristic of the silicon planar transistor.

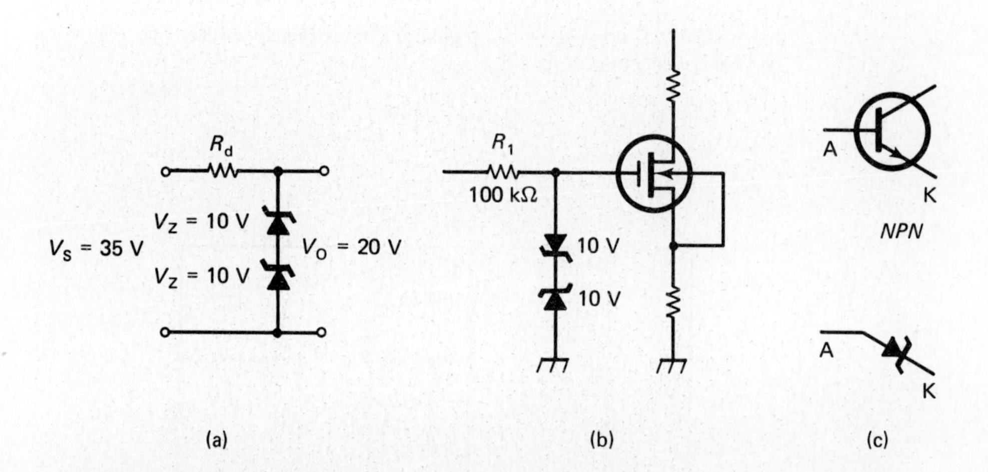

Figure 15-19 (a) Two 10-V zeners in series make a 20-V zener. (b) Series reverse-directed zeners limit input voltage to protect an FET. (c) The base-emitter junction of a silicon planar transistor forms a low-quality zener diode. Typically, $V_Z = 7$ V.

CHAPTER SUMMARY

1. Regulator diodes begin to conduct sharply at reverse voltage V_Z. True zener regulators ($V_Z < 7$ V) have a negative temperature coefficient (higher temperature, lower V_Z). Avalanche diodes ($V_Z > 7$ V) have a positive coefficient. Both types are loosely called "zeners." Both types conduct at 0.6 V in the forward direction.

2. The basic regulator circuit uses a dropping resistor R_d in series with the zener. The load is connected across the zener and V_L is held at V_Z as long as the zener is conducting. Excess voltage from V_{IN} is taken by R_d.

$$R_d = \frac{V_{IN} - V_Z}{I_L + I_Z} \tag{15-1, 2}$$

3. Avalanche diodes of $\frac{1}{2}$-W rating have a dynamic resistance which is generally less than 10 Ω for $V_Z = 10$ V, 40 Ω for $V_Z = 40$ V, and so on. The basic regulator circuit greatly reduces input ripple V_{in}.

$$V_o = V_{in} \frac{r_z}{R_d} \tag{15-5}$$

4. Load regulation for the basic regulator circuit is

$$\text{load regulation} \approx \frac{r_z}{R_L} \tag{15-6}$$

provided that the diode is operated at $\frac{1}{10}$ or less of its rated power. With higher power dissipation, self-heating causes regulation to rise to several percent.

5. A zener regulator permits V_O to remain stable (low ΔV_O) in spite of large changes in the input line voltage (high ΔV_{IN}).

$$\text{line regulation } \frac{\Delta V_O / V_O}{\Delta V_{IN} / V_{IN}} \approx \frac{V_{IN} r_z}{V_Z R_d}$$

again provided that zener self-heating is negligible.

6. In more advanced regulators a series-pass transistor is placed between the source and load to regulate V_O. The transistor is turned on more when comparison of V_O and V_Z shows that V_O is too small. This system permits variable V_O, higher power levels, and better regulation.

7. Current-limiting circuits use a low-value resistor in series with the output-current path to develop base-emitter voltage for a shut-down transistor. When $I_O R = 0.6$ V the series-pass transistor is not permitted to turn on more.

8. IC regulators use an op amp to compare V_O to V_{REF}. The popular type *723* contains a 7-V reference source, an op amp, a 100-mA series-pass transistor, and a shut-down transistor in one package.

9. Three-terminal IC regulators provide fixed output voltages and require no external components except a single 0.1-μF capacitor.

10. Switching regulators provide better efficiency than linear (series-pass) regulators. They typically operate between 5 and 50 kHz, and use a switching transistor, fast-switching diode, toroid or pot-core inductor, and tantalum filter capacitor.
11. High-frequency supplies provide the advantages of transformer coupling with a transformer that is smaller and lighter than would be possible at the 60-Hz line frequency. The ac line is rectified and filtered directly, switched at perhaps 20 kHz by a transistor, and then put through a high-frequency transformer for final rectification and filtering.

QUESTIONS AND PROBLEMS

15-1. A regulator diode has $V_Z = -12$ V, $r_{z(\text{typ})} = 5\ \Omega$, and $r_{z(\text{max})} = 15\ \Omega$. The current I_Z changes from -3.0 to -3.5 mA. What is the maximum change in V_Z to be expected?

15-2. In the circuit of Fig. 15-2, $R_L = 1.0$ kΩ and $R_d = 240\ \Omega$. Find I_Z.

15-3. In Problem 15-2, what is the minimum value of V_S that will keep the zener conducting at $I_{Z(\text{min})} = 1$ mA?

15-4. In Problem 15-2, $V_{s(\text{ripple})} = 0.15$ V p-p, and $r_z = 8\ \Omega$. Find $V_{o(\text{ripple})}$ in p-p terms.

15-5. In Problem 15-2, find the increase in V_O if R_L is removed, assuming that thermal effects are negligible. What minimum power rating of the diode would justify this assumption?

15-6. The diode in Problem 15-2 has $r_z = 8\ \Omega$. V_S changes from 20 V to 18 V. Verify that the diode is still conducting and find the change in dc output voltage V_O, neglecting temperature effects.

15-7. In a series-transistor regulator similar to that of Fig. 15-6, $V_{IN} = 42$ V dc average, $V_{\text{ripple}} = 8$ V p-p, $R_d = 1.5$ kΩ, $V_Z = 27$ V, the transistor is silicon, and $I_{R(L)} =$ 150 mA. Find the minimum transistor β that will keep the diode conducting with $I_{Z(\text{min})} = 1$ mA, even at the ripple valleys.

15-8. In Problem 15-7, find V_o ripple in p-p terms.

15-9. In Problem 15-7, transistor β_{min} is 50. Find the minimum V_{IN} dc average that will keep the regulator diode conducting at the ripple valley.

15-10. In the variable regulator circuit of Fig. 15-8, $R_B = 10$ kΩ total and is set to midrotation, $V_Z = 36$ V, and $\beta = 60$. Find V_O, first when I_L is 1 mA, and then when I_L is increased to 25 mA.

15-11. In the amplified regulator of Fig. 15-9, $V_Z = 10$ V, $R_1 = 300\ \Omega$, $R_2 = 700\ \Omega$, $I_L =$ 120 mA, and the transistors are silicon with $\beta = 100$. Find V_O.

15-12. In Problem 15-11, minimum V_{IN} is 22 V dc. Find the largest value of R_B that will permit full regulated V_O.

15-13. In the circuit of Fig. 15-10, $V_Z = 7$ V, $R_1 = 1.8$ kΩ, and $R_2 = 2.5$ kΩ. The transistors are silicon. Find V_O.

15-14. In Fig. 15-11(b), find the value of R_2 necessary to limit I_O to 300 mA.

15-15. For the op-amp regulator of Fig. 15-12(b), find $V_{O(max)}$ and $I_{O(max)}$.

15-16. Find $V_{O(max)}$ and $I_{O(max)}$ for the IC regulator of Fig. 15-13.

15-17. What advantage do switching regulators have over linear regulators?

15-18. Sketch a partial schematic diagram of the switching regulator of Fig. 15-15(b) showing the complete current path through L (a) when pin 6 of the IC is switched low; (b) when pin 6 is switched high.

15-19. In Fig. 15-15(b), V_{SAT} of the IC is +1 V and +14 V. How high will V_O rise before the transistor is switched off? How low will V_O fall before it is switched on? Assume that the 0.1-μF capacitor shunting R_{in} has been removed.

16

SPECIAL DIODES AND TRIGGERED DEVICES

16.1 VARACTOR DIODES

As described in Section 2.8, a reverse-biased junction diode consists of the two conductive areas (*N*- and *P*- doped, respectively) separated by an insulative depletion zone. These regions form the "plates" and dielectric of a capacitor. Increasing the reverse bias widens the depletion zone and decreases the capacitance.

Diodes manufactured to exploit this variable-capacitance effect are called varactors, varicaps, or VVCs (voltage-variable capacitors). Capacitance values from a few pF to nearly 1000 pF are available. Usable capacitance for an individual device may be variable over a range as wide as 20 : 1, although 5 : 1 is more common.

Varactors have two common applications—electronic tuning and frequency multiplication.

Varactor tuning is the basis of AFC (automatic frequency control) in FM receivers and its equivalent, AFT (automatic fine tuning) in television. It is also used in radio transmitters to produce frequency modulation and it has been used in receivers to completely eliminate the bulky and vibration-sensitive mechanical variable capacitor.

An oscillator whose frequency is variable by means of a varicap is shown in Fig. 16-1. A small frequency change can be observed using an ordinary 1-A rectifier diode as a varicap. Notice that a common-collector oscillator is employed, although the output is taken from the collector. This is to minimize the RF voltage across the varicap so that capacitance does not change appreciably with ac-signal swings.

Varactor multipliers are used to produce several watts of RF at frequencies in the gigahertz range where power transistors are unavailable or are prohibitively expen-

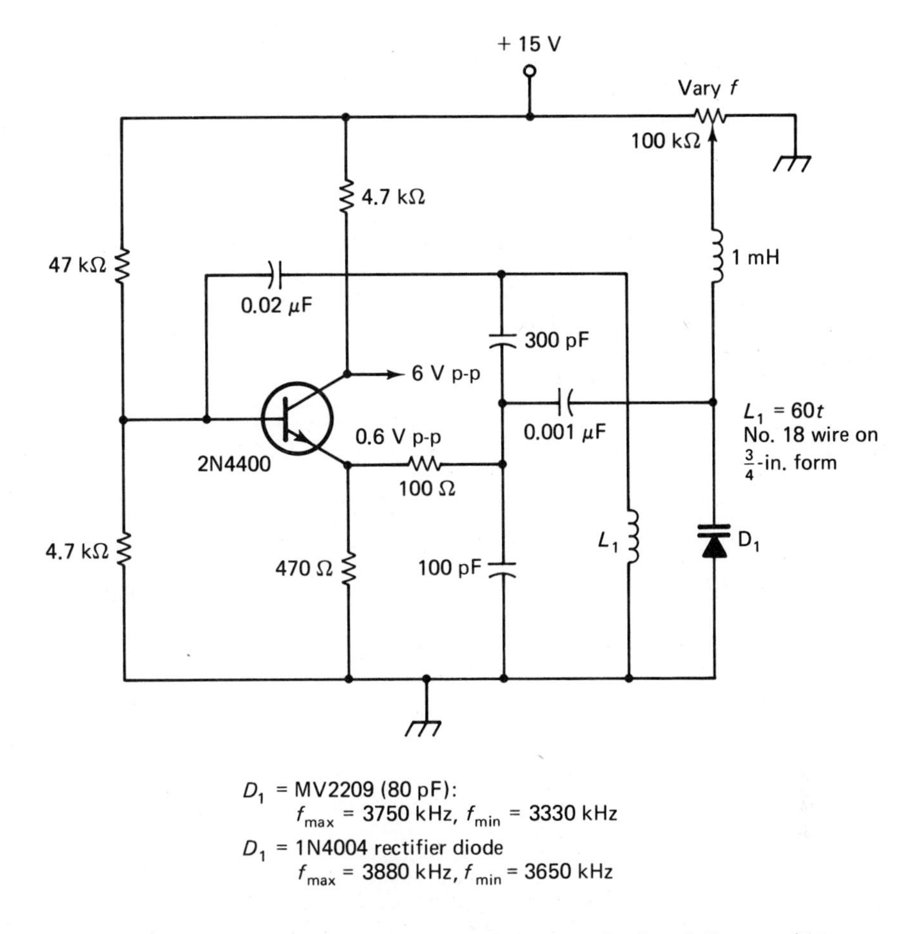

Figure 16-1 Varactor diode D_1 controls frequency of emitter-follower oscillator. Amplified output signal appears at collector.

sive. An output of 5 W at 500 MHz might be converted to 3 W at 1500 MHz, for example. A demonstration circuit at frequencies within the range of most lab instruments is shown in Fig. 16-2. The varactor builds up a waveform with flattened positive peaks because its capacitance is high at positive and low-negative bias. The negative peaks are exaggerated because the capacitance is low at high reverse bias. This distorted wave contains a strong second harmonic which is picked up by the tuned circuit L_2C_2.

16.2 TUNNEL DIODES

The tunnel or Esaki diode has a characteristic curve which exhibits *negative resistance* (downward slope) over a range typically from 0.25 to 0.4 V, as shown in Fig. 16-3(a). When biased in this region the negative resistance of the diode can be used to over-

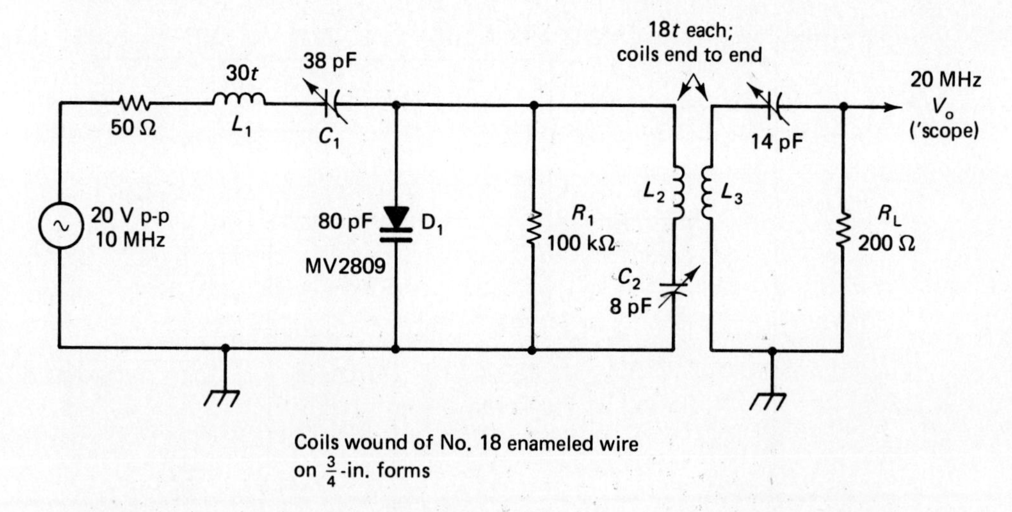

Figure 16-2 A demonstration varactor multiplier inputs 10 MHz and outputs 20 MHz.

come the positive resistance (loss) in a tuned circuit. Positive resistance causes the oscillations in an LC circuit to dampen and die away as energy is lost on each cycle. Negative resistance causes the oscillations to build up until the limits of the negative-resistance region are reached.

Oscillator: Figure 16-3(b) shows a tunnel-diode oscillator operating in the FM broadcast band. The frequency of the tuned circuit is modulated by a varactor diode

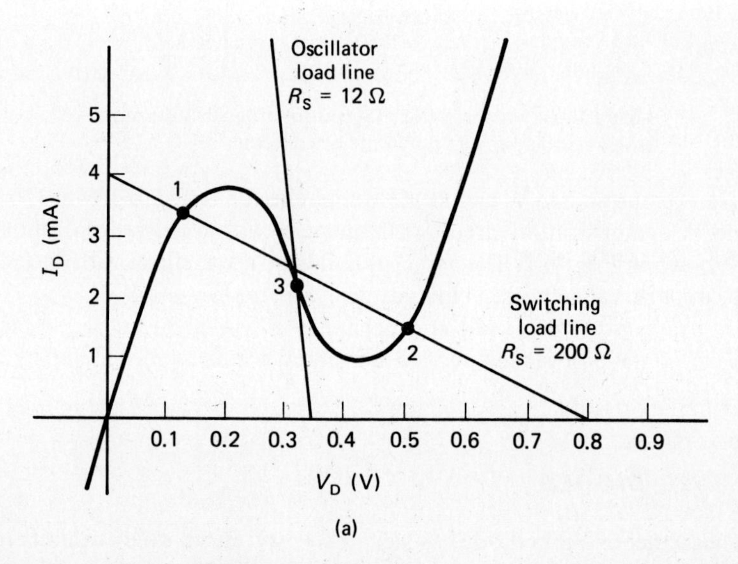

Figure 16-3 (a) Tunnel-diode characteristic curves; (b) FM wireless mike with tunnel-diode oscillator and varactor frequency modulator.

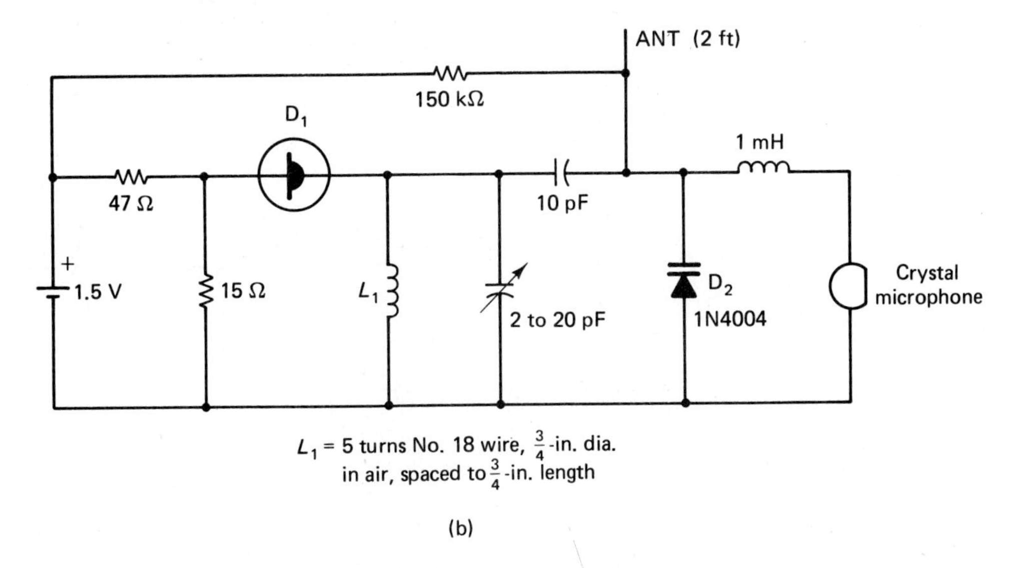

Figure 16-3 Continued

fed with audio signals from a crystal microphone. The range of this "wireless mike" is limited to about 20 ft.

Amplifier: A tunnel diode can be placed in parallel with a load resistance to effect current amplification. The signal, seeing a positive and negative resistance in parallel, sees a total resistance higher than the load.

Switch: The tunnel diode of Fig. 16-3(a) in series with 0.8 V and 200 Ω makes a two-position switch which will remain at point 1 or 2 as determined by externally applied pulses.

To each of these applications the tunnel diode brings two advantages—simplicity of circuitry and high speed. Switching times of a few nanoseconds and frequencies in the gigahertz range are common. Its disadvantages include a power limitation of a few milliwatts, and a cost several times higher than that of competitive transistors.

16.3 DISPLAY DEVICES

Light-emitting diodes: In a forward-biased *P-N* diode, hole-electron pairs are continuously being formed and recombining. The recombination releases energy, which for certain materials is manifest as visible light. The most popular of these light-emitting diodes employ gallium arsenide or gallium phosphide as the semiconductor material, and produce a red glow when forward biased at about 1.4 V and 5 to 20 mA.

Seven-segment LED displays are common in calculators, and various numerical-readout instruments. Figure 16-4(a) shows the most popular pinout for seven-segment displays. The schematic equivalent for a common-anode type is given in Fig. 16-4(b). Common-cathode displays are equally popular. The series resistors shown are required to limit the LED current to a safe value.

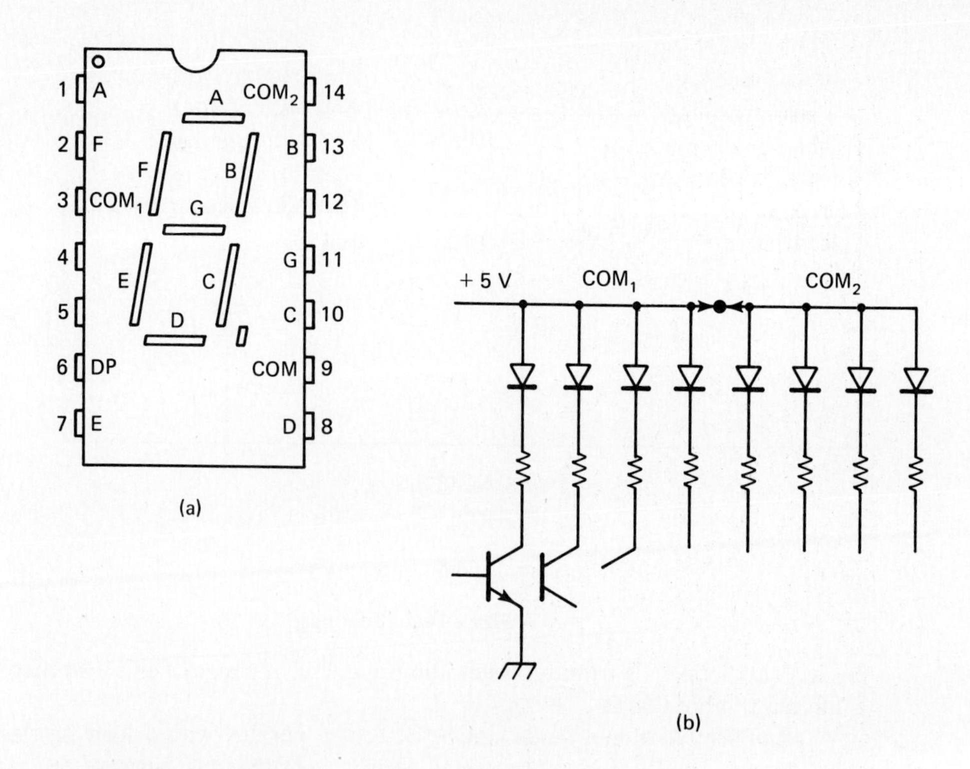

Figure 16-4 (a) Segment identification and most common pinout for LED displays; (b) representation of seven common-anode segments plus decimal, and switching method.

Liquid-crystal displays are preferred over seven-segment LED displays in many applications because they require current in the range of 100 μA per digit, rather than the 100-mA range of LED displays. They are activated by ac voltages in the range 4 to 40 $V_{p\text{-}p}$ (dc will work, but it shortens the life of the display). Liquid crystals do not generate light, but change their light-reflective properties when activated.

16.4 PHOTOSENSITIVE DEVICES

Photoconductive cells made of cadmium sulfide (CdS) or cadmium selenide (CdSe) are popular because of their low cost and ability to handle relatively large voltages and currents (250 V and 350 mW are typical limits). They have extremely slow response times—on the order of 100 ms. This slowness can be used to advantage. In control systems where most other pickup and logic devices are vulnerable to microsecond noise spikes, photoconductive cells will respond only to true signals.

Photodiodes conduct reverse currents from microamperes to a few tenths of a milliampere in response to stimulation by light. They have response times in the nanosecond range. Phototransistors and photo FETs have response times on the order of 1 μs.

Solar cells are large-area silicon *P-N* junctions which produce direct current when struck by light. Conversion efficiency in sunlight is about 10%. Under bright sunlight they can yield a power of about 50 W per square meter of exposed area. However, covering such an area requires an expenditure of several thousand dollars under presently prevailing prices. Output voltages are typically 0.4 V per cell. Response times are on the order of 5 μs.

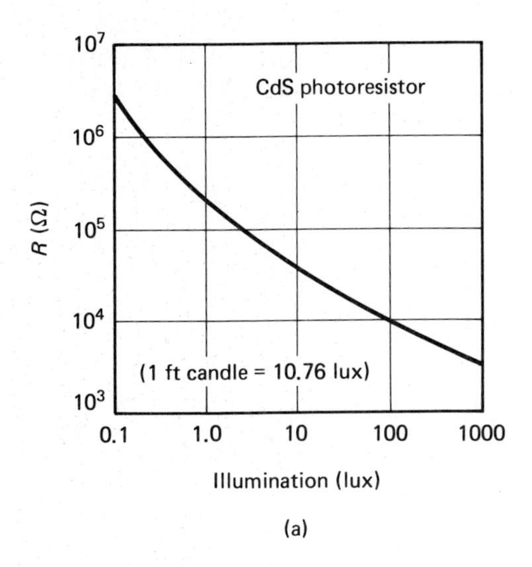

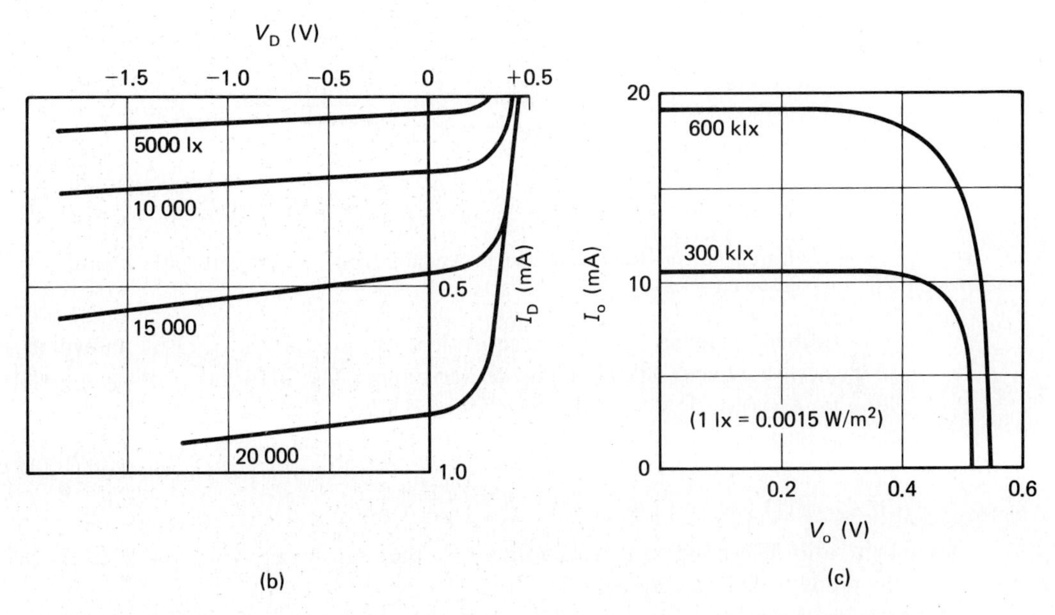

Figure 16-5 Characteristics of three popular photodetectors: (a) photoresistor; (b) photodiode; (c) solar cell.

Figure 16-5 shows characteristics for a typical photoresistor, photodiode, and solar cell.

16.5 THE UNIJUNCTION TRANSISTOR

The unijunction transistor (UJT) is a switching device used primarily as an oscillator, timer, and trigger. The name *transistor* and the symbol, so similar to the junction-FET symbol, are misleading because the UJT cannot be employed as an amplifier and has few similarities to the other "transistors" from a user's point of view.

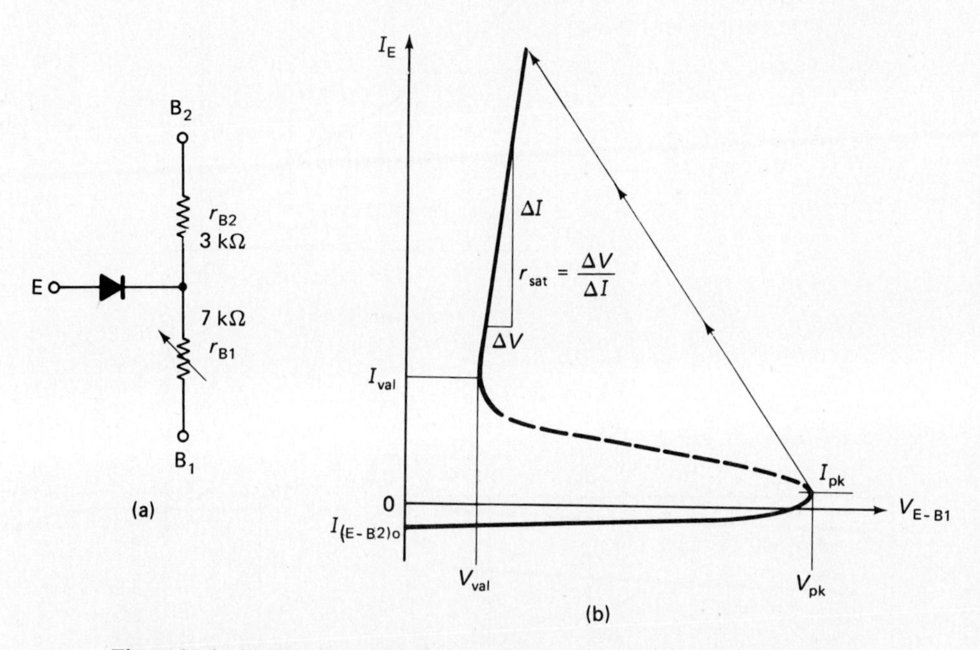

Figure 16-6 (a) Representation of a unijunction transistor; (b) UJT characteristic curves.

Intrinsic standoff ratio: An equivalent circuit for the UJT is shown in Fig. 16-6(a), with typical values of device resistance. The high value of r_{B1} prevails as long as V_E is too low to make the diode conduct. The voltage-division ratio of r_{B2} and r_{B1} is called the intrinsic standoff ratio η(eta) of the device.

$$\eta = \frac{r_{B2}}{r_{B1} + r_{B2}} \tag{16-1}$$

This ratio determines the peak voltage V_P that can be applied from E to B_1 before the diode begins to conduct.

$$V_P = V_D + \eta V_{B2-B1} \tag{16-2}$$

The instant that V_E reaches the critical value the diode injects charge carriers into r_{B1}, switching it to the low value. The high-resistance state is restored when the current through the diode drops below a minimum or valley level, I_V.

The firing voltage V_P is a fixed fraction of V_{B2-B1} for any given device. Thus we can vary the firing voltage by varying V_{B2}. The characteristic curve for a UJT at one value of V_{B2-B1} is given in Fig. 16-6(b). The graph is not to scale.

The UJT oscillator circuit of Fig. 16-7 forms the basis of most UJT applications and illustrates the device in action. The operation of the circuit is as follows:

1. R_t charges C_t toward V_{CC} following the $\tau = RC$ curve. I_E is negligible at this time.
2. When V_C (same as V_E) reaches the critical value V_P, the UJT fires and C_t discharges through E-B_1 and R_1. The discharge is many times faster than the charge because R_t is large and $r_{B1\,(on)}$ and R_1 are low.
3. When C_t discharges to the point where I_E drops below I_V, r_{B1} returns to its high-resistance value and the cycle starts over.

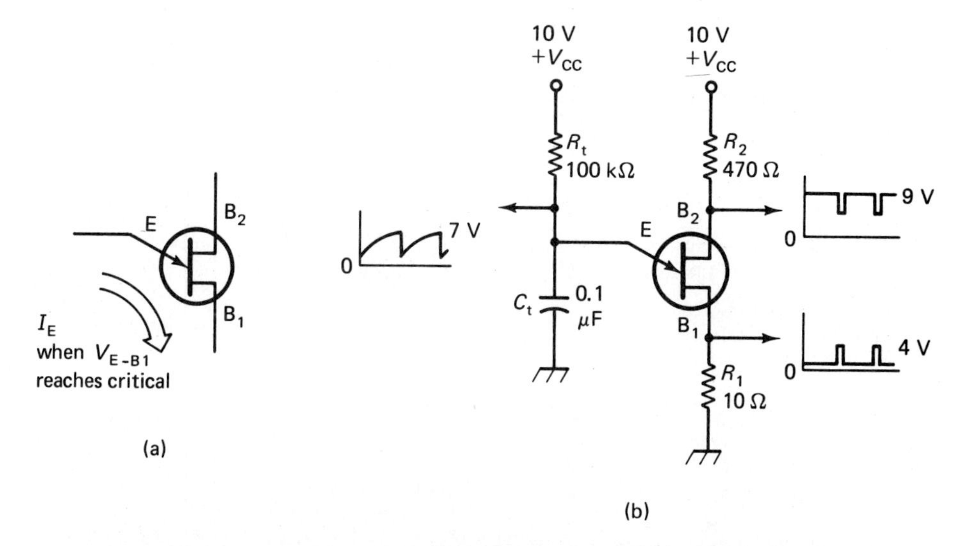

Figure 16-7 (a) Schematic symbol, lead identification, and main current path for the unijunction transistor; (b) basic UJT oscillator with representative waveforms at the three terminals.

Output signals may be taken from any of the three leads of the UJT as shown in Fig. 16-7. The emitter waveshape is roughly a sawtooth with an amplitude in the vicinity of $0.7V_{CC}$. The load on this output must have a resistance much higher than R_t or C_t will never reach the firing voltage. R_1 may be omitted if the positive-pulse B_1 output is not required. $V_{E1\,(pk)}$ is determined by the voltage division of V_P across $r_{B1\,(on)}$ and R_1.

The negative pulses at B_2 are caused by the increased current in R_2 when r_{B1} assumes its low value. R_2 is seldom omitted even if the V_{B2} output is not required, since it has two other functions:

1. R_2 prevents a V_{CC}-to-ground short in the event of a semiconductor failure.
2. R_2 can compensate for oscillator frequency drift with temperature change if chosen by the equation

$$R_2 = \frac{0.7 r_{BB}}{V_{CC}} + \frac{(1 - \eta)R_1}{\eta} \tag{16-3}$$

Limitations on the range of R_t are imposed by the peak and valley currents, I_P and I_V, shown in Fig. 16-6(b). If R_t is too high, it will never be able to deliver the required firing current I_P. If it is too low, it will continue to deliver the hold-on current I_V even after C_t has discharged. We may calculate the maximum and minimum values of R_t from the device specifications.

$$R_{t(max)} = \frac{V_{CC} - V_{P(max)}}{I_{P(max)}} \tag{16-4}$$

$$R_{t(min)} = \frac{V_{CC} - V_{V(min)}}{I_{V(min)}} \tag{16-5}$$

Of course, it is a good idea to include a safety factor by keeping R_t between half the maximum and twice the minimum value.

Example 16-1

Design a UJT oscillator to give positive pulse outputs at a 2-Hz rate. Use the following specifications: $V_{CC} = 12$ V, $\eta_{max} = 0.75$, $r_{B1} = 7$ kΩ, $r_{B2} = 3$ kΩ, $I_{P(max)} = 2\ \mu$A. Use the smallest possible value of C_t, and let $R_1 = 20\ \Omega$.

Solution

$$R_2 = \frac{0.7 r_{BB}}{V_{CC}} + \frac{(1 - \eta)R_1}{\eta} \tag{16-3}$$

$$= \frac{(0.7)(10\,000)}{12} + \frac{(1 - 0.75)(20)}{0.75} = \mathbf{590\ \Omega}$$

$$V_{B2-B1} = V_{CC}\frac{r_{BB}}{R_2 + R_{BB}} = (12)\frac{10\,000}{590 + 10\,000}$$

$$= 11.3 \text{ V}$$

$$V_{P(max)} = V_D + \eta_{max} V_{B2-B1} \tag{16-2}$$

$$= 0.6 + (0.75)(11.3) = 9.1 \text{ V}$$

$$R_{t(max)} = \frac{V_{CC} - V_{P(max)}}{I_{P(max)}} \tag{16-4}$$

$$= \frac{12 - 9.1}{2\ \mu\text{A}} = = 1.45 \text{ M}\Omega$$

For safety, let $R_t =$ **680 kΩ.**

$$\frac{9.1 \text{ V}}{12 \text{ V}} = 76\%$$

From the time-constant chart in Appendix C, 76% is reached in 1.4τ.

$$1.4\tau = \frac{1}{f} = \frac{1}{2} = 0.5 \text{ s}$$

$$\tau = 0.36 \text{ s}$$

$$C = \frac{\tau}{R} = \frac{0.36 \text{ s}}{680 \text{ k}\Omega} = \mathbf{0.53\ \mu F}$$

16.6 THE SILICON-CONTROLLED RECTIFIER

The SCR is a rectifier that will not conduct even in the forward direction until a current pulse of the required amplitude is applied to a third electrode called the gate. The advantage of this controlled turn-on is illustrated in Fig. 16-8. The total power delivered to the load is variable from $\frac{1}{2}(V_s)^2/R_L$ to zero, as indicated by the shaded areas under the sine curves. Of course, the triggering point does not normally shift over such a range in only five cycles—the sketch merely indicates the range of average load power that is possible by delaying turn-on of the SCR.

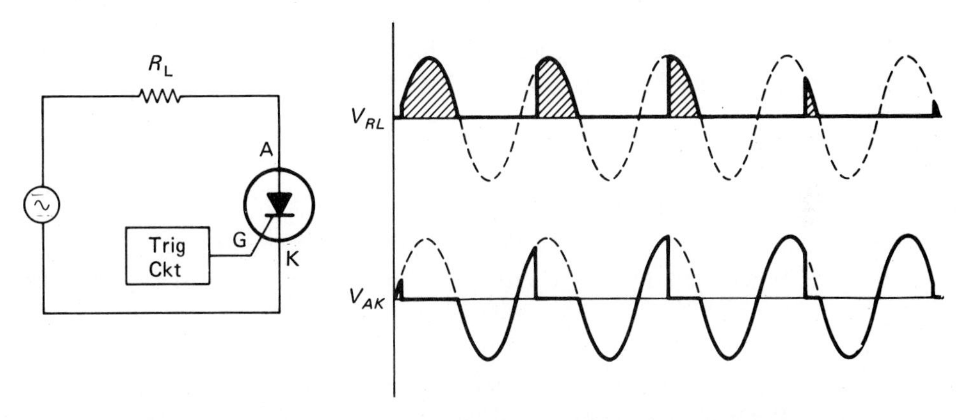

Figure 16-8 Decreasing average load power as SCR firing-delay angle progresses from 0° to 180°. Real-life progressing generally takes hundreds of cycles.

SCR characteristics: A typical medium-power SCR might be rated at 300 V, 2 A, anode-to-cathode, and might require +5 mA at 0.8 V from gate to cathode to turn on. Smaller and much larger types are available, of course.

Once an SCR has been turned on by a positive gate pulse, it continues to conduct (A to K) regardless of whether or not the gate pulse is removed or even made negative. The only way to turn it off is to remove the forward voltage V_{AK}. On an ac line this happens automatically at the end of every positive half-cycle.

SCR phase control: Many different methods are used to generate a positive gate-trigger pulse at the desired instant of the ac line cycle. A popular and very stable method is shown in Fig. 16-9. The operation of the circuit will be explained starting from the beginning of a positive half-cycle of the ac line.

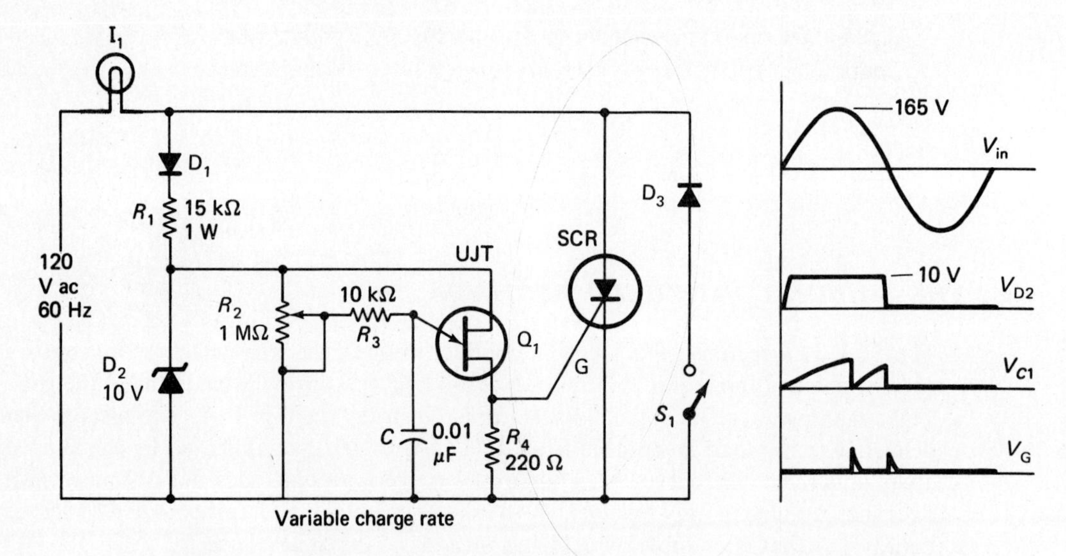

Figure 16-9 Highly stable SCR trigger using a UJT and variable charge time on C.

1. The voltage across zener diode D_2 reaches 10 V at about 8° into the cycle. D_2 holds the UJT supply at +10 V as the line voltage rises.
2. C charges toward +10 V through R_2 and R_3. The range of charge times, if the UJT fires after one time constant, is τ, where

$$\tau_{\text{min}} = R_3C = (10\ \text{k}\Omega)(0.01\ \mu\text{F}) = 0.1\ \text{ms}$$

$$\tau_{\text{max}} = (R_3 + R_2)C = (1\ \text{M}\Omega)(0.01\ \mu\text{F}) = 10\ \text{ms}$$

3. When the UJT fires, C discharges through the UJT E to B_1 and thence through the SCR G to K. This pulse turns the SCR on. Under τ_{min}, turn-on is at about 10° into the cycle, so essentially one-half of the line cycle is passed. Under τ_{max} the positive half-cycle ends (at 8.3 ms) before turn on, so none of the line is passed.
4. When the SCR fires V_{AK} drops to about 1 V, leaving the remaining line voltage across lamp I_1. The UJT supply voltage is thus removed until the start of the next positive half-cycle.
5. After 180° (8.3 ms) the line swings negative and the SCR turns off. D_1 prevents negative current from getting to the UJT and limits the dissipation of R_1.

With S_1 open R_2 varies the conduction from 0° to about 170° of the positive half cycle. Closing S_1 permits D_3 to conduct for the 180° of the negative half-cycle, extending the total range to 0 to 350° of the 360° cycle.

SCR crowbar: An overvoltage-protection circuit utilizing the "snap on and hold on" characteristic of the SCR is shown in Fig. 16-10. If the supply regulator should short, excessive voltage could be passed to the entire system causing untold damage. However, D_2 turns on SCR for any V_O greater than 5.6 V. SCR then shorts (crowbars the output), diverting excess current and blowing fuse F.

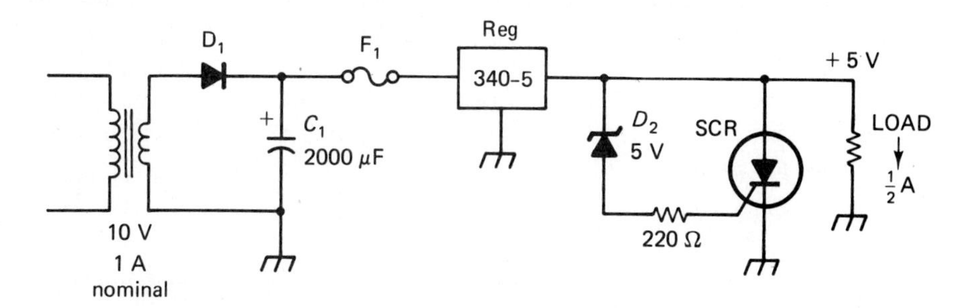

Figure 16-10 Crowbar overvoltage protection. Regulator failure turns on SCR, blowing fuse F_1.

16.7 TRIACS AND DIACS

As its name implies, a triac is a three-lead device capable of controlling ac. It functions the same way as the SCR, except that either positive or negative turn-on pulses can be applied from the gate to the "cathode"—which is now called *main-terminal 1.* Also the "anode"—now called *main-terminal 2*—passes either positive or negative current to MT_1 once fired. As with the SCR, the gate pulse turns the triac on, but the only way to turn it off is to bring "V_{AK}" (now $V_{MT2-MT1}$) to zero. Of course this happens twice each cycle on the ac line. The triac is at the heart of most commercial lamp dimmers, motor speed controllers, variable heat controls, and similar devices.

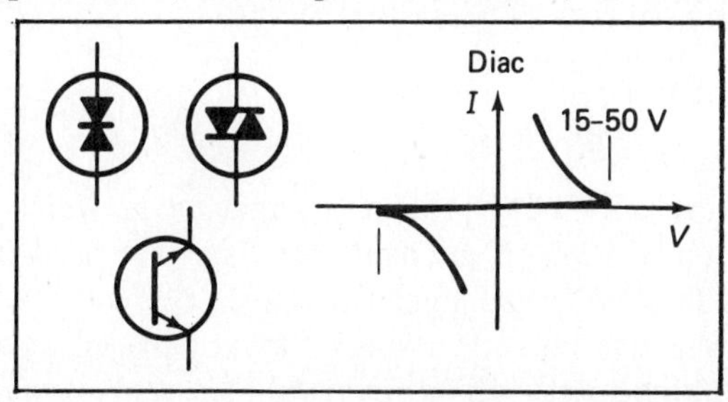

Figure 16-11 Diac characteristics and three schematic symbols currently fighting to represent it.

The **diac** is a two-terminal device used primarily in triac-trigger circuits. It conducts no current until the threshold voltage (equal in each direction) is reached. Thresholds in the range 15 to 50 V are commonly available. Once the threshold is reached the device fires and its voltage drops significantly, as illustrated in Fig. 16-11.

A **triac lamp dimmer** is shown in Fig. 16-12. C_1 is charged through $R_1 + R_2$, but V_{C1} lags the voltage applied by the line. R_3 charges C_2, but again there is a delay in the buildup of V_{C2}. These delays cannot properly be analyzed as phase shifts because the applied voltage (top of R_1) is, in general not a sine wave. However, increasing R_2 decreases the amplitude while increasing the delay of V_{C2}. The threshold of D_1 is reached near the beginning of each half-cycle (positive and negative) with R_2 at zero, resulting in nearly full conduction through I_1 and Q_1. With R_2 at maximum V_{C2} is delayed more and the line polarity reverses before threshold is reached.

L_1 and C_3 are RFI (radio-frequency interference) filters. The abrupt pulsing of line current as the triac switches on causes trouble with AM radios if these components are omitted.

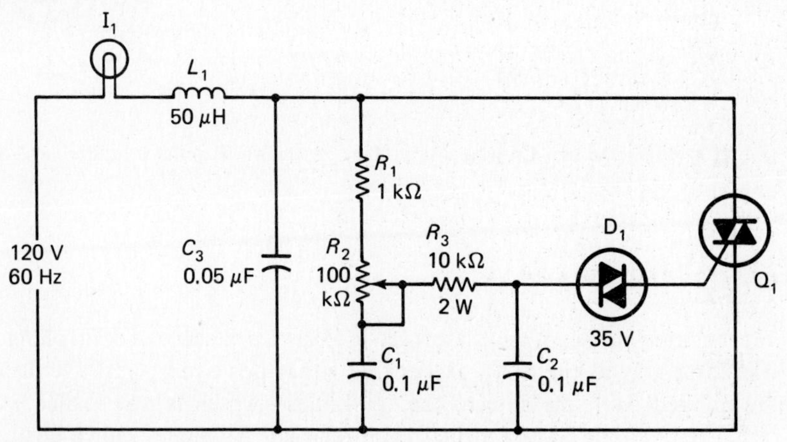

Figure 16-12 Triac lamp dimmer. C_1 and C_2 delay buildup of voltage until after the start of each half-cycle. Diac fires when V_{C2} reaches 35 V to trigger triac.

CHAPTER SUMMARY

1. Varactor diodes (varicaps or VVCs) present a capacitance which varies with reverse dc voltage. Values up to 1000 pF are available. They are used for electronic tuning and as frequency multipliers in the gigahertz range.

2. Tunnel diodes exhibit negative resistance when forward biased at about 0.3 V. They are used as low-level, high-speed oscillators, amplifiers, and switches.

3. LEDs emit a red glow when forward biased at about 1.4 V and 10 mA. They are common in seven-segment numerical displays. Liquid-crystal displays require about 1/1000 the current of LEDs.

4. Photocells (CdS or CdSe) are able to conduct large currents when illuminated but they are very slow. Photodiodes are much faster but conduct only small currents. Silicon solar cells are photovoltaic—they produce about 0.4 V per cell under bright light.

5. The unijunction transistor is used as a relaxation oscillator and as a timer. It becomes conductive from E to B_1 once the emitter voltage $V_{E\text{-}B1}$ reaches a critical level. An RC charging network feeding the emitter determines the time required to reach the firing voltage.

6. The SCR conducts conventional current from anode (A) to cathode (K) only after being triggered by a relatively smaller current pulse from gate (G) to cathode. Once triggered it remains on until V_{AK} drops back to zero. With a suitable trigger circuit (often using a UJT) the SCR can provide continuously variable power to a load by switching on for only a portion of each ac cycle.

7. The triac is similar to the SCR except that it can be triggered to conduct in either direction, so the full ac cycle can be passed. It is often triggered by a diac, which is a bidirectional breakdown device.

QUESTIONS AND PROBLEMS

16-1. Are varactor diodes normally forward or reverse biased? Do higher bias voltages increase or decrease capacitance?

16-2. In the circuit of Fig. 16-1 is higher frequency obtained with the potentiometer wiper to the left or to the right? Explain.

16-3. In the tunnel-diode oscillator of Fig. 16-3(b), could the amplitude of the signal be increased by changing the battery from 1.5 V to 6 V? Explain.

16-4. In Fig. 16-4(b) what should be the value of each resistor to limit LED current to 15 mA per segment? Assume that $V_{CE(sat)} = 0.1$ V for each transistor.

16-5. What is the advantage of liquid-crystal over LED displays?

16-6. Of the three photodevices whose curves are given in Fig. 16-5, which has the fastest response time? Which is the slowest?

16-7. A silicon UJT has $V_P = 4$ V when V_{B1-B2} is 5.5 V. Find η.

16-8. Find the output frequency of the UJT oscillator of Fig. 16-7 if $\eta = 0.75$.

16-9. In Problem 16-8, the UJT has $V_{V(min)} = 0.8$ V and $I_{V(min)} = 0.5$ mA. What is the highest frequency that can be generated using the given value of C_t?

16-10. How is an SCR turned on? How is it turned off? Can it be placed in a partially conductive state like a transistor?

16-11. An SCR is in series with a 120-V ac line and a 50-Ω load. The SCR is triggered on at the 90° point of each positive half-cycle. What is the power delivered to the load?

16-12. In Fig. 16-9, when is the power dissipation in R_1 maximum; lamp bright or lamp dim? Calculate $P_{R1(max)}$.

16-13. What is the advantage of the triac over the SCR?

16-14. Why is a UJT suitable for triggering an SCR but unsuitable for triggering a triac? What characteristic of a diac makes it suitable as a triac trigger?

17

CIRCUIT EXAMPLES

This chapter contains a number of diagrams for complete electronic instruments together with detailed analyses of the circuits contained therein. Most of the circuits have been taken from commercial products and some are intended as useful construction projects for the student. All of the circuits have been chosen to illustrate real-life applications of the concepts and analysis techniques presented in the earlier chapters.

It must be emphasized that this chapter is not light reading. Allow plenty of time for each instrument and study the circuits one section at a time. See if you can duplicate the analyses on your own, looking only at the schematic.

17.1 HIGH-VOLTAGE DC POWER SUPPLY (FIG. 17-1)

This supply was designed for use with a small radio transmitter. It contains three separate rectifier and filter circuits, the first being a full-wave voltage doubler consisting of D_1 through D_4, C_1, and C_2. The second supply is a half-wave voltage doubler with a pi-section filter, consisting of D_5, D_6, C_4, C_5, and L_1. A switch selects transformer secondary voltages of 125 or 95 V, allowing a choice of dc output voltages. The third supply is a low-current negative-voltage supply using a half-wave rectifier and pi-section filter with a resistor in place of the filter choke in the interest of economy.

We will assume that we wish to duplicate this supply, meeting the specifications given. The main problem will be to determine the specifications of the transformer.

High-voltage section: The no-load and full-load dc output voltages are given, so the winding-resistance requirements of the transformer will be determined.

$$V_{s(NL)pk} = \frac{V_{O(NL)DC} + 4V_D}{2} = \frac{820 + 2.4}{2} = 411 \text{ V}$$

$$V_{s(NL)rms} = (411)(0.707) = 291 \text{ V}$$

This is slightly greater than the 282-V measurement shown at no-load on the diagram, partly because it does not include the drop due to the small current in bleeder resistors

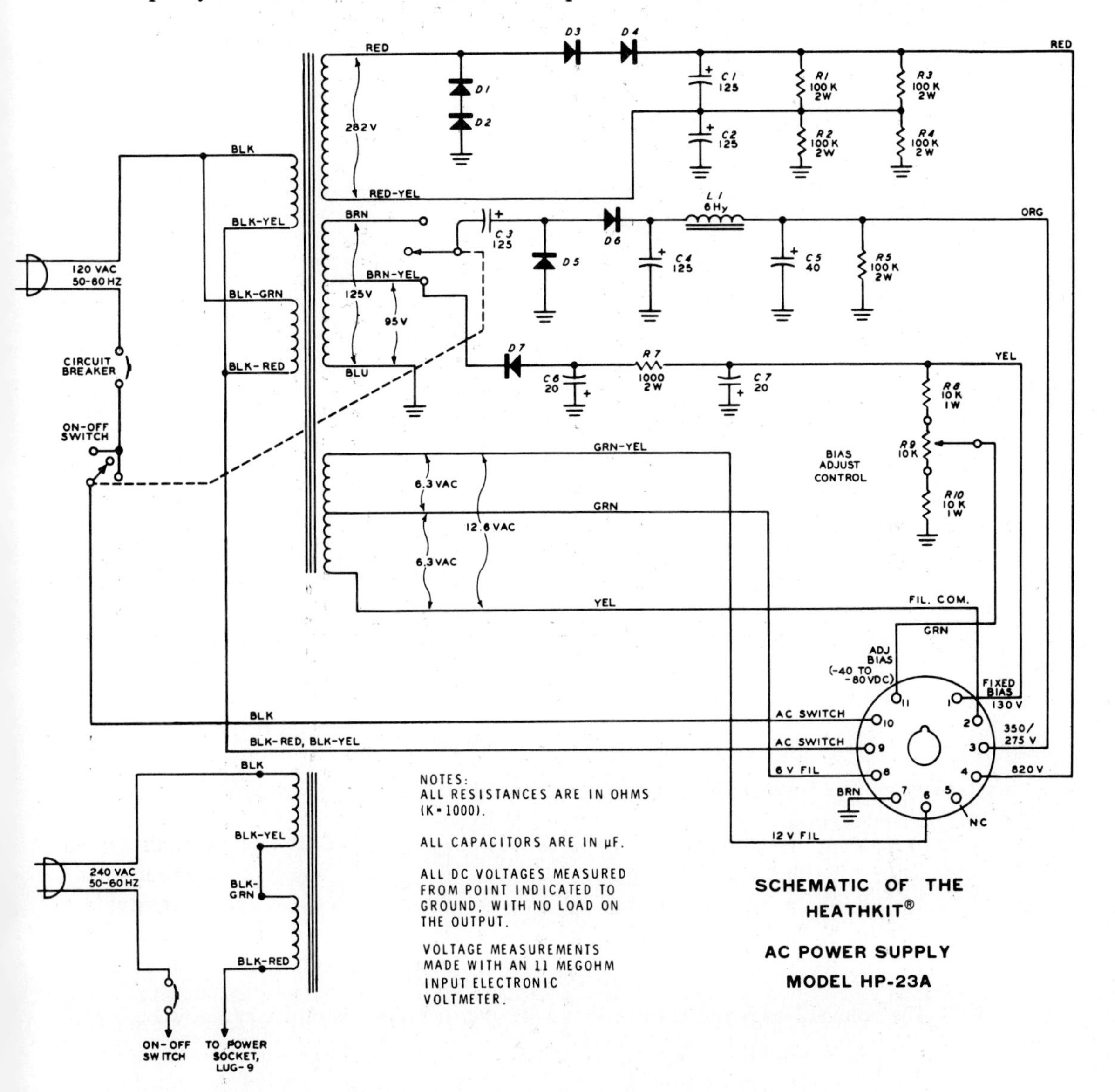

Figure 17-1 (a) Schematic diagram and (b) specifications of a power supply for a medium-power radio transmitter. (Courtesy Heath Co., Benton Harbor, Mich.)

SPECIFICATIONS

HIGH VOLTAGE SUPPLY

Output Voltage........................	820 volts DC, no load. 700 volts DC at 250 mA.
Effective Output Capacitance.............	62.5 μF.
Ripple............................	Less than 1% at 250 mA.
Duty Cycle..........................	Continuous up to 150 mA. 50% at 300 mA.

LOW VOLTAGE SUPPLY (high tap)

Output Voltage........................	350 volts DC, no load. 300 volts DC at 150 mA (with 100 mA load on High Voltage).
Ripple............................	Less than .05% at 150 mA.
Duty Cycle..........................	Continuous up to 175 mA.

LOW VOLTAGE SUPPLY (low tap)

Output Voltage........................	275 volts DC, no load. 250 volts DC, at 100 mA (with 100 mA load on High Voltage).
Ripple............................	Less than .05% at 150 mA.
Duty Cycle..........................	Continuous up to 175 mA.

OTHER OUTPUT VOLTAGES

Fixed Bias..........................	-130 volts DC, no load. -100 volts DC, at 20 mA.
Adjustable Bias......................	-80 to -40 volts DC, at 1 mA maximum.
Filaments...........................	6.3 volts AC at 11 amperes. 12.6 volts AC at 5.5 amperes.

GENERAL

Power Requirements..................	120/240 volts AC, 50/60 Hz, 350 watts maximum.
Dimensions..........................	9" long x 4-3/4" wide x 6-3/4" high.
Net Weight..........................	16 lbs.

Figure 17-1 (Continued)

R_1 to R_4. The peak-to-peak ripple and peak voltage out under load are next determined from the specifications.

$$V_{o(rip)p\text{-}p} = (V_{O(DC)AV})(\text{ripple factor})(2.828) \tag{3-3}$$
$$= (700)(1\%)(2.828) = 20 \text{ V}$$
$$V_{O(DC)PK} = V_{O(DC)AV} + \tfrac{1}{2}V_{o(rip)p\text{-}p}$$
$$= 700 + \tfrac{1}{2}(20) = 710 \text{ V}$$

The voltage loss permitted in the transformer-winding resistance can now be found.

$$V_{O(DC)PK} = 2(V_{s(NL)pk} - V_{r(w)} - 2V_D)$$
$$V_{r(w)} = V_{s(NL)pk} - 2V_D - \tfrac{1}{2}V_{O(DC)PK}$$
$$= 411 - 2 - \tfrac{1}{2}(710) = 54 \text{ V}$$

The transformer current pulses are estimated as $4I_L$ for a full-wave rectifier. Bleeder resistors R_1 to R_4 draw 7 mA in addition to the 250-mA I_L.

$$r_w = \frac{V_{r(w)}}{4I_O} = \frac{54}{4(0.257)} = 53\ \Omega$$

The true transformer turns ratio is found.

$$n = \frac{V_{s(NL)}}{V_p} = \frac{291}{120} = 2.42$$

The actual winding resistances would most efficiently be apportioned half to the secondary and half reflected from the primary.

$$r_s = \tfrac{1}{2} r_w \approx \mathbf{27\ \Omega}$$

$$r_{p1} = \frac{r_{p(refl)}}{n^2} = \frac{26}{2.42^2} = \mathbf{4.4\ \Omega}$$

Additional primary current caused by the low-voltage supplies will lower the required r_{p1} slightly.

The filter-capacitor requirement can be checked.

$$C = \frac{I_O t}{V_{rip(p\text{-}p)}} = \frac{(257\text{ mA})(8.3\text{ ms})}{20\text{ V}} = 107\ \mu\text{F}$$

This formula is a bit conservative because the capacitors are not discharging for the *entire* 8.3-ms half-cycle, but increasing C_1 and C_2 to 220 μF each (110 μF total) would increase our confidence of meeting the 1% ripple specification.

At no-load, each bleeder resistor R_1 to R_4 sustains $\frac{1}{2}V_O$ or 410 V. Their power ratings must each exceed

$$P_R = \frac{V^2}{R} = \frac{410^2}{100\,000} = 1.7\text{ W}$$

At no-load, D_3 and D_4 in series sustain a reverse voltage of $V_{s(pk)} + V_{C1}$, which is about $2V_{s(pk)}$ or 411 V. Similarly, D_1 and D_2 sustain $V_{s(pk)} + V_{C2}$. A safety factor of at least 2 should be allowed, so *each* diode should be rated above 411 V. 600-V, 1-A diodes are inexpensive and widely available.

The low-voltage supply is a half-wave voltage doubler with a pi-section choke filter. We will determine values for r_{s2} and r_{ch} and confirm the ripple spec at high-output tap.

$$V_{s(NL)pk} = \frac{V_{O(NL)DC} + 2V_D}{2} = \frac{350 + 1.2}{2} = 176\text{ V}$$

$$V_{s(NL)rms} = (176)(0.707) = 124\text{ V}$$

This agrees well with the 125-V measurement given on the schematic. Bleeder R_5 takes 3.5 mA in addition to the 150-mA I_L.

$$V_{C4(rip)p\text{-}p} = \frac{It}{C} = \frac{(153.5\text{ mA})(16.7\text{ ms})}{125\ \mu\text{F}} = 20\text{ V}$$

$$V_{O(FL)DC} = 2V_{s(NL)pk} - V_{r(w2)} - 2V_D - \tfrac{1}{2}V_{C4(rip)p\text{-}p} - V_{r(ch)}$$

$$V_{r(w2)} + V_{r(ch)} = 2V_{s(NL)pk} - 2V_D - \tfrac{1}{2}V_{C4(rip)p\text{-}p} - V_{O(FL)DC}$$
$$= 2(176) - 2 - \tfrac{1}{2}(20) - 300 = 40 \text{ V}$$

We will presume to apportion this drop 30 V to r_{w2} and 10 V to r_{ch}, since r_{w2} carries charging pulses estimated as $7I_O$.

$$r_{w2} = \frac{V_{r(w2)}}{7I_O} = \frac{30}{7(0.1535)} = 28\ \Omega$$

The actual secondary resistance should be half of this. The other half should be reflected from additional primary resistance in parallel with the 4.4 Ω found previously.

$$r_{s2} = \tfrac{1}{2}r_{w2} = 14\ \Omega$$
$$r_{p2} = \frac{r_{p(refl)}}{n^2} = \frac{14}{(124/120)^2} = 13\ \Omega$$
$$r_{p(t)} = r_{p1} \,||\, r_{p2} = 4\,4 \,||\, 13 = \mathbf{3.3\ \Omega}$$

Since there are two primaries in parallel in the 120-V connection, each winding should have a resistance of 6.6 Ω.

$$V_{o(rip)p\text{-}p} = V_{C4(rip)p\text{-}p}\frac{X_{C5}}{X_{L1}}$$
$$= 20\frac{1}{(2\pi \times 60)^2(40\ \mu F)(6)} = 0.59 \text{ V}$$
$$\text{ripple factor} = \frac{V_{O(rip)p\text{-}p}}{V_{O(FL)DC}(2.828)} = \frac{0.59}{(300)(2.828)} = 0.07\ \%$$

This formula is quite conservative, since it treats the sawtooth ripple at C_4 as a sine wave. The 0.05% specification is probably not unreasonable.

Analysis of the *LV* supply at low tap is left to the student.

Total power consumed is approximately $V_{O(NL)}I_O$ for each of the three supplies, plus filament power.

$$P_T = (820)(0.257) + (350)(0.153) + (130)(0.02) + (6.3)(11)$$
$$= 336 \text{ W}$$

If the line current were sinusoidal, its rms value would be

$$I = \frac{P}{V} = \frac{336}{120} = 2.8 \text{ A}$$

However, it is sure to contain pulses at the voltage peaks when the first filter capacitors charge, making its rms value considerably higher (maybe two or three times higher). Therefore, one cannot simply multiply 2.8 A by a safety factor (say 150%) to obtain the required circuit-breaker rating. Analysis would be quite difficult, so the recommended procedure is to measure I_{line} at full load with a true-rms ammeter, and then apply the 50% overcurrent factor.

17-2 VARIABLE REGULATED BENCH SUPPLY (FIG. 17-2)

This variable transistor regulated supply is simple enough to be constructed by the average student, yet its performance rivals that of many commercial supplies. It will deliver a maximum of 20 V at 1 A with a ripple less than 3 mV rms. Electronic current limiting at 10 mA, 100 mA, or 1 A can be selected to protect both the supply and the circuit powered by it in the event of a short circuit.

In constructing the supply, it is essential that the transformer center tap and the positive side of C_1 be connected directly together by a wire with no other connections intervening. The remaining "common" connections can then be made from the positive side of C_1. The last paragraph of Section 3.16 explains the reason for this. It is also essential to heat-sink Q_3 to the instrument cabinet through an electrically insulative washer or to a separate finned heat sink.

Transformer and main rectifier: D_1, D_2, and C_1 form the main rectifier and filter. The output voltage capability at the ripple valley must be 20 V even if $I_O = 1$ A through the 0.5-Ω resistor R_8, $V_{CE(sat)}$ of Q_4 = its maximum of 1.1 V, and C_1 is at its worst case of −20%.

$$V_{rip(C1)p\text{-}p} \frac{I_O t}{C_1} = \frac{1\text{A} \times 8.3\text{ ms}}{1600\ \mu\text{F}} = 5.2\text{ V}$$

$$\begin{aligned} V_{s(FL)pk} &= V_O + V_{R8} + V_{CE(sat)} + V_{rip(p\text{-}p)} + V_D \\ &= 20 + 0.5 + 1.1 + 5.2 + 1.0 = 27.8\text{ V} \end{aligned}$$

The transformer must be capable of delivering this full-load voltage at each side of the center tap at minimum line voltage, which we will specify as 105 V ac. Transformer secondary current for a full-wave rectifier is about $4I_L$, or 4 A.

$$\frac{1.414\,V_{s(NL)rms}}{2} - 4I_L r_w = V_{s(FL)pk}$$

$$0.707 V_{s(NL)rms} - 4r_w = 27.8$$

$$r_w = \frac{0.707 V_{s(NL)rms} - 27.8}{4}$$

Obviously, a variety of $V_{s(NL)}$ and r_w combinations are possible. Three reasonable ones are given below. Note that r_w is the effective winding resistance of one-half of the secondary plus reflected primary resistance.

$V_{s(NL)rms}$	r_w
42 V	0.5 Ω
48 V	1.5 Ω
56 V	2.9 Ω

At normal 120-V line, each of these voltages will be 120/105 or 1.14 times the listed value. The low-voltage low-resistance transformer will be larger, heavier, and more expensive. The high-voltage high-resistance transformer, though smaller and less expensive, will waste power and tend to overheat.

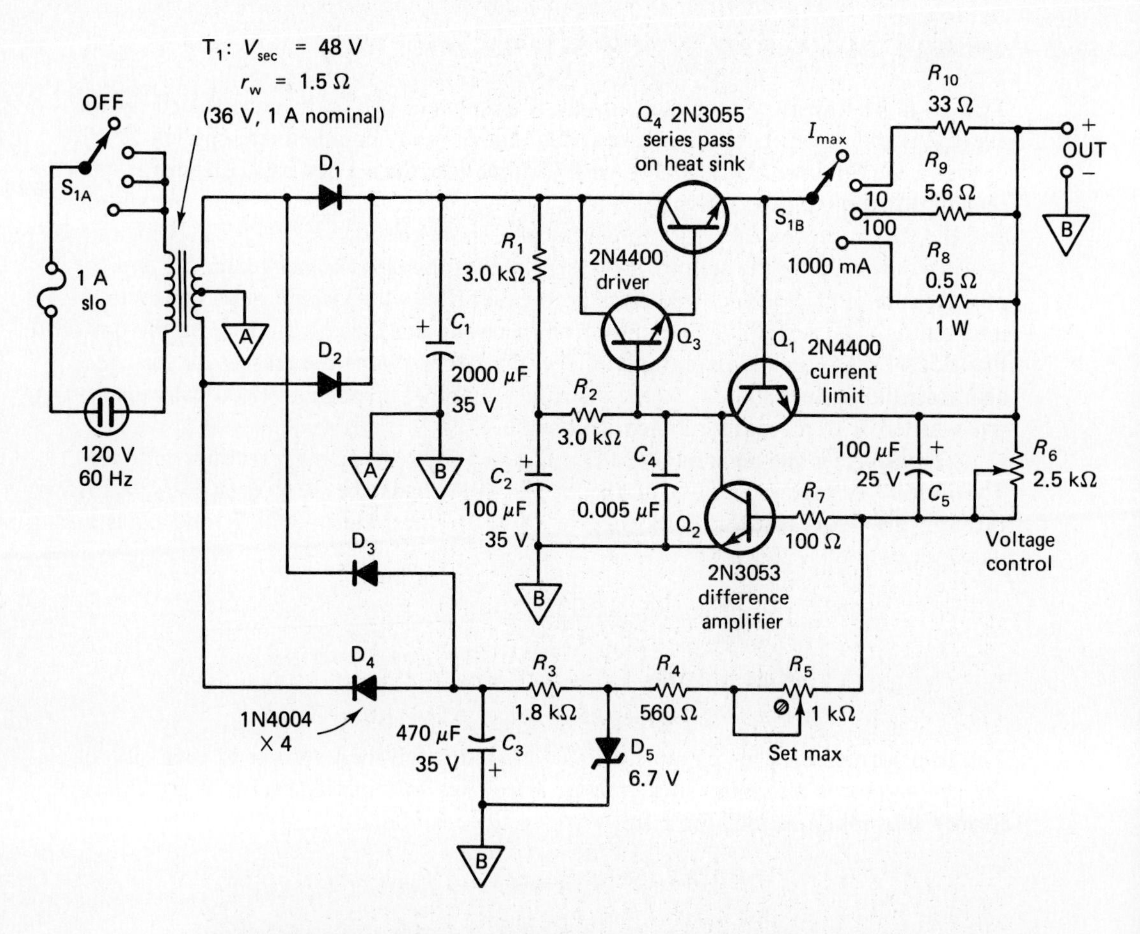

Figure 17-2 A 0- to 20-V, 1-A regulated power supply suitable for student construction. Adjustable current limiting is featured.

Amplifier and pass transistor: R_1 and R_2 feed the base of Darlington pair Q$_3$–Q$_4$. Assuming that $\beta_{\min} = 50$ each, the current required is

$$I_{3B} = \frac{I_1}{\beta_3\beta_4} = \frac{1\text{ A}}{(50)(50)} = 0.4\text{ mA}$$

C_2 filters the driving voltage to the average between the peak and valley of V_{C1} (26.8 and 21.6 V, respectively), which is 24.2 V. The base-voltage requirement is $V_O + V_{R8} + V_{4BE} + V_{3BE}$, or 21.7 V. Thus a voltage difference of 24.2 V − 21.7 V or 2.5 V across $R_1 + R_2$ must supply the 0.4-mA base drive.

$$R_1 + R_2 = \frac{V_{C1(\text{avg})} - V_{3B}}{I_{3B}} = \frac{2.5\text{ V}}{0.4\text{ mA}} = 6.25\text{ k}\Omega$$

C_2 is chosen to have a reactance several hundred times lower than R_1 to keep ripple to a minimum.

$$C_2 = \frac{1}{2\pi fX} = \frac{1}{2\pi(120)(15)} = 88\ \mu\text{F}$$

C_4 presents a low reactance at high frequencies and prevents spurious oscillations.

R_6 balances the negative current from R_3 and R_4 with a positive current from V_{O1} and in addition supplies base current to Q_2. Base voltage for Q_3, supplied by R_1 and R_2, is limited when Q_2 turns on. At high V_{line} and low I_L, V_{C1} may be 50 V or more (depending on the transformer selected). Taking $V_O = 10$ V, so R_6 is set to one-half of maximum resistance:

$$I_{R1,2} = \frac{V_{C1} - V_{3B}}{R_1 + R_2} = \frac{50 - 11.2}{6.0\ \text{k}\Omega} = 6.5\ \text{mA}$$

$$I_{2B} = \frac{I_{R1,2}}{\beta_2} = \frac{6.5}{50} = 0.13\ \text{mA}$$

R_6 must supply this base current with a minimum of extra voltage drop.

$$R_{6(1/2)} = \frac{\Delta V}{I_{2B}} = \frac{200\ \text{mV}}{0.13\ \text{mA}} = 1.5\ \text{k}\Omega$$

The full value of R_6 should be 3 kΩ or less.

Load regulation at $V_O = 10$ V is calculated by assuming that I_L changes to 1 A, requiring $I_{3B} = 0.4$ mA. This decreases I_{2C} by 0.4 mA and I_{2B} by $0.4/50 = 0.008$ mA.

$$V_O = \Delta I_{2B} R_6 = (0.008\ \text{mA})(1.5\ \text{k}\Omega) = 12\ \text{mV}$$

This ΔV_O is due to R_6. Additional ΔV_O must be anticipated as lowered transformer voltage reduces I_{D5}, cooling the zener.

C_5 feeds any output ripple around R_6 directly to Q_2. Q_2 then adjusts I_{3B}, hence I_L, to keep V_O constant. Ripple reduction is thus considerably improved. R_7 protects Q_2 in the event that a voltage source (such as a charged capacitor) is connected to V_O while R_6 is set to zero resistance. The choice of the relatively hefty 2N3053 over the less expensive 2N4400 for Q_2 is a result of this same consideration.

Reference supply: D_5 is chosen near 7 V because zeners in this range generally have the lowest voltage drift with temperature change. We want to minimize I_Z to avoid self-heating, but we must ensure that the zener conducts even under worst-case conditions of low line, maximum I_L, −20% C_3 (375 μF), and +10% V_Z (7.4 V). At minimum line voltage, letting $I_Z = 1$ mA:

$$I_{R4,5} = I_{R6} = \frac{V_{O(\text{MAX})}}{R_{6(\text{MAX})}} = \frac{20\ \text{V}}{2.5\ \text{k}\Omega} = 8\ \text{mA}$$

$$V_{C3(\text{PK})} = V_{C1(\text{PK})} = 26.8\ \text{V}$$

$$V_{C3(\text{rip})\text{p-p}} = \frac{I_{R3}t}{C_3}$$

$$= \frac{(1 + 8)\text{mA}(16.7\ \text{ms})}{376\ \mu\text{F}} = 0.40\ \text{V}$$

$$V_{R3\,(\mathrm{min})} = V_{C3\,(\mathrm{val})} - V_Z = 26.8 - 0.40 - 7.4 = 19\text{ V}$$

$$R_3 = \frac{V}{I_{R3}} = \frac{19}{9} = 2.1\text{ k}\Omega$$

In 10% tolerance, 1.8 kΩ would be the highest safe value. 2.0 kΩ could be used in 5% tolerance. The power in D_5 is now determined at high line and minimum V_O, using the 56-V center-tapped transformer.

$$V_{C3} = \left(\frac{V_{\mathrm{hi}}}{V_{\mathrm{low}}}\right)\tfrac{1}{2}(V_{s\,(\mathrm{NL})})(1.414) - V_D$$

$$= \left(\frac{130\text{ V}}{105\text{ V}}\right)\frac{(56)(1.414)}{2} - 1 = 48\text{ V}$$

$$I_{R3} = \frac{V_{C3} - V_{Z\,(\mathrm{max})}}{R_{3\,(\mathrm{min})}} = \frac{48 - (110\% \times 6.7)}{90\%(1.8)} = 25\text{ mA}$$

$$I_Z = I_{R3} - I_{R4,5} = 25 - 8 = 17\text{ mA}$$

$$P_Z = I_Z V_{Z\,(\mathrm{max})} = (17)(7.4) = 126\text{ mW}$$

A 300-mW zener would do, but a 1-W or 2-W unit would offer better line and load regulation. Assuming that $r_{z\,(\mathrm{max})} = 10\ \Omega$, the ripple at D_5 is

$$V_{\mathrm{rip}\,(D5)} = V_{\mathrm{rip}\,(C3)}\frac{r_z}{R_3} = 0.40\frac{10}{1800} = 2.2\text{ mV p-p}$$

Without C_5 this ripple would be multiplied by V_O/V_Z and appear at V_O. With $X_{C5} = 13\ \Omega$, ripple is much less.

$R_4 + R_5$ have -6.7 V at their left end (D_5) and $+0.6$ V at their right end (V_{BE} of Q_2). They carry essentially the same current as R_6, so at maximum V_O of 20 V,

$$\frac{V_O - V_{2BE}}{R_6} = \frac{V_Z + V_{2BE}}{R_4 + R_5}; \qquad \frac{20 - 0.6}{2500} = \frac{6.7 + 0.6}{R_4 + R_5}$$

$$R_4 + R_5 = 940\ \Omega$$

We choose $R_4 = 560\ \Omega$ and $R_5 =$ a 1-kΩ potentiometer to center on this value.

Current-limit circuit: Resistors R_8, R_9, and R_{10} must develop V_{BE} of 0.6 V when I_L reaches the selected limit. These resistors also carry the 8-mA current I_{R6}.

$$R_8 = \frac{V_{BE}}{I_L + I_{R6}} = \frac{0.6\text{ V}}{1.008\text{ A}} = 0.6\ \Omega$$

$$P_{R8} = I^2R = (1^2)(0.6) = 0.6\text{ W}$$

$$R_9 = \frac{V_{BE}}{I_L + I_{R6}} = \frac{0.6\text{ V}}{0.108\text{ A}} = 5.6\ \Omega$$

$$R_{10} = \frac{V_{BE}}{I_L + I_{R6}} = \frac{0.6\text{ V}}{0.018\text{ A}} = 33\ \Omega$$

Total power consumption is approximately $V_{s\,(\mathrm{pk})}\ I_O$ or about (50 V)(1 A) =

50 W. Line current would then be

$$I_{avg} = \frac{P}{V} = \frac{50}{120} = 0.42 \text{ A}$$

However, I_{rms} is likely to be quite a bit larger since it is pulsating, not sinusoidal. A true-rms ammeter should be used to find I_{line} at maximum V_{line} and I_L, and the fuse selected about 50% above this value.

The heat sink required for Q_4 can be determined once P_{Q4} is found. P_{Q4} is maximum at high line when $I_L = 1$ A and $V_O = 0$.

$$V_{Q4(CE)avg} = V_{C1(avg)} \frac{V_{line(hi)}}{V_{line(low)}} - V_{R8}$$

$$= (24.2)\frac{130}{105} - 0.6 = 29.4 \text{ V}$$

$$P_{Q4} = IV = (1)(29.4) = 29.4 \text{ W}$$

The 2N3055 has $T_{J(max)} = 200°\text{C}$ and $\theta_{JC} = 1.52°\text{C/W}$. Assuming that $T_{A(max)} = 50°\text{C}$ and a 0.35°C/W case-to-sink insulator,

$$\theta_{SA} + \theta_{CS} + \theta_{JC} = \frac{T_J - T_A}{P} \tag{11-17}$$

$$\theta_{SA} = \frac{200 - 50}{29.4} - 0.35 - 1.52 = 3.2 \text{ °C/W}$$

Figure 11-23 shows that a 25-in.^{2} heat sink is required.

If you build the power supply of Fig. 17-2 you may find to your delight that it can be "opened up" to 1.5 A, or that 1000 μF will suffice for C_1, or that a lighter transformer can be used. This is fine for individual projects, but for mass-produced and marketed instruments the worst-case situation is bound to come up eventually—lowest line voltage, highest load current, low-beta transistors, and so on—all in the same unit. It is therefore essential to hold firmly to the mathematically determined specifications and component values. The cost to the company, in service expense and loss of reputation, is not worth the temporary gain to be achieved by cutting corners on the worst-case design.

17.3 A PORTABLE CASSETTE RECORDER

Figure 17-3 shows a portable cassette recorder which illustrates the interaction of many of the circuits described earlier in this text. Since the diagram is fairly extensive an orientation and overview may be helpful.

Two ground lines extend leftward from the bottom of the 8-Ω speaker and the battery + side, respectively. The −9-V V_{CC} line starts at the center tap of T_2 and proceeds leftward through decoupling networks (see Fig. 3-12) R_{21}–C_{12}, R_{17}–C_{11}, and R_{11}–C_1.

There are five separate circuits to the recorder:

1. Power supply, T_3, D_2, D_3, and the unmarked filter capacitor. When the line is plugged in, relay coil RY throws contacts S_3 to the supply output, which must be about -9 V. With the supply deenergized, S_3 switches V_{CC} to the -9-V battery.
2. Tr_1 and Tr_2 comprise a direct-coupled preamplifier, similar to Fig. 10-2(b).
3. Tr_3 is a transformer-coupled driver, nearly identical to that of Fig. 11-3.
4. Tr_4 and Tr_5 form a push-pull class-B output amplifier (see Fig. 11-6) feeding a speaker in the playback mode and a record head RPH in the record mode.
5. Tr_6 is a Hartley oscillator operating in the ultrasonic (about 40-kHz) range. The circuit is a cross between Figs. 13-2 and 13-4. It has been found that adding this ultrasonic "bias" signal to the audio signal during recording greatly increases the linearity of the signal impressed on the magnetic tape. C_{19} couples the bias signal to the record head.

S_{1-1} and S_{1-2} connect the tape head to the amp input on playback, but connect the microphone to the input and the tape head to the output (long transformer secondary) on record. R_{25} and R_{26} establish the head current on record. Above 1.6 kHz C_{16} has a reactance lower than R_{25}, and additional record current reaches the tape head. This preemphasis of the audio highs is universally used in recording to mask noise, which tends to be concentrated in the higher frequencies. Deemphasis is used in playback to restore equal amplification of highs and lows. S_{1-4} connects the speaker to the short transformer secondary on playback and substitutes a 10-Ω dummy load on record. S_{1-5} switches V_{CC} to the bias oscillator on record and also applies dc to the erase head EH, which is positioned just before the record head in the tape path. Thus the old program is erased while the new one is being recorded.

The following transformer data were obtained by measurement of actual components. Data on center-tapped windings is for one-half winding.

	N_S/N_P	r_P (Ω)	r_S (Ω)
T_1	0.44	200	85
T_2 (upper)	0.67	5	1
T_2 (lower)	3.3	5	65
T_3	0.085	200	4
T_4	20	0	0
$R_{RY} = 900$ Ω			
$R_{RPH} = 300$ Ω			
$R_{EH} = 300$ Ω			

The unmarked power-supply filter is 470 μF. The motor filter is 330 μF, and the motor draws about 100 mA. We will analyze each of the five circuits in turn.

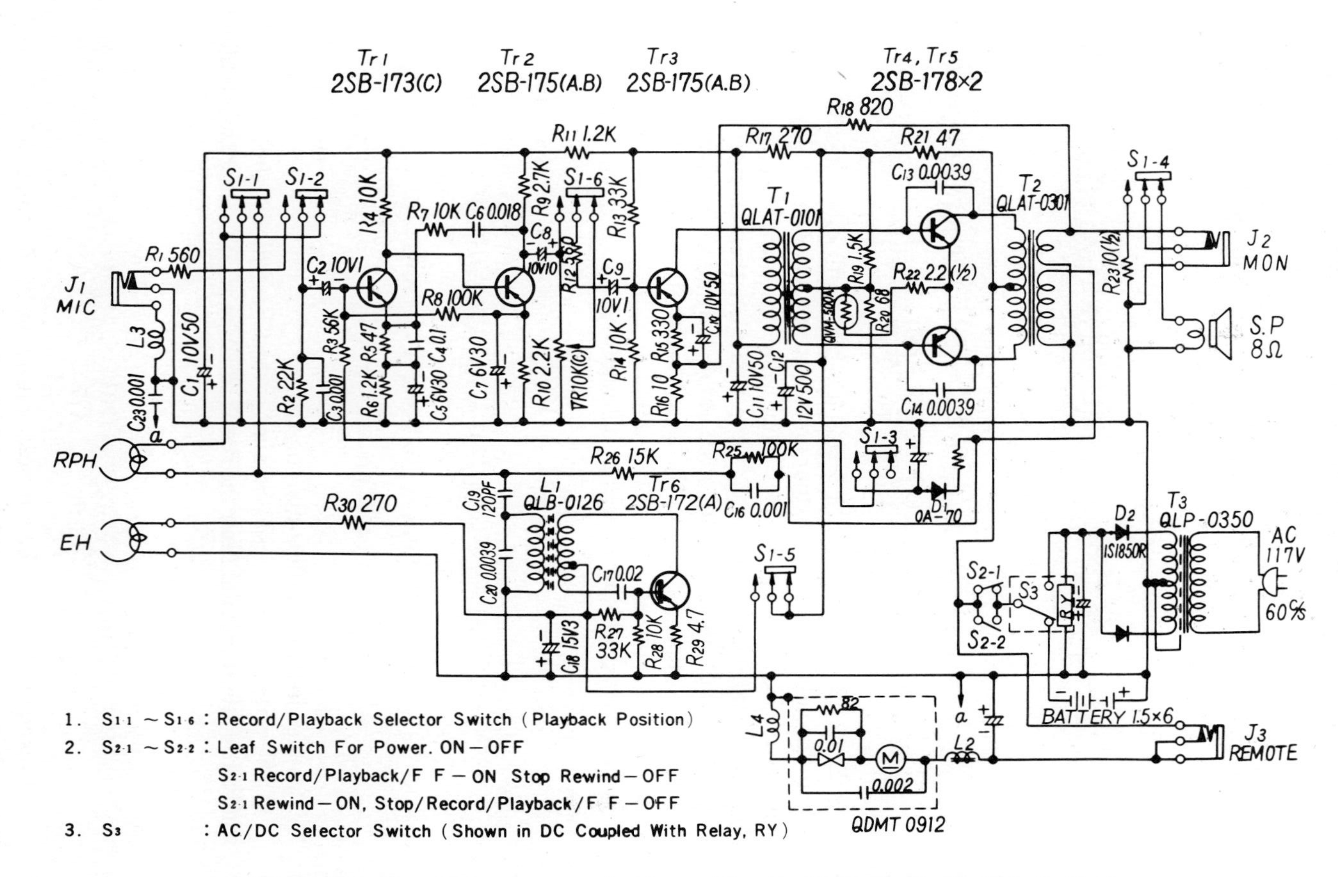

Figure 17-3 Portable cassette recorder/player. (Courtesy Radio Shack, Fort Worth, Tex.)

Power supplies: Assuming a 400-mW audio output and a 60% overall efficiency, the amplifier current drain would be

$$I_{AMP} = \frac{P_O}{60\%(V_{CC})} = \frac{0.4}{(0.6)(9)} = 74 \text{ mA}$$

Admittedly this is a shot in the dark, but it is not likely to be off by more than $\pm 50\%$. We now find the worst-case ripple.

$$I_{EH} = \frac{V_{CC}}{R_{EH} + R_{30} + R_{21}} = \frac{9}{300 + 270 + 47} = 15 \text{ mA}$$

$$I_{RY} = \frac{V_{CC}}{R_{RY}} = \frac{9}{900} = 10 \text{ mA}$$

$$I_L = I_{AMP} + I_{MOT} + I_{EH} + I_{RY} = 74 + 100 + 15 + 10 = 200 \text{ mA}$$

$$V_{rip(p\text{-}p)} = \frac{I_L t}{C_{PF} + C_{MF}} = \frac{(200 \text{ mA})(8.3 \text{ ms})}{470\ \mu\text{F} + 330\ \mu\text{F}} = 2.1 \text{ V}$$

This is 8% of V_{CC}, which seems too high until one realizes that it is applied only to the collectors of the output amp. The output signal of this amp is $V_{in}A_v$, and depends on V_{CC} only to the small extent that I_C depends on V_{CE} of the transistors. R_{21}–C_{12} filters the ripple to 0.12 V p-p (0.5%) at the collector of Tr_3 and bases of Tr_4 and Tr_5, but ripple here still does not directly affect the signal. Base-bias resistor R_{13} does add ripple directly to the signal, but at this point R_{17}–C_{11} has reduced it by an additional factor of X_C/R to 0.02%.

The average dc output voltage is found next.

$$V_s = V_{line}\left(\frac{N_S}{N_P}\right) = (120)(0.085)(1.414) = 14.4 \text{ V pk}$$

$$r_w = r_S + r_P\left(\frac{N_S}{N_P}\right)^2 = 4 + (200)(0.085)^2 = 5.4\ \Omega$$

$$V_{CC(AV)} = V_{s(pk)} - V_D - 4I_L r_w - V_{rip(pk)}$$

$$= 14.4 - 1.0 - (4)(0.20)(5.4) - \frac{2.1}{2} = 8.1 \text{ V}$$

This voltage increases at less-than-maximum I_L.

The primary is not fused because even if both diodes shorted, the transformer winding resistance would limit the result to a slow cook rather than a blinding flash.

$$R_T = r_P + \frac{2r_S}{(2n)^2} = 200 + \frac{8}{0.17^2} = 477\ \Omega$$

$$P = \frac{(V_{line})^2}{R_T} = \frac{120^2}{477} = 30 \text{ W}$$

Preamplifier: Tr_1 is biased near saturation by R_8 since there is no R_{B2} to fix base voltage. This bias point is not a problem because the signal swings from the few millivolts of tape-head input are not large. Assuming that $V_{1CE} = 1.0$ V,

$$V_{1E} = (V_{CC} - V_{CE})\left(\frac{R_5 + R_6}{R_4 + R_5 + R_6}\right) = (9 - 1)\left(\frac{1247}{11\,247}\right) = 0.9 \text{ V}$$

$$V_{1C} = V_{1E} + V_{1CE} = 0.9 + 1.0 = 1.9 \text{ V}$$

$$I_{1C} = \frac{V_{1E}}{R_5} + R_6 = \frac{0.9}{1247} = 0.72 \text{ mA}$$

The transistors are germanium, so $V_{BE} \approx 0.2$ V.

$$V_{2E} = V_{C1} - V_{BE} = 1.9 - 0.2 = 1.7 \text{ V}$$

$$V_{1B} = V_{1E} + V_{BE} = 0.9 + 0.2 = 1.1 \text{ V}$$

Now we find I_{1B}, β_1, and V_{2C}.

$$I_{1B} = I_{R8} = \frac{V_{2E} - V_{1B}}{R_8} = \frac{1.7 - 1.1}{100 \text{ k}\Omega} = 0.006 \text{ mA}$$

$$\beta_1 = \frac{I_{1C}}{I_{1B}} = \frac{0.72}{0.006} = 120$$

$$I_{2C} = \frac{V_{2E}}{R_{10}} = \frac{1.7}{2.2} = 0.77 \text{ mA}$$

$$V_{2C} = V_{CC} - I_{2C}R_9 = 9 - (0.77)(2.7) = 6.9 \text{ V}$$

Lower values of β_1 will increase V_{1C} and decrease V_{2C}. Continuing with the ac analysis, and assuming that $\beta = 100$ for all transistors:

$$r_{2j} = \frac{0.03}{I_{2C}} = \frac{0.03 \text{ V}}{0.77 \text{ mA}} = 39\ \Omega$$

$$r_{2in} = \beta_2 r_{2j} = 100 \times 39 = 3900\ \Omega$$

$$A_{v2} = \frac{R_9 \| R_{10} \| (R_{12} + r_{3in})}{r_{2j}}$$

$$\approx \frac{2.7 \| 2.2 \| 2}{39} = \frac{750}{39} = 19$$

We have assumed that $r_{3in} \approx 1.5$ kΩ for the moment.

$$r_{1j} = \frac{0.03}{I_{C1}} = \frac{0.03}{0.72} = 42\ \Omega$$

$$A_{v1} = \frac{R_4 \| r_{2in}}{r_{1j} + R_5} = \frac{10\,000 \| 3900}{42 + 47} = 32$$

$$r_{1b} = \beta_1(r_{1j} + R_5) \| R_8 = 100(42 + 47) \| 100 \text{ k}\Omega = 8.2 \text{ k}\Omega$$

$$Z_{in(amp)} = r_{1b} \| R_2 + R_1 \approx 8.2 \| 22 + 0.56 = \mathbf{6.5\ k\Omega}$$

$$A_{v(preamp)} = A_{v1}A_{v2} = (32)(19) \approx 600$$

C_2, C_5, C_7, and C_9 each have the potential to set the low-frequency limit of the amplifier. A check shows that $X_{C7} = r_{2j}$ at 136 Hz. The other three capacitors hold out to substantially lower frequencies. C_3 and C_6 limit the gain at ultrasonic frequencies.

In the record mode S_{1-3} switches in additional negative base bias, which is obtained from the output signal at the long T_2 secondary through D_1 and the filter capacitor next to S_{1-3}. When this rectified voltage exceeds V_{1B} (at $V_{2s} \approx 3.4$ V p-p) R_3 begins to bias Tr_1 actually into saturation. The result is not distortion because the input signal from the microphone is so small. Rather, A_{v1} is reduced as $V_{in(mic)}$ increases, so V_{2s} is held at about 3.4 V p-p (max) regardless of overload microphone input. Look at Fig. 6-8 to see how A_v depends upon spacing of the I_B lines on the transistor curves. Then look at Fig. 4-7(b) to see how these lines bunch together as the transistor saturates. This automatic volume control (AVC) ensures proper record setting without adjustments. S_{1-6} switches in manual volume control on the play-back mode.

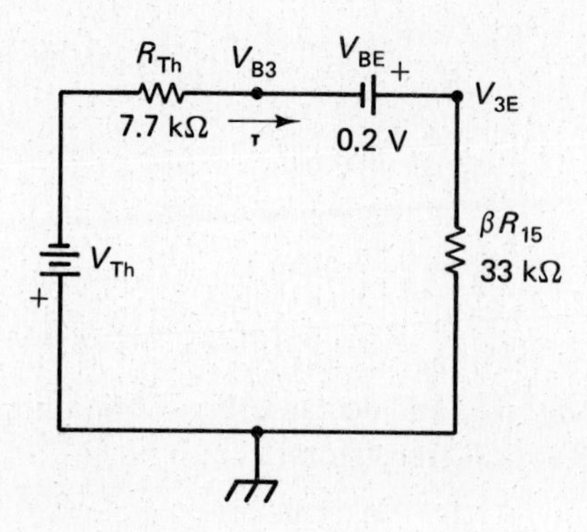

Figure 17-4 Equivalent base-bias circuit for Tr3 in Fig. 17-3.

Driver amp: R_{14} is much greater than $10R_{15}$, so R_{13} and R_{14} cannot be treated as an unloaded voltage divider. The Thévenin equivalent is shown in Fig. 17-4.

$$V_{Th} = V_{CC}\frac{R_{14}}{R_{13} + R_{14}} = 9\frac{10}{33 + 10} = 2.1 \text{ V}$$

$$R_{Th} = R_{13} \| R_{14} = 33 \| 10 = 7.7 \text{ k}\Omega$$

$$I_{3B} = \frac{V_{Th} - V_{BE}}{R_{Th} + \beta R_{15}} = \frac{2.1 - 0.2}{7.7 + 33} = 0.047 \text{ mA}$$

$$I_{3E} = I_{3B} = (100)(0.047) = 4.7 \text{ mA}$$

$$V_{3E} = I_{3E}R_{15} = (4.7 \text{ mA})(330) = 1.6 \text{ V}$$

$$r_{3j} = \frac{0.03}{I_{3E}} = \frac{0.03}{4.7 \text{ mA}} = 6.4 \ \Omega$$

The load resistance for the driver is r_b of the push-pull stage. Assuming that an average I_E for Tr_4 (or Tr_5) is 30 mA, we find r_{4b} and A_v of the driver stage.

$$r_{4j(av)} \approx \frac{0.03}{I_E} \approx \frac{0.03}{0.03} = 1 \ \Omega$$

$$r_{4b} = \beta(r_j + R_{22}) = 100(1 + 2.2) = 330 \ \Omega$$

$$A_{v(b3\text{-}b4)} = \frac{r_{b4}}{n_1 r_{j3}} = \frac{330}{(0.44)(6.4)} = 117 \qquad (11\text{-}5)$$

This last figure assumes that R_{16} is shorted—we will take care of R_{16} shortly.

Push-pull amp: Thermistor QVM-500A in parallel with R_{20} makes the bias point difficult to determine. We may reasonably assume that V_{R22} is no more than 0.1 V, so $I_{IDLE} \leq 45$ mA. Voltage gain (base to 8-Ω speaker) and $V_{o(p\text{-}p)max}$ are determined.

$$A_{v(b4\text{-}RL)} = \frac{R_L}{n_2(r_{j4} + R_{22})} = \frac{8}{0.67(1 + 2.2)} = 3.6 \qquad (11\text{-}5)$$

$$V_{o(p\text{-}p)max} = \frac{2R_L(V_{CC} - V_{CE(sat)})}{n_2(R_{22} + r_P) + \dfrac{r_S + R_L}{n_2}} \qquad (11\text{-}8)$$

$$= \frac{2(8)(9 - 0.5)}{0.67(2.2 + 5) + \dfrac{1 + 8}{0.67}} = 7.4 \text{ V}$$

$$P_{o(max)} = \frac{V^2}{R_L} = \frac{(7.4/2.828)^2}{8} = 0.85 \text{ W}$$

This level of output, if sustained, would reduce the V_{CC} even below the 8.1 V determined in part 1.

The open-loop gain from base of Q_3 to the speaker is $A_{v(b3\text{-}b4)} A_{v(b4\text{-}RL)} = (117)(3.6) = 420$. However, R_{18} and R_{16} supply voltage-derived, series-applied negative feedback (see Sections 10.4 and 10.5) with $B = R_{16}/(R_{16} + R_{18}) = 10/830 = 0.012$. Actual closed-loop gain is, therefore,

$$A_{vc} = \frac{A_{vo}}{1 + A_{vo}B} = \frac{420}{(1 + 420)(0.012)} = \mathbf{70} \qquad (10\text{-}10)$$

The input impedance of stage 3 is thus increased.

$$Z_{in(c)3} = R_{13} \| R_{14} \| \beta_3 r_{j3}(1 + A_{vo}B) \qquad (10\text{-}12)$$

$$= 33 \text{ k}\Omega \| 10 \text{ k}\Omega \| (100)(6.4)(1 + 420 \times 0.012)$$

$$= 2.6 \text{ k}\Omega$$

Our estimate of 1.5 kΩ for this value was a little low. Actual A_v of stage 2 is therefore a bit higher than first predicted.

Total amplifier gain, from tape head to spaker, is $A_{v(preamp)} A_{vc} = (600)(70)$ or 42 000 at maximum volume. Maximum V_o of 7.4 V p-p could be produced by $V_{in} = 7.4/42\,000$ or 0.18 mV p-p.

Bias oscillator: Tr_6 is biased on by voltage divider R_{27}–R_{28}. C_{19} places a reactance of 33 kΩ in series with the record-head load, which has an unknown, but probably much lower, impedance. C_{20} is reflected back to the primary as $C_{20}n^2$ or $(0.0039)(6)^2 = 0.14\ \mu$F. This trick permits the use of a smaller resonating capacitor. Primary inductance required is

$$L_{1P} = \frac{1}{4\pi^2 f^2 C} = \frac{1}{4\pi^2(40 \text{ kHz})^2(0.14\ \mu\text{F})} = 110\ \mu\text{H}$$

17.4 A TRANSISTOR INTERCOM

The intercom amplifier of Fig. 17-5 contains several additional audio circuits covered earlier in the text, and illustrates how multistage circuits can be made more simple if viewed as a unit rather than as a cascade of separate stages.

The power supply consists of T_1, D_1, D_2, filter capacitor C_{14}, and decoupling network R_{22}–C_6, which feeds the first stage and the base bias of the second stage. Q_6, Q_5, and Q_4 comprise a complementary-symmetry output stage with direct-coupled driver, similar to Fig. 11-13. Q_3 is an inverted-ground common-emitter amplifier similar to Fig. 7-6(b). Q_1 and Q_2 form a differential amplifier somewhat like that of Fig. 10-4.

A differential input stage is used because the remote microphones (actually speakers) may be located over 100 ft from the master-station amplifier. These long input lines will pick up 60-Hz noise from nearby power lines, causing hum in the output. Shielded cable is an expensive solution. Ordinary twisted-pair wire can be used if neither wire is grounded. Each wire will then pick up identical noise with respect to ground (a large common-mode noise signal) so the difference noise signal between lines (differential input) will be near zero. The differential amplifier rejects common-mode signals but amplifies differential signals. The microphone is connected differentially, of course.

The output stages will be analyzed first so the loads on the driving stages will be known. C_{13} couples the output to the load, which is an 8-Ω speaker. $X_{C13} = R_L + R_{23}$ at 150 Hz, so f_{low} cannot be lower than this. The amplifier output impedance was assumed to be negligibly low because voltage-derived negative feedback (R_{18}–R_{17}) is used (see Section 10.5).

Assuming a $\frac{1}{4}$-W audio output across the 8-Ω load we find that r_{5j} and r_{6j} are nearly negligible in determining the A_v of the output stage.

$$I_L = \sqrt{\frac{P}{R}} = \sqrt{\frac{0.25}{8}} = 0.177 \text{ A}$$

$$r_{6j} = \frac{0.03}{I_L} = \frac{0.03}{0.177} = 0.17\ \Omega$$

$$A_{v(5,6)} = \frac{R_L}{R_L + R_{23} + r_j} = \frac{8}{8 + 2.7 + 0.17} = 0.74 \qquad (11\text{-}10)$$

Base bias is fed to Q_6 from V_{CC} through R_{21} and R_{20}, which also are the collector resistors for Q_4. Taking the 12.9-V reading of V_{B6} on faith for the moment (we will verify it later):

$$I_{R20,21} = \frac{V_{CC} - V_{6B}}{R_{21} + R_{20}} = \frac{27 - 12.9}{1800 + 1800} = 3.9 \text{ mA}$$

At zero signal I_{6B} is negligible, so

$$I_{R(19)} = I_{R20,21} = 3.9 \text{ mA}$$

$$V_{R(19)} = IR = (3.9 \text{ mA})(330) = 1.29 \text{ V}$$

$$V_{R(23+24)} = V_{R19} - 2V_{BE} = 1.29 - 1.20 = 0.09 \text{ V}$$

$$I_{E(6,5)\text{idle}} = \frac{V_{R23+R24}}{R_{23} + R_{24}} = \frac{0.09}{5.4} = 17 \text{ mA}$$

Maximum output, assuming that $\beta = 100$ and that V_{CC} can hold at 27 V under full load, is

$$V_{o(\max)} = \frac{V_{CC}\beta R_L}{(R_{20} + R_{21}) + \beta(R_{23} + R_L)} = \frac{(27)(100)(8)}{(3600) + (100)(2.7 + 8)} \quad (11\text{-}11)$$

$$= 4.6 \text{ V p-p} = 1.6 \text{ V rms}$$

$$P_o = \frac{V^2}{R_L} = \mathbf{0.33 \text{ W}}$$

R_{20} is bootstrapped for ac (see Section 7.2), so the load on the collector of Q_4 appears to have the value

$$Z_{in(6)} = R_{20}\frac{r_{6j} + R_{23} + R_L}{r_{6j}} \Bigg\| \beta(r_{6j} + R_{23} + R_L) \quad (7\text{-}4)$$

$$= (1800)\frac{0.17 + 2.7 + 8}{0.17} \Bigg\| 100(0.17 + 2.7 + 8)$$

$$= 115 \text{ k}\Omega \,\|\, 1.1 \text{ k}\Omega = 1.1 \text{ k}\Omega$$

Driver stages: The gain of Q_4 is determined as $R_{c(\text{line})}/r_j$:

$$r_{4j} = \frac{0.03}{I_E} = \frac{0.03}{I_{R19}} = \frac{0.03}{3.9 \text{ mA}} = 7.7\ \Omega$$

$$A_{v4} = \frac{R_{c(\text{line})}}{r_{4j}} = \frac{Z_{in(6)}}{r_{4j}} = \frac{1100}{7.7} = \mathbf{140}$$

$$Z_{in(4)} = \beta r_{4j} = (100)(7.7) = 770\ \Omega$$

The collector resistor for Q_3 (R_{16}) is connected directly across the Q_4 base-emitter, so it obviously has about 0.6 V.

$$I_{R(16)} = I_{3E} = \frac{V_{4BE}}{R_{16}} = \frac{0.6}{2700} = 0.22 \text{ mA}$$

$$r_{3j} = \frac{0.03}{I_{E3}} = \frac{0.03}{0.22 \text{ mA}} = 136\ \Omega$$

$$A_{v3} = \frac{R_{16} \,\|\, Z_{in4}}{r_{3j}} = \frac{2700 \,\|\, 770}{136} = 4.4$$

$$r_{3b} = \beta r_{3j} = (100)(136) = 13.6 \text{ k}\Omega$$

This is open-loop gain and input resistance assuming the Q_3 emitter is at signal ground. The overall open-loop gain from base of Q_3 to load is

$$A_{v(3,4,6)} = A_{v3}A_{v4}A_{v(5,6)} = (4.4)(140)(0.74) = 460$$

SCHEMATIC OF THE HEATHKIT®
TWO-STATION INTERCOM
MODEL GD-140

AMPLIFIER CIRCU

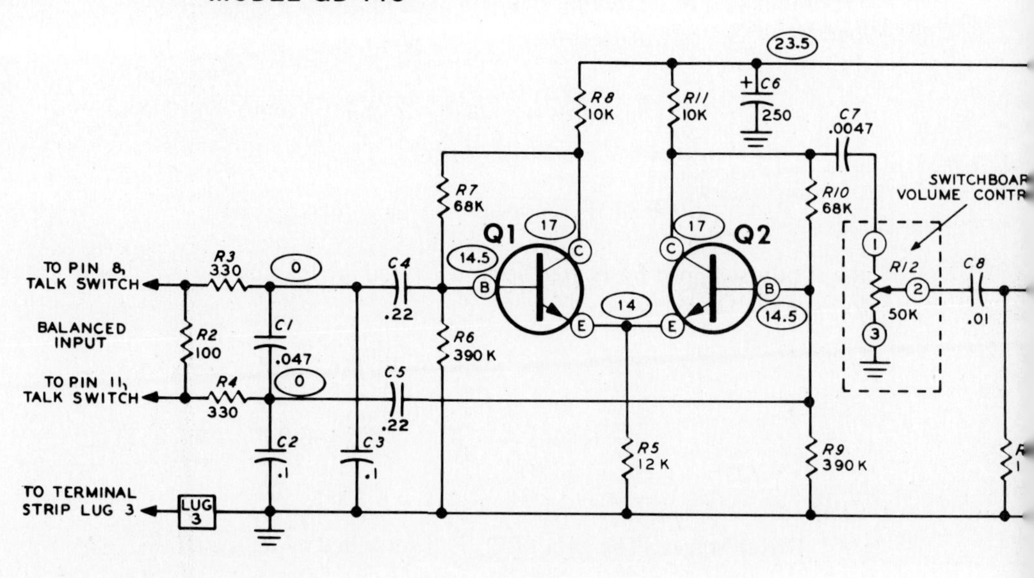

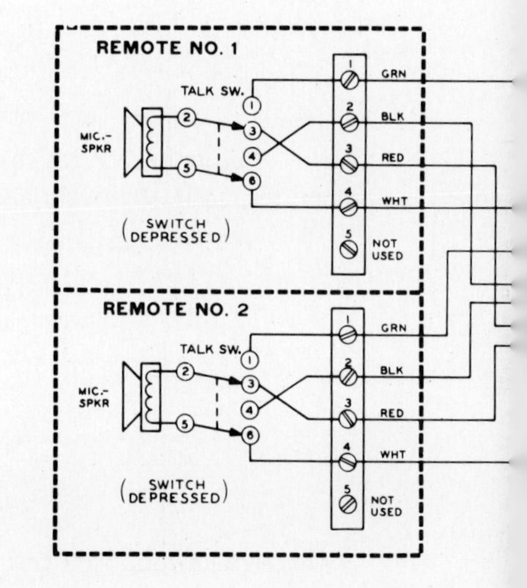

Figure 17-5 Intercom amplifier featuring differential input for noise rejection. (Courtesy Heath Co., Benton Harbor, Mich.)

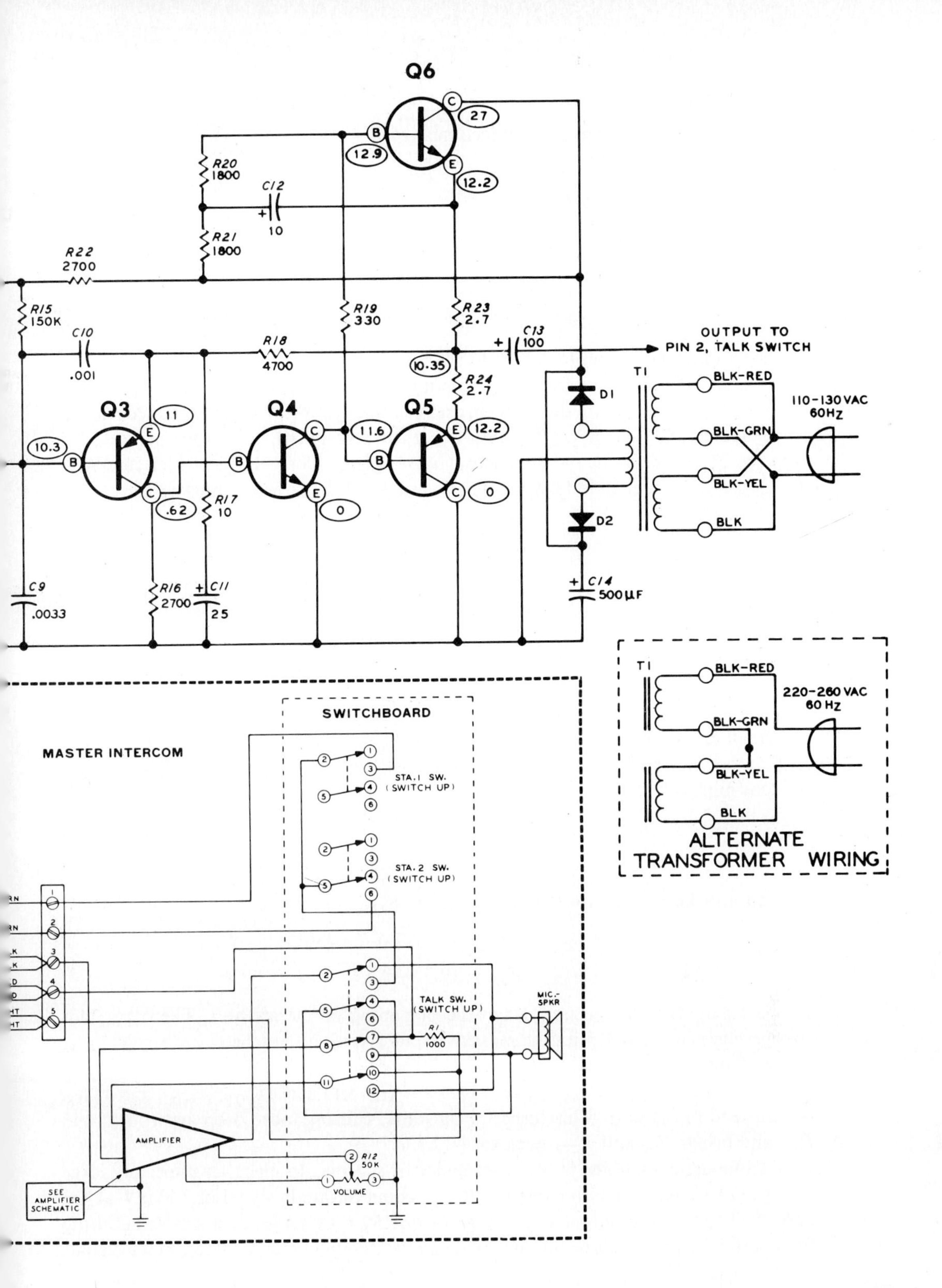
Q6
C
27
B
12.9
E
12.2
R20
1800
C12
10
R21
1800
R22
2700
R15
150K
C10
.001
R18
4700
R19
330
R23
2.7
C13
100
OUTPUT TO
PIN 2, TALK SWITCH
10.35
R24
2.7
Q3
Q4
Q5
11
10.3
11.6
12.2
.62
R17
10
0
0
D1
D2
T1
BLK-RED
BLK-GRN
BLK-YEL
BLK
110-130 VAC
60 HZ
C9
.0033
R16
2700
C11
25
C14
500 μF
T1
BLK-RED
BLK-GRN
BLK-YEL
BLK
220-260 VAC
60 HZ
ALTERNATE
TRANSFORMER WIRING
MASTER INTERCOM
SWITCHBOARD
STA. 1 SW.
(SWITCH UP)
STA. 2 SW.
(SWITCH UP)
TALK SW.
(SWITCH UP)
MIC.-
SPKR
R1
1000
AMPLIFIER
R12
50K
VOLUME
SEE
AMPLIFIER
SCHEMATIC

R_{18} and R_{17} provide series-applied feedback with

$$B = \frac{R_{17}}{R_{17} + R_{18}} = \frac{10}{10 + 4700} = 0.0021$$

$$A_{vc} = \frac{A_{vo}}{1 + A_{vo}B} = \frac{460}{1 + (0.0021)(460)} = \mathbf{230} \qquad (10\text{-}10)$$

$$Z_{in3} = R_{15} \| R_{13} \| r_{b3}(1 + A_{vo}B) \qquad (10\text{-}12)$$

$$= 150 \| 120 \| 13.6(1 + 460 \times 0.0021) = \mathbf{19\ k\Omega}$$

C_{11} has a reactance equal to R_{17} at 635 Hz, so below this frequency B is higher and A_{vc} is lower. If better low-frequency response is required, C_{11} should be increased.

Q_3, Q_4, and $Q_{5,6}$ are dc coupled, and the bias point for them all is determined by voltage divider R_{15}–R_{13}. (R_{22}, with C_6, is for decoupling, and drops perhaps 3 V from the 27-V supply; R_{14} is small enough to neglect.) The voltage-divider current is 40 times I_{3B}, so negligible loading occurs and V_{3B} can be determined directly.

$$I_{3B} = \frac{I_{3E}}{\beta} = \frac{0.22}{100} = 0.0022 \text{ mA}$$

$$I_{R(15,13)} = \frac{V_{CC} - V_{R(22)}}{R_{15} + R_{13}} = \frac{27 - 3}{150 + 120} = 0.089 \text{ mA}$$

$$V_{3B} = V_{CC} = V_{R(22)} \frac{R_{13}}{R_{13} + R_{15}} = (24)\frac{120}{270} = 10.7 \text{ V}$$

$$V_{3E} = V_{3B} + V_{3(EB)} = 10.7 + 0.6 = 11.3 \text{ V}$$

The small emitter current for Q_3 comes from V_{CC} through $Q_{6(CE)}$, R_{23}, and R_{18}. The drop across R_{18} and hence the center-point voltage at the junction of R_{23} and R_{24} are determined.

$$V_{R(18)} = I_{E3}R_{18} = (0.22)(4.7) = 1.0 \text{ V}$$

$$V_{CTR} = V_{E3} + V_{R18} = 11.3 + 1 = 12.3 \text{ V}$$

V_{6B} can now be found with the knowledge of V_{R23}.

$$V_{6E} = V_{CTR} + V_{R(23)} = 12.3 + 0.045 \approx 12.3 \text{ V}$$

$$V_{6B} = V_{6E} + V_{BE} = 12.3 + 0.6 = 12.9 \text{ V}$$

This is exactly the voltage assumed at the beginning of the analysis. The other bias voltages agree quite well with the measurements given on the schematic, except V_{CTR}, which seems to be in error.

If V_{CTR} attempts to rise above +12.3 V, $V_{R(18)}$ and hence $I_{R(18)}$, $I_{R(16)}$, and $V_{R(16)}$ will all tend to increase. This turns Q_4 on more, causing more drop across R_{20} and R_{21}, and brings V_{B6} and V_{CTR} back to +12.3 V.

Differential amplifier Q_1–Q_2 uses collector self-bias to place an unusually large voltage (14 V) across emitter resistor R_5. A common-mode signal of 140 mV at the bases will thus cause only a 1% change in I_{R5}. R_5 thus serves as a simple subtitute for a current source. This technique is called *longtailing*. Under common-mode signals the two collector voltages track so R_8 and R_{11} are effectively in parallel, along with R_7, R_{10}, and R_{12}. $A_{v(com)} = (R_7 || R_8 || R_{11} || R_{10} || R_{12})/R_5 = 4/12 = 0.33$.

Assuming the measured value of $V_C = 17$ V, and using the analysis technique of Fig. 17-4:

$$V_{Th} = V_C \frac{R_6}{R_6 + R_7} = (17)\frac{390}{390 + 68} = 14.5 \text{ V}$$

$$R_{Th} = R_6 || R_7 = 390 || 68 = 58 \text{ k}\Omega$$

$$I_{B1} = \frac{V_{Th} - V_{BE}}{R_{Th} + 2\beta R_5} = \frac{14.5 - 0.6}{58 + 2(100)(12)} = 0.0057 \text{ mA}$$

(The factor of 2 associated with R_5 is used because Q_2 supplies half of I_{R5}.)

$$I_{E1} = \beta I_{B1} = (100)(0.0057) = 0.57 \text{ mA}$$

$$I_{R5} = 2I_{E1} = 2(0.57) = 1.14 \text{ mA}$$

$$V_{R5} = IR_5 = (1.14)(12) = 13.7 \text{ V}$$

We now proceed to find the differential gain of the input stage. The differential input is applied across two r_j resistances, while the output is taken from one R_C resistance in parallel with R_{10}, R_{12}, and Z_{in3}.

$$r_{j(1,2)} = \frac{0.03}{I_E} = \frac{0.03}{0.57 \text{ mA}} = 53 \ \Omega$$

$$A_{v(dif)} = \frac{R_{11} || R_{10} || R_{12} || Z_{in3}}{r_{j1} + r_{j2}} = \frac{(10 || 68 || 50 || 19) \text{ k}\Omega}{(53 + 53)\ \Omega} = 50$$

$$\text{CMRR} = \frac{A_{v(dif)}}{A_{v(com)}} = \frac{50}{0.33} = 152 = 44 \text{ dB}$$

The Miller effect lowers the effective resistance of R_7 and R_{10}, but R_2 shunts the overall differential Z_{in} down to 100 Ω. This is still high compared to the 8-Ω source impedance.

$$r_{in(1)} = \beta r_j = (100)(53) = 5.3 \text{ k}\Omega$$

$$Z_{in(1)} = \frac{R_7}{A_{v1} + 1} \Bigg\| R_6 \Bigg\| r_{in(1)} = \frac{68}{50 + 1} \Bigg\| 390 \Bigg\| 5.3 = 1.1 \text{ k}\Omega$$

$$Z_{in(dif)} = (Z_{in(1)} + Z_{in(2)} + R_3 + R_4) || R_2$$
$$= (1100 + 1100 + 330 + 330) || 100 = \mathbf{97\ \Omega}$$

The common-mode input impedance (each input to ground) is

$$Z_{in(com)} = \frac{R_7}{A_{v1} + 1} \Bigg\| R_6 \Bigg\| \beta(r_j + 2R_5)$$
$$= \frac{68}{50 + 1} \Bigg\| 390 \Bigg\| (100)(2)(12) = 1.3 \text{ k}\Omega$$

Checking for low-frequency cutoffs:

$$X_{C4} = Z_{in1} \text{ at } 650 \text{ Hz}$$

$$X_{C7} = R_{11} + R_{12} \| Z_{\text{in3}} \text{ at 1400 Hz (at full volume)}$$

$$X_{C7} = R_{11} + \tfrac{1}{2}R_{12} + \tfrac{1}{2}R_{12} \| Z_{\text{in3}} \text{ at 730 Hz (at half volume)}$$

$$X_{C8} = R_{12} \| R_{11} + Z_{\text{in3}} \text{ at 580 Hz (at full volume)}$$

More natural voice quality could be obtained at full volume by increasing the value of C_7. High-frequency limits are imposed by C_1, C_2, C_3, and C_9. Differentially, C_1 appears in parallel with the series combination of C_2 and C_3.

$$X_{C(1,2,3)} = R_3 + R_4 \text{ at 2.5 kHz}$$

$$X_{C9} = R_{14} + R_{12} \| R_{11} \text{ at 5.2 kHz}$$

In common-mode, X_{C2} and X_{C3} each present a reactance 27 kΩ to ground at 60 Hz. The source impedance of the stray capacitive pickup is likely to be several megohms.

Overall voltage gain of the amplifier at maximum volume is

$$A_{v(T)} = (A_{vc})(A_{v(\text{dif})})\frac{Z_{\text{in(com)}}}{Z_{\text{in(com)}} + R_3 + R_4}$$

$$= (230)(50)\frac{1300}{1300 + 330 + 330} = \mathbf{7600}$$

17.5 AN OP-AMP FUNCTION GENERATOR

The performance of this function generator (Fig. 17-6) may not rival that of commercial units, nor even that of LSI (large-scale integration) chips selling for $5.00, but it is quite serviceable and it provides three distinct examples of op-amp applications.

Integrator and switch: U_1 and U_2 serve as an integrator and a hysteresis switch, respectively, as in Fig. 12-18. Saturation-level output from U_2 is about ±9 V. The input to R_4 needed to switch the output of U_2 is

$$V_{\text{IN}(2)} = \pm V_{\text{SAT}}\frac{R_8}{R_7} = \pm(9)\frac{10}{22} = \pm 4.1 \text{ V} \qquad (12\text{-}5)$$

Breakdown diodes D_1 and D_2 limit the square-wave output to approximately ±5 V. At maximum frequency (R_1 wiper at bottom) the input resistor to integrator U_1 is simply R_3. On the lowest frequency range the U_1 output will run up a 4.1-V ramp in a time

$$t = \frac{V_{\text{O}(1)}}{V_{\text{IN}(1)}}R_3C_1 = \frac{V_{\text{IN}(2)}}{V_{\text{IN}(1)}}R_3C_1 = \frac{4.1}{5}(15 \text{ k}\Omega)(0.1 \,\mu\text{F}) = 1.23 \text{ ms}$$

This is one run from zero to peak, or one-fourth of a cycle.

$$T = 4t = 4(1.23) = 4.92 \text{ ms}$$

$$f_{\text{ih}} = \frac{1}{T} = \mathbf{203 \text{ Hz}}$$

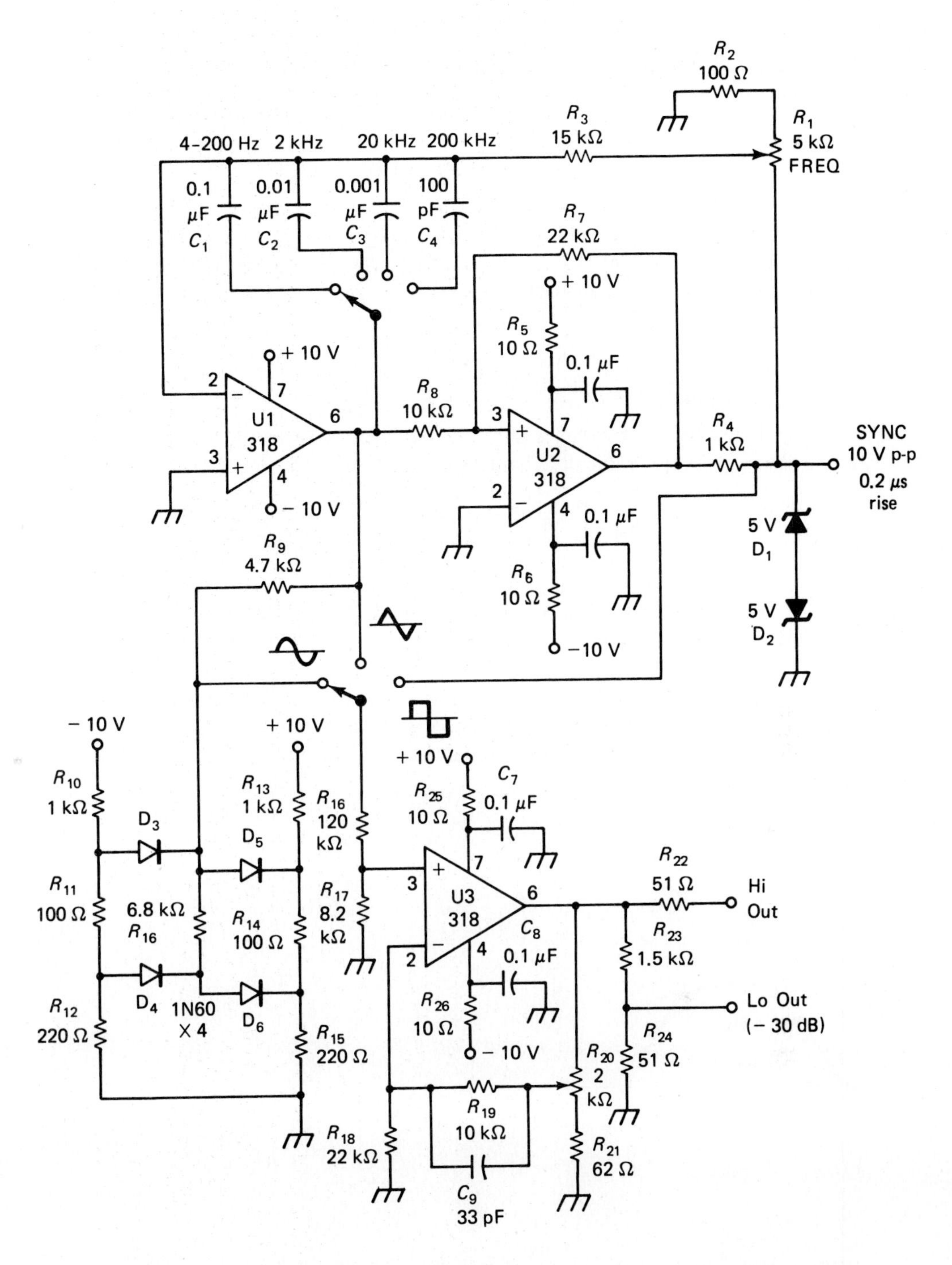

Figure 17-6 A function generator suitable for student construction. Three distinct op-amp applications are illustrated.

With the R_1 wiper at top, $V_{IN(1)}$ is reduced by the factor

$$F = \frac{R_2}{R_1 + R_2} = \frac{100}{5000 + 100} = 0.0196$$

$$f_{low} = Ff_{hi} = (0.0196)(203) = 4.0 \text{ Hz}$$

Sine shaping: R_9 through R_{15} and D_3 through D_6 comprise a sine-shaping network which places progressively heavier loads on the triangle wave as it approaches its peaks. This is illustrated in Fig. 17-7. The first break point (D_6 on positive, D_4 on negative half-cycles) occurs at

$$V_{B1} = \pm V_{CC}\frac{R_{15}}{R_{15} + R_{14} + R_{13}} \pm V_D = (\pm 10)\frac{220}{220 + 100 + 1000} \pm 0.20$$
$$= \pm 1.87 \text{ V}$$

The division of voltages above this value is in the ratio $R_{16}/(R_{16} + R_9)$ or 0.59, so the slope is not radically changed from the triangle wave. The second break is at

$$V_{B2} = \pm V_{CC}\frac{R_{15} + R_{14}}{R_{15} + R_{14} + R_{13}} \pm V_D = (10)\frac{220 + 100}{220 + 100 + 1000} \pm 0.20$$
$$= \pm 2.62 \text{ V}$$

Voltages above this value are loaded by R_9 voltage dividing against the Thévenin equivalent $R_{Th} = R_{13} \| (R_{14} + R_{15}) = 242\ \Omega$. The division ratio here is $242/(242 + 1000) = 0.20$, so the slope is flattened considerably. The sine-wave output is only slightly more than 2(2.62) or 5.2 V p-p. A few hours spent playing with the values of R_8 through R_{15} can produce a sine-wave output with 1% harmonic distortion. Typical distortion without adjustment is 2%. Commercial function generators typically specify 0.5%.

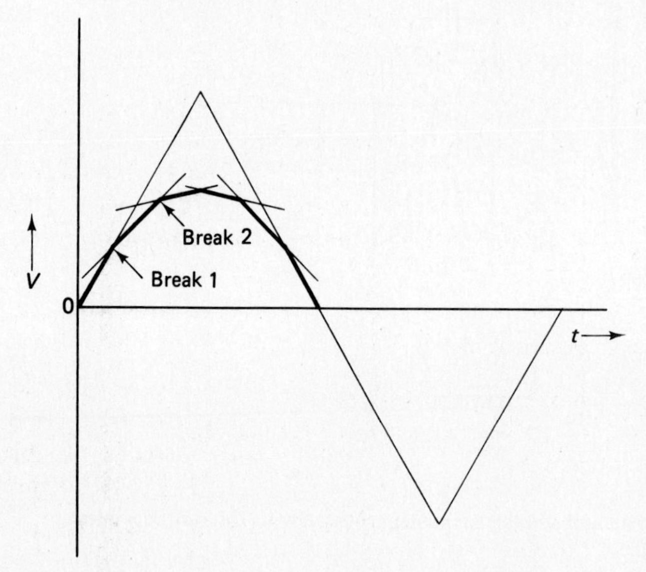

Figure 17-7 The sine wave is shaped from the triangle wave.

Amplifier: Voltage divider R_{16}–R_{17} reduces the selected waveform by a factor of 0.064 before applying it to noninverting amplifier U_3. Input levels at pin 3 are: square, 0.64 V p-p; triangle, 0.52 V p-p; and sine, 0.39 V p-p.

The minimum gain of U_3 (wiper of R_{20} at top) is

$$A_v = \frac{R_{18} + R_{19}}{R_{18}} = \frac{22 + 10}{22} = 1.45 \tag{12-12}$$

Sine-wave output at this gain is (0.39 V)(1.45) or 0.57 V p-p. With the wiper of R_{20} at bottom, negative feedback is reduced by the factor $(R_{20} + R_{21})/R_{21}$ or 33, and A_v is increased by this factor, to 48. The maximum sine-wave output is thus (0.39 V) (48) or 18.7 V p-p at no load. R_{22} maintains a 50-Ω output resistance.

R_{23} and R_{24} provide a 30-times lower output voltage (620 mV to 19 mV p-p sine wave) while maintaining a 50-Ω output impedance. C_9 is used to increase feedback at high frequencies. This compensates for stray capacitance from the wiper of R_{20} to ground which tends to decrease feedback at high frequencies. Some experimenting with its value may be necessary to produce clean high-frequency square waves.

Type-*318* op amps are used because of their high-frequency response. Type *741s* can be used for generating signals up to 10 kHz or so, and the *RC* decoupling networks at the supply pins can be omitted in this case. For both types, I_O is limited to about 15 mA, so full output voltage cannot be developed across load resistances lower than about 1000 Ω.

17.6 AN ELECTRONIC FLASH

Figure 17-8 shows an electronic flash unit used to illuminate the graticule lines for an oscilloscope camera. Here is an overview of its operation:

1. Oscillator Q_{40} chops the 6-V supply into ac which is stepped up by T_{45} and rectified by CR_{45} to 200—450 V on C_{52}.
2. Flash tube V_{50} discharges C_{52} when a 5-kV pulse from T_{50} triggers it. This pulse is initiated when Q_{64} is turned on, discharging C_{65} into the gate of SCR Q_{48}.
3. U_{10A} is a hysteresis switch which stops oscillator Q_{40} when the voltage on C_{52} reaches the desired level.
4. U_{10B} senses when the intensity level is reduced below a previously set level and reduces the charge on C_{52} through R_{17}.

We will now examine each of these circuits in detail.

Oscillator: When power is applied through S_{60}, base current builds up in Q_{40} through inductive winding 5–2 and R_{42}.

$$I_{B(40)\max} = \frac{V_{CC} - V_{BE}}{R_{42}} = \frac{6 - 0.6}{750} = 7.2 \text{ mA}$$

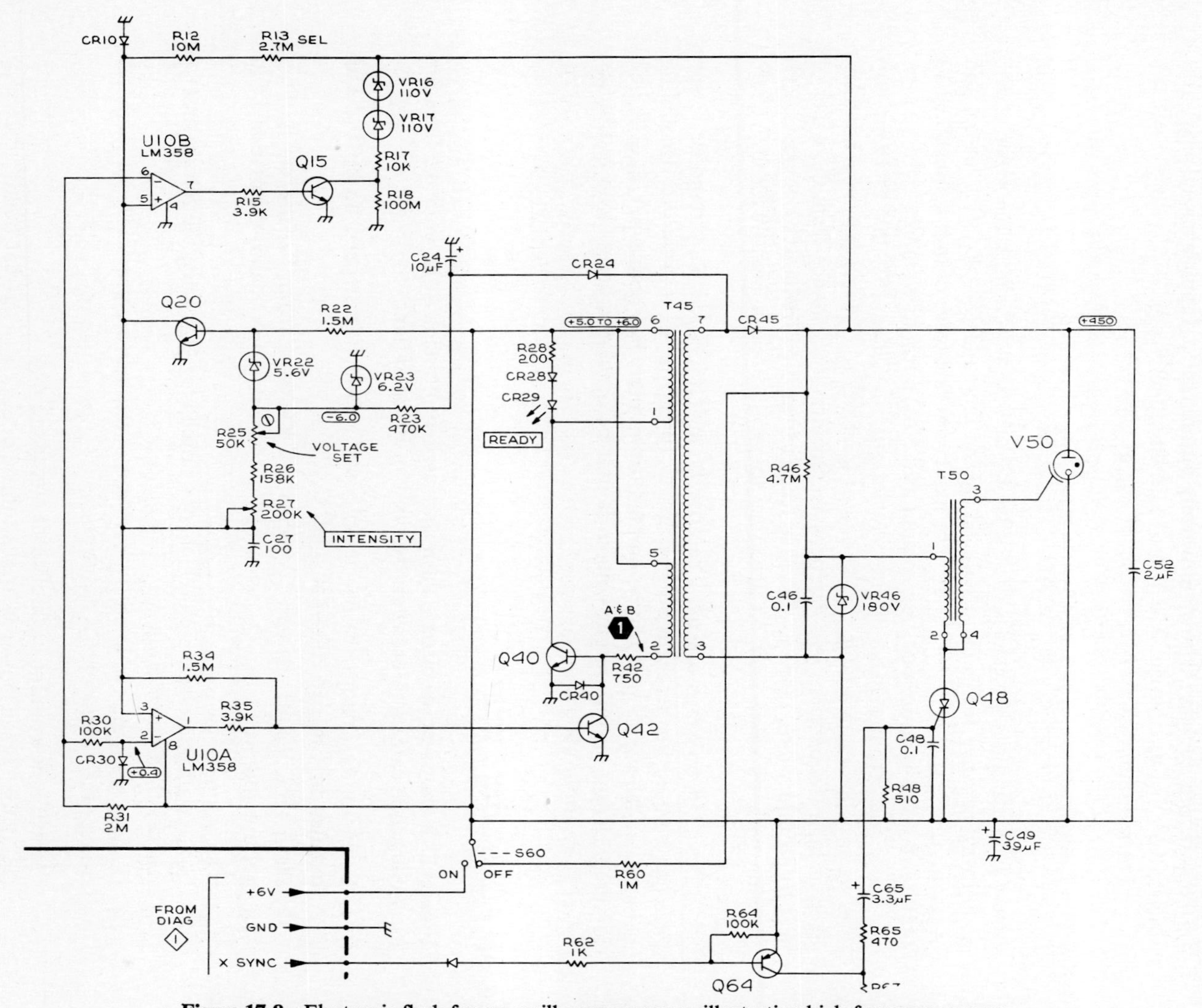

Figure 17-8 Electronic flash for an oscilloscope camera, illustrating high-frequency power supplies, SCR switching, op-amp differential comparators, and switching transistors. (Tektronix C-5C camera, Courtesy Tektronix, Inc., © 1979, Beaverton, Ore.)

This brings the collector of Q_{40} near ground and places an initial 6 V across R_{28}, CR_{28}, and CR_{29}.

$$I_{LED(max)} = \frac{V_{CC} - V_{CR28} - V_{CR29} - V_{CE(sat)}}{R_{28}}$$

$$= \frac{6.0 - 0.6 - 1.2 - 0.2}{200} = 20 \text{ mA}$$

Current also begins to rise in inductive winding 6–1. This induces a voltage in winding 5–2 (positive at 2), which adds to the turn-on current of Q_{40}. In about 5 μs the core of T_{45} saturates and the induced voltage is lost, so I_C decreases. The decreasing current in winding 6–1 induces a *reverse* voltage in winding 5–2, subtracting from the base turn-on current. Once the winding currents have returned to zero the process starts over, continuing at about a 10-kHz rate.

During the current turn-off (called flyback) high negative voltages are induced in all windings. CR_{28} protects the LED and CR_{40} protects the base of Q_{40} during this time. CR_{45} rectifies the high voltage at the secondary and charges C_{52} to as much as 450 V. The energy stored in C_{52} is

$$W_{stored} = \tfrac{1}{2}CV^2 = \tfrac{1}{2}(2\ \mu\text{F})(450)^2 = 0.2 \text{ J}$$

For comparison, 0.2 J would light a 40-W lamp for 5 ms.

Flash circuit: R_{46} charges C_{46} to 180 V. The time constant is 0.47 s, so there is no appreciable delay. Meanwhile C_{65} is charged to 6 V (positive as shown) through R_{48}, R_{65}, and R_{67}. The time constant here is only 36 ms. When the shutter-control circuit grounds point X SYNC, inverted-ground transistor Q_{64} turns on, discharging C_{65} through R_{65} and the gate of SCR Q_{48}.

$$I_{GATE(pk)} = \frac{V_C - V_{GK}}{R_{65}} = \frac{6 - 0.6}{470} = 11.5 \text{ mA}$$

The voltage-rise time constant is limited by C_{48} to prevent triggering by noise spikes.

$$\tau_{rise} = R_{65}C_{48} = (470)(0.1 \times 10^{-6}) = 47\ \mu\text{s}$$

The duration of the trigger pulse is approximately

$$\tau_{fall} = R_{65}C_{65} = (470)(3.3 \times 10^{-6}) = 1.6 \text{ ms}$$

The firing of the SCR discharges C_{46} through the primary of T_{50}. Because of the high turns ratio, a 5-kV pulse is generated at the secondary, firing the flash tube.

Intensity control: CR_{24} rectifies the smaller negative pulses from T_{45} during the *on* period of Q_{40}. C_{24} filters them to dc and R_{23} and VR_{23} regulate this dc to -6.2 V. R_{25}, R_{26}, and R_{27} apply this as one input to hysteresis switch U_{10A}. Let us assume that the string is set to 200 kΩ. $V_{C(52)}$ is applied through R_{12} and R_{13} as the other input. R_{31}, R_{30}, and CR_{30} keep the reference (inverting input) at about $+0.6$ V, so the signals are in the active region between 0 and $+6$ V. When V_{52} rises sufficiently, pin 1 of U_{10A} goes high (about $+5.5$ V) and Q_{42} turns on. This grounds the base of Q_{40}, turning it off, and charging stops. Q_{42} can sink a current far more than I_{B40}.

$$I_{42C} = \beta_{42}\frac{V_{(U10A)1} - V_{42(BE)}}{R_{35}} = (100)\frac{5.5 - 0.6}{3900} = 126 \text{ mA}$$

R_{34} feeds back +0.6 V when Q_{42} turns on to provide snap action on/off at either end of a hysteresis zone. With V_{42B} at 0.6 V, $I_{R(34)} = 0$ and $I_{R(27)} = I_{R(12)}$ at the lower trigger point.

$$\frac{V_{VR(23)} - V_{REF}}{R_{25,26,27}} = \frac{V_{C(52)} - V_{REF}}{R_{12} + R_{13}}$$

$$\frac{6.2 - 0.6}{200\ \text{k}\Omega} = \frac{V_{C(52)} - 0.6}{10\ \text{M}\Omega + 2.7\ \text{M}\Omega}$$

$$V_{C(52)} = \mathbf{365\ V}$$

With V_{42B} at 0 V, $I_{R(34)}$ adds to the negative current directed to U_{10A} pin 3, and the upper trigger point is

$$\frac{V_{VR23} - V_{REF}}{R_{25,26,27}} + \frac{V_{REF} - V_{B(42)}}{R_{34}} = \frac{V_{C(52)} - V_{REF}}{R_{12} + R_{13}}$$

$$\frac{6.2 - 0.6}{200\ \text{k}\Omega} + \frac{0.6 - 0}{1.5\ \text{M}\Omega} = \frac{V_{C(52)} - 0.6}{10\ \text{M}\Omega + 2.7\ \text{M}\Omega}$$

$$V_{C(52)} = \mathbf{361\ V}$$

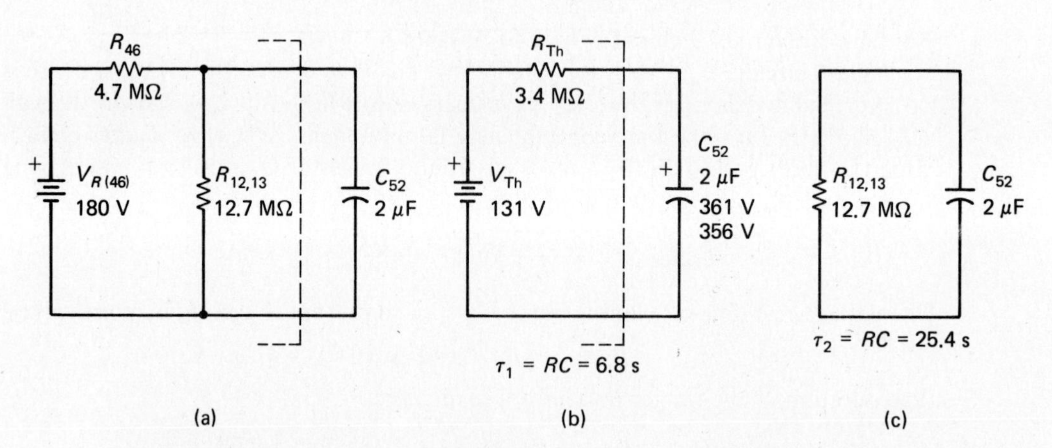

Figure 17-9 (a) Discharge path for storage capacitor C_{52}; (b) Thévenin equivalent with VR_{46} conducting; (c) with VR_{46} nonconducting.

$V_{C(52)}$ discharges from 361 V toward a V_{Th} of 131 V (a 230-V drop) through an R_{Th} of $R_{46} \| R_{12,13}$, as shown in Fig. 17-9(b). If discharge were allowed to continue, Fig. 17-9(c) would apply after VR_{46} turned off. However, after a ΔV of only 5 V (361 to 356 V), U_{10A} and Q_{42} turn the oscillator back on, recharging C_{52}. The time required for this drop is

$$\frac{\Delta V}{V_{DISCH}} = \frac{\Delta t}{\tau_1}; \qquad \frac{5}{230} = \frac{t}{6.8}; \qquad t = 150\ \text{ms}$$

LED CR_{29} is on when the oscillator runs, so it flashes at about 1/(150 ms) or 7 Hz when charging is complete.

Charge reduction of C_{52} if the intensity setting is reduced is provided by U_{10B} and Q_{15} through R_{17}. Discharge is toward 220 V ($VR_{16} + VR_{17}$) with a time constant

$$\tau_{disch} = R_{17}C_{52} = (10\ k\Omega)(2\ \mu F) = 20\ ms$$

R_{30} keeps the reference input of U_{10B} slightly more positive than that of U_{10A}, so a substantial decrease ΔV must be made before discharge begins.

$$VR_{30} = (V_{CC} - V_{CR30})\frac{R_{30}}{R_{30} + R_{31}}$$

$$= (6 - 0.6)\frac{100\ k\Omega}{100\ k\Omega + 2\ M\Omega} = 0.26\ V$$

$$V_{C(52)} = V_{R(30)}\frac{R_{12} + R_{13}}{R_{25,26,27}}$$

$$= (0.26\ V)\frac{12.7\ M\Omega}{200\ k\Omega} = 16.5\ V$$

System startup: When S_{60} is first turned on there is no negative voltage on C_{24} and the system may "hang up" with positive output from pin 1 of U_{10A} supplying positive input to pin 3. Q_{42} would then hold Q_{40} off and the charging cycle could not begin. To prevent this, Q_{20} is turned on by R_{22} until the oscillations have charged C_{24}, whereupon VR_{22} turns Q_{20} off. With Q_{20} turned on, pin 3 of U_{10A} cannot be more positive than 0.1 V, whereas pin 2 rises immediately to +0.4 V, ensuring a negative output at pin 1.

APPENDIX A
POPULAR TRANSISTOR CASES (SIDE AND BOTTOM VIEWS)

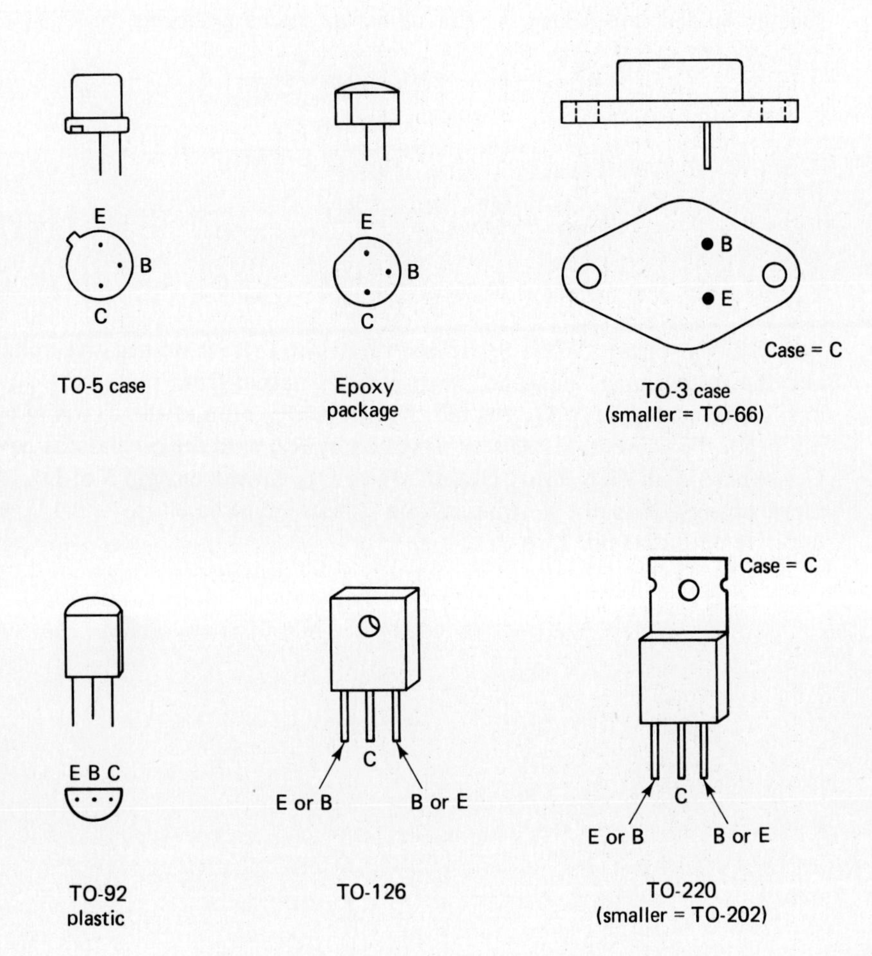

APPENDIX B
SCHEMATIC SYMBOLS

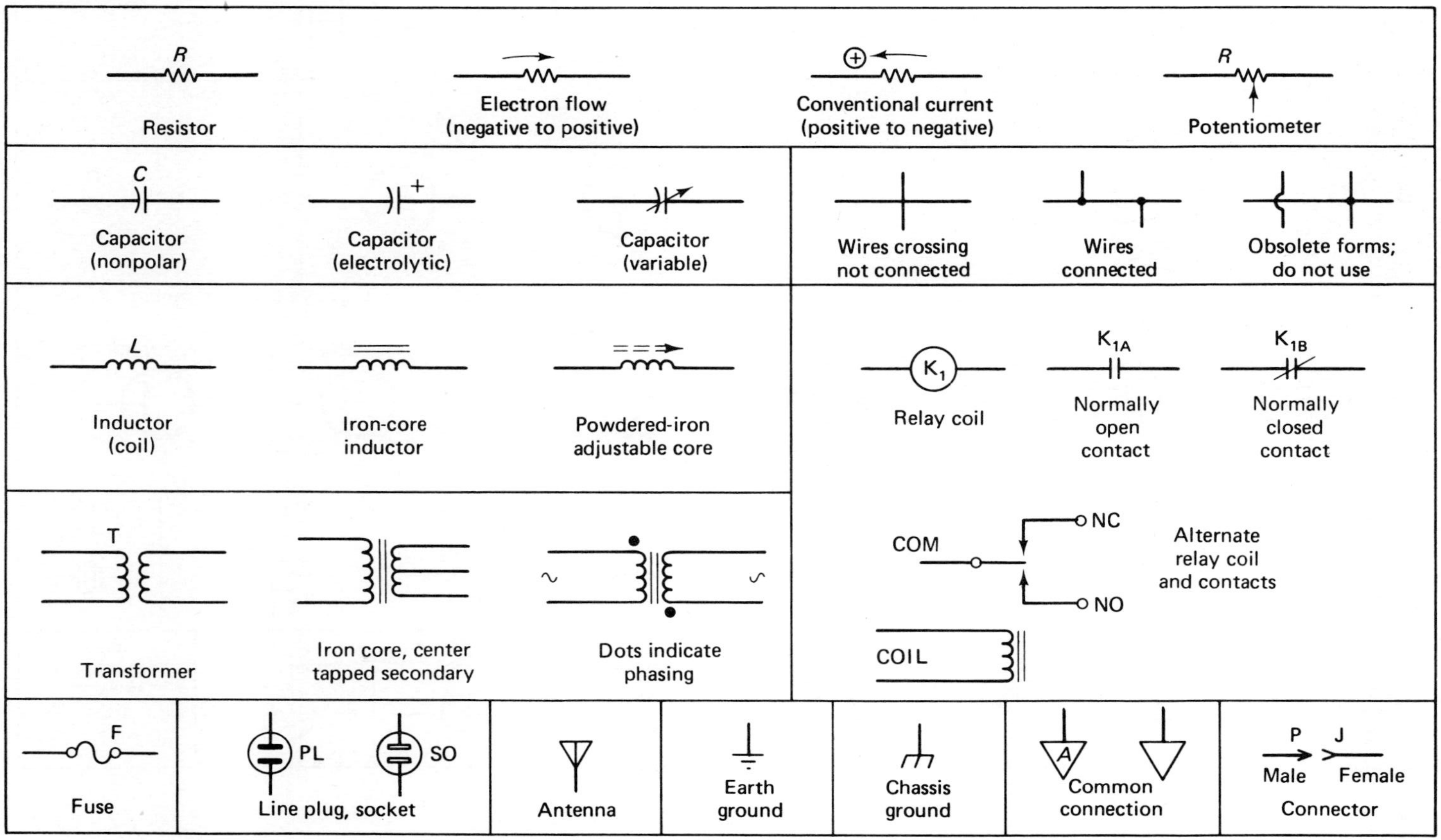

R
Resistor

Electron flow
(negative to positive)

Conventional current
(positive to negative)

R
Potentiometer

C
Capacitor
(nonpolar)

+
Capacitor
(electrolytic)

Capacitor
(variable)

Wires crossing
not connected

Wires
connected

Obsolete forms;
do not use

L
Inductor
(coil)

Iron-core
inductor

Powdered-iron
adjustable core

K_1
Relay coil

K_{1A}
Normally
open
contact

K_{1B}
Normally
closed
contact

T
Transformer

Iron core, center
tapped secondary

Dots indicate
phasing

COM
NC
NO
COIL
Alternate
relay coil
and contacts

F
Fuse

PL
SO
Line plug, socket

Antenna

Earth
ground

Chassis
ground

A
Common
connection

P
J
Male
Female
Connector

APPENDIX B
SCHEMATIC SYMBOLS (Continued)

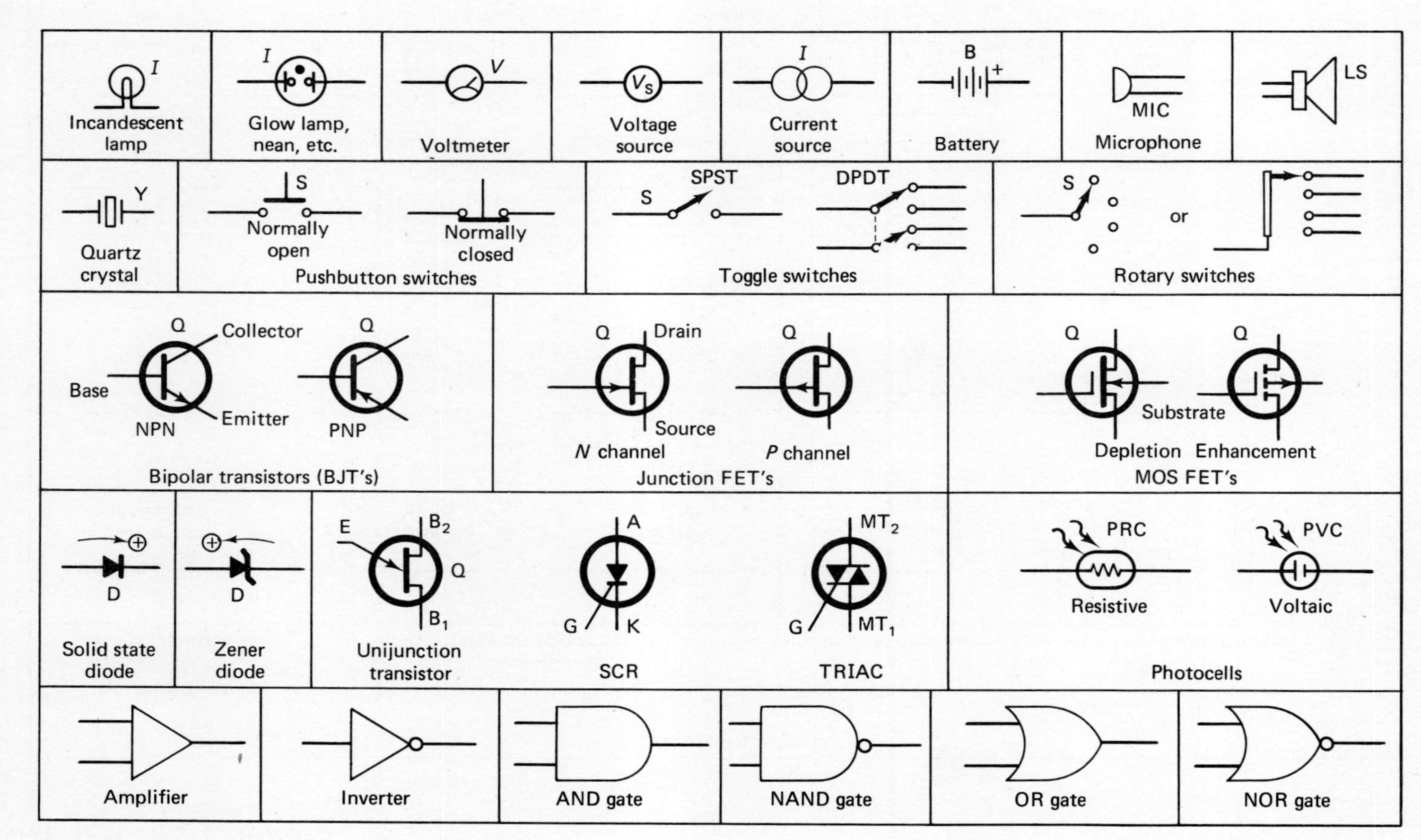

APPENDIX C
REACTANCE CHART

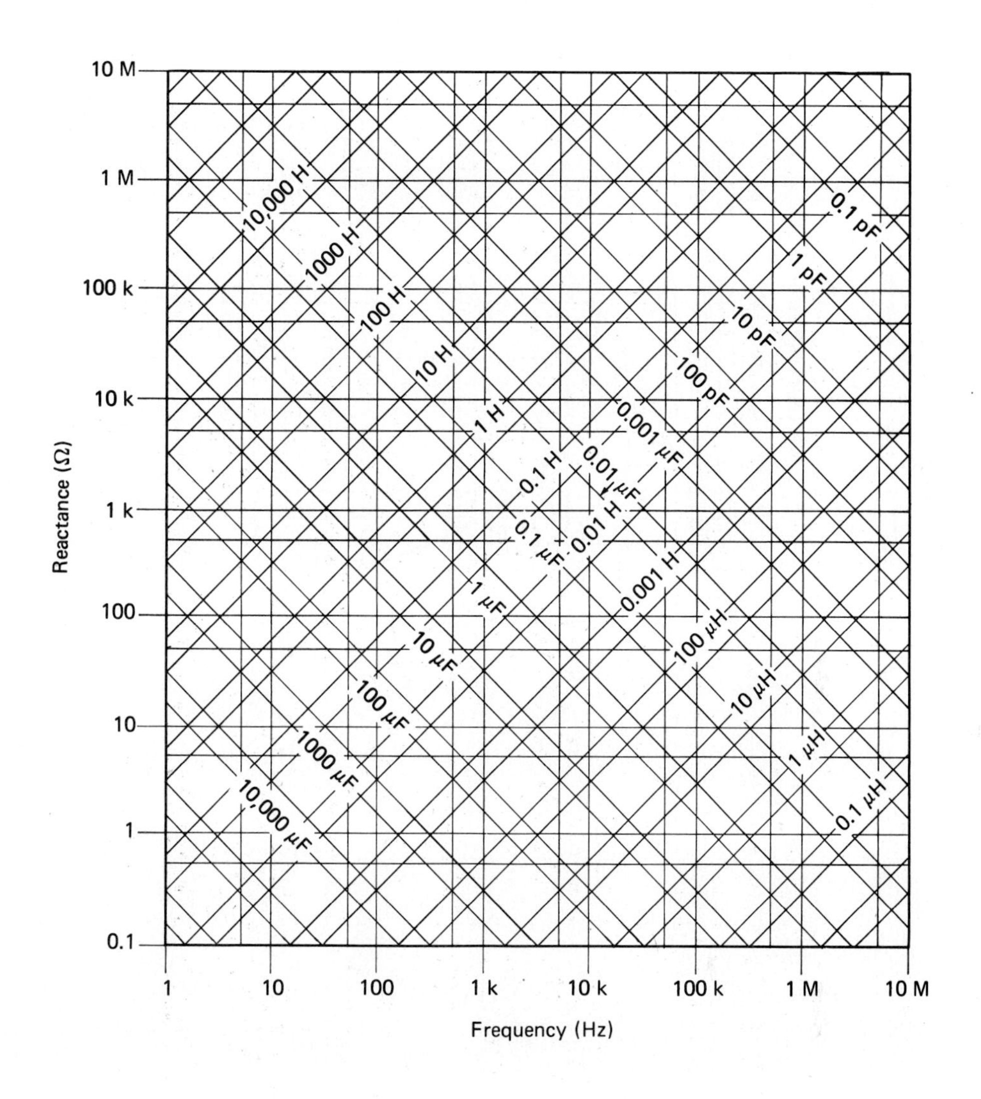

APPENDIX D
UNIVERSAL TIME-CONSTANT CHART

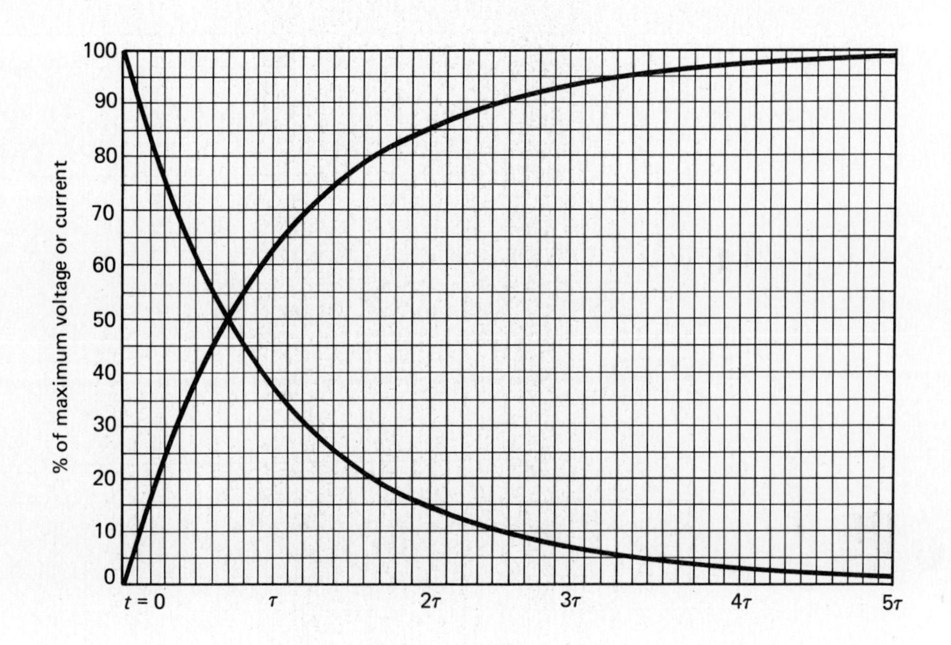

APPENDIX E
DERIVATION OF EQUATIONS

Equation (3-8): Maximum input capacitance for a given diode surge-current rating.

Assuming a specified surge time of 20 ms and a linear charging of the capacitor:

$$CV = It \qquad (3\text{-}2)$$

$$C = \frac{It}{V} = \frac{0.02 I_{\text{surge}}}{V_{s(\text{pk})}}$$

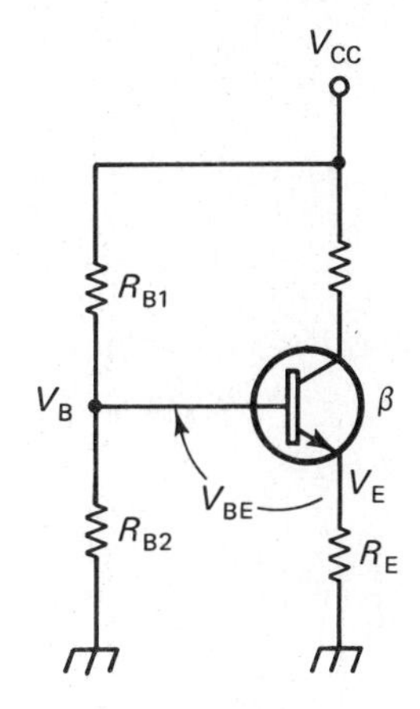

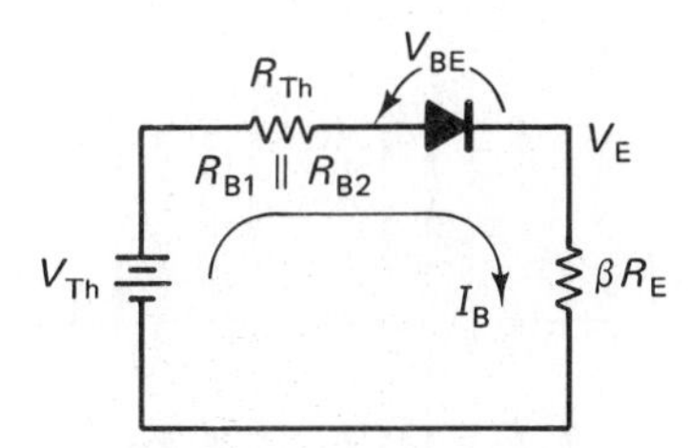

Figure E-1

Equation (6-21): Bias stability factor for voltage-divider bias circuit (see Fig. E-1).

$$S = \frac{I_{E(\min)}(\text{at } \beta_{\min})}{I_{E(\max)}(\text{at } \beta \longrightarrow \infty)}$$

$$I_{E(\max)} = \frac{V_{Th} - V_{BE}}{R_E}$$

(since $V_{R(Th)} \longrightarrow 0$ when $\beta \longrightarrow \infty$)

$$I_{E(\min)} = \beta I_B = \frac{\beta(V_{Th} - V_{BE})}{R_{Th} + \beta R_E}$$

$$S = \frac{I_{E(\min)}}{I_{E(\max)}} = \frac{\beta R_E}{R_{Th} + \beta R_E}$$

$$= \frac{\beta_{\min} R_E}{R_{B1} \| R_{B2} + \beta_{\min} R_E}$$

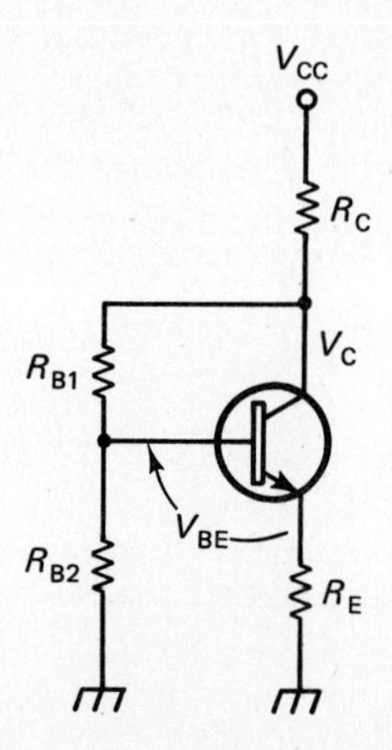

Figure E-2

Equation (7-6): Ideal V_C for collector-self-bias circuit (see Fig. E-2). Assuming that $\beta \longrightarrow \infty$ so $I_B \longrightarrow 0$ and $I_E = I_C$:

$$V_C = V_{CC} - I_C R_C$$

$$I_C = \frac{V_B - V_{BE}}{R_E} = \frac{V_C \dfrac{R_{B2}}{R_{B1} + R_{B2}} - V_{BE}}{R_E}$$

$$V_C = V_{CC} - \left(\frac{V_C R_{B2}}{R_{B1} + R_{B2}} - V_{BE}\right)\frac{R_C}{R_E}$$

$$V_C + V_C \frac{R_C R_{B2}}{R_E(R_{B1} + R_{B2})} = V_{CC} + \frac{R_C V_{BE}}{R_E}$$

$$V_C = \frac{V_{CC} + \dfrac{R_C V_{BE}}{R_E}}{1 + \dfrac{R_C R_{B2}}{R_E(R_{B1} + R_{B2})}} = \frac{V_{CC} R_E + R_C V_{BE}}{R_E + \dfrac{R_C R_{B2}}{R_{B1} + R_{B2}}}$$

Equation (8-12): FET source self-bias circuit. We term $V_{GS(off)}$ the pinchoff voltage V_P.

$$V_{GS} = I_D R_S = V_P\left(1 - \sqrt{\frac{I_D}{I_{DSS}}}\right) \quad (8\text{-}3)$$

$$R_S = \frac{V_P}{I_D} - \left(\frac{V_P}{\sqrt{I_{DSS}}}\right)\frac{1}{\sqrt{I_D}}$$

$$V_P \frac{1}{\sqrt{I_D}^2} - \frac{V_P}{\sqrt{I_{DSS}}}\left(\frac{1}{\sqrt{I_D}}\right) - R_S = 0$$

This is a quadratic equation of the form $ax^2 + bx + c = 0$ with the term $x = 1/\sqrt{I_D}$. By the quadratic formula:

$$x = \frac{-b \pm \sqrt{b^2 - 4ac}}{2a}$$

$$\frac{1}{\sqrt{I_D}} = \frac{\dfrac{V_P}{\sqrt{I_{DSS}}} \pm \sqrt{\dfrac{V_P^2}{I_{DSS}} + 4V_P R_S}}{2V_P}$$

$$= \frac{1}{2\sqrt{I_{DSS}}} \pm \sqrt{\frac{1}{4I_{DSS}} + \frac{R_S}{V_P}}$$

$$I_D = \left(\frac{1}{\dfrac{1}{2\sqrt{I_{DSS}}} \pm \sqrt{\dfrac{1}{4I_{DSS}} + \dfrac{R_S}{V_P}}}\right)^2$$

$$= \frac{I_{DSS}}{\left(0.5 \pm \sqrt{0.25 + \dfrac{I_{DSS}R_S}{V_P}}\right)^2}$$

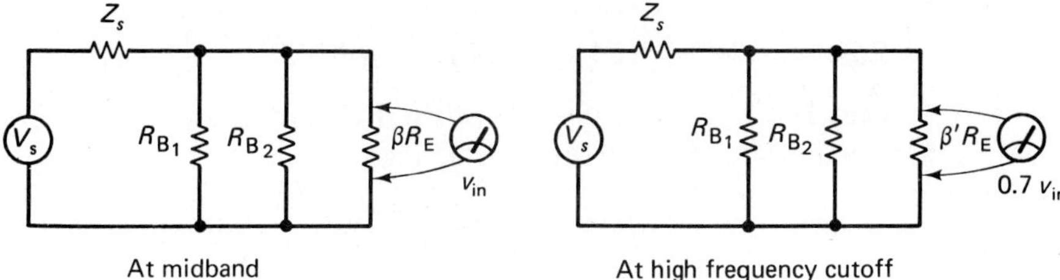

Figure E-3

Equation (9-19): High-frequency beta (β') for cutoff in a common-emitter amplifier (see Fig. E-3).

Let R_B indicate $R_{B1} \| R_{B2}$:

$$V_{in} = V_s \frac{(\beta R_E) \| R_B}{Z_s + (\beta R_E) \| R_B}$$

$$0.7V_{in} = V_s \frac{(\beta' R_E) \| R_B}{Z_s + (\beta' R_E) \| R_B} = 0.7V_s \frac{(\beta R_E) \| R_B}{Z_s + (\beta R_E) \| R_B}$$

$$\frac{\beta' R_E R_B}{(\beta' R_E + R_B)\{Z_s + [\beta' R_E R_B/(\beta' R_E + R_B)]\}} = 0.7 \frac{\beta R_E R_B}{(\beta R_E + R_B)\{Z_s + [\beta R_E R_B/(\beta R_E + R_B)]\}}$$

$$Z_s\beta'\beta R_E + \frac{\beta'\beta^2 R_E^2 R_B}{\beta R_E + R_B} + Z_s\beta' R_B + \frac{\beta'\beta R_E R_B^2}{\beta R_E + R_B}$$
$$= 0.7Z_s\beta\beta' R_E + \frac{0.7\beta(\beta')^2 R_E^2 R_B}{\beta' R_E + R_B} + 0.7Z_s\beta R_B + \frac{0.7\beta\beta' R_E R_B^2}{\beta' R_E + R_B}$$

$$0.3Z_s\beta\beta' R_E + 0.3\beta\beta' R_E R_B + Z_s\beta' R_B = 0.7Z_s\beta R_B$$

$$\beta' = \frac{0.7Z_s\beta R_B}{0.3Z_s\beta R_E + 0.3\beta R_E R_B + Z_s R_B}$$

$$\frac{1}{\beta'} = \frac{0.3Z_s\beta R_E}{0.7Z_s\beta R_B} + \frac{0.3\beta R_E R_B}{0.7Z_s\beta R_B} + \frac{Z_s R_B}{0.7Z_s\beta R_B}$$

$$\beta' = \frac{0.7}{(0.3R_E/R_B) + (0.3R_E/Z_s) + (1/\beta)}$$

Equation (10-19): Q of a parallel-resonant circuit with series winding and parallel load resistances.

$$Q_{coil} = \frac{X}{r_w}; \qquad R_p = Q_{coil}X$$

$$Q_{ckt} = \frac{R_p \| R_L}{X} = \frac{Q_{coil}X \| R_L}{X} = \frac{\frac{X^2}{r_w} \Big\| R_L}{X}$$

$$= \frac{\frac{X^2}{r_w}R_L}{X\left(\frac{X^2}{r_w} + R_L\right)} = \frac{XR_L}{X^2 + R_L r_w} = \frac{1}{\frac{X}{R_L} + \frac{1}{Q_{coil}}}$$

Equation (11-5): Voltage gain of a transformer-coupled amplifier.
I_c is ac collector signal current, I_L is ac load current, and $n = N_S/N_P$.

$$I_c = \frac{V_{in}}{r_j + R_{E1}}$$

$$I_L = \frac{I_c}{n} = \frac{V_{in}}{n(r_j + R_{E1})}$$

$$V_L = I_L R_L = \frac{V_{in}R_L}{n(r_j + R_{E1})}$$

$$A_v = \frac{V_L}{V_{in}} = \frac{R_L}{n(r_j + R_{E1})}$$

Equation (11-6): Optimum bias point for a transformer-coupled class-A amplifier (see Fig. E-4).

At turn-off, I_Q goes to zero:

$$V_{p(pk)} = I_Q\frac{r_S + R_L}{n^2}$$

At the Q point:

$$V_{CE} = V_{CC} - I_Q(r_P + R_{E1} + R_{E2})$$

and

$$V_p = 0$$

At turn-on, I_Q doubles through r_P and R_{E1}:

$$V_{p(pk)} = V_{CC} - V_{CE(sat)} - I_Q R_{E2} - 2I_Q(r_P + R_{E1})$$

At optimum I_Q the swings to turn-on and turn-off are equal. Combining the two euqations for $V_{p(pk)}$:

$$I_Q \frac{r_S + R_L}{n^2} = V_{CC} - V_{CE(sat)} - I_Q(R_{E2} + 2r_P + 2R_{E1})$$

$$I_Q = \frac{V_{CC} - V_{CE(sat)}}{\frac{r_S + R_L}{n^2} + R_{E2} + 2R_{E1} + 2r_P}$$

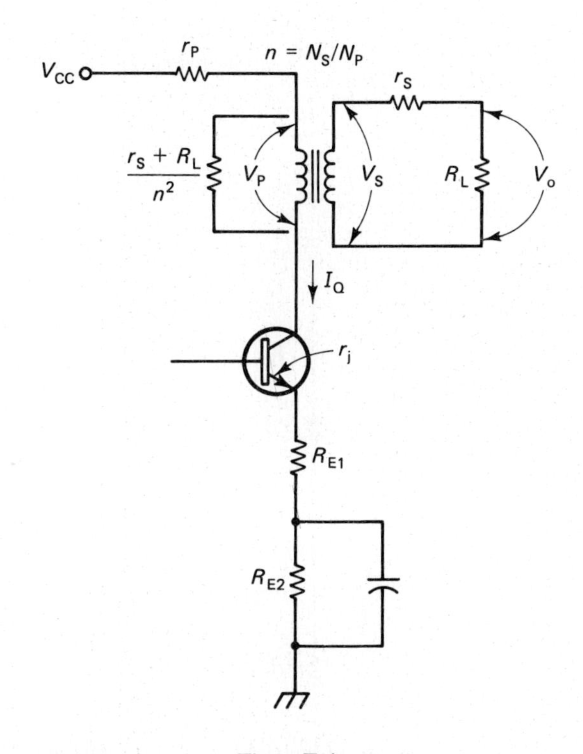

Figure E-4

Equation (11-7): Maximum peak output for an optimally biased transformer-coupled class-A amplifier.

$$V_{R(L)pk} = I_{s(pk)}R_L$$

$$= \frac{I_{p(pk)}}{n}R_L$$

$$= \frac{I_Q R_L}{n}$$

Equation (11-8): Maximum output for a class-B transformer-coupled amplifier (see Fig. E-4).

From cutoff where $I_C = 0$ to saturation, V_{CE} swings from V_{CC} to $V_{CE(sat)}$. The reflected resistance divides this voltage with R_P and R_E.

$$V_{p(pk)} = (V_{CC} - V_{CE(sat)})\frac{\dfrac{r_S + R_L}{n^2}}{r_P + \dfrac{r_S + R_L}{n^2} + R_E}$$

$$V_{R(L)pk} = n\left(\frac{R_L}{r_S + R_L}\right)V_{p(pk)}$$

$$= n\left(\frac{R_L}{r_S + R_L}\right)\frac{(r_S + R_L)(V_{CC} - V_{CE(sat)})}{n^2\left(r_P + \dfrac{r_S + R_L}{n^2} + R_E\right)}$$

$$= \frac{(V_{CC} - V_{CE(sat)})R_L}{n(r_P + R_E) + \dfrac{r_S + R_L}{n}}$$

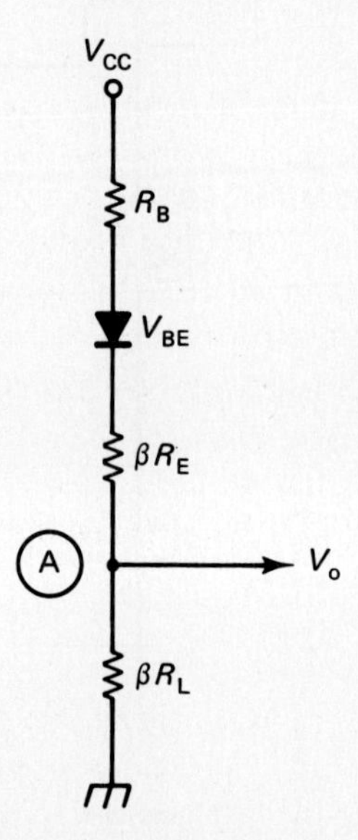

Figure E-5

Equation (11-11): Maximum output for a complementary-symmetry amplifier (see Figs. 11-12 and E-5).

Point A is at $\frac{1}{2}V_{CC}$ with no signal. At maximum positive signal R_B turns Q_2 on.

$$V_{o(pk)} = (V_{CC} - V_{BE})\frac{\beta R_L}{R_B + \beta R_E + \beta R_L}$$

$$= \frac{(V_{CC} - V_{BE})\beta R_L}{R_B + \beta(R_E + R_L)}$$

APPENDIX F
REVIEW OF ELECTRIC-CIRCUIT CONCEPTS

F.1 Thévenin's Theorem

This theorem states that any circuit composed entirely of voltage sources and resistors, no matter how complex, behaves the same as a simple series circuit containing only one voltage source and one resistor. The only problem is to find the values of voltage and resistance for that simple equivalent circuit.

The rules for finding the Thévenin equivalent values are:

1. Pick two points of the circuit to be the output terminals. These will generally be the two ends of a component for which it is desired to find the voltage, current, or power.
2. Remove the component (or components) connected between these two terminals, and analyze the remaining circuit to find the voltage between the two points with the "load" open-circuited. This is the voltage of the Thévenin equivalent circuit. (Removing the load will generally modify the circuit in such a way that two resistors will appear in series, although they were not so connected previously. Combining these resistors then leads to the solution.)
3. Mentally replace all voltage sources in the circuit by short circuits, and determine the resistance seen looking back into the two output terminals. This is the Thévenin equivalent resistance.
4. Analyze the Thévenin equivalent circuit with the load components reconnected at the output terminals.

Example F-1

Find the voltage across R_3 in the circuit of Fig. F-1(a).

Solution The output terminals are the top and bottom terminals of R_3. Since R_3 is the load element, we remove it and find the voltage at the open-circuited output terminals:

$$V_{R2} = V_1 \frac{R_2}{R_1 + R_2} = (1.5)\frac{2\text{ k}\Omega}{1\text{ k}\Omega + 2\text{ k}\Omega} = 1\text{ V}$$

$$V_{Th} = V_{OUT} = V_{R2} - V_2 = 1.0 - 0.6 = \mathbf{0.4\ V}$$

To find R_{Th}, we consider V_1 and V_2 to be short circuits in Fig. F-1(a). Looking back into the circuit from the output terminals with R_3 still removed, we see R_2 in parallel with R_1.

$$R_{Th} = R_2 \,||\, R_1 = 2\text{ k}\Omega \,||\, 1\text{ k}\Omega = \mathbf{667\ \Omega}$$

The Thévenin equivalent circuit with the load R_3 reconnected is shown in Fig. F-1(b). It is now a simple matter to find the voltage across R_3.

$$V_{R3} = V_{Th}\frac{R_3}{R_{Th} + R_3} = (0.4)\frac{1\text{ k}\Omega}{0.667\text{ k}\Omega + 1\text{ k}\Omega}$$

$$= \mathbf{0.239\ V}$$

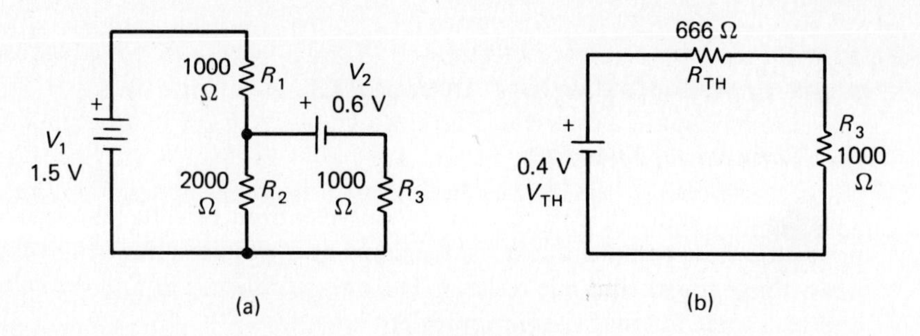

Figure F-1

F.2 Time Constant

When a capacitor is charged or discharged through a resistor, the voltage vs. time and the current vs. time graphs follow a definitely shaped curve which is called a negative exponential curve. This same curve is followed by the voltage and current in

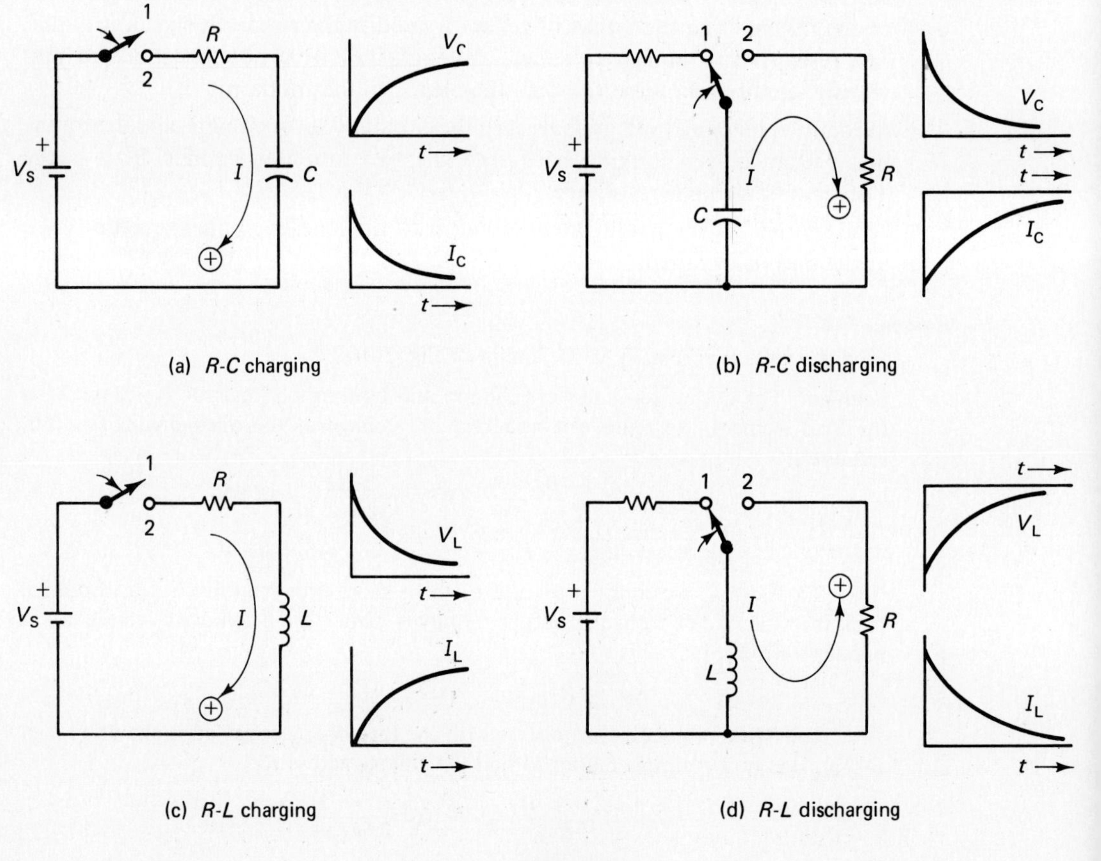

Figure F-2

a simple *RL* circuit. Figure F-2 summarizes the general shapes of the negative exponential curves for charging and discharging *RC* and *RL* circuits.

Notice that in all parts of Fig. F-2 there is only one basic negative-exponential-shaped curve—sometimes rising and sometimes falling. Notice also that the rise or decay starts out at a rapid rate but slows down to almost no change at all after a certain time. Theoretically, this rise or fall never stops completely—it just gets slower and slower until no further change can be measured.

Appendix D gives an expanded view of a negative exponential curve. If the rate of decay were constant at the initial value, the curve would drop to zero in a length of time designated by τ. Actually, the rate of decay diminishes as soon as the decay starts, so, in fact, the curve drops only to 37% of its full value in time τ. For *RC* and *RL* circuits, this time τ is called the *time constant* and can be easily calculated:

$$\tau = RC$$

$$\tau = \frac{L}{R}$$

where τ is in seconds, R is in ohms, C is in farads, and L is in henrys.

The general statement can be made that one time constant is the length of time that it takes for the voltage or current in a circuit to fall (or rise) 63% of the way to its final value. Thus in Fig. F-2(b), the capacitor's voltage will drop from V_S to 37% of V_S in the time τ. After 2τ it will drop to 37% of its value at τ, and after 3τ it will drop to 37% of its value at 2τ. The following table summarizes the values of the negative exponential for both rising and falling curves:

	Rising (%)	Falling (%)
$t = 0$	0	100
$t = \tau$	63	37
$t = 2\tau$	86	14
$t = 3\tau$	95	5
$t = 4\tau$	98.1	1.9
$t = 5\tau$	99.3	0.7

As can be seen from the table, the curve never reaches its "final" value, but after 5τ it has gone more than 99% of the way, and we often consider the action to be essentially complete after 5τ.

A region of special interest in the figure of Appendix D is that part of the curve which lies between $t = 0$ and $t = 0.2\tau$. In this interval the negative exponential curve is nearly identical with a straight line from 100% at $t = 0$ to 0% at $t = \tau$. Thus it can be predicted with reasonable accuracy that in 0.1τ the curve will decay by 10%, to 90% of its full value, and that a decay to 80% of full value (20% decay) will take 0.2τ. This approximation is used extensively in Chapter 3 on power supplies.

Example F-2

The capacitor in Fig. F-2(b) is charged to 18 V by the battery in position 1. Approximately how long will it take for this voltage to decay 0.6 V when the switch is thrown to position 2? $R = 15\ \text{k}\Omega$ and $C = 5\ \mu\text{F}$.

Solution We want to determine the time to decay to 0.6/18 or 3% of full value. The curve of Appendix D indicates that a decay to 3% will take about 3.5τ. For the R and C values given,

$$\tau = RC = (15 \times 10^3)(5 \times 10^{-6}) = 75 \times 10^{-3}$$

$$3.5\tau = 3.5(75 \times 10^{-3}) = 263 \times 10^{-3} = \mathbf{263\ ms}$$

Example F-3

What value resistor is needed to completely discharge a 1000-μF capacitor in 4 s? How much time will be required for it to discharge to 90% of full voltage through this resistor?

Solution Complete discharge is assumed to occur after 5τ. If $5\tau = 4$ s, then $\tau = \frac{4}{5} =$ 0.8 s.

$$\tau = RC$$

$$R = \frac{\tau}{C} = \frac{0.8}{1000 \times 10^{-6}} = \frac{0.8}{10^{-3}} = \mathbf{800\ \Omega}$$

One time constant is 0.8 s, and a decay of 10% (to 90%) will take approximately 10% of one time constant.

$$t_{90\%} = 0.1\tau = (0.1)(0.8) = \mathbf{0.08\ s}$$

F.3 Peak and RMS Values of AC

An ac waveform is not delivering its peak power all through the cycle. In fact, at the 0° and 180° points, the power delivered by an ac waveform is zero. To compensate for these "down" portions of the cycle, the *peak* ac voltage must be considerably higher than the overall *effective* voltage. In fact, an ac wave with an effective voltage of 100 V must have a *peak* voltage of $100 \times \sqrt{2}$, or about 141 V. To put it another way, a light bulb connected to an ac source of 100 V effective (141 V peak) would be equally bright when connected to a 100-V dc source. This *effective* or *equivalent-to-dc* measure of ac is commonly called *rms* after the mathematical technique (root-mean-square) used to derive it. The rms system is the standard way of measuring an ac sine wave, and unless specifically stated otherwise, any reference such as "the 120-V ac line" implies rms measure.

On an oscilloscope it is most convenient to measure from the top to the bottom of a wave (peak to peak), and a 100-V rms wave would thus be measured as 282 V p-p. To convert from one of these systems to the other, the formulas given with Fig. F-3 may be used.

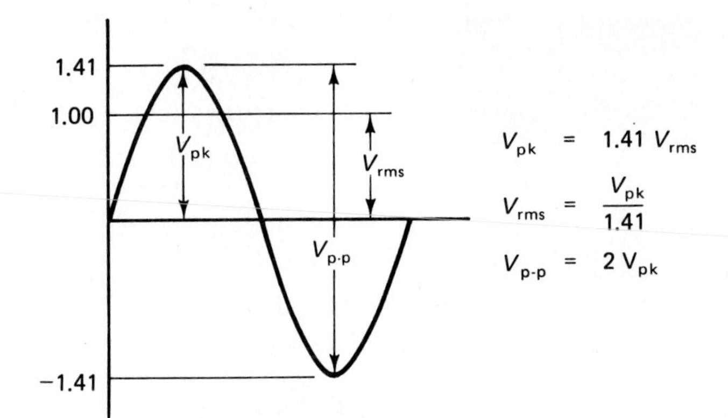

Figure F-3

A few precautions concerning these systems of measure may be in order:

1. The ratio of peak to rms value is 1.41 for *sine waves only*. Other waveforms have different ratios entirely.
2. All power calculations must be done in the rms system of measure, unless *instantaneous* power is *really* what is desired (which is not often the case).

Example F-4

What is the *peak* value of the 120-V ac line?

Solution

$$V_{pk} = 1.41\,V_{rms} = 1.41 \times 120$$
$$= \mathbf{169\ V}$$

Example F-5

An oscilloscope is used to measure a voltage of 17 V p-p across a 47-Ω resistor. What is the power being dissipated in the resistor?

Solution

$$V_{pk} = \frac{V_{p\text{-}p}}{2} = \frac{17}{2} = 8.5\text{ V}$$

$$V_{rms} = \frac{V_{pk}}{1.41} = \frac{8.5}{1.41} = 6.03\text{ V}$$

$$P = \frac{V^2}{R} = \frac{6.03^2}{47} = \mathbf{0.77\ W}$$

APPENDIX G
SELECTED DEVICE SPECIFICATIONS

1N4001 thru 1N4007

Designers▲Data Sheet

"SURMETIC"▲ RECTIFIERS

. . . subminiature size, axial lead mounted rectifiers for general-purpose low-power applications.

Designers Data for "Worst Case" Conditions

The Designers▲ Data Sheets permit the design of most circuits entirely from the information presented. Limit curves — representing boundaries on device characteristics — are given to facilitate "worst case" design.

LEAD MOUNTED SILICON RECTIFIERS

50-1000 VOLTS DIFFUSED JUNCTION

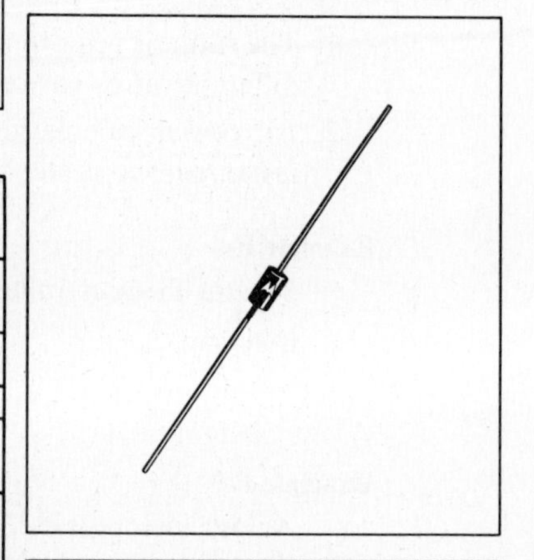

*MAXIMUM RATINGS

Rating	Symbol	1N4001	1N4002	1N4003	1N4004	1N4005	1N4006	1N4007	Unit
Peak Repetitive Reverse Voltage Working Peak Reverse Voltage DC Blocking Voltage	V_{RRM} V_{RWM} V_R	50	100	200	400	600	800	1000	Volts
Non-Repetitive Peak Reverse Voltage (halfwave, single phase, 60 Hz)	V_{RSM}	60	120	240	480	720	1000	1200	Volts
RMS Reverse Voltage	$V_{R(RMS)}$	35	70	140	280	420	560	700	Volts
Average Rectified Forward Current (single phase, resistive load, 60 Hz, see Figure 8, $T_A = 75°C$)	I_O	1.0							Amp
Non-Repetitive Peak Surge Current (surge applied at rated load conditions, see Figure 2)	I_{FSM}	30 (for 1 cycle)							Amp
Operating and Storage Junction Temperature Range	T_J, T_{stg}	-65 to +175							°C

*ELECTRICAL CHARACTERISTICS

Characteristic and Conditions	Symbol	Typ	Max	Unit
Maximum Instantaneous Forward Voltage Drop ($i_F = 1.0$ Amp, $T_J = 25°C$) Figure 1	v_F	0.93	1.1	Volts
Maximum Full-Cycle Average Forward Voltage Drop ($I_O = 1.0$ Amp, $T_L = 75°C$, 1 inch leads)	$V_{F(AV)}$	–	0.8	Volts
Maximum Reverse Current (rated dc voltage)	I_R			μA
$T_J = 25°C$		0.05	10	
$T_J = 100°C$		1.0	50	
Maximum Full-Cycle Average Reverse Current ($I_O = 1.0$ Amp, $T_L = 75°C$, 1 inch leads	$I_{R(AV)}$	–	30	μA

*Indicates JEDEC Registered Data.

MECHANICAL CHARACTERISTICS

CASE: Void free, Transfer Molded
MAXIMUM LEAD TEMPERATURE FOR SOLDERING PURPOSES: 350°C, 3/8" from case for 10 seconds at 5 lbs. tension
FINISH: All external surfaces are corrosion-resistant, leads are readily solderable
POLARITY: Cathode indicated by color band
WEIGHT: 0.40 Grams (approximately)

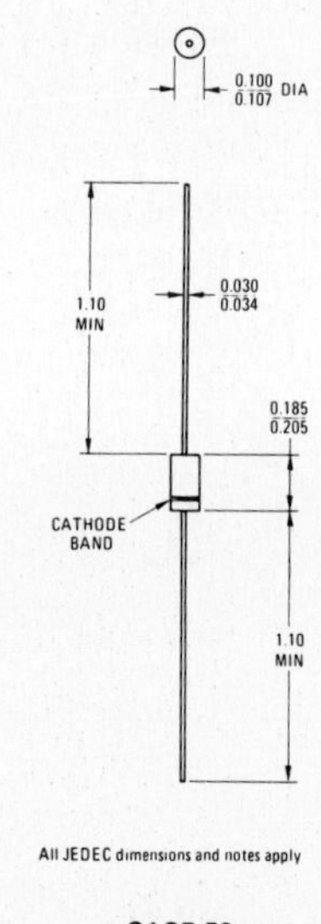

All JEDEC dimensions and notes apply

CASE 59
DO-41

1N4001 THRU 1N4007

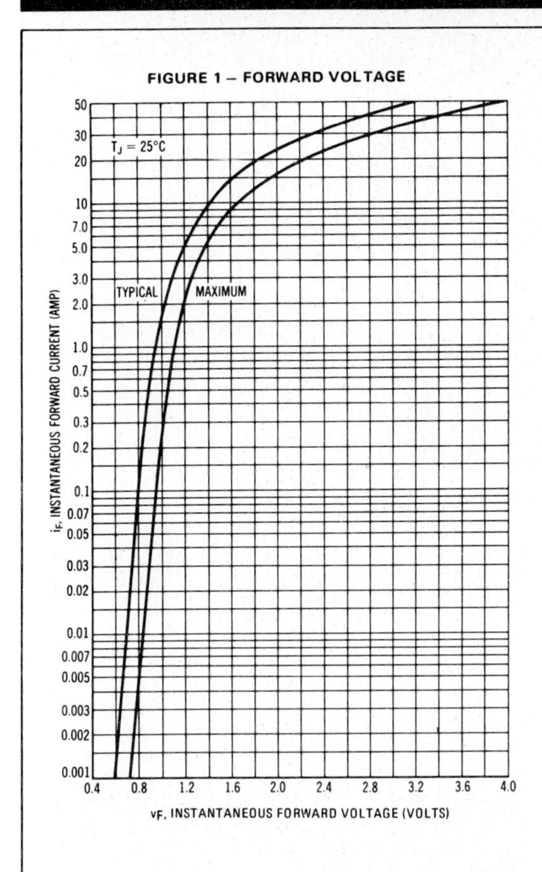

FIGURE 1 – FORWARD VOLTAGE

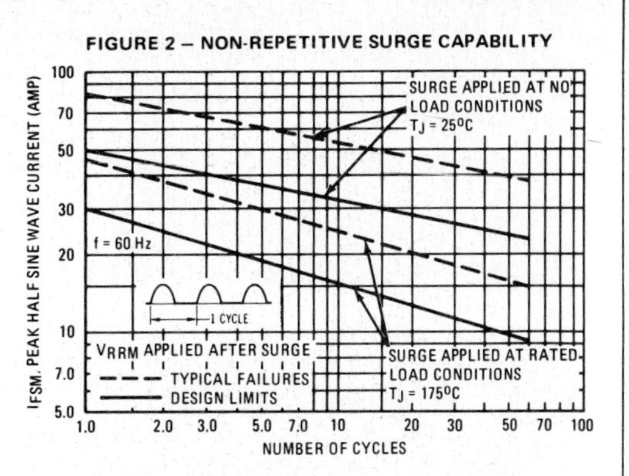

FIGURE 2 – NON-REPETITIVE SURGE CAPABILITY

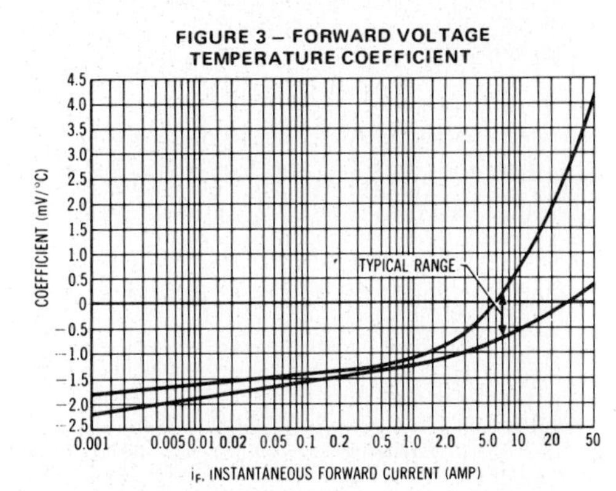

FIGURE 3 – FORWARD VOLTAGE TEMPERATURE COEFFICIENT

TYPICAL DYNAMIC CHARACTERISTICS

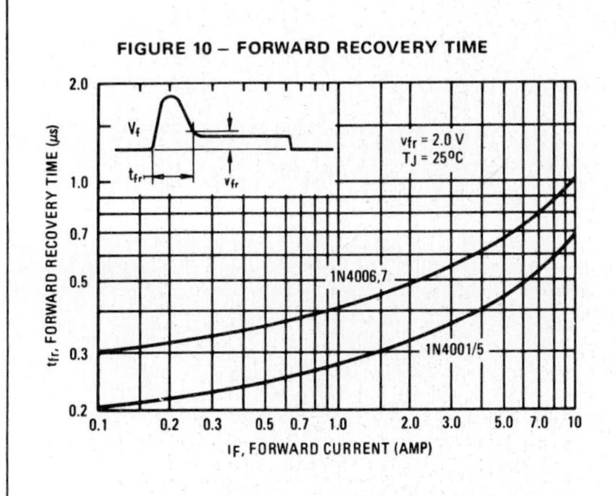

FIGURE 10 – FORWARD RECOVERY TIME

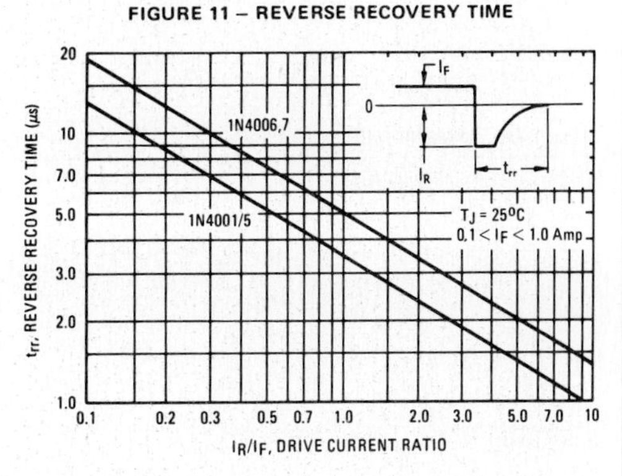

FIGURE 11 – REVERSE RECOVERY TIME

MOTOROLA *Semiconductor Products Inc.*

(Courtesy of Motorola Semiconductor Products, Inc.)

2N4123
2N4124

NPN SILICON SWITCHING & AMPLIFIER TRANSISTORS

FEBRUARY 1966 — DS 5143

NPN SILICON ANNULAR* TRANSISTORS

. . . designed for general purpose switching and amplifier applications and for complementary circuitry with PNP types 2N4125 and 2N4126.

- General Purpose Switching from 10 mA to 100 mA
- Amplifier Applications from Audio to > 100 MHz
- Wide-Band Audio Noise Figure = 5-6 dB maximum
- Low-Leakage, High Stability Annular Structure
- One-Piece, Injection-Molded Unibloc† Package for High Reliability

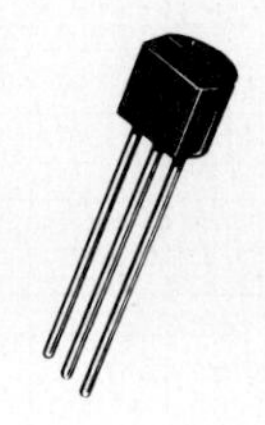

MAXIMUM RATINGS

Characteristic	Symbol	2N4123	2N4124	Unit
Collector-Emitter Voltage	V_{CEO}	30	25	Vdc
Collector-Base Voltage	V_{CB}	40	30	Vdc
Emitter-Base Voltage	V_{EB}	5		Vdc
Collector Current	I_C	200		mAdc
Total Device Dissipation @ T_A = 60°C	P_D	210		mW
Total Device Dissipation @ T_A = 25°C Derate above 25°C	P_D	310 2.81		mW mW/°C
Operating and Storage Junction Temperature Range	T_J, T_{stg}	-55 to +135		°C

THERMAL CHARACTERISTICS

Characteristic	Symbol	Max	Unit
Thermal Resistance, Junction to Ambient	θ_{JA}	0.357	°C/mW

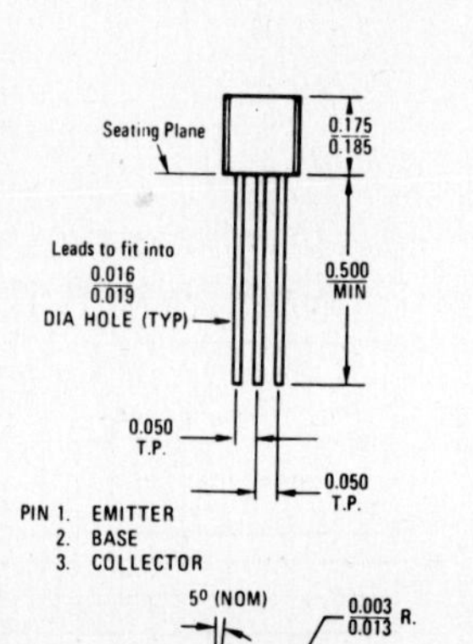

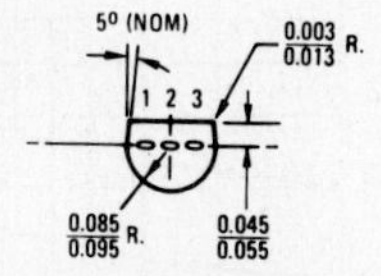

To convert inches to millimeters multiply by 25.4

All JEDEC dimensions and notes apply

CASE 29-01
TO-92

*Annular Semiconductors patented by Motorola Inc.
†Trademark of Motorola Inc.

A SUBSIDIARY OF MOTOROLA INC

(*Courtesy of Motorola Semiconductor Products, Inc.*)

2N4123 / 2N4124

ELECTRICAL CHARACTERISTICS (T_A = 25°C unless otherwise noted)

Characteristic		Fig. No.	Symbol	Min	Max	Unit
OFF CHARACTERISTICS						
Collector-Emitter Breakdown Voltage* (I_C = 1 mAdc, I_E = 0)	2N4123 2N4124		BV_{CEO}*	30 25	– –	Vdc
Collector-Base Breakdown Voltage (I_C = 10 μAdc, I_E = 0)	2N4123 2N4124		BV_{CBO}	40 30	– –	Vdc
Emitter-Base Breakdown Voltage (I_E = 10 μAdc, I_C = 0)			BV_{EBO}	5	–	Vdc
Collector Cutoff Current (V_{CB} = 20 Vdc, I_E = 0)			I_{CBO}	–	50	nAdc
Emitter Cutoff Current (V_{BE} = 3 Vdc, I_C = 0)			I_{EBO}	–	50	nAdc
ON CHARACTERISTICS						
DC Current Gain* (I_C = 2 mAdc, V_{CE} = 1 Vdc)	2N4123 2N4124	9	h_{FE}*	50 120	150 360	–
(I_C = 50 mAdc, V_{CE} = 1 Vdc)	2N4123 2N4124			25 60	– –	
Collector-Emitter Saturation Voltage* (I_C = 50 mAdc, I_B = 5 mAdc)		10, 11	$V_{CE(sat)}$*	–	0.3	Vdc
Base-Emitter Saturation Voltage* (I_C = 50 mAdc, I_B = 5 mAdc)		11	$V_{BE(sat)}$*	–	0.95	Vdc
SMALL SIGNAL CHARACTERISTICS						
High-Frequency Current Gain (I_C = 10 mAdc, V_{CE} = 20 Vdc, f = 100 MHz)	2N4123 2N4124		$\|h_{fe}\|$	2.5 3.0	– –	–
Current-Gain — Bandwidth Product (I_C = 10 mAdc, V_{CE} = 20 Vdc, f = 100 MHz)	2N4123 2N4124		f_T	250 300	– –	MHz
Output Capacitance (V_{CB} = 5 Vdc, I_E = 0, f = 100 kHz)		1	C_{ob}	–	4	pF
Input Capacitance (V_{BE} = 0.5 Vdc, I_C = 0, f = 100 kHz)		1	C_{ib}	–	8	pF
Small-Signal Current Gain (I_C = 2 mAdc, V_{CE} = 1 Vdc, f = 1 kHz)	2N4123 2N4124	5	h_{fe}	50 120	200 480	–
Noise Figure (I_C = 100 μAdc, V_{CE} = 5 Vdc, R_S = 1 kohm, Noise Bandwidth = 10 Hz to 15.7 kHz)	2N4123 2N4124	3, 4	NF	– –	6 5	dB

SWITCHING CHARACTERISTICS

Characteristic		Fig. No.	Symbol	Typ	Unit
Delay Time	V_{CC} = 3 Vdc, $V_{EB(off)}$ = 0.5 Vdc,	2	t_d	24	ns
Rise Time	I_C = 10 mAdc, I_{B1} = 1 mAdc	2	t_r	13	ns
Storage Time	V_{CC} = 3 Vdc, I_C = 10 mAdc,	2	t_s	125	ns
Fall Time	I_{B1} = I_{B2} = 1 mAdc	2	t_f	11	ns

*Pulse Test: Pulse Width = 300 μs, Duty Cycle = 2%

FIGURE 1 — CAPACITANCE

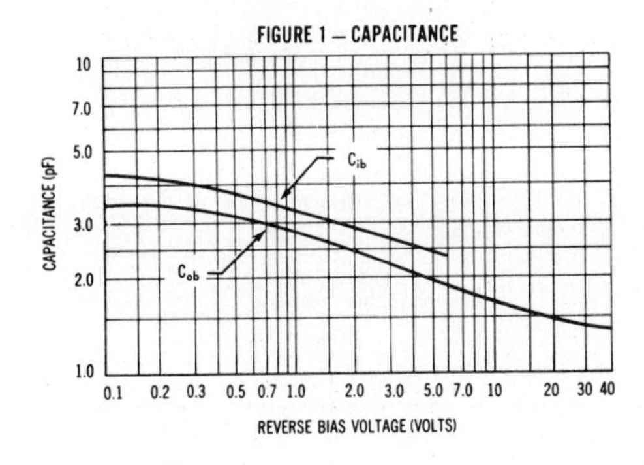

FIGURE 2 — SWITCHING TIMES

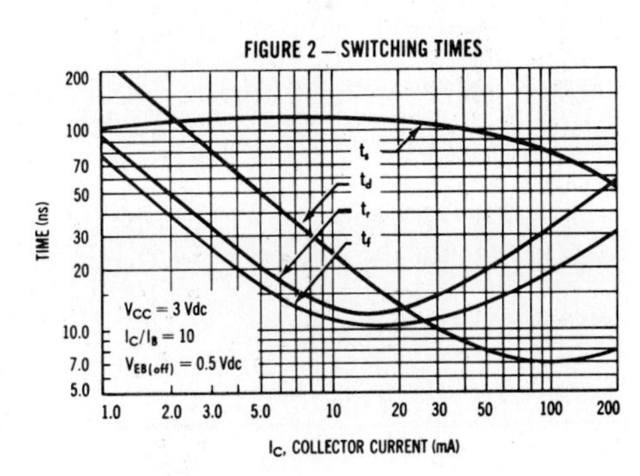

MOTOROLA Semiconductor Products Inc. 

(*Courtesy of Motorola Semiconductor Products, Inc.*)

RCA

Solid State Division

Power Transistors

2N3053
40389 40392

40392 H-1375

2N3053S

40389 H-1468

2N3053L H-1380

General-Purpose, Medium-Power Silicon N-P-N Planar Transistors

For Small-Signal Applications
In Industrial and Commercial Equipment

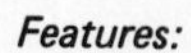

Features:

- Maximum safe-area-of-operation curve
- Forward- and reverse-bias operation without second breakdown
- Low leakage current

These devices are available with either 1½-inch leads (TO-5 package) or ½-inch leads (TO-39 package). The longer-lead versions are specified by suffix "L" after the type number; the shorter-lead versions are specified by suffix "S" after the type number.

RCA-2N3053 is a silicon n-p-n planar transistor useful up to 20 MHz in small-signal, medium-power applications. Type 40389 is a 2N3053 with a factory-attached diamond-shaped mounting flange.

Applications:

- Audio amplifiers
- Controlled amplifiers
- Power supplies
- Power oscillators

MAXIMUM RATINGS, *Absolute-Maximum Values:*

		2N3053	40389 40392	
COLLECTOR-TO-BASE VOLTAGE	V_{CBO}	60	60	V
COLLECTOR-TO-EMITTER SUSTAINING VOLTAGE:				
With external base-to-emitter resistance (R_{BE}) = 10 Ω	$V_{CER}(sus)$	50	50	V
With base open	$V_{CEO}(sus)$	40	40	V
With base-emitter-junction reverse-biased	$V_{CEV}(sus)$	60	60	V
EMITTER-TO-BASE VOLTAGE	V_{EBO}	5	5	V
COLLECTOR CURRENT	I_C	0.7	0.7	A
TRANSISTOR DISSIPATION:	P_T			
At case temperatures up to 25°C		5	7 (40392)	W
At free-air temperatures up to 25°C		1	3.5 (40389)	W
At temperatures above 25°C		See Figs.1, 2, and 3		
TEMPERATURE RANGE:				
Storage and operating (Junction)		← −65 to +200 →		°C
LEAD TEMPERATURE (During soldering):				
At distance ≥ 1/32 in. (0.8 mm) from seating plane for 10 s max.		← 235 →		°C

(*Courtesy of RCA Solid State Division*)

ELECTRICAL CHARACTERISTICS, *at Case Temperature (T_C) = 25°C unless otherwise specified*

Characteristics	Symbol	TEST CONDITIONS: DC Collector Voltage V		DC Emitter or Base Voltage V		DC Current mA			LIMITS: Types 2N3053 40389 40392		Units
		V_{CB}	V_{CE}	V_{EB}	V_{BE}	I_C	I_E	I_B	Min.	Max.	
Collector-Cutoff Current	I_{CBO}	30					0		—	0.25	μA
Emitter-Cutoff Current	I_{EBO}			4		0			—	0.25	μA
DC Forward-Current Transfer Ratio	h_{FE}		10			150[a]			50	250	
Collector-to-Base Breakdown Voltage	BV_{CBO}					0.1	0		60	—	V
Emitter-to-Base Breakdown Voltage	BV_{EBO}					0	0.1		5	—	V
Collector-to-Emitter Sustaining Voltage: With base open	$V_{CEO}(sus)$					100[a]		0	40	—	V
With external base-to-emitter resistance (R_{BE}) = 10 Ω	$V_{CER}(sus)$					100[a]			50	—	V
Base-to-Emitter Saturation Voltage	$V_{BE}(sat)$					150		15	—	1.7	V
Collector-to-Emitter Saturation Voltage	$V_{CE}(sat)$					150		15	—	1.4	V
Small-Signal, Forward Current Transfer Ratio (At 20 MHz)	h_{fe}		10			50			5	—	
Output Capacitance	C_{ob}	10					0		—	15	pF
Input Capacitance	C_{ib}			0.5		0			—	80	pF
Thermal Resistance: Junction-to-Case	$\theta_{J\text{-}C}$								35(max.) 2N3053		°C/W
									25(max.) 40392		°C/W
Junction-to-Free Air	$\theta_{J\text{-}FA}$								175(max.) 2N3053		°C/W
									50(max.) 40389		°C/W

[a]Pulsed; pulse duration = 300μs, duty factor = 1.8 %.

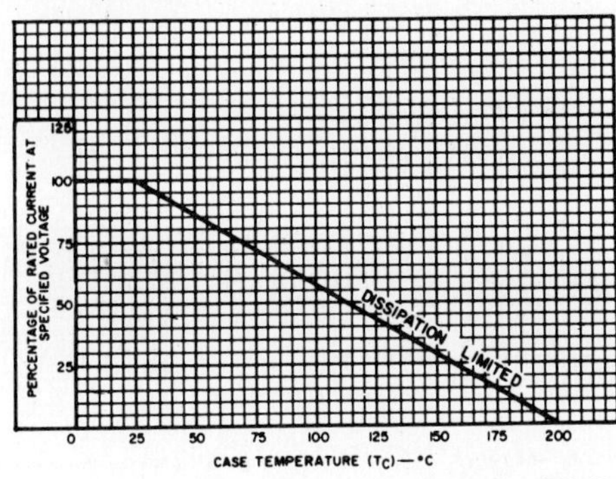

Fig.1 — Derating curve for type 2N3053. 92LS-1469

(*Courtesy of RCA Solid State Division*)

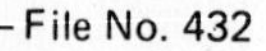

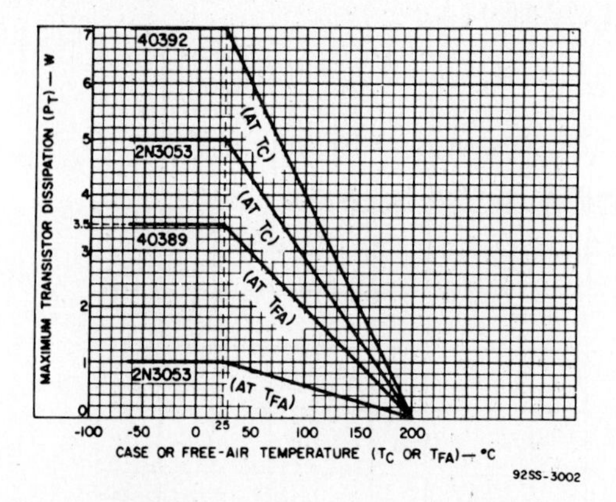

Fig. 3 - Dissipation derating curves for all types.

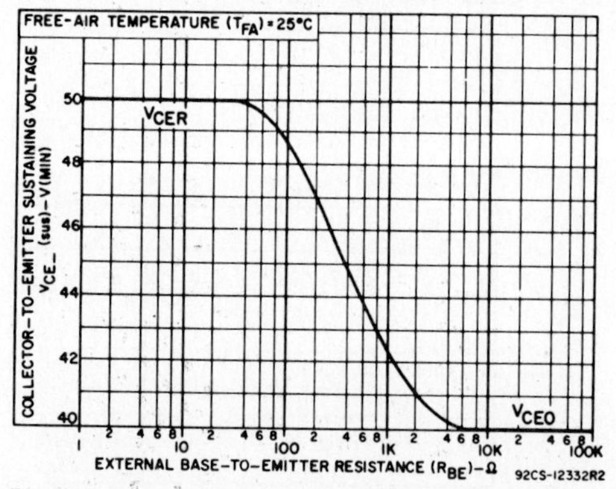

Fig. 4 - Sustaining voltage vs. base-to-emitter resistance for all types.

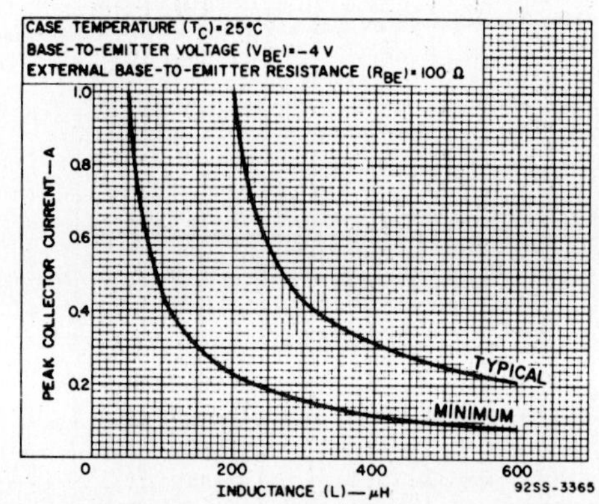

Fig. 5 - Reverse-bias, second-breakdown characteristics for all types.

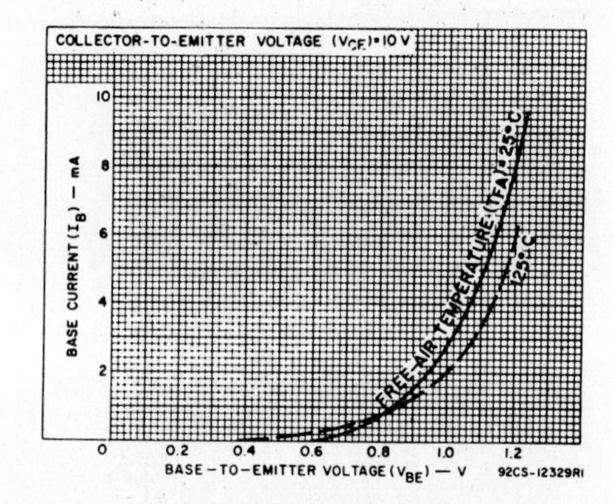

Fig. 6 - Typical dc-beta characteristics for all types.

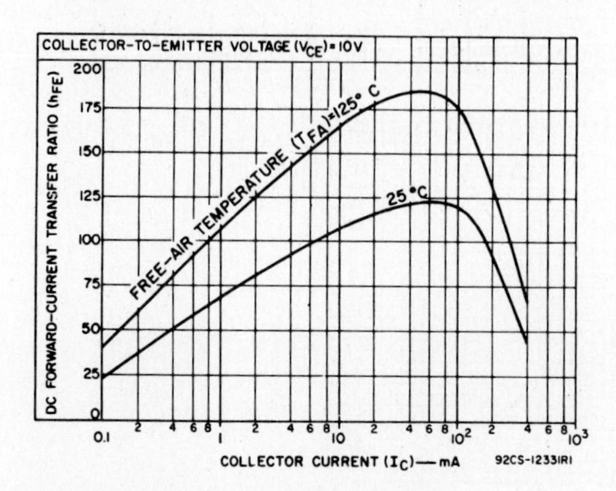

Fig. 7 - Typical input characteristics for all types.

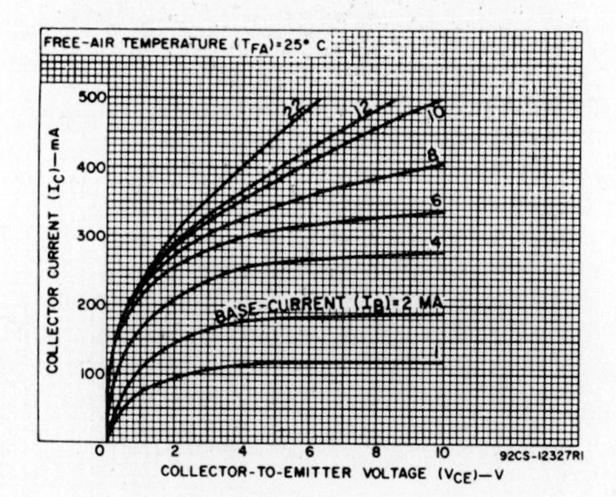

Fig. 8 - Typical output characteristics for all types.

(*Courtesy of RCA Solid State Division*)

TYPE 2N3819
N-CHANNEL PLANAR SILICON FIELD-EFFECT TRANSISTOR

SILECT† FIELD-EFFECT TRANSISTOR

For Industrial and Consumer Small-Signal Applications

- **Low C_{rss}: ≤ 4 pf** • **High y_{fs}/C_{iss} Ratio (High-Frequency Figure of Merit)**
- **Cross Modulation Minimized by Square-Law Transfer Characteristics**

mechanical data

This transistor is encapsulated in a plastic compound specifically designed for this purpose, using a highly mechanized process‡ developed by Texas Instruments. The case will withstand soldering temperatures without deformation. This device exhibits stable characteristics under high-humidity conditions and is capable of meeting MIL-STD-202C method 106B. The transistor is insensitive to light.

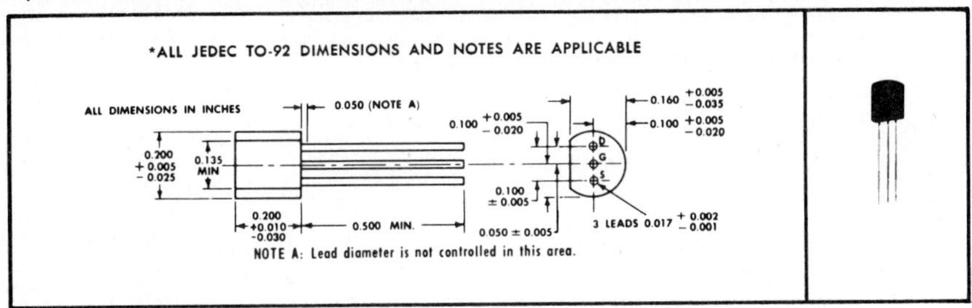

***absolute maximum ratings at 25°C free-air temperature (unless otherwise noted)**

Drain-Gate Voltage	25 v
Drain-Source Voltage	25 v
Reverse Gate-Source Voltage	−25 v
Gate Current	10 ma
Continuous Device Dissipation at (or below) 25°C Free-Air Temperature (See Note 1)	360 mw
Storage Temperature Range	−65°C to 150°C
Lead Temperature 1/16 Inch from Case for 10 Seconds	260°C

***electrical characteristics at 25°C free-air temperature (unless otherwise noted)**

	PARAMETER	TEST CONDITIONS	MIN	MAX	UNIT
$V_{(BR)GSS}$	Gate-Source Breakdown Voltage	$I_G = -1\ \mu a$, $V_{DS} = 0$	−25		v
I_{GSS}	Gate Cutoff Current	$V_{GS} = -15$ v, $V_{DS} = 0$		−2	na
		$V_{GS} = -15$ v, $V_{DS} = 0$, $T_A = 100°C$		−2	μa
I_{DSS}	Zero-Gate-Voltage Drain Current	$V_{DS} = 15$ v, $V_{GS} = 0$, See Note 2	2	20	ma
V_{GS}	Gate-Source Voltage	$V_{DS} = 15$ v, $I_D = 200\ \mu a$	−0.5	−7.5	v
$V_{GS(off)}$	Gate-Source Cutoff Voltage	$V_{DS} = 15$ v, $I_D = 2$ na		−8	v
$\|y_{fs}\|$	Small-Signal Common-Source Forward Transfer Admittance	$V_{DS} = 15$ v, $V_{GS} = 0$, f = 1 kc, See Note 2	2000	6500	μmho
$\|y_{os}\|$	Small-Signal Common-Source Output Admittance	$V_{DS} = 15$ v, $V_{GS} = 0$, f = 1 kc, See Note 2		50	μmho
C_{iss}	Common-Source Short-Circuit Input Capacitance	$V_{DS} = 15$ v, $V_{GS} = 0$, f = 1 Mc		8	pf
C_{rss}	Common-Source Short-Circuit Reverse Transfer Capacitance	$V_{DS} = 15$ v, $V_{GS} = 0$, f = 1 Mc		4	pf
$\|y_{fs}\|$	Small-Signal Common-Source Forward Transfer Admittance	$V_{DS} = 15$ v, $V_{GS} = 0$, f = 100 Mc	1600		μmho

NOTES: 1. Derate linearly to 150°C free-air temperature at the rate of 2.88 mw/C°.
2. These parameters must be measured using pulse techniques. PW ≈ 100 msec, Duty Cycle ≤ 10%.

*Indicates JEDEC registered data.
†Trademark of Texas Instruments
‡Patent Pending

(*Courtesy of Texas Instruments, Inc.*)

μA741A • μA741E

FREQUENCY COMPENSATED OPERATIONAL AMPLIFIER

FAIRCHILD LINEAR INTEGRATED CIRCUITS

GENERAL DESCRIPTION – The μA741A and E are high performance monolithic Operational Amplifiers constructed using the Fairchild Planar* epitaxial process. They are intended for a wide range of analog applications. High common mode voltage range and absence of "latch-up" tendencies make the μA741A and E ideal for use as voltage followers. The high gain and wide range of operating voltage provides superior performance in integrator, summing amplifier, and general feedback applications. Electrical characteristics are identical to MIL-M-38510/10101.

- **NO FREQUENCY COMPENSATION REQUIRED**
- **SHORT-CIRCUIT PROTECTION**
- **OFFSET VOLTAGE NULL CAPABILITY**
- **LARGE COMMON-MODE AND DIFFERENTIAL VOLTAGE RANGES**
- **LOW POWER CONSUMPTION**
- **NO LATCH UP**

ABSOLUTE MAXIMUM RATINGS

Supply Voltage	±22V
Internal Power Dissipation (Note 1)	
Metal Can	500 mW
DIP	670 mW
Flatpak	570 mW
Differential Input Voltage	±30V
Input Voltage (Note 2)	±15V
Storage Temperature Range	−65°C to +150°C
Operating Temperature Range	
Military (741A)	−55°C to +125°C
Commercial (741E)	0°C to +70°C
Lead Temperature (Soldering, 60 seconds)	300°C
Output Short Circuit Duration (Note 3)	Indefinite

EQUIVALENT CIRCUIT

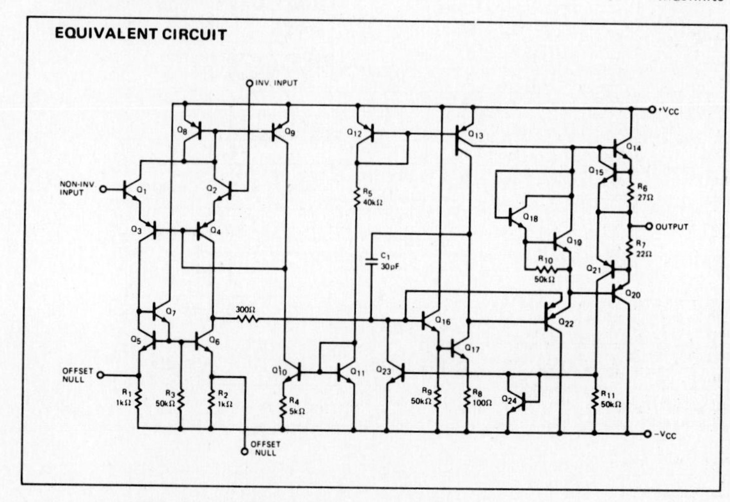

CONNECTION DIAGRAMS
8-LEAD METAL CAN
(TOP VIEW)
PACKAGE OUTLINE 5B

ORDER INFORMATION

TYPE	PART NO.
741A	741AHM
741EC	741EHC

14-LEAD DIP
(TOP VIEW)
PACKAGE OUTLINE 6A

ORDER INFORMATION

TYPE	PART NO.
741A	741ADM
741EC	741EDC

10-LEAD FLATPAK
(TOP VIEW)
PACKAGE OUTLINE 3F

ORDER INFORMATION

TYPE	PART NO.
741A	741AFM

*Planar is a patented Fairchild process.

(*Courtesy of Fairchild Semiconductors*)

FAIRCHILD LINEAR INTEGRATED CIRCUITS • μA741A • μA741E

741E

ELECTRICAL CHARACTERISTICS (V_S = ±15V, T_A = 25°C unless otherwise specified)

PARAMETERS (see definitions)		CONDITIONS		MIN.	TYP.	MAX.	UNITS
Input Offset Voltage		$R_S \leq 50\Omega$			0.8	3.0	mV
Average Input Offset Voltage Drift						15	μV/°C
Input Offset Current					3.0	30	nA
Average Input Offset Current Drift						0.5	nA/°C
Input Bias Current					30	80	nA
Power Supply Rejection Ratio		V_S = +10, −20; V_S = +20, −10V, R_S = 50Ω			15	50	μV/V
Output Short Circuit Current				10	25	35	mA
Power Dissipation		V_S = ±20V			80	150	mW
Input Impedance		V_S = ±20V		1.0	6.0		MΩ
Large Signal Voltage Gain		V_S = ±20V, R_L = 2kΩ, V_{OUT} = ±15V		50			V/mV
Transient Response	Rise Time				0.25	0.8	μs
(Unity Gain)	Overshoot				6.0	20	%
Bandwidth (Note 4)				.437	1.5		MHz
Slew Rate (Unity Gain)		V_{IN} = ±10V		0.3	0.7		V/μs
The following specifications apply for 0°C ≤ T_A ≤ 70°C							
Input Offset Voltage						4.0	mV
Input Offset Current						70	nA
Input Bias Current						210	nA
Common Mode Rejection Ratio		V_S = ±20V, V_{IN} = ±15V, R_S = 50Ω		80	95		dB
Adjustment For Input Offset Voltage		V_S = ±20V		10			mV
Output Short Circuit Current				10		40	mA
Power Dissipation		V_S = ±20V				150	mW
Input Impedance		V_S = ±20V		0.5			MΩ
Output Voltage Swing		V_S = ±20V,	R_L = 10kΩ	±16			V
			R_L = 2kΩ	±15			V
Large Signal Voltage Gain		V_S = ±20V, R_L = 2kΩ, V_{OUT} = ±15V		32			V/mV
		V_S = ±5V, R_L = 2kΩ, V_{OUT} = ±2 V		10			V/mV

VOLTAGE OFFSET NULL CIRCUIT

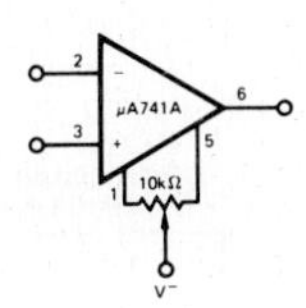

TRANSIENT RESPONSE TEST CIRCUIT

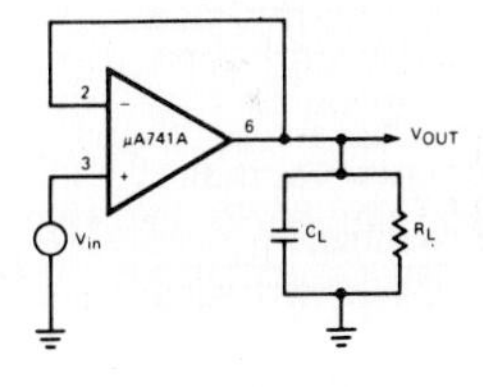

(Courtesy of Fairchild Semiconductors)

FAIRCHILD LINEAR INTEGRATED CIRCUITS • μA741A • μA741E

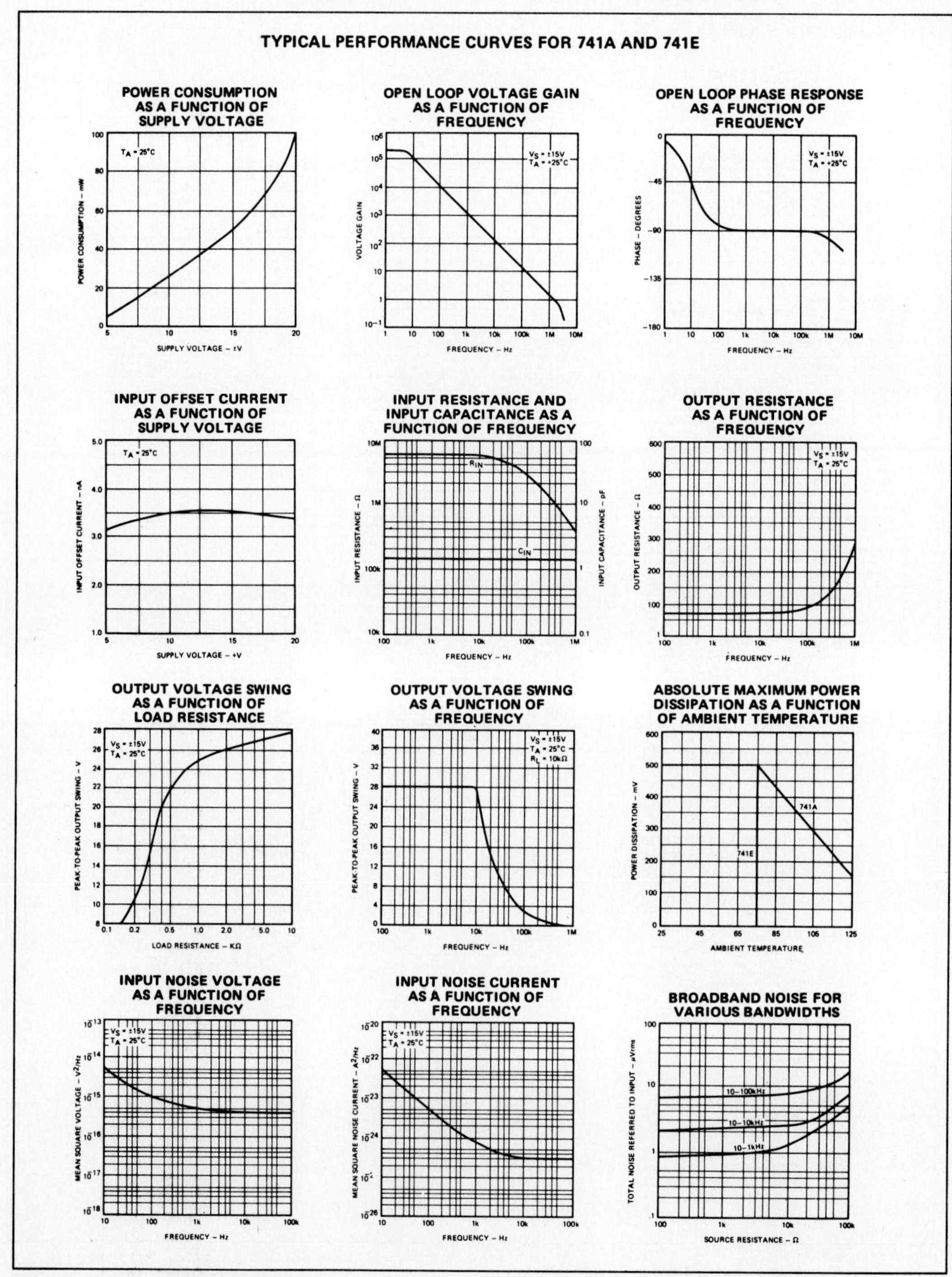

(Courtesy of Fairchild Semiconductors)

APPENDIX H
SYMBOL IDENTIFICATION

Page numbers reference the first appearance of each symbol in the text. For a general discussion of symbol usage, see pages 5 through 7.

A—ampere, p. 11
A (subscript)—ambient, p. 240
A—anode, p. 331
A_v—voltage gain, p. 97
B—base, p. 57
B—battery, p. 71
B—feedback ratio, p. 198
B_w—bandwidth, p. 204
c (subscript)—closed-loop, p. 198
C—capacitance, capacitor, p. 26
C—collector, p. 57
°C—degree Celsius, p. 16
ch (subscript)—choke, p. 37
chg (subscript)—charging, p. 30
$CMRR$—common-mode rejection ratio, p. 194
com (subscript)—common-mode, p. 194
CT—center-tapped, p. 34
D—diode, p. 10
D—drain, p. 147
dB—decibel, p. 165
dif (subscript)—differential, p. 191
dischg (subscript)—discharging
E—emitter, p. 57
f—frequency, p. 27
f_c—carrier frequency, p. 294
f_m—modulating frequency, p. 294
f_r—resonant frequency, p. 204
f_T—unity-gain frequency, p. 176
F—farad, p. 28
F—fuse, p. 46
°F—degree Fahrenheit
G—giga (JIG-uh), $\times 10^9$, p. 14
G—gate, p. 147
h_{fe}—ac current gain, p. 59
h_{FE}—dc current gain, p. 59
Hz—hertz, cycle per second, p. 10
I—lamp, p. 11
I—current, p. 79
I_{CBO}—collector-to-base leakage, p. 61

I_{CEO}—collector-to-emitter leakage, p. 62
I_{DSS}—zero-bias drain current, p. 148
J (subscript)—junction, p. 240
k—kilo, $\times 10^3$, p. 11
K—relay, p. 89
K—cathode, p. 331
L—inductance, inductor, p. 36
L (subscript)—load, p. 25
LED—light-emitting diode, p. 77
m—milli, $\times 10^{-3}$, p. 11
M—mega, $\times 10^6$, p. 174
n—nano, $\times 10^{-9}$, p. 62
n—turns ratio N_S/N_P, p. 30
N—number of turns, p. 30
o (subscript)—open-loop, p. 198
o (subscript)—output, p. 30
p (subscript) parallel, p. 237
P (subscript)—primary, p. 30
P—power, p. 67
PIV—peak inverse voltage, p. 12
PRC—photoresistive cell, p. 79
PRV—peak reverse voltage, p. 12
Q—transistor, p. 65
Q—charge, p. 27
Q—quiescent point, p. 104
Q—quality factor, p. 204
r (subscript)—resonant, p. 204
r_b—ac resistance from base to ground, p. 98
r_c—dynamic resistance of collector, $\Delta V_{CE}/\Delta I_C$, p. 102
r_d—dynamic resistance of drain, V_{DS}/I_D, p. 154
r_j—dynamic resistance of emitter junction, p. 95
r_L—ac resistance from output terminal to ground, p. 141
r_P—primary-winding resistance, p. 30
r_S—secondary-winding resistance, p. 30
r_s—source resistance, $1/y_{fs}$, p. 154
r_w—Thévenin equivalent of all winding resistance, p. 30
r_z—dynamic zener resistance, p. 301
R—resistance, resistor, p. 11
R_b—bias-compensating resistor, p. 258
R_B—base resistor, p. 59
$R_{c(line)}$—ac resistance through which collector current passes, p. 102
R_d—dropping resistor, p. 301
R_D—drain resistor, p. 152
$R_{e(line)}$—ac resistance through which emitter current passes, p. 102

R_f—filter resistor, p. 38
R_F—feedback resistor, p. 81
R_t—timing resistor, p. 264
R_T—total resistance
s—second, p. 16
s (subscript)—source, p. 6
s (subscript)—series, p. 237
S—siemens, reciprocal ohm, p. 150
S—switch, p. 46
S—source (FET), p. 147
S—stability factor, p. 109
S—slew rate, p. 255
S (subscript)—secondary, p. 30
S (subscript)—sink, p. 240
sat (subscript)—saturation, p. 62
sc (subscript)—short-circuit, p. 311
Si—silicon, p. 18
t—time, p. 13
t_r—risetime, 10% to 90%, p. 182
T—transformer, p. 25
T (subscript) total, p. 204
V—volt (unit of voltage), p. 11
V—voltage (quantity), p. 11
V_D—diode forward voltage, p. 11
V_{FL}—full-load voltage, p. 29
$V_{GS(off)}$—gate-source voltage for $I_D \rightarrow 0$, p. 148
V_{NL}—no-load voltage, p. 29
V_O—output voltage, p. 30
V_{rip}—ripple voltage, p. 29
V_s—secondary voltage, p. 26
V_{Th}—Thévenin voltage, p. 350
V_{TH}—threshold voltage, p. 150
V_Z—zener voltage, p. 300
V (subscript)—valley, p. 328
W—watt, p. 12
X—reactance, p. 37
y_{fs}—forward transconductance, p. 148
Z—impedance, p. 98
Z (subscript)—zener or regulator, p. 300
α (alpha)—current gain I_C/I_E, p. 136
α_{dB}—power ratio expressed in decibels, p. 165
β (beta)—current gain I_C/I_B, p. 58
Δ (delta)—small change in a quantity, p. 64
η (eta)—efficiency ratio P_o/P_{in}, p. 214

η—UJT voltage ratio, p. 328
θ (theta)—thermal resistance, p. 240
μ (mu)—micro, $\times 10^{-6}$, p. 22
π (pi)—3.14159, p. 37
τ (tau)—time constant, p. 27
Ω (omega)—ohm, p. 11
$\approx$—approximately equal to, p. 42
∞—infinity, p. 11
$\rightarrow$—yields, approaches, p. 11
$\|$—parallel resistances, p. 99
$\gg$—much greater than, p. 87
$\leq$—equal to or less than, p. 86

INDEX

O

P

Q

R

S